39.00

W9-CUE-474

Building Trades DICTIONARY

Leonard P. Toenjes

AMERICAN TECHNICAL PUBLISHERS, INC.
HOMEWOOD, ILLINOIS 60430

1 2 3 4 5 6 7 8 9-89-9 8 7 6 5 4 3 2 1

Printed in the United States of America

Library of Congress Cataloging-in-Publication Data

Toenjes, Leonard P., 1953–
    Building trades dictionary

    1. Building—Dictionaries.   I. Title.
TH9.T585   1989   690'.03'21   88-7904
ISBN 0-8269-0403-3

Contents

Acknowledgments

The author and publisher are grateful to the following companies and organizations that provided photographs and technical assistance.

Advertising Communications, Inc.
Aero-Motive
Alvarado Manufacturing
American Institute of Steel Construction
American Plywood Association
American Society of Mechanical Engineers
American Welding Society, Inc.
Arbor International, Inc.
Asphalt Institute
Associated Pile and Fitting Corporation
Atlas Powder Company
Autodesk, Inc.
Bethlehem Steel Corporation
Black and Decker
Bon Tool Company
Brodhead-Garrett
Rodger A. Brooks, Architect
Bryant Electric
The Burke Company
Burro Crane
Carlon Meter
Carrier Corporation
Caterpillar, Inc.
Ceramic Tile Institute
Cherne Industrial, Inc.
Cincinnati Electrical Tool, Inc.
Circle Seal Controls
Cleaver-Brooks
Cooper Tools
Crain Cutter Company, Inc.
Cronkhite Industries, Inc.
CSC Scientific Company, Inc.
David White Instruments, Division of Realist, Inc.
Dayton Electric Manufacturing Company
Deere and Company
Delphi Body Works
Delta International Machinery Corporation
Dresser Industries, Inc.
Elkay Manufacturing Company
The Foxboro Company
Gammans Industries, Inc.
The Garlinghouse Company
Garrett Wade Company, Inc.
Gemtor, Inc.
General Electric Company
Georgia-Pacific Corporation
Goldblatt Tool Company
GOMACO Corporation
Hearlihy and Company
The Heil Company
Henry Valve Company

Hoffman Engineering Company
Ideal Industries, Inc.
Indco, Inc.
Intergraph Corporation
International Masonry Institute
International Union of Bricklayers and Allied Craftsmen
ITW Ramset/Red Head
H.B. Ives, a HARROW Company
Jamesbury Corporation
Kaiser Cement Corporation
Klein Tools, Inc.
Koh-I-Noor Rapidiograph, Inc.
Kohler Company
Laser Alignment, Inc.
The Lietz Company
Milwaukee Electric Tool Corporation
Modern Engineering Company
Montgomery Elevator
Morrison's Concrete and Equipment, Inc.
The Mueller Company
E.L. Mustee and Sons, Inc.
National Roofing Contractors Association
NIBCO
Owens Corning Fiberglas Corporation
Paxton-Patterson
Porter Cable
Portland Cement Association
Red Devil, Inc.
The Ridge Tool Company
Rosedale Products, Inc.
Safway Steel Products, Figgie International, Inc.
Semler Industries
A.O. Smith
Snap-on Tools Corporation
Square D Company
Stanley Tools
The L.S. Starrett Company
Stockham Valves and Fittings
Sunshine Rooms® Inc.
Super Radiator Coils
Symons Corporation
Thomas and Betts Corporation
Trus Joist Corporation
Truth, Inc.
United States Gypsum Corporation
Von Duprin, Inc.
Wacker Corporation
Wade Division, Tyler Pipe
Walsh Construction Company of Illinois
Watts Regulator
Wendy's International, Inc.

Introduction

BUILDING TRADES DICTIONARY is a comprehensive reference source for persons working in the building trades or related fields. Building trades, basic engineering, and architectural terms are presented in a highly illustrated, concise format. Emphasis is placed on construction tools, materials, and processes used in the construction industry.

A separate section of the book, Contractors' Terms, defines terms encountered by supervisors, contractors, and other persons involved in planning and development of construction projects. Terms that apply to estimating, bidding, labor relations, labor law, building codes, and bonding and licensing are included. The Appendices provide a comprehensive reference to plane and solid geometric figures, conversions, trade organizations, and print components and symbols.

USING BUILDING TRADES DICTIONARY

BUILDING TRADES DICTIONARY is composed of several elements, including terms, definitions for the terms, synonyms, illustration references, and/or references to similar terms.

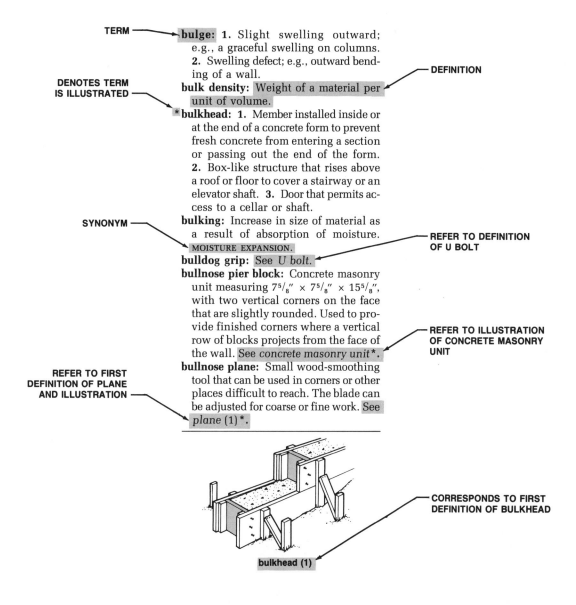

TERM ——→ **bulge: 1.** Slight swelling outward; e.g., a graceful swelling on columns. **2.** Swelling defect; e.g., outward bending of a wall. ←—— DEFINITION

DENOTES TERM IS ILLUSTRATED ——→ **bulk density:** Weight of a material per unit of volume.

＊**bulkhead: 1.** Member installed inside or at the end of a concrete form to prevent fresh concrete from entering a section or passing out the end of the form. **2.** Box-like structure that rises above a roof or floor to cover a stairway or an elevator shaft. **3.** Door that permits access to a cellar or shaft.

SYNONYM ——→ **bulking:** Increase in size of material as a result of absorption of moisture. MOISTURE EXPANSION.

bulldog grip: See *U bolt*. ←—— REFER TO DEFINITION OF U BOLT

bullnose pier block: Concrete masonry unit measuring $7^5/_8''$ × $7^5/_8''$ × $15^5/_8''$, with two vertical corners on the face that are slightly rounded. Used to provide finished corners where a vertical row of blocks projects from the face of the wall. See *concrete masonry unit*＊. ←—— REFER TO ILLUSTRATION OF CONCRETE MASONRY UNIT

REFER TO FIRST DEFINITION OF PLANE AND ILLUSTRATION ——→ **bullnose plane:** Small wood-smoothing tool that can be used in corners or other places difficult to reach. The blade can be adjusted for coarse or fine work. See *plane (1)*＊.

CORRESPONDS TO FIRST DEFINITION OF BULKHEAD

bulkhead (1)

Aaron's rod: Ornament or molding having a straight rod with a leaf design or scrollwork emerging on either side at regular intervals.

abacus: Round or square slab forming the uppermost section of a capital.

abamurus: Buttress or secondary supporting wall.

abate: To remove or shape material to form a relief design.

abatement: Amount of waste wood produced by sawing or planing timber to size.

abat-jour: Sloped opening in a roof or wall designed to admit light and deflect it downward; skylight.

abatson: Device used to deflect sound waves downward.

abattoir: Livestock slaughterhouse.

abatvent: Louver that keeps wind from entering a building, but allows light and air circulation.

* **abbreviation:** Letter or group of letters representing a term or phrase. Used on a print to conserve space and drafting time.

A-block: Hollow concrete masonry unit with one closed and one open end with a web in the center. Commonly used at wall openings.

Abrams' law: Rule stating that the strength of concrete is determined by the ratio of the amount of water to the amount of cement in the mix, providing the mix is a workable consistency.

abrasion: 1. Wearing away of a surface by friction. 2. Reducing of dimensions of a material by grinding.

abrasion resistance: Ability of a material to resist being worn away by friction or grinding.

abrasive: Hard natural or synthetic material used to wear or polish a surface by friction. Includes emery, sand, diamond, crushed garnet, quartz, pumice, and tripoli.

abrasive paper: Paper or cloth backed with an abrasive, such as emery or crushed garnet, that is glued to one surface. Used to smooth or polish a surface.

abreuvoir: Void between stones in a wall or between two arch stones that is filled with mortar.

ABS: See *acrylonitrile butadiene styrene*.

absolute humidity: Weight of water vapor per unit volume of air. Expressed as grams per cubic centimeter, grams per cubic foot, or pounds per cubic foot.

absolute pressure: Pressure equal to the sum of gauge pressure and atmospheric pressure.

absolute volume: 1. Volume that a solid material displaces including all permeable and impermeable voids, but excluding the spaces between particles. 2. Volume that a fluid occupies.

absolute zero: Hypothetical temperature at which all thermal motion ceases, approximately $-273.2°C$ or $-459.7°F$.

absorber: 1. Component of a solar energy system that collects radiant energy and converts it to thermal energy. 2. Device used to absorb refrigerant vapors in the low-pressure side of a refrigeration system.

absorption: 1. Drawing of gases or vapors into permeable pores of porous material. Results in physical and/or chemical changes in the material. 2. Conversion of radiant energy into other energy forms.

absorption bed: Large pit containing coarse aggregate and distribution pipe through which the septic tank effluent may seep into the surrounding soil.

* **absorption field:** System of trenches containing coarse aggregate and distribution pipe through which the septic tank effluent may seep into the surrounding soil. ABSORPTION TRENCH.

absorption rate: Rate at which weight of water is drawn into a masonry unit in one minute. Expressed as grams or ounces of water per minute.

absorptive lens: Filtered lens designed to decrease glare and reflected, stray light.

absorptivity: Comparison of surface radiant energy to amount of energy absorbed by a black surface at similar temperatures. Used for classification of a solar surface.

abut: To join along an edge or end but not overlap.

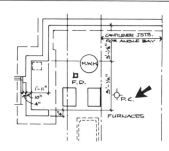

abbreviation

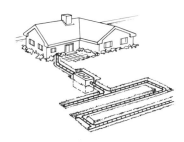

absorption field

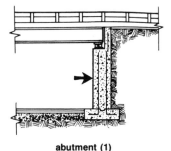

abutment (1)

acme thread

* **abutment: 1.** Structure that supports the end of a bridge or arch. **2.** Anchorage for bridge suspension or prestressing cables. **3.** Side of an earth bank that supports a dam.

abutment piece: Lowest structural member that receives and distributes the load of an upright or strut; e.g., sole plate.

abutting joint: 1. Joint formed by fastening two members end-to-end without overlapping. See *butt joint*. **2.** Wood joint formed by fastening two members, grain perpendicular to one another, without overlapping.

acanthus: Decorative ornament shaped like leaves of an acanthus plant; with tall, slender pointed leaves. Used in Corinthian capitals.

accelerator: 1. Chemical substance that increases the rate of hydration and/or strength development of concrete, mortar, or grout. **2.** Substance that decreases setting time of plaster. **3.** Centrifugal pump in the return duct of a central heating system. Used to increase air flow.

accent lighting: Directional illumination designed to emphasize an object or design.

access: 1. Approach to a room or building; a corridor between rooms. **2.** Opening in floors, walls, or ceilings used to install, inspect, and repair concealed equipment.

access connection: Roadway used to enter or exit a highway.

access court: Approach to a dwelling within a group of residential structures.

access door: Door used to enclose an area that houses concealed equipment. ACCESS PANEL.

access flooring: See *pedestal floor*.

accessible: 1. Capable of being removed or exposed without damage to the building structure or finish. **2.** Capable of being reached quickly.

accessory to corner: Physical objects, such as trees or rock formations, that are close to surveying corners and used as future reference points.

access stair: Stairway used to gain access from one level to another, but is not a required exit stairway.

accolade: Decorative design consisting of two ogee curves intersecting over the center of a door or window.

accordion door: Door constructed of a series of narrow, vertical slats joined with straps or hinges. Supported by rollers in an overhead track. Edges of doors butt against one another when door is closed, and fold face-to-face when open. See *door**.

accouple: To join or couple.

accouplement: 1. Timber tie or brace. **2.** Placing columns in pairs.

accretion: Change in a shoreline resulting from reliction (recession) or alluvium (deposition of soil).

accumulator: 1. Pump used to maintain consistent pressure in a power-operated jacking system. Consists of a small piston utilizing hydraulic pressure and a large pump utilizing pneumatic pressure. **2.** Storage chamber for liquid refrigerant on the low-pressure side of a system. SURGE DRUM, SURGE HEADER.

acetone: Organic flammable solvent that evaporates quickly. Used in paint remover, lacquer, and thinner.

acetylene: Colorless gas formed from a mixture of calcium carbide and water. Commonly combined with air or oxygen to form a combustible gas. Stable under low pressure; unstable if compressed to more than 15 pounds per square inch.

achromatic color: White light without hue.

acid: 1. Water-soluble compound capable of forming hydrogen-concentrated salts. Acidity (acid content) is expressed as pH, denoting the hydrogen ion concentration. Affects strength characteristics of concrete when present in sufficient amounts. **2.** See *muriatic acid*.

acid core solder: Solder wire or bar with flux core.

* **acme thread:** Thread design with tapered teeth in which the widest portion is at the root of the thread. Angle between the sides of the thread and an axis perpendicular to the screw thread is 29°. Used for feed screws on equipment such as milling machines or

lathes.

acorn nut: Nut with a hexagonal base and domed top that covers the end of the screw threads. Used as a decorative nut or where protection is required for the end of the threads. See *nut**.

acoustical board: Manufactured product used to absorb sound waves and prevent passage of sound. Available in flat sheets.

acoustical material: Material used to absorb sound waves and prevent passage of sound. Includes tile, expanded foam, plastic, or material primarily composed of mineral, wood, cork, metal, or vegetable fibers.

acoustical plaster: Plaster containing vermiculite or other porous material; used to reduce reflection of sound waves.

acoustical tile: Small square or rectangular piece of acoustical board, approximately $1/2$". Applied to ceilings or walls to prevent passage of sound.

acoustics: 1. Study of sound, including the generation, transmission, and effects of sound waves. **2.** Overall characteristics of a room that affect the way the quality of sound is perceived within the room. Characteristics include reverberation, extraneous noises, volume of the original sound, and size and shape of the room.

acre: Land measure equal to 43,560 square feet or 4840 square yards.

acre-foot: Volume of a material required to cover an acre to a 1'-0" depth.

acropodium: Raised platform used to support a statue.

acroter: 1. Small pedestal placed on the apex or at the base of a pediment to support a statue or other ornament. ACROTERIA, ACROTERIAN. **2.** Statue or ornament placed on a pedestal.

acrylic plastic: Soft, non-crystalline thermoplastic with optical clarity, excellent weather resistance, and good shatter resistance. Used for glazing, lights, hardware, window frames, and lighting fixtures.

acrylic plastic glaze: Synthetic, shatterproof material available in sheets for use on windows in high-breakage areas.

acrylic resin: Thermoplastic resin formed from esters of acrylic acid. Transparent, strong, and chemical-resistant. Used in paint, caulking, and sealants.

acrylonitrile-butadiene-styrene (ABS): Plastic pipe and fittings used for sanitary drainage and vent piping systems, and above- and below-grade storm water drainage.

actinic glass: Yellow tinted glass used to filter ultraviolet and infared light.

activated alumina: Drying agent manufactured from aluminim oxide.

activated charcoal: Carbonized organic material used to absorb odors from the air or remove color in solution.

activated rosin flux: Rosin or resin-base flux used to increase wetting by the solder.

active earth pressure: Horizontal force of the earth applied against a vertical surface such as a wall.

active leaf: The portion of a hinge that moves.

active solar energy system: System in which solar energy is obtained from the sun and distributed throughout the system by using fans or blowers. Primary components are collectors, storage unit, fan or blower, and an auxiliary heating system. ACTIVE SYSTEM.

actuator: Automatic device used to open and close a damper or valve.

acuminated: Finished to a point at the top.

acute angle: Angle less than 90°.

adamant plaster: Quick-setting plaster finish coat.

* **adapter:** Fitting used to match the size or characteristics and connect two or more components. Used in plumbing, HVAC, and electrical trades.

addition: 1. Expansion to an existing structure. 2. Material blended into cement during manufacture to aid in processing or to modify the properties of concrete.

additive: See *addition, admixture.*

adherence: Property of unlike particles that causes them to stick together.

adhesion: 1. Bonding members together using an adhesive. 2. Physical and chemical attraction between molecules.

adhesive: Substance used to bond two surfaces together. Includes cement, paste, mastic, and glue. Available in solid, liquid, and semiliquid forms.

adhesive bond: Attractive force between an adhesive and base material.

adiabatic curing: Maintenance of consistent heat in mortar or concrete during the hardening process.

adiabatic process: Process in which heat neither enters or leaves a closed system.

adit: Entrance to a building or mine used for ventilation and drainage.

adjacent: Adjoining or touching.

adjustable clamp: Clamp in which the opening between the jaws can be adjusted to fit the workpieces.

adjustable key: Key for a sliding door lock.

adjustable pipe hanger: Pipe support consisting of a beam clamp, adjustable rod, and adjustable ring.

adjustable tie: Heavy gauge wire reinforcement device installed in masonry cavity wall to join adjacent wythes of masonry in which the bed joints are at different heights. See *tie* (1)*.

adjustable wall furring bracket: 20-gauge galvanized steel strips with corrugated edges used to attach metal furring to masonry.

adjustable wire stripper: See *wire stripper.*

adjustable wrench: Wrench with one fixed jaw and one movable jaw that is adjusted to the desired position with a thumbscrew in the head of the tool. See *wrench**. CRESCENT® WRENCH.

* **adjusting lever:** Narrow piece of metal in a wood-shaping plane used to adjust the angle of the plane.

admixture: Material, other than water, aggregate, fiber reinforcement, and cement, used as an ingredient in a batch of concrete or mortar. Used to add color, control strength, shorten or lengthen setting time, or modify temperature range of the mix. Incorporated into the mix before or during mixing.

adobe: Aluminous clay used to form unfired brick. Primarily used in western United States.

adsorbent: Material that extracts a substance from a solid, liquid, or gaseous material without a chemical or physical change.

adsorption: Process of extracting a substance from a solid, liquid, or gaseous material by causing the substance to adhere to an adsorbent. Chemical or physical change is not present in the adsorbent.

advancing slope grouting: Process in which grout is moved horizontally through preplaced aggregate with a suitable injection method.

adze: Long-handled cutting tool with a thin arched blade that is perpendicular to the handle. Used for rough dressing timber. ADZ.

* **adze eye:** Design of the head of a hammer in which opening that receives the handle is extended to provide a larger bearing surface for the handle.

adze-eye hammer: Hammer with an extended head that provides a larger bearing area for the handle.

aedicula: Door or window with a pilaster or column on each side and a pediment across the top.

aerated concrete: Lightweight concrete formed with gas-forming or foaming admixtures that form a cellular structure. Used as a retardant of sound transmission.

aeration: Process of naturally or mechanically introducing air into a substance.

aerator: Mechanical device used to introduce air into water, soil, or sewage.

aerial survey: Survey of a large area of land based on aerial photographs. Predetermined reference points are marked on the ground to guide the photographer.

aerograph: Spray gun used to apply paint. Paint passes from an attached container to a small pneumatic nozzle where it is discharged as a fine mist.

aerosol paint: Paint packaged in a pressurized can with a nozzle designed to atomize the paint. Pressure is provided by compressed liquefied gas.

A-frame: 1. Structure with a gable roof extending to the foundation. 2. A-shaped steel frame on casters which supports a chain fall or hoist from a connecting beam. Beam extends between the A-shaped members. 3. Guyed derrick with an A-shaped mast.

aft batter: Vertical position of a pile with the bottom of the pile closest to the crane. IN BATTER.

after cooler: System used to remove heat from compressed gas in a refrigeration system.

adapter

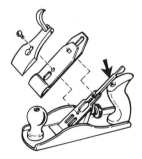

adjusting lever

adze eye

after flush: Water that drains from a toilet before the valve is closed.

* **aggregate:** Granular material, such as gravel, sand, vermiculite, or perlite, that is added to cement paste to form concrete, mortar, or plaster. Graded as fine, coarse, lightweight, and heavyweight. Comprises 60% to 80% of the volume of concrete.

aggregate gun: Equipment used to manually apply drywall textured materials to gypsum drywall. Consists of a material hopper that stores and discharges material and a spray gun that supplies compressed air to the nozzle. Material is atomized during application.

aggregate seeder gun: Hopper attached to an electric motor with an oscillating gear attachment. Aggregate feeds into the hopper by gravity and the sponge oscillates to push aggregate onto the surface.

aging: Natural process in which the characteristics of a material change with time.

agitation: 1. Gentle motion applied to concrete in a plastic state to prevent segregation of the ingredients. 2. Mixing or moving a substance by air or mechanical means.

agitator: 1. Device used to retain plasticity of concrete, plaster, and gypsum finish material. 2. Device used to create a turbulent motion in confined fluid or powder.

aiguille: Slender instrument used to bore holes in stone or other masonry material.

air: Gas primarily composed of oxygen and nitrogen with traces of other gases.

air acetylene welding: Welding process in which metal is joined using heat obtained from a gas flame or flame produced from the combustion of acetylene with air, with or without applying pressure, and with or without filler material.

air balancing: Adjusting a duct system to provide equal air distribution.

air break: Plumbing piping arrangement in which a drain discharges into another fixture at a level below the flood rim.

air brick: 1. Hollow or perforated brick used to ventilate a structure. 2. Perforated metal box with similar dimensions to a brick; used to ventilate a structure.

airbrush: Device used to apply a fine mist of paint using compressed air.

air carbon arc cutting: Process of severing metal using heat obtained from a carbon arc to melt the metal and removing molten metal with a stream of air.

* **air chamber:** Short section of supply pipe that extends beyond the last fitting; used to prevent water hammer.

air change: Volume of air in an enclosed area that is replaced by fresh air. Expressed as a ratio of the volume of air exchanged to the total volume of the enclosed area.

air circulation rate: Rate at which air is exchanged in an enclosed area. Expressed as the volume of air in the room divided by unit time.

air cleaner: Device designed to remove airborne impurities such as dust, fumes, gases, and vapors from the air. Includes air washers, air filters, electrostatic precipitators, and charcoal filters.

air compressor: See *compressor*.

air conditioner: Equipment used to control humidity and temperature and remove airborne impurities in an enclosed area. Consists of a cooling coil or evaporator, electric compressor, and condenser that work together to cool the air. Ranges from small window-mounted units to large roof-mounted units.

air conditioning: Process used to control humidity, temperature, cleanliness, and/or distribution of air within an enclosed structure.

air conditioning condenser: Air- or water-cooled unit used to condense hot, compressed refrigerant gases.

air content: Volume of air voids in cement paste, concrete, or mortar, exclusive of the pores in aggregate. Expressed as a percentage of the total volume of the paste, mortar, or concrete.

air-cooled slag: Product resulting from the solidification of molten blast furnace slag that is cooled under atmospheric conditions. Slag is excavated, crushed, and sorted for commercial applications such as concrete and bituminous aggregate.

* **air curtain:** Continuous stream of air circulated across an opening and used as a barrier against thermal transmission and contaminants.

* **air diffuser:** Air distribution outlet in a supply duct used to deflect and mix air. Commonly fitted with vanes or louvers.

air drain: Flue or passageway around the outside of a foundation wall that prevents earth from laying directly against the wall and that conveys fresh air to the wall to prevent dampness.

air-dried: Dried naturally, without using artificial heat.

air-dried lumber: Lumber that has been naturally dried to a moisture content ranging from 12% to 15%.

air duct: Sheet metal or fiberglass duct used to transfer air from a fan or motor to an air diffuser or register.

air entrainment: Process in which minute air bubbles, ranging in size from .01″ to .001″, are mixed in a concrete or mortar mix. Improves workability and frost resistance of the mix. In specifications, designated by the letter A following the concrete type designation.

air escape: Device used to automatically discharge excess air from a water supply pipe. Consists of a ball cock that opens a discharging valve when enough air has collected, and closes it to prevent loss of water.

air flue: Flue fitted with a valve that is built into a chimney stack to exhaust air from a room.

* **air gap:** Vertical distance from the outlet of a supply pipe, such as a faucet, to the flood level rim of a fixture or receptacle into which it discharges.

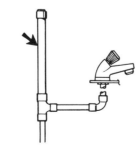
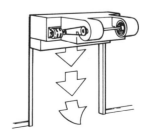

aggregate   air chamber   air curtain   air diffuser

air gap

aligning punch

air grate: Fixed grate through which air is brought into, or exhausted from, a structure for ventilating purposes. Commonly built into the soffit or wall of a structure.

air gun: Paint spray gun. See *aerograph*.

air hammer: **1.** Pneumatic equipment used to drive piles. **2.** Pneumatic percussion tool used to break hard surfaces such as concrete. Chisel or hammer is inserted into the tip.

air handling unit: Prefabricated air conditioner assembly used to treat air before distribution into an air conditioned area.

airless spraying: Process of applying liquid finish by forcing it through specially designed openings in a spray gun under pressure. Minimizes fogging and spray rebound.

air lift: Cleaning of loose debris out of a caisson or pipe pile by introducing compressed air at the bottom of the pile.

air lock: **1.** Stoppage of water flow in a low-pressure supply system caused by air that is trapped in the pipe. **2.** Enclosed area used to isolate an air conditioned space from another area. **3.** Area between rooms with different pressures, such as the entrance to an air-supported structure.

air main: Supply pipe that connects a compressor to a branch line in a pneumatic system.

air plug: Removable stopper that is screwed onto a watertight manhole or scuttle. Used when conducting a hydraulic or pneumatic test of a waste system.

air pocket: Undesirable void in concrete filled with air. Caused by improper placement and consolidation of the mix.

air receiver: Storage tank of an air compressor.

air ring: Perforated manifold in a shotcrete nozzle used to introduce air into the concrete.

air-seasoned lumber: See *air-dried lumber*.

air slaking: Absorption of moisture by quicklime or cement, which chemically decomposes the material.

air space: Cavity in a wall or between various structural members; e.g., between a masonry veneer and a wood-framed wall.

air splitter: Blades extending across the interior of an air duct that divide one air stream into several air streams.

air-supported structure: Non-bearing structure supported by air pressure. Pressure inside the structure is slightly greater than atmospheric pressure.

air terminal: See *lightning arrester*.

air throw: Distance that an air stream travels before the velocity diminishes below a given speed.

airtight construction: Construction in which the structure does not allow air to pass through.

air transformer: Air compressor attachment that removes oil, moisture, and debris from the air line. Commonly used with spray guns.

air trap: See *trap* (1).

air vessel: Enclosed air chamber that uses the compressibility of air to provide better water flow and minimize water hammer.

air void: Space in concrete paste, mortar, or concrete occupied by air. Larger voids than entrained air.

air washer: Enclosed chamber in which air is drawn or forced through a spray of water and passed over electrically charged, wet plates or through conductors to filter; regulates the heat and humidity of the air. May be used as a humidifier or dehumidifier.

airway: Space between thermal insulation and roof sheathing that permits air movement.

aisle: Passageway between upright members or fixtures.

ajarcara: Ornamental brickwork relief.

alabaster: Fine-grained gypsum, white or lightly tinted. Carved for mantel ornaments or other decorative construction.

alameda: Shaded walkway.

albarium: White lime used for stucco.

alburnum: See *sapwood*.

alclad: Aluminum product coated with an aluminum or aluminum alloy for improved corrosion resistance. ALU-MINUM CLAD.

alcohol: Fast-drying, flammable fluid. Ethyl, butyl, amyl, and isopropyl alcohol are used as paint thinners. Denatured alcohol is used as a cleaning solvent.

alcove: Recessed area that opens into a larger room.

alette: **1.** Wing of a building. **2.** Buttress. **3.** Pilaster-like abutment of an arch on either side of a large column or pilaster that supports an entablature.

alga: Small aquatic plant present in stagnant water or on aggregate that decreases the final strength of concrete.

alidade: Basic surveying instrument consisting of a telescope mounted on a column and an index for measuring angles.

align: To arrange in a line.

* **aligning punch:** Long tapered tool used to align bolt or rivet holes in structural steel.

alignment: Formation of a line through two or more points.

aliphatic resin glue: Cream-colored, ready-mixed glue used to bond wood or other porous material. Has high bonding strength and good thermal and solvent resistance, but poor moisture resistance.

alkali: Water-soluble salt including sodium or potassium. Naturally occurring in concrete and mortar. Reaction with aggregate or soil may cause expansion of the concrete.

alkali soil: Soil with a pH value of 8.5 or higher.

alkyd plastic: Self-extinguishing thermosetting plastic with good thermal and electrical resistance and low impact strength. Used in paint, lacquer, and molded electrical parts where temperature does not exceed 400°F.

alkyd resins: Synthetic thermosetting resin used as a binding agent in paint, lacquer, varnish, and adhesives.

Allen screw: See *hex socket screw*.

Allen wrench: See *hex wrench*.

alley: **1.** Narrow roadway extending behind buildings that provides access to the rear of a building. **2.** Covered passage along the side a house that permits access directly to an inner court or yard without entering the house.

all heart: Heartwood throughout; free of sapwood.

all-purpose wire tool

all rowlock wall

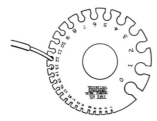

American wire gauge

alligatoring: Rough texture of a painted surface created by the application of another coat applied before the base coat has completely dried.

alligator wrench: See *pipe wrench.*

allowable load: Maximum load an object or assembly can support within specified safety tolerances.

allowable span: Maximum distance a structural member can be unsupported and maintain structural integrity.

allowable stress: Maximum stress an object or assembly can withstand within specified safety tolerances.

alloy: Composition of two or more metals or one metal and a nonmetallic material.

alloy steel: Composition of steel with other metals that yields desirable qualities.

all-purpose framing anchor: Light-gauge galvanized metal member with preformed perforations. Used to join and reinforce the joints of wood framing members such as studs, plates, and rafters. See *anchor**.

* **all-purpose wire tool:** Hand tool with two jaws that can cut wire and small bolts, strip insulation from various size wire, and crimp wire.

* **all rowlock wall:** Masonry wall in which two courses of brick in the rowlock stretcher position alternate with one course of brick in the rowlock position.

all-thread nipple: Short pipe fitting externally threaded its entire length.

alluvial deposit: Soil, sand, gravel or other material deposited by flowing water, especially during times of floods. Not stable enough for structural foundations. ALLUVIAL SOIL, ALLUVIUM.

all-weld-metal test specimen: Test sample composed wholly of weld metal.

altar: Table or elevated structure in the worship area of a church.

altar rail: Rail along the front of an altar. Separates the chancel from the body of the church.

alteration: Modification of an existing structure, including HVAC, electrical, mechanical, or structural, but does not enlarge the overall dimensions of the structure.

alternate lay: Wire rope pattern in which the strands are alternately wound in a clockwise and counterclockwise direction around the core.

alternating current (AC): Electric current that reverses direction at regular intervals of time and that has alternately positive and negative values. Average value of the current over time is zero.

alternating tread stairway: Stairway in which treads are attached to supporting sides at alternating heights. Supporting sides are set at a 50° to 70° angle from the horizontal.

altitude: Perpendicular distance from a specified elevation.

alto-relievo: High relief sculpture in which at least one-half of a figure projects from the face of the surface on which it is carved.

alum: Chemical compound used to accelerate setting time in plaster.

alumina: See *aluminum oxide.*

alumina zirconia: Abrasive material used on metal to remove heavy stock and high-pressure grinding.

aluminize: To apply a surface coating of aluminum by spraying or dipping.

aluminous cement: See *calcium sulfate cement.*

aluminum: Lightweight, noncorrosive, nonmagnetic metal. Good thermal and electrical conductivity. Commonly used in alloy form for additional strength. Used for siding, flashing, exterior ornamentation, and other exterior building products.

aluminum bronze: Alloy of copper and 3% to 11% aluminum. Good corrosion resistance.

aluminum conduit: Aluminum tubing for electrical conductors.

aluminum foil insulation: Insulation in which blanket insulation or gypsum board is covered on one or both sides with aluminum foil or coating.

aluminum molding: See *molding.*

aluminum nail: Lightweight, stain- and corrosion-resistant nail.

aluminum oxide: Synthetic product primarily composed of crystallized bauxite. Commonly used as a wood and metal abrasive, and as a component of brick and tile. ALUMINA.

aluminum paint: Paint containing aluminum paste. Good thermal and corrosion resistance.

ambient noise: Total amount of sound in a given area.

ambient temperature: Temperature of the surrounding air.

ambit: Boundary line.

ambo: Large pulpit or desk in a church.

ambry: Storage area recessed in a wall. ALMERY.

ambulatory: Covered passage or sheltered walk immediately inside the walls surrounding a building.

American bond: See *common bond.*

American National Thread Series: See *Unified Thread Series.*

American standard channel: Structural steel member in which the outside of the flanges and web are perpendicular to one another, forming a C shape. Inner flange surfaces are at approximately 16.67° angle from the outside of the flanges. See *structural steel*.*

American standard pipe thread: Standard pipe thread used for connecting gas, steam, and water pipes. Adjoining sides of the threads are at a 60° angle to each other.

* **American wire gauge:** Standard diameter designation for wire of electrical conductors or the thickness of aluminum, copper, and brass sheet material. Higher number indicates smaller diameter or thickness. AMERICAN STANDARD WIRE GAUGE.

amino plastic: Thermosetting plastic with good electrical insulation and abrasion-resistant qualities.

ammeter: Device used to measure electric current in a circuit. Expressed in amperes.

ammonal: Inexpensive explosive used in a dry blast hole.

ammonia: Colorless, pungent gas used in liquefied form as a refrigerant. Soluble in water or alcohol.

ampacity: Current-carrying capacity of an electrical conductor without exceeding its temperature rating. Expressed in amperes.

ampere: Measure of electric current. Rate at which 1 volt of electricity flows through a conductor with 1 ohm of resistance. AMP.

ampere-hour: Amount of electric current that flows in one hour.

ampere-hour meter: Device used to measure and register the electric current that flows in one hour.

amphiprostyle: Having a row of columns along the front and back, as applied to a structure, but not along the sides.

amphitheater: Elliptical, circular, or semicircular arena with rows of seats in an oval or circular pattern.

amplitude: Maximum displacement from the mean position.

analysis: Process of reducing a problem or compound to its primary components or elements; e.g., soil analysis.

analyzer: 1. Monitoring device used to measure the force applied to a pile when driving it. 2. Device in the high-pressure side of an absorption system used to increase the concentration of refrigerant in vapor entering the condenser.

* **anchor:** 1. To secure a structural member in position. 2. Device used to reinforce a structural member by securing it to another member; device used to secure two or more members together. 3. Metal device used to fasten structural members to concrete or masonry. 4. To lock stressed tendons in a prestressed concrete member in position

to maintain predetermined stress. 5. To attach precast panels to a wall or foundation. 6. Special metal form used to fasten together timbers or masonry. 7. Egg-shaped ornament alternating with a dartlike tongue used to enrich a molding. EGG-AND-DART MOLDING, EGG-AND-TONGUE MOLDING, EGG-AND-ANCHOR MOLDING.

anchorage: 1. Device used to secure stressing tendons in position after stressing a concrete member. 2. Device used to permanently fasten a precast panel to a wall or foundation.

anchor block: Block of wood replacing a brick in a masonry wall. Used as an attachment surface for partitions or fixtures.

anchor bolt: Steel rod threaded at one end used to secure structural members to concrete or masonry. Commonly formed in an L or J shape.

anchor dart: See *anchor* (7).

anchor hook: Hook attached to the lower end of a block-and-tackle. Used without a safety assembly.

anchor knot: Knot used to fasten a rope to a post or ring. BUCKET HITCH.

anchor nail: See *toenail.*

anchor pile: 1. Vertical support member that is driven into the ground behind a retaining wall. Used for securing a tieback rod or cable. 2. See *reaction pile.*

anchor plate: Flat perforated piece of metal used to support a hanger, such as a pipe hanger. Adhesive is applied to the back of the plate and pressed against the mating surface to force the adhesive around the perimeter and through perforations to form a key.

anchor tie: See *tie* (1).

ancon: 1. Projection on a concrete masonry unit. 2. Vertical bracket or corbel used to support a cornice or entablature.

anechoic room: Room with walls that do not reflect sound waves.

anemometer: Device used to calculate air velocity.

angle: 1. Geometric figure formed by two lines extending from the same point. 2. Angular difference in direction between two lines with the same origin. Expressed in degrees, minutes, and seconds, or radians. 3. L-shaped section of rolled, drawn, or extruded metal. See *structural steel*.

ANCHORS

General-purpose anchors

 expansion sleeve

 molly

 plastic

 rawl plug

 screw

 self-drilling

 stud bolt

 toggle bolt

Brick anchors

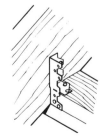 bent strap

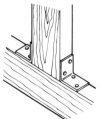 corrugated

 dovetail

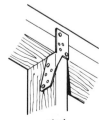 wire tire

Framing anchors

 all-purpose framing

butt joint

rafter

stud

angle bar: 1. Vertical bar between the faces of a bow or bay window. **2.** See *angle iron*.

angle bead: 1. See *corner bead*. **2.** Molded wood, metal, or plastic strip used to provide protection along the corner of a wall.

angle block: See *glue block*.

angle board: Board used as a guide when planing multiple members to similar angles. Desired angle is cut in angle board prior to planing.

angle bond: Metal tie used to bond the corners of masonry walls together.

angle brace: 1. Brace fastened across the inside corner of framework to add rigidity; e.g., light-gauge metal strap fastened across the corners of a wood-framed wall. **2.** Hand tool used to drill in an area without sufficient space to operate a brace.

angle bracket: Support member with two faces, commonly at a 90° angle to one another. Web may extend between the faces for added strength.

angle cleat: Small angle iron member used to fasten a structural member to structural steel.

angle closer: See *closer*.

angle dozer: Earth-moving equipment with a blade fixed at an angle to push earth forward and to the side.

angle drill: Electric drill in which the boring tool is at a 90° angle to the motor. Used to drill holes in inaccessible places where a conventional electric drill cannot be used. See *drill**.

angle float: Float with two opposite edges bent to a 90° angle and used to finish concrete or plaster in a corner. Made of plexiglass, stainless steel, or aluminum. INSIDE ANGLE TOOL.

angle gauge: Template used to lay out or verify angles in carpentry or masonry construction.

angle iron: 1. See *angle* (3). **2.** Perforated metal strap used to reinforce 90° corners.

angle joint: Joint between two wood members that creates a change in direction.

angle newel: Newel post located where a stairway changes direction, such as a landing.

angle of discharge: Highest included angle between the center lines of primary air stream.

angle of repose: 1. Maximum angle above horizontal at which bulk material, such as soil, will remain without sliding. **2.** Greatest angle at which bed joints of an arch will remain supported only by friction. ANGLE OF REST.

angle plane: 1. Wood smoothing tool with cutting iron positioned to form an internal angle. **2.** Hand tool used to prepare a plaster surface after brown coat has set. Removes projections and cleans inside corners.

angle plow: Hand tool used to finish inside plaster corners.

angle post: See *angle newel*.

angle rafter: See *hip rafter*.

angle shaft: Decorative member fastened to an outside corner of a structure.

angle staff: See *corner bead*.

angle tie: See *angle brace* (1).

angle valve: Globe valve in which the inlet and outlet are at 90° to one another. Less resistance to water flow than a globe valve and a 90° elbow. ANGLE STOP.

angling dozer: Earth-moving equipment with a pivoting blade used to push earth forward and/or to the side.

angular flashing: Sheet metal flashing formed into an L shape with an obtuse angle. Used to divert water away from a structure and prevent leakage. See *flashing**.

anhydrous lime: See *quicklime*.

animal glue: Adhesive made from hooves, bones, and hides of animals. A good adhesive for gap filling; used in interior furniture. HIDE GLUE.

annealed wire: Soft, pliable wire used extensively for tie wires, especially for wiring concrete forms and reinforcing steel.

annealing: Process of heat treating metal in which the metal is heated and allowed to cool slowly. Used to relieve internal stress, improve machinability and cold-working characteristics, and mechanical and physical properties.

* annual ring: Growth ring visible in a horizontal cross section of a tree. Composed of springwood (light and porous wood) and summerwood (dark and dense wood). Age of a tree may be determined by counting the rings. GROWTH RING.

annular: Ring-shaped.

annular ball bearing: Component of a pivoting assembly in which the ball bearings are sealed within a cylindrical race.

annular bit: See *plug cutter*.

annular nail: See *ring-shank nail*.

annular vault: Ring-shaped barrel vault formed by connecting two concentric rings.

annulated column: Column composed of two or more shafts surrounded by a decorative retaining band. Common in clustered columns of Gothic structures.

annulet: Small band of molding encircling a capital or column.

annunciator: Electrical signaling device that indicates when a circuit is energized with an audio or visual signal.

anode: 1. Electrode through which current enters a nonmetallic material. **2.** Electrode or portion of an electrode on which a chemical reaction occurs; e.g., an anode in an electric water heater. Used to protect components from electrolysis.

anodize: To coat a metallic surface with a protective film obtained through electrolytic action.

anta: Rectangular pier or pilaster formed by a thickened end of a wall.

anta cap: Capital of an anta.

antechamber: Foyer or vestibule.

antefix: Upright ornament placed at regular intervals along the eaves or cornices of a roof to conceal the ends of roofing tile.

antepagment: Stone or stucco ornamentation around a doorway or window.

anteroom: Waiting area adjacent to a larger room.

ante-solarium: Balcony that faces the sun.

* antichecking iron: Narrow piece of flat iron driven into the end of a piece of timber to prevent it from splitting. Sharpened on one edge and bent to an S, C, or Z shape.

anticipator: Fixed or adjustable thermostat component used to regulate a heating and/or air conditioning unit. Prevents overheating or overcooling of an area beyond a predetermined temperature.

antifreeze: Substance added to liquid to lower its freezing point and prevent solidification.

antimonial lead: Alloy of lead and antimony; increases tensile strength and

annual ring

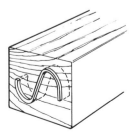

antichecking iron

hardens lead. Used for roofing and tank lining.

antiquing: Method of applying a second layer of contrasting or complimentary color of paint over a base coat. Base coat is partially exposed by combing, graining, or marbling.

anti-siphon trap: Trap that increases the resealing quality of a drainage system. Additional seal is obtained by using a large bowl on the outlet leg of a trap, thus increasing the volume of water seal.

anvil: Pile hammer component that transmits the driving force to a pile. Located directly below the ram.

anvil block: Movable piece of steel in a paving breaker or air hammer used to transfer rotating action of the motor to a reciprocating chisel or tip.

apartment: Room or group of rooms designed as a one-dwelling.

aperture: Opening in a wall for a door, window, or ventilating unit.

apex: Highest point of a structure.

apex stone: Triangular piece of masonry at the top of a gable, vault, or dome. Often highly decorated. SADDLE STONE.

apparent candlepower: Measure of illumination expressed as the candlepower of light transmitted across a specified distance.

apparent wattage: Measure of electrical power in a circuit. Determined by multiplying the voltage by the amperage.

appearance grade: Lumber grade based solely on the finished appearance. Divided into industrial, architectural, and premium grades.

appendage: See *addition*.

appentice: See *lean-to*.

appliance: Utilization equipment, generally other than industrial equipment, installed or connected to perform duties such as heating, cooking, and cooling.

appliance branch circuit: Branch circuit that supplies energy to one or more outlets. Individual lighting fixtures are not attached to the circuit.

applied molding: Trim member attached to the face of a flat surface. APPLIED TRIM.

applique: Ornamental trim member.

apprentice: Person who enters into an agreement with an employer or Joint Apprenticeship and Training Committee (JATC) for a required period of time to receive instruction and learn a trade. Related instruction is provided by a JATC. Protected by federal and some state laws and the local JATC in regard to work hours, wages, and conditions of employment.

apprenticeship: Training for a trade or skilled craft that requires a wide variety of skills and knowledge. Provided with instruction and on-the-job experience in practical and theoretical aspects of a skilled trade. Commonly lasts from three to five years.

approach: Difference in temperatures between exhausted air or fluid and the wet bulb temperature of the entering air or fluid in a water cooling device or heat exchanger.

approach ramp: 1. Sloped access for the handicapped. **2.** Access road for vehicles entering a highway.

approved: Acceptable to the authority having jurisdiction, as applied to construction or installation of materials or equipment.

appurtenance: 1. Non-structural portion of a building that is built-in, such as electrical equipment, ventilating units, and windows. **2.** Right-of-way; easement.

* apron: 1. Plain or molded finish piece immediately below a window stool. Used to conceal the rough edge of wall finish. **2.** Front gate of the body of an earth-moving scraper. **3.** Short ramp or slab with a slight pitch, such as around a garage or airport hangar. **4.** Horizontal member directly below the edge of a tabletop. Attached to the legs and underside of the tabletop. **5.** See *backsplash*.

apron piece: Beam or horizontal timber that supports the upper ends of a stringer in a stairway. PITCHING PIECE.

apron rail: See *lock rail*.

apron sink: Sink in which the front and sides extend horizontally approximately 5″ to 8″ outside of the bowl.

apron wall: Vertical exterior wall panel extending from a window sill to the top of a lower window.

apse: 1. Semicircular section of a structure that projects horizontally beyond the main structure. **2.** Vaulted section of a church behind the altar.

aquastat: Underwater device used to regulate water temperature.

aqueduct: Structure used to transport water above or below ground.

aquifer: Underground formation that supplies water to springs and wells.

arabesque: Intricate pattern of interwoven plants and geometric figures. May be painted, inlaid, or carved into a surface.

* arbor: 1. Rotating shaft of an electric tool, such as an electric saw or shaper. May be direct- or belt-driven. **2.** Area of closely planted trees and shrubs supported by latticework.

arc: 1. Sustained discharge of electricity across a gap in a circuit or between electrodes. **2.** Continuous section of a curved line.

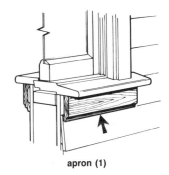

apron (1)

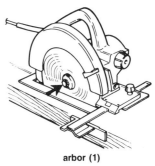

arbor (1)

arcade: 1. Arched roof or covered passageway. **2.** Series of arches supported by piers or columns.

arc blow: Deflection of an electric arc from the normal path as a result of magnetic forces.

arc cutting: Cutting process in which metal is cut with the heat of an electric arc between an electrode and the base metal.

arc gouging: Arc cutting process used to form a groove or bevel.

arch: Curved, pointed, or flat structure or structural component that spans an opening and is supported on both ends.

arch bar: Metal strap or beam used to support a flat arch.

arch brace: Curved structural member used to support a roof frame.

arch brick: 1. Wedge-shaped brick used in arch construction or other circular brickwork. COMPASS BRICK, RADIAL BRICK, VOISSOIR BRICK. **2.** Hard brick that has been overburned by being placed in contact with the fire in the kiln.

arch buttress: See *flying buttress*.

arch center: Formwork used to support a masonry arch or vault during construction. Removed after mortar has obtained sufficient strength.

arched beam: Beam with a slightly arched upper surface.

arching: Transfer of shear stress in a soil mass across an area of low shear strength to an area with a higher shear strength.

architect's scale

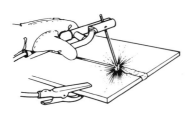

arc welding

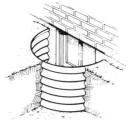

areaway

armored cable

architect: Person licensed and qualified to perform architectural services such as analyzing a construction project, creating and developing the project design, drawings, specifications, bidding requirements, and administering the project.

* **architect's scale:** Triangular-shaped scale with six graduated surfaces. Two different scales are marked on five of the surfaces and inches and fractions of an inch are marked on the other surface.

architectural area: Total interior floor area of a structure. Exterior roofed areas, such as a carport or porch, are calculated by multiplying the actual floor area by .5. Exterior unroofed areas are not included in the calculations.

architectural bronze: Alloy composed of 57% copper, 40% zinc, 2.75% lead, and 0.25% tin. Used for ornamental work.

architectural concrete: Concrete that is permanently exposed to view and requires care in selection, forming, placing, and finishing of the surface to meet certain appearance standards.

architectural door: Door classification that designates high material and appearance specifications.

architecture: Art and science of designing and constructing structures based on aesthetic and functional criteria.

architrave: Lowest part of an entablature resting on the capital of a column.

architrave cornice: Entablature consisting of the architrave and cornice.

archives: Repository for public records and other documents.

archivolt: 1. Ornamental band around the perimeter of a curved structure such as an arch. **2.** Ornamental molding on the face of an arch.

arch rib: Primary load-bearing member of a ribbed arch.

arch ring: Curved member of an arch that is the primary load-bearing member.

arch stone: See *voussoir*.

archway: Passageway under an arch.

arc length: 1. Distance between the end of an electrode and the surface of the base metal. **2.** Length of an arc measured along a curve.

arc plasma: See *plasma arc welding*.

arcuated construction: Stonework construction characterized by arches and vaults.

arc welding: Welding process in which metal is joined using heat obtained from an electrical arc, with or without the application of pressure, and with or without filler material.

area: 1. Surface measurement within two boundaries. Expressed as square units, such as square feet or square inches. **2.** Space designated for a specific purpose, such as an area drain. **3.** See *areaway*.

area drain: Drain used to collect surface water from a given area. Commonly used in a basement or depressed loading dock.

area separation wall: Fire-rated, 2″ to 4″ thick partition used to prevent the spread of fire. Extends from the foundation to or above the roof line.

* **areaway:** Open, below-grade area around a basement window or door. Provides light, ventilation, and means of access.

areaway wall: Horizontal structure built to retain and support earth around an areaway. Constructed of concrete, concrete block, brick, rubble stone, or steel. AREA WALL.

arenaceous: Primarily composed of sand.

argillaceous group: Group of material primarily composed of clay and shale. Provides alumina component of portland cement.

argon: Inert gas used for shielding gas in gas tungsten arc welding.

armature: Rotating component of a motor or generator; consists of copper wire wrapped around an iron core.

armor coat: Asphalt pavement produced by applying two or more thin layers of aggregate and asphalt.

* **armored cable:** Assembly of insulated conductors with a covering consisting of a helically wound metal strip with interlocking edges. Commonly circular in cross section, but may be oval or flat. BX.

armored concrete: See *reinforced concrete*.

armored front: Device used with a mortise lock to prevent tampering with cylinder set screws. Consists of an underplate attached to the door jamb and a finish plate fastened to the underplate.

armor plate: See *kick plate*.

array: Group of solar collectors positioned so as to maximize efficiency.

arrester: Corrosion-resistant enclosure installed over a chimney flue to reduce the hazard of flying sparks. Constructed of perforated sheets, wire, or expanded metal. SPARK ARRESTER.

arris: Edge or ridge where two surfaces intersect; e.g., edge formed where two moldings intersect.

arris fillet: Triangular wood member along the intersection of a roof and a wall and under roofing material. Used to divert rainwater away from the vertical surface. See *cant strip*.

arrisway: 1. Tile or slate laid diagonally. **2.** Squared timber cut diagonally. ARRISWISE.

articulation: Process of constructing movable joints, usually with joint pins.

artificial marble: See *artificial stone*.

artificial seasoning of wood: Removal of moisture from wood by a means other than air drying. See *kiln dried lumber*.

artificial stone: Mixture of stone chips with cement, mortar, or plaster that is seasoned for several months and polished to a smooth finish to simulate stone.

artisan: Skilled craftsperson; artist.

art metal: Metal shaped into artistic forms and used for ornamental purposes such as a door knocker.

asbestine: Silicate of magnesuim used in paint as a binding agent and to retain pigment. FRENCH CHALK.

asbestos: Non-combustible mineral fiber derived from natural hydrous magnesium silicate. Has high thermal and electrical resistance. Airborne particles pose a health hazard. Extreme care should be taken in installing and removing asbestos products.

asbestos abatement: Process of safely cleaning, removing, and disposing of asbestos from areas of potential human exposure. ASBESTOS REMOVAL.

asbestosis: Lung disease caused by extended exposure to asbestos fiber.

ash dump: Opening in the floor of a firebox or fireplace into which ashes are swept to an ashpit below.

*__ashlar:__ Squared stone with a flat surface used in foundations and for facing masonry walls.

ashlar brick: Brick with a roughly surfaced face that resembles stone.

ashlaring: Process of facing a wall with ashlar.

ashlar line: Horizontal line on the exterior face of a masonry wall.

ashlar masonry: Masonry consisting of squared stone with flat surfaces bonded with mortar.

ash pan: Metal container under a grate of a fireplace used for collecting and removing ashes.

ashpit: Cleanout compartment under a firebox or fireplace used for collecting and removing ashes.

ashpit door: Cleanout door that provides access to an ashpit for removal of ashes.

askarel: Group of nonflammable synthetic chlorinated hydrocarbons used as insulation for electrical conductors and other electrical devices. Gases produced under arcing conditions are generally noncombustible.

aspect ratio: Relationship of the longer dimension to the shorter dimension in a rectangular form or configuration; e.g., relationship between the width and depth of a rectangular duct.

asphalt: Dark-colored pitch primarily composed of natural or synthetic bitumens. Used as ingredient in paving material, waterproof roof covering, and expansion joints.

asphalt block: Paving member composed of 88% to 92% crushed stone and the remainder asphalt cement.

asphalt emulsion: Asphalt in which water has been suspended; used as an initial paving coat. As water evaporates, asphalt hardens.

asphaltic: Containing asphalt.

asphaltic cement: Material made by refining petroleum to remove water and foreign material. Contains less than 1% ash. Must be heated to fluid consistency for application. ASPHALT CEMENT.

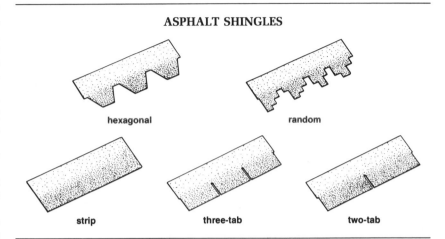

ASPHALT SHINGLES

hexagonal   random

strip   three-tab   two-tab

asphaltic mastic: Mastic composed of asphalt and filler material such as sand. Used as an adhesive, sealant, and for waterproofing purposes. ASPHALT MASTIC.

asphalt-impregnated: Saturated with an asphaltic substance to improve moisture resistance.

asphalt paint: Liquid asphaltic product used for waterproofing purposes. May be mixed with lampblack or other mineral pigments.

asphalt roofing: See *roll roofing*.

*__asphalt shingle:__ Composition shingle made of asphalt-impregnated felt surfaced with mineral granules.

asphaltum: Natural mineral pitch and a residue produced from petroleum and coal tar. Dark-colored, brittle, and glossy bitumen.

aspiration: 1. Introduction of room air into the air stream discharging from a diffuser. **2.** Process of removing a material by means of a vacuum.

aspirator: Device that draws a stream of liquid or gas through it by means of negative pressure produced by the flow of a liquid through an orifice.

assembly drawing: Drawing showing the entire machine or structure with all detailed parts in their operating position.

assize: Cylindrical block of stone that forms part of a columm.

associate dimensioning: Process used by a computer-aided design and drafting system whereby the software program automatically changes some dimensions to reflect revisions that the operator has made to a dimensioned member.

*__astragal:__ Molding consisting of one or two pieces; used to cover the opening between the stiles of a pair of doors or casement windows. Provides a weather seal.

astreated: Decorated with star-shaped design.

athley wagon: Large trailer with a bottom dump used to transport soil and rock.

atmosphere: Mass of air surrounding the earth.

atmospheric pressure: Pressure exerted by the earth's atmosphere under standard conditions at sea level. Equal to 14.7 pounds per square inch.

atmospheric refraction: Deflection of a line of sight of a surveying instrument as a result of atmospheric conditions.

atomic hydrogen welding: Welding process in which metal is joined by heat produced by an electric arc in an atmosphere of hydrogen.

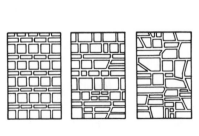

ashlar

astragal

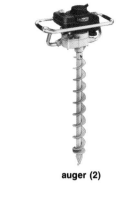

auger (2)

automatic taping machine

---

atomization: Fine spray produced by impinging a stream of air on a small stream of paint.

atrium: 1. Large hallway or lobby with galleries on three or more sides at each floor level. **2.** Area enclosed with glass or other transparent material that admits sunlight. Commonly used to house plants.

attached column: Column with three-fourths of its diameter projecting from the adjacent wall.

attachment plug: Device that when inserted in a receptacle makes connection between the conductors in an electrical receptacle and an attached flexible cord.

attenuation: Reduction in the intensity of sound due to sound energy being converted to thermal or mechanical energy.

Atterberg limits: Water content of plastic soil that defines the upper and lower boundaries between the different states of soil consistency.

attic: Area directly below the roof and above the ceiling of the upper story of a structure. GARRET.

attic louver: Screened or slotted opening used to ventilate an attic. May be rectangular or triangular.

attic ventilator: Electric or wind-driven fan used to ventilate an attic. May be thermostat-controlled.

* **auger: 1.** Tool used to bore holes up to 1″ diameter in wood. Fastened to the jaws of a brace. Lead screw advances

bit into wood. See *bit* (1)*. AUGER BIT. **2.** Rotary tool used to bore holes into soil or rock. May be manually operated or power-driven.

auger-grout pile: Pile formed by boring a hole with an auger and placing pressurized grout in the hole through the hollow center of the auger. Auger is withdrawn at a steady rate as the grout is placed.

autoclave: Pressurized chamber in which an environment of steam is used to cure concrete products and test hydraulic cement.

autogenous sealing: Natural process of sealing cracks in concrete or mortar while it is kept damp.

autogenous weld: Fusion weld made without adding filler metal.

automatic: Activated by the environment or predetermined condition.

automatic center punch: See *self-centering punch.*

automatic fire suppression system: Fire protection system in which heat or smoke activates a control component that permits a noncombustible material to be discharged onto the combustion area. Commonly used in commercial kitchens.

automatic flue damper: Automatic device between a vent and draft diverter in a furnace used to prevent the escape of excess heat.

* **automatic taping machine:** Device that applies reinforcing tape and joint compound simultaneously to flat joints or corners. AMES TAPING TOOL®, BAZOOKA.

automatic transfer switch: Electrical switch that automatically transfers a load conductor connection from one power source to another.

auxiliary switch: Electrical control operated by a main device for signaling, interlocking, and other similar purposes.

avalanche protector: Guard plate along the sides of the tracks or wheels of earth-moving equipment. Used to prevent loose material from sliding into and contacting the tracks or wheels.

average bond stress: Measure of the tensile force applied to a rebar. Determined by dividing the tensile force by the product of the perimeter and embedded length of the rebar. ANCHORAGE BOND STRESS.

aviation snips: Snips used to cut compound curves and intricate patterns in sheet metal. Available as right- and left-cutting. See *tin snips.*

avoirdupois weight: Measure of weight based on a pound equaling 16 ounces and an ounce equaling 16 drams.

avulsion: Modification of a shoreline resulting from the formation of a new stream or river channel.

awl: Short pointed instrument used to pierce holes in soft materials such as wood. Also used to lay out lines on hard, nonporous surfaces.

awl haft: Handle of an awl.

awning: Roof-like covering of canvas or metal placed over a window or entrance to provide protection against rain or sunlight. May be retractable.

awning window: Window that is hinged along the upper edge of the sash with the bottom sash swinging outward. See *window*.*

ax: Cutting tool with a long handle and two sharpened edges used to split wood and hew logs. AXE.

axhammer: Hand tool used to dress or spall rough stone. May have either two sharpened edges or one sharpened edge and one hammer face.

axial: Situated on or around an axis.

axial force: Compressive and tensile forces acting along the length of a member.

axis: Straight line or plane about which a body or member rotates.

axis of symmetry: Straight line about which a body or member is symmetrically developed and on which the center of gravity is located.

axis of weld: Straight line through the length of a weld; perpendicular to and at the center of its cross section.

axle load: Portion of the gross weight of a vehicle that is transmitted to a structure or pavement through wheels supporting a given axle.

Axminster carpet: Machine-woven carpet in which pile tufts are inserted between the warp threads and bound with a filler material. Permits intricate designs and multiple colors.

axonometric drawing: Pictorial drawing in which the angle of the axes varies with the type of representation. Types include isometric, dimetric, and trimetric. Position of object is inclined with respect to the plane of projection, and all receding lines are foreshortened.

azeotrope: Refrigerants characterized by a minimum or maximum boiling point that is lower or higher than any of the components.

azeotropic point: Temperature at which liquid boils and forms a vapor.

azimuth: Horizontal angle that is measured clockwise from the true north meridian to an object or fixed point. Expressed in degrees, minutes, and seconds.

babbitt: Anti-friction alloy of soft metal made of copper, tin, and antimony. Used in the manufacture of bushings and bearings.

back: Top exposed portion of a masonry member, slate, or tile.

back arch: Concealed arch that supports the inner portion of a wall. A lintel or other support carries the exterior of the wall.

back band: Rabbeted outer molding of the inner trim of a window or door casing. BACK BEND.

back bar: Horizontal bar in the chimney of an open fireplace on which kettles and other cooking utensils may be hung.

back blocking: Small pieces of gypsum drywall installed on the back of otherwise unreinforced single layer drywall joints. Backing pieces are held in place with joint compound.

back catch: Fastener attached to an outside wall for catching and holding a shutter, window blind, or door in an open position.

back check: Portion of a door closer that reduces the speed at which the door may be opened.

back draft damper: Flue-installed control that allows the passage of air in the flue in only one direction.

backdrop: Connection in a manhole where the branch drain enters by means of a vertical pipe.

back edging: Chipping away the biscuit below the glazed face of a brick or tile, the front itself being scribed.

back electromotive force: Voltage across the starting and common connections in a 1 phase motor. COUNTER EMF.

backer sheet: Thin layer of plastic laminate on the back of a laminated piece.

backfill: 1. Dirt, stone, or other materials used to fill the excavated trench around a foundation, plumbing piping, or other trenches and build up the ground level. **2.** See nogging.

back fillet: Edge by which a slightly projecting structural part returns to the face of the wall, as a quoin or architrave.

back-flap hinge: Hinge in two leaves, screwed to the face of a door that is not thick enough to permit the use of a butt hinge.

backflow: Flow of water or sewage in the direction opposite its normal flow.

backflow preventer: Plumbing device that stops the flow of water or sewage in the direction opposite its normal flow. BACKFLOW VALVE.

backflow valve: Plumbing device in a drain that prevents a reversal of the flow of sewage. BACKFLOW PREVENTER.

back gouging: Removal of welding and base metal from the opposite side of a partially welded joint to help complete joining when further welding is done on this side.

*__back hearth:__ Fireproof floor under a fireplace grate upon which the fire is built.

*__backhoe:__ Piece of excavating equipment with a loading bucket in front, which is drawn inward during operation.

backing: 1. Rubble or broken stones used at back of masonry facing. **2.** Wood strips nailed at the meetings of walls and ceilings to provide solid corners for nailing finish materials. **3.** Bevel on top edge of a hip rafter that allows roofing boards to fit the top of a rafter. **4.** Paper, cloth, fiber, or a combination of these materials to which abrasive materials are attached for making coated abrasives (sandpaper). **5.** Material such as metal, carbon, and gas used to support a joint during welding. **6.** Materials that form the back of a carpet. **7.** First coat of plaster applied to lath.

8. Excavation material used for backfill against a retaining wall. **9.** Coursed masonry laid above the extrados of an arch. **10.** Removing portion of the rear of a member to enable it to fit tightly against another surface.

backing board: Gypsum drywall installed in a suspended ceiling; serves as an attachment surface for acoustical tile.

backing brick: Lower-quality brick used for constructing the inner part of a brick wall.

backing up: Laying inside masonry units after the exterior of the wall has been laid.

back iron: Steel plate that reinforces the cutting iron of a wood plane and breaks the shavings.

back joint: Rabbeted section of masonry that accepts a wood nailer.

back lintel: Horizontal masonry supporting member on the rear of a wall.

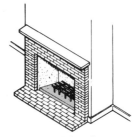

back hearth

backhoe

back nut: Female threaded fastener with one end shaped to accept a grommet. Used to form a watertight seal.

back observation: See *backsight*.

back of an arch: Extrados of arch or vault.

back plastering: 1. Installation of lath and plaster in the stud space midway between the outside sheathing and inside lath and plaster of an exterior wall to provide a double air space for insulation. **2.** Application of a ³/₈″ thick mortar coat on the back of a masonry facing tier for purposes of moisture-proofing and airproofing. PARGING.

backplate: Piece used on the inside of a door for framing an opening for a letter drop.

back pressure: Air pressure in pipes that is greater than atmospheric pressure. BACK SIPHONAGE.

back puttying: Forcing of putty into space left between edges of the rebate and glass on the back of the sash after face puttying.

backsaw: Cutting hand tool with stiffened blade. Used in cabinet work as a bench saw. See *saw*.

back seam: Carpet seam made while the carpet is turned face down.

backset: Horizontal distance from a door edge to the center line of the knob or keyhole. SETBACK.

backsight: 1. Vertical distance from the bench mark to the line of sight. **2.** Point for determining the elevation and/or angular position of a surveying instrument.

back siphonage: See *backflow, back pressure*.

*** backsplash:** Narrow vertical piece of material, usually waterproof, installed at the meeting of a cabinet top and the rear wall.

back splice: Means of preventing the unraveling of an end of rope. A crown knot is tied and the ends are given several extra tucks.

backup: 1. Part of a masonry wall behind the exterior facing; consists of one or more wythes of brick or other masonry material. **2.** Material installed prior to application of a sealant or caulking to reduce amount of sealant required and prevent sagging.

back vent: Ventilation pipe attached to a waste pipe on the sewer side of its trap to prevent siphonage of waste or excess fluid.

backwater flap: Hinged metal flap that prevents flow of liquid in the wrong direction.

*** backwater valve:** Automatic valve placed in a house drain or sewer lateral to keep sewage from backing up into plumbing fixtures during heavy rain and flooding.

badger: 1. Implement used to clean out excess mortar at the joints of a drain after it has been laid. **2.** See *badger plane*.

badger plane: Wide wooden rabbet plane having a skew mouth with iron flush at one side. Used for planing into a corner.

badigeon: Cement or paste of suitable material used to fill holes or cover defects in stone or wood.

baffle: 1. Artificial surface, usually a plate or wall, for deflecting, retarding, or regulating flow of fluids or gases. **2.** Portable screen that absorbs or deflects sound to improve its quality or characteristics.

bag of cement: Measure of portland cement equal to 94 pounds.

bag plug: Inflatable drain stopper used to block lower end of a drain while testing plumbing.

bag trap: S-shaped plumbing fitting with inlet and outlet in alignment.

bague: Ring of a plate of an annulated column.

bahut: Rounded upper course of masonry on a wall or parapet.

baked: Paint finish exposed to temperatures exceeding 150°F, creating a hard finish.

balanced circuit: Electrical circuit with the same load in each ungrounded conductor and no load on the neutral conductor or imbalance between the ungrounded conductors.

balanced door: Opening covering that is equipped with a counterbalance to facilitate opening and closing.

balanced load: Electrical load connected so that current taken from the ungrounded conductors of a three-wire system are equal.

balanced sash: Double-hung window member that can be easily raised because of a system of springs or weights that acts as a counterbalance.

balanced step: Any one of a series of winding treads so arranged that small ends of the treads are not much narrower than the parallel treads or fliers.

balancing fitting: Movable control valve that increases and decreases fluid flow but does not act as a shut-off. BALANCING VALVE.

balcony: 1. Platform enclosed by a railing or balustrade; projects from the face of an inside or outside wall of a building. **2.** Gallery in a theater.

balk: 1. Large squared timber or beam. **2.** Low earth ridge marking a boundary line.

ballast: 1. Autotransformer that delivers a specific voltage and limits the electric current required by a gas lamp, such as a fluorescent lamp. **2.** Layer of coarse stone or gravel on which concrete or timbers are placed.

ball bearing: Bearing in which inner and outer edges turn upon loose, hardened steel balls in a channel inside the bearing.

ball bearing hinge: Knuckled hinge equipped with ball bearings that prevent wearing at the joints and ensure easy and quiet operation of doors. Designed for heavy, oversized doors and inside doors of hospitals, schools, and office buildings. BALL BEARING BUTT.

*** ball catch:** Door-fastening hardware in which a spring-controlled metal ball projects through a smaller hole to engage a sunken strike plate. BULLET CATCH.

ball check valve: Fluid control device that permits the flow of liquid in one direction while stopping flow in the opposite direction.

ball cock: Self-regulating faucet that is opened or closed by the rising or falling of a hollow ball floating on the water surface. Used to supply water to a tank.

ballflower: Ornamental carving resembling a ball enclosed within three flower petals.

balling up: Globules of molten brazing material resulting from lack of wetting of the base metal.

ball joint: Plumbing connection in which a spherical object is held in a cuplike

backsplash

backwater valve

ball catch

shell. Allows for movement in many directions, except in the direction of the axis.

balloon: Ball or globe crowning a pillar, pier, or other similar structural member.

* **balloon framing:** Residential building construction in which one-piece studs extend from the first floor line or sill to the roof plate. Joists for upper floors are nailed to the sides of studs and receive additional support from ledger boards.

ball peen hammer: Striking tool with the end opposite the striking face shaped in a half-round design. Used by metalworkers and stonemasons. See *hammer*.

ball points: Measuring tool used for determining the center of a hole. A semicircular end is fitted on a straight shaft, creating a seat for dividers or trammel points.

ball reamer: Hand tool used for finishing concave portion of a ball joint.

ball test: 1. Method of determining consistency of concrete by dropping a 6″ diameter ball into concrete in a plastic state. BALL PENETRATION TEST. 2. Method of determining roundness of a drain by dropping a ball of a given diameter through the drain.

ball valve: 1. Water valve that regulates flow in a tank by means of a floating ball. The ball fits into a spherical seat to control fluid flow. 2. See *ball cock*.

* **baluster:** One of a series of slender vertical members of a balustrade; an upright support for the railing of a stairway.

baluster shaft: Column of a baluster.

balustrade: Railing consisting of a series of small columns connected at the top by a coping; a row of balusters on which a rail is mounted.

band: 1. Horizontal decorative feature of a wall, such as a flat frieze or fascia, usually with a projecting molding at upper and lower edges. The flat portion between the moldings may be ornately decorated or flat and unadorned.

2. Steel strap that holds lumber, brick, or other building materials together during transportation. 3. To decorate with a band, strip, or stripe.

bandage: 1. Strengthening band or strip of stone or iron. 2. Metal band placed around a tower or dome to prevent spreading and provide added strength.

band board: Floor joist set on top of the sill around the perimeter of a wood-framed floor. HEADER JOIST, RIM JOIST.

band clamp: Strip of steel or canvas attached to a cinching device for holding irregularly shaped objects in position. See *clamp*.

band course: Horizontal row of brick or stone around a column or around a building; serves as an ornamental feature and belt course.

banding: 1. Steel straps fitted tightly at the driving end of wood pile to prevent splitting during driving. 2. Steel straps stretched around material, such as brick, to hold them together during shipping.

band saw: Electric-powered cutting equipment with a continuous blade passing around two wheels and secured by roller guides. Used for irregular-shaped cuts in wood and metal. See *saw*.

bandshell: Bandstand with a sounding board at the rear; shaped like an open portion of a sphere.

banister: 1. Handrail. 2. See *baluster*.

banjo bar: Large U-shaped bar held against a rivet while forming its head.

* **banjo taping machine:** Holder and semiautomatic applicator for paper tape used to cover seams in drywall. For wet or dry application of paper tape.

bank: Soil that rises above an excavation or trucking level.

banker: Workbench on which bricklayers and stonemasons work when shaping arches or other construction requiring shaped materials.

bank measure: Calculation of soil and/or rock volume prior to blasting or excavation.

bank of transformers: Electrical trans-

formers located at one place and connected to the same circuit.

bank plug: Surveyor's wooden stake; driven with top approximately 24″ above ground level. A string is attached and stretched between stakes to measure grade.

bank run: Excavated stone not graded for size. Contains sand, gravel, and stone of various sizes. BANK GRAVEL.

bank sand: Fine stone material with relatively sharp edges. Used to form strong bonding in plaster.

bar: 1. Slender strip of wood that separates and supports panes in a window sash. 2. Round, deformed piece of steel rod used to reinforce concrete. See *reinforcing bar*. 3. Any solid metal product with a simple geometric cross section. 4. Ridge or mound of unconsolidated material at an elevation just below high water level.

barbed: With jagged edges or points.

* **barbed dowel pin:** Short headless nail with a sharp, pointed lower end and sharp edges between ends. Used largely for fastening mortise-and-tenon joints in a window sash and for door work.

bar chair: See *chair*.

bar clamp: Tool consisting of a long flat piece of metal with two clamping jaws and a tightening device. Used primarily by woodworkers. See *clamp*.

bare conductor: Uninsulated electrical wire.

barefaced tenon: Tenon that is shouldered on one side only.

barefoot joint: Joint set up and fastened without a mortise and tenon; e.g., in balloon framing, post or stud butting against another and held in position by toenailing.

bargeboard: Decorative piece of lumber that covers projecting portion of a gable roof. GABLEBOARD, VERGEBOARD.

barge couple: See *barge rafter*.

barge course: 1. Part of roof tiling that usually projects beyond principal rafters or bargeboards along the sloping edge of a gable roof. 2. Course of brick laid on edge to form coping of a wall.

balloon framing

baluster

banjo taping machine

barbed dowel pin

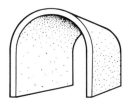

barrel vault

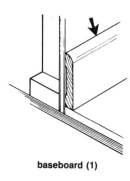

baseboard (1)

barge rafter: Gable end roof member that supports a decorative end member; may project beyond the face of gable end wall.

barge spike: Large square nail with a chisel point; used in heavy timber construction.

barge stone: One of the stones that forms sloping edge of a gable roof.

bar handle: Horizontal rod mounted on one or more brackets for opening a door.

bariom plaster: Mixture of barite aggregate with gypsum plaster or portland cement. Used on x-ray room walls to reduce radiation transmission.

bar joist: Flat steel truss with an open web. Used for floor and roof framing. See *joist*.

bark: Rough outer covering of tree trunks. Sometimes used for decorative effect or landscape ground cover.

bark pocket: Patch of bark nearly or wholly enclosed in wood. See *defect*.

bar list: Bill of materials showing all quantities, sizes, lengths, and bending dimensions of rebars.

bar mat: Steel reinforcement of two or more layers of rods set perpendicular to each other and tied or welded together.

barn boards: Softwood lumber commonly used for barn siding.

barn door hanger: Device consisting of a sheave mounted in a frame fastened to a sliding door that travels on an overhead track, or rail, which carries heavy sliding doors.

barn door latch: Heavy thumb latch.

barn door pull: Large handle attached to heavy doors.

barn door roller: Roller assembly for a barn door or any other extra-heavy door that travels on a floor-mounted track or rail.

barn door stay: Device consisting of a small roller, sometimes mounted on a screw or spike, for guiding heavy sliding doors.

barn siding: Overlapping boards used as an exterior covering of a building.

bar number: Designation of diameter of rebars. Approximate bar diameter given in eighths of an inch; e.g., a #5 bar is approximately $5/_8$″ in diameter.

barometer: Device for measuring atmospheric pressure.

barracks: Simple building that provides temporary living and/or sleeping quarters.

barrage: Low dam for controlling water level in a stream.

barrel: 1. Measure of cement equal to 4 U.S. bags, a total of 376 pounds. 2. Liquid measure equal to $31^1/_2$ gallons. 3. Portion of a pipe in which bore and wall thickness are uniform.

barrel arch: Arch that resembles a segment of a barrel with greater length than span.

barrel bolt: Round piece of metal for fastening a door or window sash; made to slide into a cylindrical socket or barrel.

barrel drain: Drain that is cylindrical in shape.

barrel hitch: Knot used in hoisting materials.

barrel nipple: Short piece of pipe with an outside thread on each end and a bare middle section. BARREL FITTING.

barrel roof: Roof similar in shape to the interior of a barrel.

* **barrel vault:** Semi-cylindrical arch or roof with parallel abutments and same section throughout.

barricade: Obstruction designed to deter vehicular or pedestrian traffic.

barrow: See *wheelbarrow*.

bar sash lift: Bent rod forming a handle fastened at each end of a sash with screws. Used as a handle to raise and lower a sash.

bar size section: Hot-rolled steel member, such as an angle, a bar, or channel, with a maximum cross-sectional dimension of 3″.

bar spacing: Placement of rebars measured from center-to-center.

bar support: Device, usually of formed wire, that supports, holds, and spaces rebars. See *chair*.

basal angle: Angle at the base of a structural member.

bascule: Member that pivots on an axis and is counterbalanced at one end.

base: 1. Lowest part of a wall, pier, monument, or column; the lower part of a complete architectural design. 2. Supporting ground or aggregate for a roadway. 3. See *baseboard*. 4. Lowest point in a vertical plumbing stack. 5. Main chemical ingredient in paint.

base bead: See ground.

base block: See *plinth block*.

* **baseboard:** 1. Finish trim member covering lower edge of a finished wall where wall and floor meet. 2. Board forming base of an object.

baseboard heating unit: Long, narrow radiant heating unit installed at base of interior walls. Hot water, steam, or electricity provides heat.

base cabinet: Lower kitchen cabinet with countertop.

base coat: 1. Substance containing equal parts of raw linseed oil and turpentine applied as a base for wood finish or stain; does not obscure woodgrain. 2. Initial coat of finish over unfinished work.

base course: 1. Lowest masonry row of a wall or pier; foundation row of masonry that supports the masonry structure. 2. Layer of material installed on subgrade or subbase in preparation for pavement.

base court: 1. Outer, lower, or interior court of a castle or mansion. 2. Rear courtyard of a farmhouse.

base elbow: Plumbing fitting with an integral flange that acts as a supporting piece.

base flashing: Sheet metal strip installed at lowest meeting point of a vertical and a horizontal surface. Used to provide moisture protection. See *flashing*.

baseline: 1. Established reference axis from which measurements are taken when laying out building lines, property lines, street lines, or other working lines. 2. Line formed by intersection of ground plane and picture plane in a perspective drawing. 3. Line extending due east and west that serves as surveyor's reference for positioning north or south.

base load generation: Large, steady electrical load produced by the largest and most efficient electrical generator within the system.

basement: Story or stories of a building below the main floor and partially or completely below ground level.

basement space: See *crawl space*.

basement window: Opening, commonly of an awning or hopper design with a

basket strainer (2)

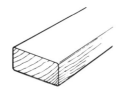

bastard sawed lumber

batch plant

batten seam

metal frame and sash, for use in basement walls. Windows are enclosed in an areaway where below grade.

base metal: 1. Metal to be welded, soldered, plated, or cut. PARENT METAL. 2. Metal material that oxidizes. 3. Major ingredient in an alloy.

base molding: 1. Trim member above plinth of a wall, pillar, or pedestal. Part between shaft and pedestal, or if no pedestal, part between shaft and plinth. 2. Trim member at top of a built-up baseboard. See *molding**.

base of a column: Lower part of a vertical support upon which shaft rests. Part between shaft and plinth. Sometimes the base is composed of all lower members of the column and the plinth.

base plate: 1. Plate or steel slab upon which a column or scaffold section stands. 2. Part of a surveying instrument that supports the lower end of footscrews used in instrument leveling.

base screed: See *ground*.

base sheet: Asphalt-saturated and coated composition roofing used in built-up roof systems.

base shoe: Strip of trim molding at baseboard and floor. See *molding**.

base shoe corner: Square member at wall corners and floor that eliminates mitered base shoe joints.

base tape: Highly accurate distance-measuring tool used in surveying.

base trim: 1. Finish at baseboard at lower part of an inside wall. 2. Lower part of a column that may consist of various members that make up the base as a whole, including an ornate pedestal and other decorative parts.

basic stress: See *stress*.

basil: See *bezel*.

basin: Open vessel that is the portion of a plumbing fixture for holding liquid.

basin fittings: All plumbing trim attached to an open vessel for holding liquid, including faucet, trap, and waste.

basin wrench: Plumbing hand tool with a long shank and ratcheted jaws for reaching into inaccessible locations. See *wrench**.

basket guard: Protective device over a table saw blade.

***basket strainer:** 1. Removable perforated drain fitting for a sink. 2. Fluid strainer in which a removable perforated basket retains undesirable solid material.

basket weave: Masonry pattern in which alternate units form a checkerboard design. See *bond* (1)*.

bas-relief: Sculpture in low relief with figures projecting only slightly from the background face.

basso-relievo: Bas-relief in which figures are projected from a background but not detached from it.

bastard file: Fine or medium grain flat file used for smoothing metal. See *file**.

bastard masonry: 1. Rough ashlar stones. BASTARD ASHLAR. 2. Thin facing stones.

***bastard sawed lumber:** Hardwood lumber in which annual rings are at 30° to 60° with the surface; midway between true quarter-sawn and true plain-sawn.

bastard tuckpointing: Filling of masonry joints with a special mortar to form a wider ridge along joint center than in true tuckpointing. See *tuckpointing*.

bat: 1. Piece of broken brick. 2. See *batten*.

batch: Amount of concrete, mortar, or plaster mixed at one time.

batching: Measuring ingredients for a batch of concrete or mortar by weight or volume and placing them in the mixer.

batch meter: Device that controls mixing time by locking a discharge mechanism. The mixer cannot be discharged until the preset mixing time has expired on the batch meter.

***batch plant:** Site for storage and mixing of all concrete components. Various amounts of cement, sand, aggregate, water, and other materials are mixed and blended to create concrete.

batement light: Window with an angular bottom rather than a horizontal bottom; used to fit an arch or sloping member.

bathroom: Area with a basin and one or more of the following: toilet, tub, or shower.

bathseat: Bench installed in a bathtub.

bath trap: P-shaped pipe fitting, sometimes made with a lesser depth than other plumbing traps. Used to prevent sewer gas from entering living areas.

bathtub: Plumbing fixture designed for bathing. May be built-in, corner, free-standing, island, bench, leg, recessed, sitz, sunken, right-hand, or left-hand.

batt: See *batt insulation*.

batted work: Stone with vertical or oblique parallel scoring. BROAD TOOLED WORK.

batten: 1. Narrow strip of board used to cover cracks between boards, most commonly joints in vertical exterior siding. See *board-and-batten*. 2. Thin narrow strip of board nailed across the surface of one or more boards to prevent warping. 3. Cleat. 4. Squared timber of special size used for flooring. Usually measures 7″ wide, 2$\frac{1}{2}$″ thick, and 6′-0″ or more long. 5. Narrow wood strip nailed across rafters to provide a base for wood shingles, slate, or clay tile roofing. 6. Narrow wood strip applied to act as a support for lath and plaster.

batten door: Door made of vertical sheathing boards reinforced with strips of board nailed horizontally. Nails are clinched on opposite side.

battening: Application of wood strips to which lath or other material may be attached.

batten plate: Spacer piece to hold component parts of a member the correct distance apart.

***batten seam:** Metal roofing joint in which the meeting of adjoining sheets is formed around a narrow wood strip.

batten strap: Metal strip, usually of copper. Used for securing a batten or nailing strip to top edge of a copper roof gutter.

batter: Receding upward slope; the backward inclination of a timber or wall that inclines away from a person facing it.

batterboard: Piece of lumber nailed horizontally to stakes driven near each proposed corner of an excavation. Notches are made or nails driven into batterboards to mark work lines such as a building corner or footing width lines. Strings are stretched between batterboards for transferring reference points. Batterboards may also be set to a specified elevation, with elevation written on batterboard.

batter brace: Inclined support at end of a truss that gives added strength and structural stability.

battered foundation: Structural support consistng of a wall with one vertical and one sloping face. Wall is wider at bottom than at top. See *foundation*.

battering wall: Vertical, continuous structure with a sloping face. Used for sustaining pressure of the weight of water or a bank of earth. Same as retaining wall.

batter pile: Foundation support driven at an angle to give both horizontal and vertical support. SPUR PILE.

batter post: See *batter brace*.

batter rule: Instrument with a measuring device or frame and a plumb bob and line by which the slope of a wall is regulated during construction.

battery: 1. Storage cell connected in

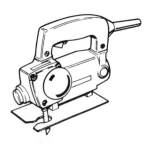

batterboard

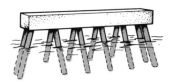

batter pile

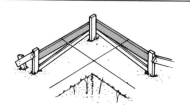

bayonet saw

series or parallel as a source of electric current. 2. Blasting machine.

battery of fixtures: Two or more similar plumbing fixtures that discharge into a common waste or branch pipe.

batting tool: Chisel measuring 3″ to 4½″ wide. Used for forming a pattern in stone of parallel alternating grooves and ridges.

batt insulation: Material used to inhibit transmission of temperature and/or sound; usually composed of fiberglass with or without a thin facing material, and made in relatively small units for convenience in handling and applying. Sizes vary from 3″ to 7″ or more in thickness, 15″ to 23″ in width, and usually 48″ in length. The facing materials may be kraft paper, metal foil, or plastic sheets.

battlement: 1. Low wall composed of a series of solid parts alternating with open spaces. 2. Low railing at the edge of a platform or bridge. 3. Indented or notched wall of a parapet. 4. Series of solid parts alternating with open spaces used as an ornamental design on furniture.

batture: Elevation of a river bed.

bay: 1. Projection outward from a wall, with sides usually at a 22½° or 45° angle. A window installed in such a projection is a bay window. 2. Window with its usual setting or framing, as jambs and window backing. 3. Space or division of a wall within a building between two rows of columns, piers, or other architectural members, or the space between a row of columns and a bearing wall. 4. Compartment in a barn where grain on the stalk and hay may be stored. 5. Coastal indentation. 6. Relatively small area of decking or roofing in which concrete is placed all at one time.

bayonet saw: Small electric saw with an interchangeable reciprocating blade. Used to cut intricate designs and curves in wood, metal, or plastic. SABER SAW, ELECTRIC JIGSAW.

bayonet saw blade: Short straight blade with teeth on one side used for cutting metal or wood. Lengths vary from 3″ to 8″ and widths from ³/₁₆″ to ³/₄″. Tang of blade is commonly reduced in width to fit in a bayonet saw. See *blade*.

bayonet socket: Lamp socket with two lengthwise slots and a 90° turn at bottom of each slot. The lamp base has two pins that slide in socket slots. The lamp is held in the socket by twisting slightly when pins reach bottom of slots.

bay stall: Window seat built into the opening of a bay window.

bay window: Three-walled window projecting outward from a wall of a structure at an angle. See *window*.

bazooka: Automatic tool for applying joint compound to drywall.

bead: 1. Circular or semicircular molding. See *beading, corner bead, screen bead.* QUIRK BEAD. 2. Layer of metal deposited on base metal during the welding process as the electrode melts. WELD BEAD. 3. Narrow strips of caulking, glue, or other crack fillers applied with a nozzle or tip.

bead and butt: Framing for panel work in which panels are flush with frame and have beads run on adjoining edges with a sticker machine. Beads run with grain of the wood and butt against the rail.

bead and reel: Decorative molding consisting of a small sphere and one or two circular disks, alternating singly or in pairs with oblong, olive-shaped beads. REEL AND BEAD MOLDING.

bead chain: String of alternating metal spheres and wires used as pulls for items such as electrical switches and key chains.

beaded dovetail seam: Sheet metal seam used to join a cylindrical member to a flat member. A bead is formed around one end of the cylinder to support the flat member and tabs are bent over to secure the flat member in position. See *seam* (1).

beaded joint: Meeting of two members with a rounded cut on the edge of one or both adjoining members to help conceal the joint.

beader: Tool for cutting semicircular shapes in wood and other soft material.

beadflush: Panel surrounded by a semicircular shape worked in the edges of the frame so that frame, bead, and panel are flush along their front faces.

beading: 1. Type of molding or rounded projecting band; used as a decorative feature. See *bead*. 2. Depression of various shapes made in sheet metal.

beading bit: Router bit with short flat projections on each side; used to form a convex shape, such as for rounding corners or forming an ornamental edge. See *bit* (2).

beading machine: Device used for making depressions in sheet metal.

beading tool: See *bead plane*.

bead molding: Small, convex molding which has a profile greater than 180°.

bead plane: Wood-shaping hand tool used to cut semicircular shapes.

beadwork: Ornamental molding cut with a beading tool. BEADING.

beak: 1. Slight, continuous projection ending in an arris or a narrow fillet.

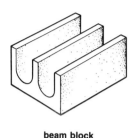

beam block

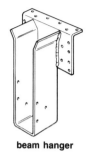

beam hanger

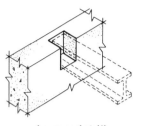

beam pocket (1)

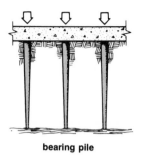

bearing pile

2. Part of a drip from which rainwater is diverted. **3.** Crooked end of a bench holdfast.

beakhead: Drip mold on extreme lower edge of the lowest member of a cornice. See *beak molding.*

beaking joint: Meeting of a number of adjacent heading joints in a straight line.

beak molding: Trim member with a downward projecting part, the outline somewhat resembling a bird's beak, to make a drip for rainwater and prevent water from running to the face of the wall below.

beam: Horizontal structural member made of a single member of concrete, timber, stone, iron, or other structural material and installed horizontally to support loads over an opening.

beam anchor: See *joist hanger.*

beam-and-slab floor construction: Reinforced concrete floor system in which a solid slab is supported by reinforced concrete beams. Concrete for beams and slabs is usually placed monolithically.

* **beam block:** Hollow concrete masonry unit with short webs or a channel for placement of horizontal reinforcement embedded in mortar or concrete.

beam bolster: Continuous bar support used to hold up reinforcing in beam bottoms.

beam ceiling: Type of construction in which horizontal or angular structural members of the ceiling are exposed to view. Beams may be either true (actual structural members) or false (boxed frames installed to give appearance of beams).

beam clamp: Holding device used to secure pipe hangers to horizontal supporting members.

beam fill: Masonry or concrete used to fill spaces between joists or between a basement or foundation wall and the framework of a structure. This provides fire stops in outside walls for checking fires that start in the basement of a building.

* **beam hanger:** Wire, strap, or other hardware device that supports formwork from structural members. BEAM SADDLE.

* **beam pocket: 1.** Opening in a wall or girder which allows for intersecting beams. **2.** Opening in a column or girder to which forms for intersecting beams will frame in.

beam schedule: Working drawing list showing number, size, and placement of beams used in a structure.

beam test: Method of determining concrete flexural strength by measuring the strength of an unreinforced beam.

bearer: 1. Small member used primarily to support another member or structure; e.g., the ledger in balloon framing that supports second floor joists. **2.** Horizontal member of a scaffold that supports the platform. **3.** Any horizontal structural member supporting a load.

bearing: 1. Portion of a beam or truss resting upon a support; that part of any member of a building that rests upon its supports. **2.** Compass reading used to indicate angle in degrees and minutes from line extending north to south; e.g., N35°E denotes a point which is 35° east of north. Reading approaches 90° in four quadrants. **3.** Part in which a shaft or pivot revolves. Types include anti-friction, needle, pilot, and solid.

bearing capacity: Maximum unit pressure any material will withstand before failure or deformation.

* **bearing pile:** Deep foundation member that supports loads placed on it by being driven to a point where the tip of the pile rests on a solid base.

bearing plate: Flat piece of steel placed under a heavily loaded truss, beam, girder, or column to distribute the load so pressure will not exceed bearing strength of the supporting member.

bearing stratum: Soil or rock layer that supports a building load.

bearing value: Load that soil will sustain without substantial deformation.

bearing wall: Continuous vertical support for floors, other walls, roofs, or other structural loads. BEARING PARTI-

TION.

becket bend: Knot for joining two ropes of different diameters; secure but difficult to untie. SHEET BEND, WEAVER'S KNOT.

bed: 1. Layer of cement or mortar in which stone or brick is embedded or against which it bears. **2.** Horizontal surface of a stone in position. **3.** Lower surface of a brick, slate, or tile. **4.** Recess formed by the mold to hold plaster ornament. **5.** Base for machinery. **6.** To set a member in place with putty or a similar material; most commonly referred to in glazing.

bedding: Filling of mortar, putty, or other substance in order to secure a firm bearing.

bedding coat: Initial layer of drywall joint compound.

bedding compound: See *joint compound.*

bedding course: Lowest layer of mortar that starts masonry construction.

bedding dot: Small section of plaster set onto the lath; acts as a screed guide.

bedding plane: Position and direction of joints between various strata or layers of rock.

bed dowel: Tubular-shaped member placed in center of a stone bed.

bed joint: 1. Horizontal joint upon which bricks rest. **2.** Radiating joints of an arch.

bed molding: 1. Trim member used where eaves of a building meet the top of outside walls. See *molding*.* **2.** Trim member, in any architectural order, used as a finish immediately beneath the corona and above the frieze. **3.** Trim member in an angle, as between projections of overhanging eaves and sidewalls.

bed of a stone: Under surface of a stone. The upper surface prepared to receive another stone is the top bed. The natural stratification of the stone is the natural bed.

bed plate: 1. Horizontal member applied on a foundation as a support for some structural part. SILL. **2.** Flat piece of metal used as a rest or support for a machine.

bed puttying: Placing a thin layer of putty or bedding compound in the rabbet of a window sash and pressing the glass onto this layer. Glazing points are driven into the wood, and the sash is face-puttied. The window is then turned over, and excess putty or glazing compound is cleared away by running a putty knife around the perimeter of the glass opening. BED IN PUTTY.

bedrock: Solid rock that underlies any superficial geological formations. Provides a firm foundation on which to erect a heavy structure.

bedroom: Sleeping area.

bed stone: Large foundation stone; e.g., stone used to support a column.

bed timber: Large wood member serving as a foundation or support for other work.

beeswax: Soft material applied to moving metal parts, especially screws, to reduce friction.

beetle maul: Large mallet used for driving materials that could be damaged by a sledgehammer.

Belfast truss: See *bowstring truss.*

belfry: Tower in which a bell is hung; separate or attached bell tower of a church.

Belgian truss: Prefabricated roof framing member with sloping upper chords joined to the bottom chord by a series of inclined struts. Struts meet at highest point of the truss and at all other connections. See *truss*.*

bell: 1. Metal device, usually cup-shaped, mechanically equipped for producing a ringing sound when in operation; e.g., an electric doorbell. 2. Part of a pipe enlarged for a short distance to receive the end of another pipe of the same diameter in order to make a joint. HUB, SOCKET. 3. Enlarging an excavation to provide additional bearing surface at the bottom of an otherwise narrow hole.

*bell and spigot joint: Joint for cast iron plumbing pipes, each length of pipe being made with an enlarged or belled end and a plain or spigot end. Spigot end of one length fits into bell end of the next length, with the joint made tight by caulking. SPIGOT AND SOCKET JOINT.

belled: Having an enlarged end shaped like a bell; often used at bottom of concrete piers or caissons to provide additional bearing.

bell hanger's bit: Long, slim wood bit used for drilling through frame of a building when installing doorbells.

bell metal: Alloy made of 75% to 80% copper and 20% to 25% tin.

bellows: Device with an accordion design for converting air or fluid pressure variation into movement.

bellows seal: Gasket made of a flexible corrugated metal member with one end attached to a ring fastened to a shaft and the other end attached to a disk that is pushed against a housing. BELLOWS EXPANSION JOINT.

bell push: Button that is pushed to ring a bell.

bell trap: Inverted bell-shaped plumbing device that prevents sewer gas backup. Installation is no longer allowed by the National Plumbing Code.

bell wire: Low-current electrical conductor with insulation rated below 50 volts.

belt: 1. Band course, or courses, of brick or stone projecting from a brick or stone wall; usually placed in line with window sills. May be molded, fluted, plain, enriched at regular intervals, or of a different type of brick or stone. 2. See *belt abrasives.* 3. Continuous strap for transferring motion and power from pulley to pulley or as a conveyor material.

belt abrasives: Coarse or fine sanding materials attached to a continuous backing for use on a belt sander.

belt conveyor: Endless strap that passes over pulleys or rollers, providing a track on which loose materials or small articles are carried from one point to another.

belt course: 1. Layer of stone or molded work carried at same level across or around a building. 2. Decorative feature, such as a horizontal band around a building or column.

belt dressing: Substance used to prolong belt life or improve its frictional grip and prevent slipping.

belt lacing: Narrow strips of rawhide used to lace belts together. Not applied to wire hooks and other types of fasteners used to hold ends of belts together.

belt loader: Excavation machine with an auger that digs and loads material onto a conveyor to remove excavated material.

belt sander: Portable or stationary electric tool with a continuous abrasive belt. A flat plate above the abrasive holds it in contact with the work for smoothing.

belvedere: Rooftop pavilion area.

bench: 1. One level or step of excavation. 2. Long, continuous seat for more than one person.

bench apron: Covering board or facing at front of a workbench.

bench brake: Large device for bending and shaping sheet metal.

bench dog: Wood or metal peg placed in a hole near end of a workbench to prevent work from slipping out of position or off the bench.

benched foundation: See *stepped footing.*

bench grinder: Electric tool with two rotating stone wheels, one mounted at each end of the motor. Used for sharpening plane irons and chisels.

bench hook: 1. Hook-shaped device used to prevent work from slipping on a workbench. 2. Flat timber or board with cleats nailed on each side and one on each end to prevent slipping, which might cause damage to workbench top.

benching: 1. Sloping, steplike excavation to control erosion or tie new fill to an existing slope. BENCH TERRACE. 2. Groove at bottom of a concrete waste pipe that carries small amounts of flow. 3. Concrete placed on a steep slope to prevent erosion or failure of the slope. 4. Concrete base in a trench for supporting pipe.

*bench mark (BM): Stable reference point marked with elevation above mean sea level from which differences of elevations are measured. The U.S. Geological surveys provide bench marks at intervals across the U.S. with elevations related to mean sea level.

bench plane: Wood-smoothing hand tool used on a workbench. Types are jack, smooth, fore, and block planes.

bench rule: 1'-0" long rule marked with

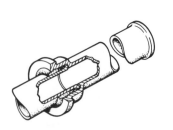

bell and spigot joint   belt sander   bench mark

divisions of $1/_8''$ on one side and $1/_{16}''$ on other side.

bench sander: Power tool similar to a belt sander but attached to a workbench. Used to sand small pieces.

bench stop: Adjustable metal device usually notched and attached near one end of a workbench to hold a piece of work.

bench table: 1. Course of projecting stones forming a stone seat around walls at the base of a building, such as a large church. **2.** Projecting course around base of a pillar sufficient to form a seat.

bench vise: Clamping device attached to edge or side of a workbench to hold work in place.

benchwork: Work done with hand tools at a workbench.

bend: 1. Short piece of curved pipe, as an elbow, used to connect two adjacent straight lengths of conduit. **2.** Intertwining ends of two ropes of different diameters to form one continuous length of rope.

bending: Deformation of a straight member into a curved shape. Compression, tension, and sliding shear occur. Compression occurs along inner edge; tension occurs along outer edge; and sliding shear occurs at inside of the bending member.

bending moment: Rotating of a structural member about an axis. For example, the midsection of a beam has a tendency to bend downward.

bending pin: Tool used to shape, expand, or straighten lead pipes, fittings, and sheet lead. BENDING IRON.

bending schedule: Written list and description of reinforcing steel rods and shapes required for a reinforced concrete structure.

bend test: Method of checking strength of a weld or piece of steel by changing the shape by 180° without heating.

benefication: Improvement of chemical or physical properties of a raw material, such as aggregate, or intermediate product by removing undesirable components or impurities.

bent: Framework transverse to length of a structure usually designed to carry lateral as well as vertical loads. A self-supporting frame having at least two legs and usually placed at right angles to length of the structure it supports; e.g., columns and cap supporting the spans of a bridge are bent.

bent bar: Reinforcing steel rod formed to a prescribed shape such as a column tie, hook bar, stirrup, or truss bar.

bent cap: Concrete beam or block extending across and encasing heads of piles or columns to make up the top of a bent for the bridge span above.

bent nose plier: See *curved jaw pliers.*

bentonite: Highly absorptive and compressible clay material tamped into place to restrict water seepage, or suspended in water slurry to keep earth from falling into an excavation.

bent strap anchor: Corrugated masonry anchor with a dovetailed end that connects to the flange of a steel column. Other end is twisted to be set into a horizontal mortar joint. See *anchor*.*

bentwood: Curved lumber shaped by steaming or boiling and bending to a form.

benzene: Liquid chemical used as a solvent or cleaning fluid.

berliner: Terazzo topping with large and small pieces of marble paving.

berm: 1. Raised earth embankment. **2.** Shoulder alongside a paved road. **3.** Terrace or shelf in a sloping excavation wall.

Bethell process: Wood preservative process of pressure-treating wood with creosote.

bevel: One side of a solid body inclined in relation to another side, with the angle between the two sides being either greater or less than a right angle; a sloping edge.

bevel board: Guide piece used in framing a roof or stairway to lay out angles. PITCH BOARD.

beveled edge: Slight depression along long edges of a sheet of drywall to allow for filling with joint compound.

beveled halving: Joint with meeting surfaces cut at an angle to the plane of the timbers so they are not pulled apart by a force in their own plane.

beveled siding: Boards used for exterior wall covering that are thicker along one edge. When applied, the thicker edge overlaps the thinner edge of siding below to shed water. The lower edge is $5/_8''$ to $3/_4''$ thick, tapering to approximately $1/_8''$ at the top edge. Width of bevel siding varies from $3^1/_2''$ to $11^1/_4''$. See *siding*.*

bevel gear: Circular turning device with teeth formed on outside of a conical shape.

bevel joint: Joining of two pieces with edges shaped at angles not equal to 90°.

bevel of door: Angle of inclination that the lockset edge of a door has from the face to allow clearance at the jamb or adjoining door.

bevel of lock: Direction of the angled side of a latch bolt. Regular bevel indicates a lock that functions on an inswinging door. Reverse bevel indicates a lock for an out-swinging door.

bevel square: See *sliding T-bevel.*

bevel washer: Wedge-shaped steel hardware fitting with a hole in the center at an angle used to give flat bearing for a nut when a threaded rod passes through at an angle. BEVELED WASHER.

bevel weld: Butt joining of two pieces of metal with one or more edges shaped at an angle. Used for metal too thick for proper penetration.

bezel: Sloping cutting edge of a tool such as a chisel or drill bit. BASIL.

Bezier curve: Smooth curve defined by four points, two of which are end points. Used to design irregular, smooth curves such as driveways.

bias: 1. Line or cut at an oblique or diagonal angle. **2.** Continuous direct voltage across a PN junction that establishes required junction operating conditions.

* bibb: Faucet or tap threaded so a hose may be attached to carry water. BIB.

bibb cock: Faucet fitted with a nozzle curving downward; used as a draw-off tap. BIBB NOZZLE, STOPCOCK.

bibb valve: Draw-off tap used for domestic water supply; closed by screwing down a washered disc onto a seat in the valve body.

* bidet: Bathroom fixture similar to a sitz bath; used for bathing lower part of the body.

bi-fold door: One of a pair of doors hinged to each other side by side. One door is pivoted at both the top and bottom. Opening of doors is guided by a track across the top and sometimes the bottom of the opening. See *door*.*

bibb

bidet

bight: Central part of a strand of rope in knot tying; between the working end and standing part.

billet: **1.** Lumber with three sawn sides and one round side. **2.** Decorative molding with a series of circles or squares placed alternately with a single or double row of notches. **3.** Steel plate placed under a column to distribute weight over a broader area.

billet steel: Metal, reduced from ingots or continuously cast, that conforms to specified engineering limits of chemical composition.

bimetal: Two materials with different expansion properties that are bonded together.

bimetal overload: Thermal condition that causes a power interruption in an electrical motor circuit. The switch opens as a result of metal warping open when heated and closing when cooled.

bimetal relay: Switch activated by thermal action. See *bimetal overload.*

bin: Storage container.

binder: **1.** Timber or steel beam supporting bridging joists in a double or framed door. **2.** Cementious material such as asphalt, hydrated cement, or resins. BINDER COURSE. **3.** Fine substances that enable excavated fill material to stick together. **4.** Nonvolatile portion of paint that provides cohesion. BINDING AGENT. **5.** Bond used to tie parts of a masonry wall together. BINDER COURSE. **6.** Small diameter reinforcing rod that ties the main steel reinforcing together. **7.** Check of deposit to validate a contract.

binding: **1.** Sticking, not moving freely, as in a moving part such as a window or door. **2.** Strip sewn on the edge of

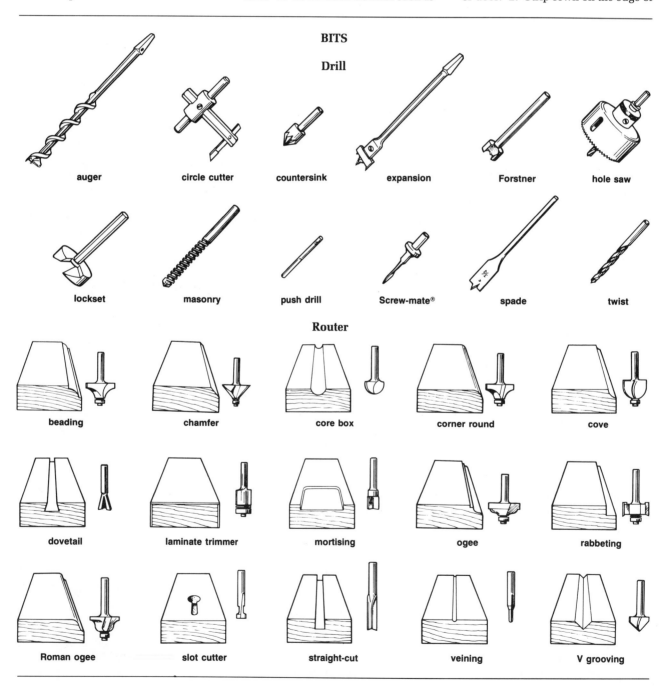

BITS

Drill

auger   circle cutter   countersink   expansion   Forstner   hole saw

lockset   masonry   push drill   Screw-mate®   spade   twist

Router

beading   chamfer   core box   corner round   cove

dovetail   laminate trimmer   mortising   ogee   rabbeting

Roman ogee   slot cutter   straight-cut   veining   V grooving

carpeting for protection or appearance.

binding beam: Horizontal structural support that carries common joists.

binding head screw: Slightly domed top of a slotted head screw with a circular shape; raised from the work by a cylindrical section. See *screw**.

binding joists: See *binder*.

binding piece: Member fastened to two opposite beams or joists to eliminate lateral deflection. STRAINING BEAM.

binding post: Setscrew or projecting member that is part of an electrical device. Used to hold conductor wire in place against a terminal.

binding rafter: See *purlin*.

binding yarn: Strands of yarn woven lengthwise into carpeting to hold pile tufts in place.

biparting door: One of a pair of doors that opens in a vertical direction; commonly used on freight elevators.

bird block: Piece of wood nailed between rafters at top wall plate of an open soffit to close the attic space. To provide ventilation, these may be drilled with holes that are covered with screening.

bird peck: Small hole or patch of distorted grain resulting from birds pecking through cells of a tree.

bird's-eye: 1. Small central spot in lumber with wood fibers arranged in an ellipse around it to give the appearance of an eye; e.g., bird's-eye maple. **2.** View from above an object, as in a bird's-eye view.

bird's mouth: Triangular notch near rafter bottom, allowing rafter to bear flat on the top wall plate.

biscuit: Unglazed tile.

bisect: To divide or cut a line, plane, angle, or solid into two equal parts.

* **bit: 1.** Hole-cutting tool for use with a brace or an electric drill. **2.** Shaping tool for use with a router. **3.** Part of a key that projects and enters a lock contacting the bolt, tumblers, or both. **4.** Point of a soldering gun or soldering iron that transfers heat and sometimes solder to the joint.

bit brace: Curved hand tool used to hold boring and drilling tools. Curved handle design gives greater leverage than a straight handle. BIT STOCK.

bitch: Steel fastening piece in which ends are bent to point in opposite directions.

bitch pot: Container holding asphaltic emulsion.

bite: Amount of overlap of a stop on a panel or glass light.

bit extension: Attachment that fits between the drill and hole-cutting tool to provide longer reach. The extension

has a long shank, tang, and jaws or set screws for holding the bit.

bit gauge: Attachment for a hole-cutting tool that stops drilling or boring at a given depth. BIT STOP.

bitting: Indentation or cut on a portion of a door-lock key that sets the tumblers.

bitty: Paint surface defect in the form of small pieces of material projecting above an otherwise smooth surface.

bitumen: Generic name for various hydrocarbon mixtures, including tars, pitches, and asphalts.

bituminous: Containing tars, pitches, or asphalts.

bituminous distributor: Mechanical device for pumping and evenly distributing hot tar, road oil, or other bituminous materials.

bituminous emulsion: Mixture of hydrocarbon materials and water with hydrocarbon materials suspended in water. Used as a waterproofing agent.

bituminous grout: Mixture of tar materials and sand. Used for filling cracks.

bituminous macadam: Mixture of tar materials and coarse aggregate. Used for paving.

black japan: Bituminous paint; used as a metal varnish.

black plate: Uncoated cold-rolled sheet steel; commonly manufactured in 12" widths.

blacktop: Asphalt pavement composed of bituminous products and aggregate.

blackwall hitch: Temporary means of attaching a load rope to a hook.

* **blade: 1.** Longest of the two arms of a framing square, usually 24" long and 2" wide. The tongue of the square forms a right angle with the blade. Rafter framing tables and Essex board measure tables are located on the blade. BODY. **2.** Serrated or razor-edged metal part fitted into or on a knife or saw for cutting. The type of metal, cutting edge, and design of the serration depend on the saw used and material cut. **3.** Part of an excavating machine that digs and pushes dirt but cannot carry it. **4.** Part of a trowel that contacts the work.

Blaine test: Method for measuring fineness of cement or other finely powdered materials by determining their permeability to air. Expressed as square meters of surface area per kilogram.

blank: 1. Piece of trim lumber cut so that waste is minimized during molding. **2.** Unfinished sheet of plate glass. **3.** Any flat, unbroken, unadorned surface.

blank arcade: See *blind arcade*.

blank door: Fixed, unopening door.

BLADES

Wood-cutting Blades

Circular

carbide-tipped

combination

crosscut

plywood

Bayonet

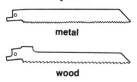

metal

wood

Concrete/Masonry Blades

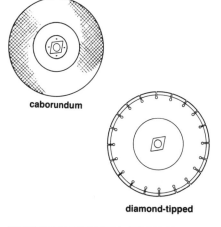

caborundum

diamond-tipped

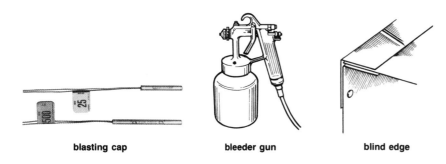

blasting cap   bleeder gun   blind edge

blanket insulation: Flexible lightweight sheet for inhibiting transmission of temperature or sound. Available in long rolls, strips, or panels with or without a vapor barrier. Insulating material is normally composed of fiberglass. Thickness varies from 2″ to 8″. Thicker types have a flange for stapling. Available in 15″ width for 16″ OC framing and 23″ width for 24″ OC framing. See *batt insulation*.

blank flange: Flattened projecting end of a pipe in which there are no bolt holes in the projecting section.

blank flue: Chamber built into one side of a fireplace and closed off at the top to conserve materials and labor and balance the weight.

blast: To loosen or remove dirt or rock with an explosive charge.

blast furnace slag: Nonmetallic by-product of iron production. Used as an ingredient in portland blast furnace slag cement and as an aggregate in lightweight concrete. Three types are air cooled, expanded, and granulated. See *slag*.

blast hole: Boring into rock or other hard material to allow for insertion of explosive materials.

blasting: 1. Cleaning or roughening a surface by the forcible projection of an abrasive material. 2. Process of loosening or removing dirt or rock with an explosive charge.

blasting agent: Material composed of fuel and an oxidizer but not classified as an explosive. Used for loosening dirt and rock.

* **blasting cap:** Metal tube with one open end containing detonating compound(s) designed to detonate from sparks or flame from a safety fuse inserted and crimped into the open end.

blasting mat: Mesh of interwoven cables designed to contain debris resulting from detonation of an explosive charge.

bleaching: Cleansing or whitening a wood finish with oxalic acid or other substances with similar properties.

bleb: Solidified small bubble in glass.

bleed: 1. To exude a liquid as in a minute leak. 2. Coloring from wood or undercoat breaking through intermediate and top coats of paint. BLEED-THROUGH. 3. Exuding of water from concrete. Water either rises to the surface or escapes through the forms, weakening the concrete. BLEEDING. 4. To remove unwanted air or fluid from hoses or passages. 5. To excrete an iron-stained liquid, as in the seams of a boiler.

bleeder: 1. Pipe that draws off liquid refrigerant in a direction parallel to the main refrigerant flow. 2. Small valve for draining fluid from a closed system.

* **bleeder gun:** Spraying tool for painting that has a continuous flow of compressed air passing through the nozzle. Material to be applied flows through the gun only when the trigger is depressed.

bleeder tile: Pipe placed in foundation walls that allows surface water accumulated by outside drains to pass into the drain on the inside of the wall. BLEEDER PIPE.

bleeding capacity: Ratio of water volume released to the volume of concrete.

bleeding of wood: Exuding of preservative from a treated timber or board.

blemish: Imperfection that mars the finished appearance of an object.

blend: 1. To combine and mix various components into one. 2. To obtain a desired shade or color by mixing two or more different color pigments. 3. To apply paint to a surface so one shade will gradually mix with another shade.

blended cement: Mixture of granulated blast furnace slag and hydrated lime, portland cement and granulated blast furnace slag, or portland cement and pozzolan.

blending valve: Control device used to mix liquid with recirculating liquid in order to obtain a desired temperature.

blind: 1. Interior covering for a window; consists of narrow slats that operate as a unit. 2. Shutter.

blind arcade: Closed series of arches used for ornamentation.

blind area: Wholly or partly covered area outside a building that keeps moisture away from walls.

blind bond: Setting of masonry units so that headers are concealed.

blind casing: Rough trim member for a door or window frame.

blind catch: Shutter fastener hardware consisting of a hooklike piece used for holding the shutter in place when in a closed position.

blind door: Louvered door. See *blank door*.

* **blind edge:** Piece of sheet metal shaped to cover fasteners and raw edges of sheet metal. FALSE EDGE, DUTCHMAN.

blind flange: Flattened projecting end on a pipe that closes the end of the pipe and prevents material passage.

blind floor: See *subfloor*.

blind header: Short blocks of stones or ends of bricks installed to give appearance of structural headers.

blind hinge: 1. Type of pivot device made especially for use on outside blinds or shutters. 2. Concealed pivot device.

blinding: 1. Applying a filling layer of concrete on another concrete surface to fill surface voids or create a more workable finished surface. 2. Compaction of earth directly above a tile drain to keep dirt out of the tile. 3. Clogging in a screen.

blind joint: Joining of materials so that the joint cannot be seen.

blind miter: Joint in which two pieces are cut at equal angles and fastened with dovetails and concealed pins.

blind mortise: Notch cut into a piece of material, such as a board, but does not pass entirely through the material.

blind nailing: Driving nails so that nail heads are concealed.

blind pocket: Recessed area in a wall or ceiling that receives a window blind.

blind rivet: Pin with a small head and an expandable shank; used for joining light-gauge metal.

blind stop: Rectangular molding used in assembly of window frames.

blind valley: Perpendicular joining of two roof sections by constructing and sheathing a main roof and attaching an intersecting roof on top of the completed main roof. FALSE VALLEY.

blind vent: Plumbing fixture vent pipe ending in a wall or installed only to give the appearance of a vent pipe.

blister: 1. Defect on a painted surface that may be caused by direct heat on the surface. See *blistering*. 2. Puffing out of a plaster coat.

blistering: 1. Defect on a painted surface

resulting from one of a number of different causes, such as subjecting the surface to direct heating, which causes the painted surface to swell. Swelling may also be caused by sealing in moisture, hand smears, or by a difference in temperature between the surface and finishing materials. Also occurs in drywall taping and plastering. **2.** Raising of a thin layer of concrete soon after finishing.

blister steel: Rough-textured metal material made from wrought iron.

bloated: Swollen; enlarged beyond a normal size.

* block: 1. Piece of wood glued into interior angle of a joint or fastened between structural members to strengthen joint, provide structural support, or block air passage. **2.** See *concrete masonry unit**. **3.** Pulley in a case through which rope or cable is placed to obtain mechanical advantage. **4.** Portion of a city or town divided by streets.

* block and tackle: Combination of rope and pulleys used to gain mechanical advantage for lifting or pulling. Double and triple pulleys (blocks) are often used with the rope reeved (threaded) through the blocks in various arrangements to give mechanical advantage or to multiply the force in making power available to move a greater load than the same amount of direct power. Increase in power is equal in ratio to the number of strands supporting the load.

block beam: Flexural member composed of individual blocks joined by prestressing.

block bridging: Short pieces of wood nailed between joists that serve as stiffening braces.

block flooring: Squares of wood flooring cut from narrow, short strips of regular flooring or made in solid blocks of wood. Types of block flooring include solid unit, laminated, and slat blocks. Size and thickness vary.

blockhouse: Fortified structure used as a shelter.

block-in-course: Masonry for use in heavy engineering construction in which stones are carefully squared and finished to make close joints and faces are dressed with a hammer.

block-in-course bond: Method of fitting masonry used for uniting concentric courses of an arch by inserting transverse courses, or voussoirs, at intervals.

blocking: See *block* (1).

blocking course: Row of stone laid on top of a cornice that crowns a wall.

blockout: 1. Frame set in a concrete form to create a void in the finished concrete structure. **2.** Void created in concrete by such a form.

block plan: Drawing of organization for a building site showing outlines of existing and proposed streets and buildings.

block plane: Small wood-smoothing hand tool, 5″ to 7″ long, with a low-angle metal cutting iron held in a frame. Cutting iron bevel is placed up, instead of down as in other wood planes, and has no cap iron. Used for smoothing face or end grain. Held with one hand. See *plane* (1) *.

block sequence: Welding in which separated increments are completely or partially joined before intervening areas of the joint are welded.

block tin: Pure tin.

block yard: Manufacturing plant for production and storage of concrete masonry units.

bloom: 1. Chemical powder that appears on masonry walls, especially brick, commonly within first year after construction as a result of improper removal of cleaning acid. See *efflorescence*. **2.** Defect on a varnished surface usually caused by a damp atmosphere. **3.** Haze on glass, old painted surfaces, or rubber.

blowback: Pressure difference at which a safety valve opens to release excess pressure.

blow counter: Pile-driving crew member who monitors hammer impacts for each increment of pile advancement.

* blowdown valve: Device on lowest point of a low-pressure boiler.

blower: Multi-blade fan unit that compresses inlet air to a higher discharge pressure; commonly used with heating and cooling systems.

blow gun: Attachment for a compressed air hose that allows a strong stream of air to be emitted.

blow hole: Opening where air and gas can escape.

blowing: Popping out a plaster surface and forming a pitted surface. Caused by expansion of backing materials.

blown joint: Joining of soft metal, such as lead, made by use of a blow pipe.

blow off: 1. Sewage pipe outlet for discharging sediment or water. **2.** Controlled removal of water, steam, or other pressurized material from a closed tank or storage system.

blow off valve: Control device that opens to allow controlled release of water or steam from a pressurized closed storage system.

blowout: Rupturing of forming materials resulting from too much internal or hydrostatic pressure.

blowtorch: Portable device that uses an

block (3)

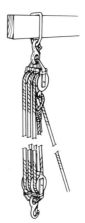

block and tackle

blowdown valve

open flame to apply intense local heat.

blub: Swelling or bulging out of newly plastered work.

blueprint: Set of drawings with dimensions and materials for a structure or building project. Included are plot plans, floor plans, elevations, details, sections, and specifications. The term blueprint is derived from the old reproduction processes, which produced a drawing copy sheet with a blue background and white lines.

blue stain: Discoloration of lumber resulting from fungus growth in unseasoned wood. Although blue stain mars the appearance of lumber, it does not seriously affect strength.

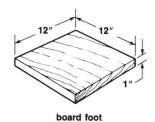

board foot

bolt cutter

---

bluestone: Grayish-blue sandstone used as a paving stone and as a building stone for window and door sills and lintels.

blue tops: Grade stakes with tops set at finished grade level.

blushing: Condition in which a bloom or gray cloudy film appears on a newly paint finished surface on hot humid days, usually caused by condensation of moisture or rapid evaporation of solvents.

board: Piece of timber sawn to a specified size, usually 1″ thick and from 4″ to 12″ wide before final planing.

board-and-batten: 1. Wall covering composed of wide boards and narrow strips. Wide boards are attached vertically with small spaces remaining. Narrow strips, or battens, are attached over spaces between boards. See *siding**. **2.** Plywood sheets with an exterior veneer applied vertically to a wall with narrow wood strips nailed at regular intervals to give appearance of true board-and-batten siding. Some plywood sheets are available with grooves routed at regular intervals. REVERSE BOARD-AND-BATTEN.

board-and-brace: Wood surface formed by alternating thick and thin boards. Thick boards are grooved on each edge and thin boards are fitted into grooves.

board butt joint: Joint in which shotcrete is blown in a sloping design to a board laid flat on an adjoining surface.

***board foot:** Piece of lumber 12″ wide, 12″ long, and 1″ thick, or its cubic equivalent.

boarding: Covering made of lumber; e.g., box sheeting of a building.

boarding-in: Process of nailing lumber to the outside studding of a house.

board knife: See *utility knife*.

board lath: 1. Thin, narrow piece of wood fastened to wall studs as a support for plaster. **2.** Lath produced in large sheets. Sometimes called plasterboard and usually intended as a base for plastering.

board measure: System of measurement for lumber with the unit of measure being 1 board foot. Quantities of lumber are designated and prices determined in terms of board feet.

board measure scale: Series of numbers found on the back of the body of a framing square. Used to calculate board feet in a piece of lumber. ESSEX BOARD MEASURE SCALE.

board rule: Measuring device used to determine the number of board feet in a quantity of lumber without calculating. BOARD MEASURE RULE, TALLY STICK.

board sheathing: Waterproof, insulating composition board made in large sheets of various dimensions. Used as base for exterior surfaces.

boardwalk: Pedestrian walking surface constructed of wood.

boast: To surface a stone with a broad chisel, boaster, and mallet.

boaster: Chisel used to smooth the surface of hard stone or remove tool marks.

boat lumber: Wide boards, 12″ to 16″ in width, of lightweight clear wood, such as cedar, redwood, or white pine.

boat spike: Long, square nail used in heavy timber construction. BARGE SPIKE.

boatswain's chair: Suspended flat bench seat, similar to a swing. Used as a sitting work platform.

bob: See *plumb bob*.

bobbin: Plumbing tool for bending and shaping lead pipe.

body: 1. Load-carrying portion of a truck or scraper. **2.** Longer portion of a framing square. BLADE.

body coat: Layer of paint between primer and finish coats. May also be final coat.

boil: 1. Wet material at bottom of an excavation. **2.** Swelling of material at bottom of an excavation caused by seepage.

boiled oil: Linseed oil heated to temperatures from 400° to 600°F with a small additional quantity of a drier, such as lead monoxide or manganese dioxide. Used to promote quick drying of newly painted surfaces.

boiler: Closed tank connected to an energy source that heats water to a high temperature. Usually connected to a pump to distribute water or steam for heating or power purposes.

boiler bushing: Pipe fitting with an outer thread and two pipe connections that are made inside and outside the boiler tank.

boiler feeder valve: Automatic control device that maintains proper amount of water in a boiler.

boiler heating surface: Part of heat-transfer apparatus that has one side in contact with fluid being heated and the other side in contact with gas or refractory being cooled. Fluid being heated forms part of the circulating system. The amount of this surface is measured on the side receiving the boiler, water walls, water floor, and water screens.

boiler horsepower: Equivalent evaporation of 34.5 pounds of water per hour at 212°F. Equal to a heat output of 33,475 Btu per hour.

boiler jacket: Insulating covering installed around outside of a boiler.

boiler plate: Steel plate from 1/4″ to 1 1/2″ thick steel that is rolled to a medium-hard texture. BOILER STEEL.

boiler rating: Measurement of heating capacity of a water- or steam-heating system. Expressed in Btu per hour.

boiler tubes: 1. Pipes that carry water or heat and gases of combustion. **2.** Pipe inserted in top of a boiler tank to direct cold water to tank bottom, keeping stored hot water from being cooled by new cold water.

boiling point: Temperature at which vapor pressure of a liquid is equal to normal atmospheric pressure.

boiserie: Ornate wood interior wall paneling.

bolection molding: Trim member of unusually large and broad convex projection; a panel molding that projects beyond the face of rails and stiles.

bollard: Stone guard attached to a corner or a freestanding stone post to protect it against damage from vehicular traffic.

bolometer: Thermal device for measuring radiant energy.

bolster: 1. Short horizontal member resting on top of a column to support beams or girders. **2.** Crosspiece on an arch centering, spanning from rib to rib. **3.** Bearing point of a truss bridge upon a pier. **4.** Wide chair for supporting rebars. **5.** Bricklayer's chisel. **6.** To reinforce.

***bolt: 1.** Cylindrical fastener, usually consisting of a piece of metal having a head or hooked end and a fully or partially threaded body. **2.** Portion of a lockset projecting from the door into the jamb to hold the door closed. **3.** Short section of a tree trunk. A short log suitable for peeling of veneer. **4.** Two or more rolls of wallpaper in a single package.

***bolt cutter:** Scissors device with large handles and hardened jaws for cutting reinforcing rods, bolts, rivets, heavy wire, and chain.

bolt rope: Manila rope of high quality. Approximately 15% stronger than #1 grade manila rope.

* **bolt sleeve:** Cylindrical tube installed around a bolt prior to concrete placement to prevent adhesion of concrete to the bolt.

* **bond: 1.** Arrangement of masonry units in a wall by lapping them one upon another to provide a sturdy structure. An inseparable mass is formed by tying the face and backing together. **2.** To stick or adhere. **3.** Surety agreement required of contractors that assures payment or performance of work. **4.** Short length of wire rope used to attach loads to a crane.

bond beam block: Horizontal concrete or masonry unit that strengthens a masonry wall and reduces cracking. Contains steel reinforcement and is filled with mortar. See *concrete masonry unit**.

bond breaker: Material, such as form oil, used to prevent adhesion or sticking of materials. For example, chemical bond breakers are applied to the inside of concrete forms to prevent newly placed concrete from sticking to forms.

BOLTS

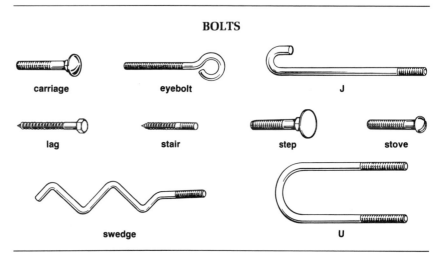

carriage eyebolt J

lag stair step stove

swedge U

bond coat: Any initial layer of material designed to improve adhesion of subsequent layers.

bond course: Row of headers or stones placed at right angles to the base of a wall.

bonded member: Prestressed concrete member in which tendons and concrete are adhered to each other. BONDED TENDON.

bonded post-tensioning: Post-tensioned concrete member in which tendons are adhered to concrete by grouting.

bonder: Masonry header. BOND BREAKER.

bonding: 1. Wire net that covers the top of a ventilating pipe or chimney. **2.** Cover for guiding and enclosing the tail end of a valve spindle. **3.** Permanent joining of metallic parts to form an electrically conductive path that has continuity and the ability to safely conduct electrical loads.

BRICK AND STONE BONDS

Brick

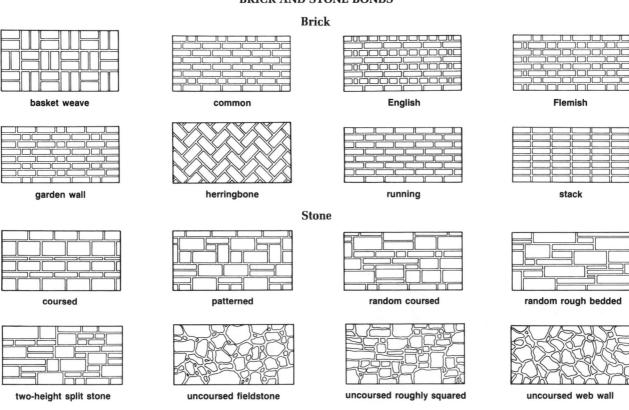

basket weave common English Flemish

garden wall herringbone running stack

Stone

coursed patterned random coursed random rough bedded

two-height split stone uncoursed fieldstone uncoursed roughly squared uncoursed web wall

bonding agent: Adhesive substance designed to fasten two members or layers of material together.

bonding conductor: Section of cable that grounds cable sheathing and metal frames of electrical equipment.

bonding jumper: Connection between portions of an electrical conductor in a circuit, piece of equipment, or main service to maintain required amps in that circuit or the necessary grounding.

bonding layer: Thin coat of mortar spread on a previously hardened concrete surface to increase adhesion of an additional layer of concrete.

bond length: Minimum amount a straight steel reinforcing rod must penetrate into concrete for secure anchoring.

bond line: Junction of weld metal and base metal or junction of two base metal parts in the absence of weld metal.

bond plaster: Type of gypsum plaster used as initial coat over concrete.

bond stone: Masonry unit placed through a wall at a right angle to its base to bind the wall together.

bond timber: Piece of lumber placed in a horizontal position in a masonry wall to help tie it together.

bones: Rocks from an aggregate base that come to the surface and separate from finer materials.

bonnet: 1. Soft covering placed over a stiff disc that is inserted in a drill or grinder and used for polishing. 2. Wire netting that covers top of vent pipes and chimneys. 3. See *drive cap.* 4. Cover for guiding and enclosing a valve stem.

bonnet tile: Semicylindrical roofing tile installed at a hip on a tile roof.

book matched: Method of joining wood veneers in which adjacent pieces of veneer from a flitch are fastened side by side, with every other sheet turned over, similar to book pages. The back of one piece joins the face of the adjoining piece. HERRINGBONE MATCHED.

* **boom:** Any heavy beam or trussed structure that is hinged at one end and carries a weight lifting device at the other; e.g., crane, jack, lattice, and live booms.

boom strap: Safety belt worn by workers working at heights on a crane or other lifting device.

booster: Auxiliary device that increases output of another primary machine.

boot: 1. Marker set behind a grade stake or hub when line of sight to grade stake or hub is obstructed. 2. Device attached to an open end piling to close the end driven into the ground and prevent it from filling with soil. 3. Projection from a beam or floor slab that supports a brick wall. 4. Transitional sheet metal piece used to join ducts of various cross-sectional shapes. 5. Metal flashing around a pipe at point of projection through a roof.

boot fitting: Sheet metal piece with one end shaped to fit a register and the other end shaped to fit a supply duct. Used for forced air heating and cooling.

boot truck: Oil truck with a spray attachment for spreading asphalt oil.

Bordeaux connection: Wire rope thimble with a permanent link.

border: Finishing strip near or along the edge of a surface.

bore: 1. Internal diameter for a pipe, cylinder, or a hole for a shaft, either machined or rough finished. 2. To drill.

bored pile: See *caisson.*

boring: Process of making holes by rotary drilling.

borrowed light frame: Window opening in an interior partition between two interior areas.

borrow site: Area from which earth is removed and hauled to another location. BORROW PIT.

borrow soil: Earth taken from another location.

boss: 1. Ornamental, projecting, knob-like block; e.g., carved keystone at rib intersections in Gothic vaulting. 2. Enlarged part of a shaft. 3. Keystone in a dome. 4. Projecting portion of a pipe, allowing for alignment and providing a point for gripping with tools. 5. See *bossing.*

bossage: Stones that are at first roughly dressed, such as corbels and quoins, then installed projecting from the surface and finish-dressed in position.

bossing: Process of shaping malleable sheet metal to conform to irregularities of a covered surface. Accomplished by tapping the metal with a bossing mallet. DRESSING.

bossing mallet: Striking tool shaped for use to process malleable sheet metal to a required form.

bossing stick: Wooden tool used to shape malleable sheet metal, such as sheet lead for tank lining.

Boston hip roof: Method of shingling used to cover the joint, or hip, of a hip roof. A double row of shingles or slate is laid lengthwise along the hip to ensure a watertight covering.

* **Boston ridge:** Finish for a roof ridge consisting of shingles saddled over the ridge and intersecting with courses of shingles on both sides of the roof.

* **Bosun's chair:** Safety harness with a rigid seat and a leather belt suspended from straps connected to a ring.

bottle brick: Hollow masonry unit manufactured in a pattern allowing for interlocking with similar units. May be set with or without reinforcing steel.

bottom bolt: Device attached on a door bottom. Friction prevents bolt from dropping until released into locking position.

* **bottom chord:** Lowest member of a truss, providing both compressive and tensile strength. May be horizontal or at an incline, depending on truss design.

bottoming: Physical removal of excavated material from bottom of an excavated area.

bottoming tap: Internal threading tool designed to cut threads the full depth of a predrilled hole. See *tap* (1).

bottom plate: Lowest horizontal member of a wood-framed wall. SOLE PLATE.

bottom rail: Lowest horizontal member of a window sash or door.

boulder: Stone material over 12″ in size.

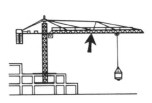

boom

Boston ridge

Bosun's chair

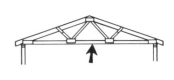

bottom chord

boulder wall: Rustic wall composed of large stones, usually undressed, and mortar.

boulevard: Wide, landscaped street.

boundary monument: Surveying marker placed on or near a boundary line to identify its location.

boundary survey: Diagram of a site. Contains dimensions, angles, compass bearings of boundary lines, and surveyor's certification.

bounding wall: Wall enclosing any area or defining a boundary.

* **Bourdon tube:** Curved device that straightens as pressure increases and bends as pressure decreases. Used to register pressure inside a pressure gauge.

bow: 1. Any part of a building that projects in the form of an arc or a polygon. **2.** Defect consisting of a flatwise deviation from an imaginary straight line from end to end of a piece. See *warp*. See *defect**.

bowl: Bucket or body of an earth-carrying scraper.

bowled floor: Floor that slopes downward toward a stage or an altar, as in a theater or church. Slope is commonly $\frac{1}{2}''$ per foot.

bowline: Knot that cannot slip or jamb. Used as a hitch, for interlocking two ropes, or for hoisting materials. See *knot**.

bow saw: Cutting hand tool for wood with blade tensioned between one open leg of an H-shaped handle or tubular frame. Blade is tightened or loosened by a cord or turnbuckled rod in the other opening of the H.

bowstring truss: Structural configuration composed of a straight bottom member and a curved, radius top member connected with struts. Used in roof and timber bridge construction. See *truss**.

bow window: Window projecting out from the face of a wall in a curved radius, normally made of several narrow, flat panes set at slight angles to each other. See *window**. COMPASS WINDOW.

box anchor: Square metal piece that ties masonry units together.

box beam: Hollow horizontal member formed like a long box. Either a structural or ornamental member, depending on materials and design.

box bolt: Door-holding hardware similar to a barrel bolt, except that it is square or flat.

box casing: Inside piece of window trim forming a blind casing. SUBCASING, INSIDE CASING.

box column: Built-up, hollow, vertical support member used in porch construction; usually square.

box connector: Electrical attachment for fastening ends of a cable to a box.

box construction: Method of cabinet-building that does not use a frame. NON-FRAME CONSTRUCTION.

* **box cornice:** Overhang trim completely enclosed with shingles, fascia, and soffit. CLOSED CORNICE.

box culvert: Enclosed drainage ditch with a square or rectangular cross section. Top, bottom, and sides are formed with reinforced concrete.

box drain: Drain with a flat top and bottom and upright sides; usually built in brickwork or concrete.

boxed cornice: See *box cornice*.

boxed mullion: Hollow divider that houses sash weights between two window units.

box end wrench: Metal hand tool with closed ends on a straight shaft handle. Shaped to fit specific sizes of nuts or bolt heads. Shaped ends (heads) are offset from the shaft (handle) from 15° to 45°. See *wrench**.

box frame: Window frame with openings at each side for sash weights.

box girder: 1. Horizontal support member of steel or cast iron with a hollow square or rectangular design. **2.** Beam made of wood formed like a long box. **3.** Bridge span having a top and bottom slab with two or more walls forming one or more rectangular spaces.

box gutter: Horizontal wood trough cut into the roof and lined with galvanized iron, tin, or copper to make it watertight. CONCEALED GUTTER.

box-head window: Window designed with a pocket above the window to allow sash to fit in the wall and leave window opening completely clear.

boxing: Continuation of a fillet weld around a corner as an extension of the principal weld.

boxing up: Applying a rough covering, such as sheathing, to a house or other structure.

box joint: See *finger joint*.

box lewis: Hoisting device of tapered metal pieces that are inserted in holes of large masonry members.

box nail: Fastener with a flat head and a shank not as thick as that of a common nail. Sizes range from 2d to 40d. Used in wood that splits easily. See *nail**.

box out: To form an opening or pocket in a structure. See *blockout*.

box pile: Foundation support formed by welding two steel channels or other structural steel shapes along their long axis, forming a void in the middle.

box sill: Header nailed on ends of joists

Bourdon tube

box cornice

and resting on a wall plate. Used in wood-frame construction.

box stair: Enclosed staircase.

box staple: Piece on a door post or jamb into which the lockbolt engages.

box stoop: High landing that makes a quarter turn in a stair at the front of a building.

box strike: Hardware device installed in a door jamb with the opening designed to receive the bolt and an enclosed rear to prevent access.

box union: Pipe fitting with female threads and a square outside nut portion. Used to join pipes while allowing for disconnection without removal of piping. See *union*.

box wrench: See *box end wrench*.

brace: 1. Structural piece, either permanent or temporary, designed to resist weights or pressures of loads. Piece fitted and firmly fastened to others at any angle to strengthen the assembly. **2.** Support used to help hold parts of furniture in place and give strength and durability to the entire piece. **3.** Hand tool with clamping jaws that hold an auger bit and a bent handle for turning the bit; used for boring holes in wood. See *drill**.

braced framing: Heavy-timber framing in which the frame is formed and stiffened by posts and braces.

brace frame: Wood-framing in which corner posts are braced to sills and plates.

brace measure: Table that gives lengths of different hypotenuses for various right angle structural supports. Usually located on the back, center of the tongue of a framing square. BRACE SCALE, BRACE TABLE.

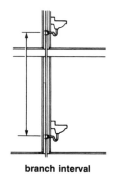

branch interval

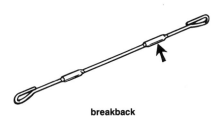

breakback

---

brace piece: Shelf of a mantel. MANTEL-PIECE.

brace scale: See *brace measure*.

brace table: See *brace measure*.

bracket: 1. Projection from face of a wall used as a support for a cornice or some ornamental feature. **2.** Support for a shelf.

bracket baluster: Upright member of a stair balustrade formed of metal that is bent to a 90° angle at the bottom and built into the string of a masonry stair.

bracket cornice: Series of exposed wall projections supporting an overhang.

bracketed stair: Staircase in which exposed step ends have a decorative scroll-bracket shape.

bracketing: 1. Wooden skeleton pieces to which lath and plaster are fastened to form a cornice surface. **2.** Shaped timber supports that form a basis for plasterwork and moldings of ceilings and parts near ceilings.

brad: Thin, small nail made of wire with uniform thickness throughout and a small head. See *nail**.

brad awl: Short straight hand tool with a handle at one end and a chisel or pointed edge at the other end along a non-tapering shank. Used to make small holes.

brad set: See *nail set*.

braided wire: Electrical conductor composed of a number of small wires twisted together or interwoven.

brake: 1. Device for holding, slowing, or stopping movement of an object. **2.** Tool for forming sheet metal.

brake horsepower: Measure of force determined by dividing running am-perage by full-load amperage listed on the equipment. This is then multiplied by the rated horsepower.

branch: 1. Inlet or outlet of a pipe fitting set at an angle with the run. **2.** Pipe into which no other branch pipes discharge. **3.** Any part of a piping system other than a main, riser, or stack.

branch circuit: Part of an electrical wiring system between the final set of circuit breakers or fuses and the fixtures and receptacles.

branch control center: Assembly of circuit breakers for protection of branch circuits feeding from the main electrical service.

branch cutout: Holder for a branch circuit fuse.

branch drain: Drain pipe between a gulley, soil pipe, or sanitary fitting and the main drain pipe.

branch ell: Elbow pipe having a back outlet in line with one of the run outlets. HEEL OUTLET ELBOW.

* branch interval: Length of plumbing waste stack approximately one building story tall (not less than 8'-0") to which the horizontal plumbing branches from one story of the building are connected.

branch joint: Plumbing pipe connection in the shape of a "T" or "Y" in which the connection is wiped with solder.

branch line: 1. Air supply line between controller and controlled device. **2.** Air duct between controller and actuator. **3.** Water supply pipe connecting one or more fixtures to the water main, riser, or another branch line.

branch pipe: Plumbing pipe having one or more branches.

branch vent: Pipe connected between a plumbing fixture and the waste vent stack to allow each fixture to drain freely.

brandering: Nailing of furring strips to girders, joists, or a solid surface to hold finish materials away from wall surface.

brandreth: 1. Wooden framework for support; e.g., a stand for a cask. **2.** Fence or railing around a well. BRANDRITH.

brashness: Condition of wood characterized by low resistance to shock and an abrupt failure across the grain without normal splintering.

brass: Alloy of copper and zinc used for decorative hardware.

brass flare fitting: Plumbing connection in which the pipe joint is made watertight by the compression of a flared, soft metal pipe end. Compression takes place when an adjoining nut is tightened around this pipe end.

brass pipe wrench: Plumbing tool with a strap designed to hold and turn soft pipe without crushing the pipe. BRASS PIPE VISE.

braze: To solder with an alloy, such as brass, that cannot or is very difficult to change into a liquid state as compared to common solder. A metal rod with a lower melting point than the metals being joined (base metals) is used as an electrode when joining the metals. The filler metal is attracted to the base metals by capillary action. The base metals do not melt.

break: 1. Lapse in continuity; any projection from the general surface of a wall; an abrupt change in direction. **2.** To shatter or fail.

breakaway clip: L-shaped light-gauge aluminum member used to fasten area separation walls to floor and roof framing. Clips dissolve in the event of fire, allowing the walls to move.

* breakback: Weakened point in a concrete form snap tie that allows the end projecting beyond the wall face to be removed by striking and breaking.

breaker: 1. See *circuit breaker*. **2.** Relatively good insulating material used between the liner and outer shell of a refrigeration unit. **3.** Rock-crushing machine.

breaker panel: Electrical box that contains circuit breakers.

breakfast room: Small eating area adjacent to a kitchen.

break ground: First operations performed when excavation is begun for a new building.

break in: Cutout in a masonry wall for inserting a wood member.

breaking joints: Staggering joints of materials in consecutive rows to improve strength and appearance by avoiding continuous seams.

breaking point: Point of failure of a member or a joint.

breaking radius: Full extent of curvature to which a material can be bent without breaking.

break iron: See *plane iron cap*.

break line: Drawing line used for an object too long to show on a sheet or that does not require full length to be shown. Styles of break lines include those with a Z in the middle, an irregular zigzag pattern, or a figure-eight shape for cylindrical objects.

breakwater: Wall projecting into the ocean to prevent damage to a harbor or shoreline caused by wave action.

breast: 1. Portion of a wall between window stool and floor. **2.** Underside of a member. **3.** Projecting portion of a wall.

breast board: See *lagging.*

breast drill: Hand tool used for drilling holes in wood or metal. A hand-turned crank transmits power through bevel gears to the drill chuck. Force is applied through a breastplate at the rear of the tool.

breast of a window: Backing of the recess and parapet under a window sill; composed of masonry.

breastsummer: **1.** Heavy timber, or summer, placed horizontally over a large opening. **2.** Beam flush with a wall or partition it supports. **3.** Lintel over a large window that supports the superstructure above it.

breather plug: Removable cap for venting a space that is otherwise airtight.

breccia: Stone composed of angular fragments embedded in a finer material.

breeching: Vent exhaust area for moving products resulting from a fuel-fired combustion chamber to a vent or chimney.

breeching fitting: Y-shaped fitting in which the flow of two pipes is mixed into one.

breeze: Finely divided materials from coke production; used as an additive in cement.

breezeway: Covered passage, open at each end, that passes through a house or between two structures, increasing ventilation and adding an outdoor living effect.

*** brick:** Rectangular block used for building or paving purposes. Bricks are made from clay or clay mixture molded into blocks that are then hardened by drying in the sun or baking in a kiln. American-made bricks average $2^1/_2'' \times 3^3/_4'' \times 8''$ or $2'' \times 4'' \times 8''$ in size. Brick types include adobe, arch, bottle, breeze, buff standard, building, economy, engineered, facing, fire, floor, gauged, jumbo, Norman, paving, Roman, salmon, SCR, and sewer.

brick anchor: Corrugated thin metal piece attached to a backing wall and embedded into brick mortar joints. Used to hold masonry veneer to the backing wall surface. See *anchor*.

brick and brick: Method of bricklaying in which bricks are stacked without mortar; mortar is applied later only to fill surface irregularities.

brickbat: Pieces of broken bricks.

brick beam: Lintel made of brick with iron straps.

brick cement: Waterproofed masonry compound used with every type of brick, concrete brick, tile or stone masonry, and also in stucco work.

brick construction: Building with ex-terior bearing walls built of brick or a combination of brick and tile masonry.

brick corbel: Projection from the face of a wall formed of brick.

brick facing: See *brick veneer.*

brick gauge: Standard heights for laying brick courses.

brick hammer: See *bricklayer's hammer.*

bricklayer: Person who builds structures of stone, brick, concrete block, and other masonry materials. BRICK MASON.

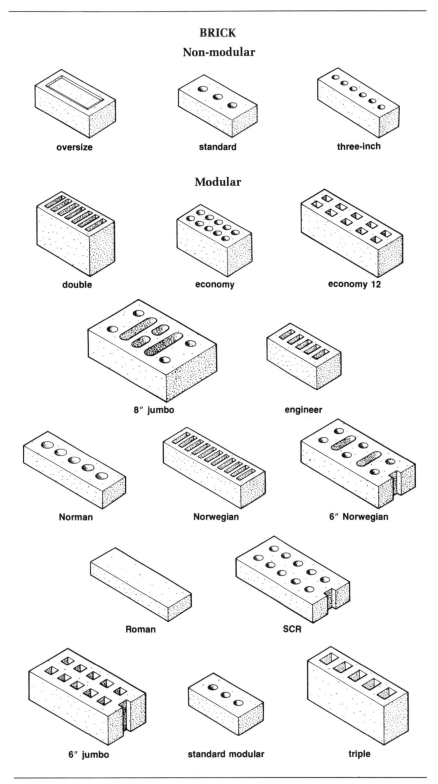

BRICK

Non-modular

oversize   standard   three-inch

Modular

double   economy   economy 12

8″ jumbo   engineer

Norman   Norwegian   6″ Norwegian

Roman   SCR

6″ jumbo   standard modular   triple

brick set　　　　brick tongs　　　　brick veneer

bricklayer's hammer: Hand tool used by bricklayers for dressing or breaking brick. It has both a sharpened peen and hammer head. See *hammer**. BRICK HAMMER.

brick mason: See *bricklayer.*

brick molding: Wooden trim member installed around door and window frames where they meet a brick wall surface. See *molding**.

brick nogging: Brickwork that fills spaces between studs or timbers in a wood-framed wall or partition. BRICK-AND-STUD WORK.

brick pier: Detached mass of masonry that serves as a vertical support.

brick seat: Ledge on a wall or footing that supports masonry.

***brick set:** Tool used to cut bricks when exact surfaces are required. A brick hammer is used to force the chisel-like brick set into the brick.

***brick tongs:** Hand tool for picking and carrying several bricks at once. A handle is attached to a self-clamping set of rods to hold and carry up to 11 bricks.

brick trimmer: Arch built of brick between vertical supports at either side of an opening. Constructed to the thickness of an upper floor to support a hearth and guard against fire.

brick trowel: Flat, triangular-shaped trowel used by bricklayers for picking up mortar and spreading it on each course of brick or on individual bricks.

***brick veneer:** Brick facing applied to the wall surface of a frame structure or other type of structure. Metal ties in mortar joints bond the facing material to the frame structure. A small air space is normally provided between the facing and the framed wall. BRICK FACING.

brick whistle: See *weephole.*

***brickwork:** Anything constructed of brick regardless of the position.

bridge: **1.** Structure erected over a depression or stream, as over a highway, chasm, or river, to provide a passageway for vehicles or pedestrians. **2.** Structure similar in form or use to a bridging joist. **3.** To span, as to bridge between two openings. **4.** Electric current-carrying wire to a blasting cap.

bridgeboard: Support or stringer of a stair; a notched board for supporting risers and treads of a wooden stairway.

bridge crane: Hoisting machine mounted on an overhead track or rail.

bridge deck: Pavement surface on a bridge.

bridge measure: Numbers on a framing square denoting hypotenuses of various triangles. These hypotenuses are in relation to a triangle with one 12 ″ leg and one leg denoted by the inch marking on the square for the unit rise of a common rafter. Hypotenuses for a hip or valley rafter are in relation to a triangle with one 17″ leg and one leg denoted by the inch marking on the square for the unit rise.

bridge stone: Flat stone placed across a depression to act as a level walkway.

***bridging:** Wood or metal pieces fastened between timbers, such as floor joists, to distribute loads and strengthen the structure. Cross bridging consists of two small pieces installed at an angle between members. Solid bridging consists of one-piece wood blocks.

bridging joist: Beam or horizontal support that rests on binding joists and supports flooring.

bridle hitch: Two, three, or four lines joined at a common lifting hook or point. Each leg is attached to the object to be lifted in order to equally distribute the load.

bridle joint: A reverse mortise-and-tenon joint. Instead of a tongue and mortise being cut in the centers of two adjoining pieces, two tongues are cut along the outside edges of the tenoned piece, and two corresponding mortises are cut at the edges of the other piece to receive them.

brindled brick: Masonry material with a brown, mottled surface.

brindle iron: Iron hanger that supports joists and beams. STIRRUP.

brine: Liquid heat transfer agent in a refrigeration system; usually a salt solution that remains liquid and has a flash point above 150° Fahrenheit.

BRICK POSITIONS

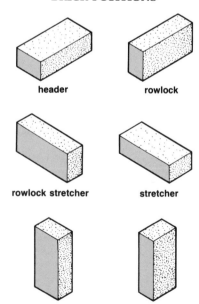

header　　　　rowlock

rowlock stretcher　　　　stretcher

sailor　　　　soldier

BRIDGING

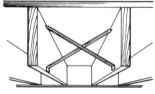

metal cross

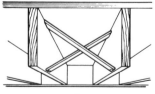

wood cross

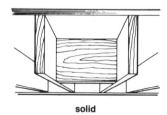

solid

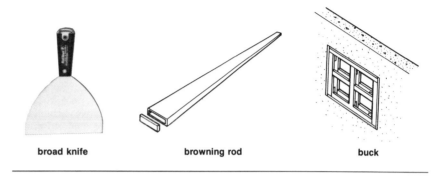

broad knife   browning rod   buck

British thermal unit (Btu): Quantity of heat required to raise temperature of 1 pound of pure water 1° Fahrenheit at or near the temperature of maximum density of water (39° Fahrenheit).

broach: 1. To enlarge or ream a hole in stone. **2.** To rough-dress a stone. **3.** Pointed structure; e.g., a spire. **4.** Half-pyramid sections that form the transition from a square base to an octagonal roof.

broached work: Broad grooves that give a finish to a building stone; made by dressing the stone with a punch.

broaching: Removal of stone by drilling a row of holes close to each other and knocking out sections of stone between holes.

broach post: Vertical timber tie used to connect the ridge and tie beam of a roof. Shaped at the bottom to allow for bearing of the two angular struts that support middle points of rafters. KING-POST.

broad ax: Hand tool for rough-dressing timber; has a wider blade than normal.

broadcast: To throw or toss material onto an open surface.

* broad knife: Hand tool with a wide flexible blade for applying and smoothing joint compound on drywall. Also used for scraping off old paint and wallpaper.

broadloom: Seamless carpeting woven from 6′ to 18′ wide.

brob: Wedge-shaped spike for bracing and securing the end of a timber where it butts against another timber.

brocade: Carpet with a raised surface pattern formed by heavy twisted yarn.

broken joint: Meeting point not in a straight line; adds strength to the structure. BREAKING JOINT.

broken range: Design of masonry construction with otherwise continuous courses interrupted at various intervals.

broken tape measurement: Lineal surveying distance calculated by taking a series of short measurements and adding them.

bronze: Alloy of copper and tin; sometimes applied to other alloys that do not contain tin.

bronzing: 1. Paint surface defect in the form of chalking resulting from exposure to the elements. **2.** Application of bronze to another surface.

broom: 1. To press layers of roofing material to a bituminous surface for achieving a complete bond between layers. **2.** To spread the head of a wood pile as a result of repeated blows. **3.** See *broom finish.*

broom closet: Small recess or cabinet for storing brooms and cleaning materials.

broom finish: Slightly rough texture in concrete created by drawing the bristles of a broom across the smoothed surface of freshly troweled concrete. BRUSHED SURFACE.

brother: Sling with two or four legs.

brown coat: Layer of plaster applied with a fairly rough finish to receive the finish coat. In two-coat plasterwork, the brown coat is the base layer of plaster applied over lath. In three-coat plasterwork, the brown coat is the second layer applied over the scratch coat.

browning brush: Tool for throwing water on the surface of applied mortar to provide slip, or lubrication, for other tools.

* browning rod: Plastering straightedge with a hollow rectangular 1″ × 4″ cross-sectional shape; available in 6′ or 8′ length.

brownstone: Dark brown or reddish-brown sandstone used as a building material.

browpiece: Beam over a door opening.

brush: 1. An electric motor part of carbon or graphite that carries electric current. **2.** Tool used for applying liquids such as paint and wallpaper paste. Also used for cleaning when dry.

brush coat: Waterproofing application of one or more layers of asphalt, pitch, or commercial waterproofing on exterior of a foundation below grade line using a brush or trowel.

brushed plywood: Plywood siding in which the surface is treated by removing the softer wood from the face by abrasive action, leaving the harder grain pattern to create a more textured surface.

Brussels carpet: Carpeting made of several different colors of worsted yarn that are fixed in backing made of strong linen thread.

bubbler: Mouthpiece of a drinking fountain.

* buck: Framing around an opening in a wall; e.g., a door buck encloses the opening in which a door is placed.

bucket: Scoop-shaped attachment on an excavating machine.

bucket trap: Valve for eliminating air and condensed moisture from pipes and radiators without allowing steam to escape. As a container is filled with condensation, buoyancy is lost and it sinks, opening the valve. Condensation is released and refilled with air to return to a floating position.

bucking: Cutting a tree trunk into lengths for lumber.

buckle: To twist or bend out of shape, become deformed or distorted under a compressive load.

buckling: 1. Compressive failure of a structural member. **2.** Shrinking of a proxylin lacquer film over an oil-base paint undercoat when full drying time has not been allowed for the undercoat.

buck opening: See *buck.*

buck saw: See *span web saw.*

buff: To polish, grind, or clean to a smooth surface and high shine.

buffalo box: See *curb box.*

buffer: Blasted rock or a metal plate that limits scattering of rock from the next blast.

buggy: Manual or powered vehicle used to transport fresh concrete from the mixer to the location where it is to be placed. GEORGIA BUGGY.

bugle head screw: Threaded fastener with a flat head design. Underside of the head is shaped like the bell of a bugle. Commonly used on drywall screws. See *screw*.*

builder's acid: Mixture of one part hydrochloric acid and four parts water; used to remove mortar stains from brick.

builder's hardware: See *finish hardware.*

builder's level: Telescope-like instrument incorporating a spirit level that mounts on a tripod; used to establish straight work lines and elevations. See *surveying*.*

builder's staging: Scaffolding formed of square timbers strongly braced together; used in handling heavy materials.

builder's tape: Measuring device, usually 50' or 100' in length, with a narrow strand of steel or fabric marked with dimensions and wrapped around a spool in a case.

building: 1. Structure that stands alone or is separated from adjoining structures by fire division walls and fire doors. 2. Assembling materials and putting them together to form a structure; the act of one who builds.

building area: Total ground space covered by each building and accessory building. Does not include uncovered entrances, terraces, and steps.

building block: Hollow rectangular block of burnt clay, terra-cotta, concrete, cement, or glass construction material.

building brick: See *common brick*.

building drain: Part of the piping of a plumbing drainage system that receives discharge from soil, waste, and other stacks inside a building and conveys it to the building sewer. Three types are *sanitary* (waste), *storm* (surface water), and *combined* (both waste and surface water.) HOUSE DRAIN.

building envelope: 1. Elements that enclose a conditioned space and through which thermal energy is transmitted. 2. Outer portion of a building.

building line: Limit on a lot beyond which the building code forbids construction of a building. Boundary on a building site within which walls of the building must be confined.

building main: Pipe that supplies fresh water to a structure; connected from supply source to first branch of the water distribution system.

building material: Material used for construction.

building paper: Felt material saturated with bitumen to form a waterproof sheet. Applied between sheathing and siding, as an undercoating for stucco applied to wire mesh, and as an undercovering on roofs below shingles to protect against leaks.

building power line carrier: Method of transmitting control signals through existing electrical wiring by means of altering or superimposing on existing waveforms.

building service equipment: Mechanical and electrical equipment and fixtures, including piping and wiring, that provide services essential for making a structure habitable.

building sewer: Part of a drainage system that receives materials from the building drain and conveys them to a public sewer or other disposal point. Three types are *sanitary* (waste), *storm* (surface water), and *combined* (waste and surface water.)

building site: Area occupied by a building, including open yard areas.

building stone: Stone that may be used in construction of a building, such as limestone, sandstone, granite, and marble.

building trades: Skilled and semi-skilled workers in the construction industry, such as carpenters, concrete finishers, electricians, floorlayers, glaziers, heavy equipment operators, HVAC technicians, ironworkers, laborers, masons, painters, pipefitters, plumbers roofers, and teamsters.

building trap: Plumbing fitting that traps fluid that acts as a stopper to prevent sewer gas backup.

buildup: 1. Welding operation in which surfacing metal is deposited to the required dimension. 2. Application of successive layers of shotcrete.

built-in: Part of the permanent structure; e.g., built-in range, built-in oven.

built-up: Composed of two or more parts fastened together to act as a single unit.

built-up beam: Horizontal support constructed by fastening two or more boards or timbers together for additional strength.

built-up column: Vertical support composed of more than one piece.

built-up girder: See *built-up beam*.

built-up roofing: Covering normally used on flat roofs composed of several layers of bituminous materials applied in an overlapping manner and sealed to provide a waterproof covering. Commonly covered with a thin layer of gravel.

bulb: 1. Gas enclosure portion of an electric lamp. 2. Glass portion of an electric lamp assembly. 3. Portion of a thermal-sensing system placed into a variable media.

bulb bar: Steel rod that is thickened toward one edge.

bulb tee: Metal member with a cross-sectional T shape. Ends of the T are enlarged to form a bulbous edge.

bulb well: Socket used in conjunction with an immersion thermostat.

bulge: 1. Slight swelling outward; e.g., a graceful swelling on columns. 2. Swelling defect; e.g., outward bending of a wall.

bulk density: Weight of a material per unit of volume.

* **bulkhead:** 1. Member installed inside or at the end of a concrete form to prevent fresh concrete flow from coming into a section or out the end of the forms. 2. Box-like structure that rises above a roof or floor to cover a stairway or an elevator shaft. 3. Door that permits access to a cellar or shaft.

bulking: Increase in size of material as a result of absorption of moisture. MOISTURE EXPANSION.

bulldog grip: See *U bolt*.

* **bulldozer:** Tractor with caterpillar tread that is used to push dirt and debris for clearing an area.

bullet catch: Spring-loaded hardware device with a ball partially projecting from one end. Ball fits into a mortise or hole in a jamb for holding a door closed. See *ball catch*.

bullet-resistant glass: Laminated material of glass and transparent resin sheets bonded with heat and pressure.

* **bull float:** Tool with a large flat piece of aluminum, magnesium, or wood attached with a rotating pivot to a long handle. Used for smoothing concrete flatwork.

bull floating: Smoothing and leveling of high spots and voids left in concrete flatwork after screeding. This is the first stage in final finishing of concrete flatwork after screeding; sometimes used as a substitute for darbying.

* **bull-headed tee:** Plumbing fitting in the shape of a T in which the branch size

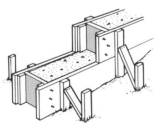

bulkhead (1)

bulldozer

bull float

bull-headed tee

is larger than the straight run size. BULLHEAD TEE.

bull header: Brick having one rounded corner, usually laid with the short face exposed to form a sill under and beyond a window frame; also used as a quoin or around doorways.

bullnose: Convex rounding of any object; e.g., a brick having one rounded corner or a stair step with a rounded end used as a starting step.

bullnose block: Concrete masonry unit measuring $7^5/_8'' \times 7^5/_8'' \times 15^5/_8''$, with one vertical corner on the end that is slightly rounded. See *concrete masonry unit**.

bullnose pier block: Concrete masonry unit measuring $7^5/_8'' \times 7^5/_8'' \times 15^5/_8''$, with two vertical corners on the face that are slightly rounded. Used to provide finished corners where a vertical row of blocks projects from the face of the wall. See *concrete masonry unit**.

bullnose plane: Small wood-smoothing tool that can be used in corners or other places difficult to reach. The blade can be adjusted for coarse or fine work. See *plane* (1)*.

bullnose step: Flat tread that is rounded at the outer corner and extends beyond the stringer to form a semicircular platform for placement of the starting newel post.

bull pin: Tapered, pointed piece of metal used in structural ironwork to align holes for driving a drift pin before bolting the adjoining pieces.

bull point: Pointed steel bar used for making holes in stone.

bull's-eye arch: Circular opening, such as a circular or an oval window.

bull stretcher: Brick with one corner rounded and laid with the long face exposed; e.g., quoin.

bulwark: Defensive structure; solid wall or rampart for protective purposes.

bumper: 1. Knoblike doorstop attached to a wall as a protection from a swinging or sliding door; device used as an obstruction for a swinging or sliding door. 2. Device that cushions an impact point. 3. Worker who molds handmade brick.

bumping: Process of raising or shaping flat metal so as to form ornaments for cornice work, curved moldings, sheet metal balls, and covers for various objects. Performed using a raising hammer and raising block.

bundled bars: Group of one to four steel reinforcing bars tied into contact with each other.

bungalow: One-story house with low, sweeping lines and a wide veranda. The attic may be finished as a second story.

bungalow siding: Bevel horizontal lap siding 8" or wider.

buoyant foundation: Building support made of a reinforced concrete raft designed to float on the soil or water that supports it and the superstructure.

burden: 1. Distance from a drilled hole in the earth to ground surface. 2. Volume of rock or other material moved by explosives in a drilled hole. 3. Loose material overlying bedrock.

burglar alarm: Device fastened to a building opening that sounds an alarm or alerts law enforcement officials when disturbed. Activated by thermal, light, or acoustical stimulus.

burl: Abnormal growth on tree trunk often in the form of a flattened hemisphere. Considered a decorative feature in woodwork and veneer.

burlap: Fabric made of coarsely woven jute, hemp, or flax. Commonly used to hold moisture in the curing of concrete by wetting and laying it on the concrete surface.

burner: Device that conveys fuel to the final point of combustion. Types include air-atomizing, atmospheric, conversion, dual-fuel, forced draft, injection, intermittent ignition, mechanical-atomizing, natural draft, pressure-atomizing, power, rotary-atomizing, steam-atomizing, vaporizing, and yellow flame.

burner register: Series of vanes that directs air flow through a combustion chamber.

burner throat: Sleeve located at the burner exit; the beginning location for combustion.

Burnett's process: Infusion of timber with chloride of zinc as a preservative.

burning: Cutting metal with an oxyacetylene torch.

burning a joint: Tooling mortar joints after mortar has become too hard, leaving a black mark.

burning in: Fastening the edge of lead flashing into a masonry wall by turning it into a dovetail groove that has been cut in the wall, then filling the groove with molten lead and caulking.

burnish: To polish or produce an edge by friction.

burnisher: Hardened steel hand tool used for finishing, polishing, or final sharpening of metalwork or tool edges by friction. The burnisher is held against a revolving metal piece or rubbed over a tool edge, which receives a smooth, polished surface as a result of compression of the outer layer of metal.

burr: 1. Ragged, sharp, projecting edge.

2. Nut with screw threads. 3. Lump of brick that has fused together during burning.

* **busbar:** 1. Aluminum or copper strip on a switch and panelboard from which all electrical circuits are tapped. BUS. 2. Large aluminum or copper bar to which main electrical feeders or circuits are connected. BUS.

* **bush hammer:** Hand tool with a serrated face of several rows of diamond-shaped points; used to roughen hardened concrete surfaces.

bushing: 1. Threaded piece or plug for connecting two pipes of different sizes in the same line. The plug that serves as a reducing adapter is designed to receive a pipe of smaller diameter than that of the pipe into which it is screwed. See *conduit fitting**. 2. Insulating tube or sleeve tht protects an electrical conductor where it passes through a hole. 3. Metal cylinder between a shaft and its support which reduces the rotating friction.

bush metal: Alloy of copper and tin; used for fabricating bushings.

bus wire: Conductor of electricity to a blasting cap.

butment: See *abutment*.

butt: 1. To meet without lapping. 2. Hinge of any type other than a strap hinge. 3. Short length of roofing material. 4. Thick end of a wood shingle.

butt chisel: Hand woodworking tool for mortising the leaf of a hinge into a wood door or jamb.

butt end: Thicker end of any member or tool.

busbar (2)

bush hammer

butterfly valve

butt gauge

butt end treatment: Method of applying preservatives to wood members by soaking the ends in a preservative material.

butterfly hinge: Ornamental pivoting device in which both leaves are exposed and manufactured in a decorative pattern.

butterfly nut: See *wing nut*.

butterfly roof: Two shed roofs connected at lower edges to form a low-sloped V shape.

* **butterfly valve:** Enclosed device housing a rotating disk that seats against a resilient material in the closed position. A 90° turn of the handle is required to open or close the device and control the flow of fluid.

buttering: 1. Process of spreading mortar on brick before laying. **2.** Deposit of welding metal on one or more surfaces to provide compatible weld metal for completion of the weld.

buttering trowel: Flat tool similar to but smaller than a brick trowel; used for spreading mortar on a brick before it is placed.

butt fusion: Joining thermoplastic resin materials by heating until molten and pressing together until cool.

* **butt gauge:** Marking device used to lay out depths and widths of mortises for hinges in doors and jambs.

butt hinge: Pivoting hardware secured to edge of a door and face of the jamb it meets when door is closed. Usually mortised into the door and jamb.

butt joint: Fastening together of two parts end-to-end without overlapping. See *joinery**.

butt joint anchor: Metal L-shaped device used to join two wood members that are 90° to one another. See *anchor**.

button: 1. Piece of hardwood with a short shaft and large round head. Inserted into countersunk holes to cover fastener heads. BUTTONHEAD. **2.** Part of a weld that fails in a destructive test of a weld specimen.

button-headed bolt: See *carriage bolt*.

button punching: Crimping the interlocking lap of sheet metal ducts with a dull punching tool.

buttress: Projecting structure built against a wall or building to give it greater strength and stability.

buttress cap: Trim member at top of an open stair carriage that receives the bottom end of balusters.

butt stile: Vertical frame of a door to which hinges are attached.

butt strap: Steel plate fastened to two meeting pieces that are not overlapped.

butt weld: Joining of two pieces of metal by an electric arc where the two pieces are not overlapped. See *weldment**.

butt-welded splice: Reinforcing bar splice made by welding the ends without overlapping.

buttwood: Lumber cut from base portion of a tree.

butyl stearate: Clear liquid used for dampproofing concrete.

buzzer: Electric call signal consisting of a doorbell with the hammer and gong removed; operates on the same principle as a vibrating bell.

BX: Flexible, armored electrical cable.

bypass: Secondary pipe or duct, a channel providing an alternative deflected route; usually controlled by a damper or valve and connected with a main passage for conveying fluid around an obstruction or some element of a system.

bypass door: See *sliding door*.

bypass feeder: Device that injects chemicals into boiler feedwater to prevent scale and reduce oxygen content.

bypass vent: Plumbing vent in which ends are connected below the lowest fixture and above the highest fixture; a separate pipe in addition to all other vents.

Byzantine architecture: Architectural style characterized by large domes, round arches, elaborate columns, and many decorative elements.

cabana: Small open shelter.

** **cab guard:** Metal protective device on a dump truck that extends from the front wall of the truck body over the cab.

cabinet: 1. Piece of furniture fitted with shelves, drawers, and/or doors. **2.** Flush or surface-mounted enclosure used for housing electrical devices or conductor connections. **3.** Suite of rooms for an exhibition.

cabinet file: Single-cut half-round file used to shape and smooth wood. CABINET RASP.

cabinet finish: Oiled, varnished, or polished finish on a wood surface.

cabinet lock: Spring-loaded or magnetic device used for securing cabinet doors. See *lock**.

cabinetmaker: Person skilled in layout, construction, and installation of cabinets and casework.

cabinet projection: Oblique drawing that shows a true view of the front face of an object with all receding lines drawn to one-half the length of corresponding lines in the true view.

** **cabinet scraper:** Flat steel blade with one edge sharpened and burnished to a sharp edge. Drawn across a wood surface to remove irregularities, dried finish or adhesive. Also used for final smoothing of surfaces before sanding.

cabinet window: Bay or projecting window used for displaying goods.

cabin hook: See *hook and eye*.

cable: 1. Conductor with insulation, or a stranded conductor with or without insulation. **2.** Wire rope used to lift and transport loads. **3.** See *tendon*. **4.** One of the reedings in the flutes of a column or pilaster.

cable bender: Tool used to bend heavy wire and cable in tight locations. Shaped similar to an open-end wrench with a head offset 22° from the handle.

cable box: See *cutout box*.

cable ceiling heat: Radiant heating system in which several small, thin conductors are attached to the lower surface of a gypsum drywall ceiling. Conductors are encased in a special plaster and current flows through them to produce heat.

cable cutter: Hand tool with sharp jaws used for cutting electrical cable.

cable duct: Rigid metal, plastic, or concrete enclosure in which electrical conductors are installed.

cable jacket: Protective covering over the insulation, core, or sheathing of a cable.

cable length: Surveying measurement equal to 720′-0″.

cable molding: See *cabling*.

cable pulling compound: Material used to lubricate cable and facilitate pulling the cable through conduit.

cable rack: See *cable tray*.

cable splice: Method used to join two wire ropes by interweaving the wire strands.

cable-supported construction: Structure in which cables are the primary structural component; e.g., suspension bridge.

** **cable tray:** Open metal frame used to support electrical conductors. Consists of a latticed frame on the sides and bottom.

cableway: Device used to transport material in which a wire rope is anchored at both ends and suspended to facilitate moving material from one point to another.

cabling: Spiral design molding similar to a rope. CABLE MOLDING.

cad cell: Solid-state device that senses flame.

cadmium: Metal element with good corrosion resistance.

cadmium plating: Protective cadmium coating applied to base metal. Provides good corrosion resistance.

cage: 1. Reinforcing steel assembly that is ready to be placed in position. **2.** Circular frame that limits movement of rollers in a bearing. **3.** Wire enclosure for lights used to prevent damage to the lamp.

caisson: 1. Watertight box or enclosure used for construction work below grade or water level. **2.** Raised or recessed ceiling panel. See *coffer*.

caisson pile: Cast-in-place pile formed by drilling a hole, removing the earth, and filling it with concrete.

caking: Hardening of material that is difficult to remix or soften.

cab guard

cabinet scraper

cable tray

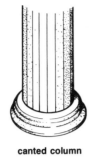

| cam (1) | canted column | cant strip (1) |

calcareous: Containing calcite or calcium carbonate.

calcimine: Inexpensive wash coating consisting of glue and powdered calcium carbonate. May be white or tinted and used on plaster or masonry surfaces.

calcine: To heat a substance to a point just below fusion temperature to remove combined water or alter its chemical composition.

calcium chloride: Concrete admixture that accelerates the setting of concrete. Increases the heat generated in the mix; used for cold-weather concreting.

calcium silicate brick: Masonry unit made primarily from cement, sand, and lime hardened by the autoclave method.

calcium silicate insulation: Hydrated calcium silicate used for pipe insulation or molded into rigid shapes.

calcium sulfate cement: Cement that depends on the hydration of calcium sulfate to provide hardness.

caliber: Inside diameter.

calibrate: To verify the graduations and incremental values of a machine or tool for accuracy and make the necessary adjustments.

caliduct: Conduit used to carry hot air, hot water, or steam in a heating system.

California bearing ratio (CBR): Measure of the bearing capacity of a foundation. Calculated as the ratio of force per unit area required for a 3 square inch circular piston at a rate of 0.05″ per minute to the force required for equal penetration of a standard crushed rock material; penetration is 0.1″.

caliper: Hand tool with one fixed and one adjustable jaw. Graduated on the body of the tool. Used to make precision inside and outside measurements such as measurements of inside and outside diameters.

calorie: Measure of heat equal to the amount of heat required to raise the temperature of 1 gram of water 1°C under 1 atmosphere of pressure.

calotte: Dome; cup-shaped ceiling.

* **cam: 1.** Eccentric wheel mounted on a rotating shaft. Used to convert rotating motion into linear non-symmetrical motion. **2.** Rotating member of a lock that moves the locking mechanism.

camber: 1. Slight convex curve of a surface that facilitates water runoff. **2.** Slight upward curve in a structural member, such as a beam or a truss, designed to compensate for deflection of the member under load.

came: Flexible lead H-shaped member that secures pieces of leaded glass together in a window.

camelback truss: Prefabricated truss in which a segmented upper chord is made of a series of straight members joined at a slight angle to each other to form an upward curve.

camp ceiling: Ceiling in which the central portion is flat and horizontal and the perimeter is sloped to the angle of the rafters.

candlepower: Measure of illumination intensity of a light source. Expressed in candelas.

candle wicking: Cotton thread used on pipe threads to make them watertight.

cane bolt: Door hardware device with an inverted-L shape. Secured to a door with brackets that allow it to travel vertically. Bottom end projects into a recess in the floor to secure the door.

canopy: 1. Roof-like structure projecting from a wall with or without support from pillars. **2.** Exterior part of an electrical lighting fixture that fits against the wall or ceiling and covers the outlet box.

cant: 1. Line or surface at an angle to another. **2.** Molding formed with plain surfaces and angles rather than curves. **3.** Squared or partially squared log that is resawn into lumber.

cant bay: Three-sided bay window; outer two sides are splayed from the face of the wall.

cant board: Wood member positioned to create a slope; e.g., on each side of a valley to support the flashing.

cant brick: Manufactured brick with one beveled side.

* **canted column:** Column with a faceted perimeter instead of flutes.

cant hook: Large, round wooden lever with an adjustable steel or iron hook near the pivot end. Used to move timber.

cantilever: 1. To rest on and project beyond a supporting member, such as a cantilevered beam or joist. **2.** Large ornamental bracket used to support a balcony or cornice.

cantilever bridge: Bridge formed by two arms that project from opposite sides of piers to form the span.

cantilever retaining wall: See *battered wall*.

canting strip: See *water table*.

cant molding: Beveled trim member.

canton: Pilaster or quoin forming a corner that projects from the face of a wall.

* **cant strip: 1.** Triangular piece of lumber, metal, or plastic placed under roofing at the intersection of a parapet wall and a roof. Facilitates water removal and weatherproofing. **2.** See *chamfer strip*.

cant window: See *bay window*.

canvas wall: Plaster wall with a canvas covering that is a base for paint and wallpaper.

cap: 1. Top member of a column, door, or molding; e.g., lintel over a door or window frame. **2.** Unthreaded fitting or fitting with female threads. Used to enclose the end of a pipe. See *pipe fitting**. **3.** See *blasting cap*. **4.** See *drive cap*. **5.** Layer of concrete placed over rock at the bottom foundation excavations. Used to prevent the rock from weathering and other damage. **6.** Smooth, flat surface bonded to the bearing surfaces of concrete test specimens to provide uniform distribution of the load during testing.

capability: Maximum load-carrying capacity of electrical generation equipment or other electrical devices under specified conditions and predetermined time. Expressed as kilovolt-amperes or kilowatts.

capacitance: Property of electrical conductors and devices that allows the storage of an electrical charge when potential differences occur between the conductors. Expressed as the quantity of electricity to the potential difference.

capacitor: Electrical device composed of two electrical conductors with an insulator between them. Used to introduce capacitance into an electrical circuit.

capacity reducer: Compressor component that regulates the capacity of a compressor without affecting operating conditions.

cap block: Concrete masonry unit with all surfaces finished except for the bottom. Bottom commonly contains voids to provide a key with the concrete. Installed at the top of a wall to produce a finished surface. See *concrete masonry unit**.

cap cable: Short tendon in a prestressed concrete member used to prestress the zone of negative bending.

* **cape chisel:** Chisel with a thin tapered face and a narrow cutting edge. Used to cut slots and grooves in deep corners of concrete, masonry, and metal.

* **Cape Cod style:** Architectural design derived from the one-story or story-and-a-half cottages built in the Cape Cod district of Massachusetts. Characterized by steeply pitched gable roofs and eave lines at approximately the same height as the first floor ceiling.

cap flashing: Sheet metal flashing wrapped around the top of a wall or other vertical projections to provide moisture protection. Installed over all other waterproofing. See *flashing**.

capillary action: Seepage of moisture through hardened concrete or mortar resulting from incomplete or faulty surface finishing. CAPILLARY FLOW.

capillary break: Open area designed to prevent capillary action of moisture.

capillary tube: Small diameter tube used to regulate the flow of refrigerant between the high and low sides of a refrigeration system.

capillary water: Water that, by means of capillary action, is above the surrounding water level or penetrates a space where water normally does not flow.

cap iron: Slotted steel plate used to reinforce and stiffen the plane iron in a hand plane. BACK IRON.

capital: Upper part of a column, pilaster, or pier.

cap molding: Uppermost member of window or door trim.

cap nut: Internally threaded fastener with a domed head. Used as a decorative exposed fastener. See *nut**.

capped butt hinge: Butt hinge with a cover that conceals the screws of each leaf. Cap is fastened to the hinge with small screws.

capped edge: Protective and reinforcing strip of sheet metal fitted to the raw edge of sheet metal.

capping: Architectural member that forms a cap.

capping brick: See *coping brick*.

capping plane: Hand plane used to round the upper surface of a wood handrail.

cap plate: See *double framing*.

cap screw: See *hex cap screw*.

cap sheet: Top layer of composition roofing on a built-up roof.

capstan: Mechanically or manually driven vertical winch used with cable or rope.

capstone: Stone used for coping.

capsule anchor: Anchoring device consisting of an individual chemical packet and threaded fastener. Packet is inserted into a predrilled hole and the fastener is inserted, breaking the packet and forming an adhesive bond.

caracole: Spiral stairway.

carbide: Hard, abrasion-resistant material used to form the cutting edges of cutting tools such as saw blades and drill and router bits. Harder than steel and provides a longer-lasting sharp edge.

carbide-tipped blade: Wood-cutting circular saw blade with carbide cutting tips, which produce a longer lasting sharp point. See *blade**.

carbolineum: Oily, dark brown substance consisting of antracene oil and zinc chloride. Used as a wood preservative.

carbon-arc cutting: Arc cutting process in which metal is severed by melting the base metal with the heat produced by an electrical arc between a carbon electrode and base metal.

carbon-arc lamp: Electric discharge lamp that uses an electric arc between two carbon electrodes to produce illumination.

carbon-arc welding: Arc welding process in which heat is produced by an electric arc between the base metal and carbon electrode. Shielding gas is not used. Pressure and filler metal may or may not be used.

carbon dioxide: Heavy, colorless gas that does not support combustion. Used as a shielding gas in arc welding.

carbon electrode: Non-filler material used in arc welding and cutting. Consists of a carbon or graphite rod coated with copper or other material.

carbon steel: Steel containing carbon and small amounts of other material.

carborundum: Abrasive material made from a combination of carbon and silicon. Used as a coating for circular blades.

carborundum blade: Circular saw blade composed of carborundum pressed into a disk. Used for cutting concrete and masonry. See *blade**.

carborundum paper: Abrasive cloth or paper in which powdered carborundum is applied to a backing material.

carborundum stone: Abrasive stone made of carborundum. Used for sharp-

cape chisel

Cape Cod style

carol (1)

ening tools such as plane irons and chisels.

carcass: Structural framework of a building or other component. CARCASE.

carcassing timber: Structural members that comprise the framework of a building.

card plate: Small metal frame on a cabinet drawer or doors used to hold a nameplate or card. CARD FRAME.

carnarvon arch: Lintel supported by corbels.

* **carol: 1.** Seat built into the recessed area of a bay window. **2.** See *carrel*.

carolitic: Ornamented with designs of branches and leaves.

carpenter: Person engaged in the construction of wood and light-gauge steel building framework, concrete formwork, and interior and exterior finish.

carpenter's level: Metal or wooden frame containing one or more clear vials with markings on and fluid in each. Bubble in each vial is used to indicate the horizontal or vertical trueness of an object when centered within the vial markings.

carpenter's square: See *framing square*.

carpentry: Building trade involving the construction of wood and light-gauge steel framework, concrete formwork, and interior and exterior finish.

carpet: Floor covering composed of natural or synthetic fibers woven through a backing material.

carpet backing: Base material on the underside of carpet that supports the carpet fibers.

carpet density: Number of pile tuft rows per inch extending lengthwise.

carpet knife: Cutting tool with a retractable double-edged blade.

carpet pad: Foam material installed between the rough floor and the carpet. Used to provide a cushion and create a soft floor surface. CARPET CUSHION.

carpet pile: Tufts of yarn that project from the face of the carpet. Ends may be cut or looped.

carpet pitch: Number of yarns across the width of a carpet. Expressed as the number of yarns per 27″ of width.

carpet strip: 1. Narrow piece of wood, metal, or rubber attached to the floor beneath a door. 2. See *tackless strip*.

* **carport:** Covered shelter with one or more open sides. Usually attached to the side of an existing structure.

carreau: Square- or diamond-shaped glass or piece of tile installed in ornamental glazing.

carport

cased opening

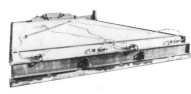

casting bed

carrel: Small room used for individual study or reading in a library. CARRELL.

carrelage: Decorative terra cotta tile.

carriage: See *stringer*.

carriage bolt: Round-headed threaded fastener with a square, finned, or ribbed shank directly below the head. Head is without recess; shank secures the bolt and prevents it from turning when torque is applied. See *bolt**. BUTTON-HEADED BOLT, CARRIAGE SCREW.

carrier: 1. Device on which wall-hung plumbing fixtures, such as sinks and lavatories, are installed. 2. Metal angle or bar that supports treads and risers in a stairway.

carrying channel: Primary support member of a suspended ceiling. MAIN T, MAIN RUNNER.

carryover: Water in steam lines that produces water hammer and creates the potential for pipe rupture.

cartoon: Design or drawing used as a detailed, full-scale model.

cartridge fuse: Low-voltage overcurrent protection device. Consists of a current-responsive element inside a fuse tube with terminals on both ends.

carvel joint: Flush joint between two adjacent boards.

carving: Ornamental designs on furniture or woodwork formed by cutting or chiseling designs in the surface.

cascade refrigerating system: Two or more cooling systems that act in series with each other. Evaporator of one system cools the condenser of the other system.

cascade sequence: Longitudinal and buildup sequence in which weld beads are deposited in overlapping layers.

case: 1. To apply one material over another, such as applying a brick veneer over a low-quality backing. 2. Box or basic cabinet. 3. Lock housing. 4. Glass enclosure in which food is displayed.

case clamp: Pneumatic or hydraulic industrial equipment used for clamping large objects.

cased frame: Wood frame for a double-hung window.

cased glass: Ornamental glass formed of two or more layers of glass, one of which commonly is a colored glass. Outer layer may be cut away to reveal the color from other layers.

cased member: Structural member or component covered with a high-grade material.

* **cased opening:** Structural opening finished with a jamb and trim, but without doors.

case-hardened timber: Timber in which the surfaces have dried too quickly, producing checks and cracks.

case hardening: Process in which the surface of steel is hardened through heating and carbonization.

casein glue: Moisture-resistant adhesive available in powdered form and mixed with cold water. Derivative of milk protein.

casement: See *casement window*.

casement adjuster: Device used to secure the sash of a casement or French window.

casement molding: Shallow concave molding similar to a cavetto or scotia.

casement window: Window with a vertically hinged sash. See *window**.

case steel: Outside layer of case-hardened steel.

casework: Assembled cabinet or case.

casing: 1. Framework around a window or door. 2. Finished lumber or molding around window and door frames. See *molding**. 3. Hollow steel pipe used to support the sides and line the hole. May be driven, drilled, or dropped into the ground.

casing bead: Metal plaster stop installed around window and door openings to provide a smooth corner and eliminate trim molding.

casing knife: Knife used to trim wall-coverings around moldings.

casing nail: Thin nail with a flared head used for finish work. See *nail**

cassoon: See *coffer*.

castellated: Having notches along the top edge.

castellated nut: Internally threaded fastener with small notches along the top edge. See *nut**. CASTLE NUT.

caster: Wheel, or set of wheels, set in a swivel frame and attached to the feet or legs of portable equipment.

casting: Impression taken from a mold made of wax, plaster of paris, or other material.

* **casting bed:** Forms and support used for forming precast concrete members.

casting plaster: Fine-textured plaster used for ornamental work, such as plaques and art statuary. Great amount of plasticity with superior surface hardness and strength.

cast-in-place: Deposited in the place where it is required to harden as part of the structure, as applied to mortar or concrete. CAST-IN-SITU.

cast-in-place pile: Concrete pile, with or without a casing, formed by placing concrete directly into its final position. See *pile* (1)*.

cast-in-situ: See *cast-in-place*.

cast iron: Iron alloy with high carbon content. Has high compressive strength.

cast iron soil pipe: Plumbing pipe manufactured from gray cast iron using a centrifugal process. Strong and corrosion-resistant. Available as hubless, single hub, and double hub pipe.

cast stone: Mortar or cement paste mixture with an aggregate of stone fragments; resembles stone when cast into a mold. Artificially colored surface material may be applied.

catalyst: Material that accelerates the reaction of chemicals without undergoing change.

catch: Device used to secure a door or gate.

catch basin: 1. Cistern used to obstruct the flow of objects that would not readily pass through a sewer. 2. Reservoir used to retain surface water. 3. Kitchen sink trap used to retain fats, grease, and oil and prevent passage into a sewer.

catenary: Shape formed by suspending a flexible cord between two points.

catenary arch: Arch that has the shape of an inverted catenary. CATENARIAN ARCH.

catface: Rough plaster surface defect caused by variations in the thickness of the base coat.

cathead: 1. See *capstan*. 2. Metal clamping device used in concrete formwork in which a pencil rod is inserted and secured with a bolt. 3. Large retention nut installed on a she bolt.

cathedral ceiling: Ceiling that slopes upward; not horizontal nor parallel with the floor.

* **catherine-wheel window:** Circular window with muntins radiating from the center similar to a wheel with spokes projecting from the rim. ROSE WINDOW, WHEEL WINDOW.

cathodic protection: Protection against electrolytic corrosion used for coating submerged tanks and reservoirs.

catspaw: 1. Knot used to form a sling in a continuous rope for hoisting heavy loads. See *knot**. 2. See *nail puller*.

catwalk: Narrow passageway providing access to inaccessible areas.

caul: 1. Tool used to shape veneer to fit a curved surface. 2. Protective sheet of metal or plywood used during the forming, shaping, and pressing process for panel products such as fiberboard, particleboard, or plywood.

caulk: 1. To fill a joint or void with a filler material. 2. See *caulking*.

caulking: Resilient material made of latex, oil, asphalt, butyl rubber, or silicone used to seal joints and cracks. Adheres to the surface and remains flexible to allow for movement of the adjoining surfaces. CAULKING COMPOUND.

* **caulking gun:** Manual or power tool used to extrude caulking through a nozzle.

* **caulking iron:** Hand tool used to compact oakum into a cast iron pipe joint that is to be sealed with molten lead.

caulking recess: Area between the hub of cast iron pipe and its mating member. Used for the application of oakum and molten lead.

caulking trowel: Hand tool with a flat, narrow blade used to smooth caulking.

causeway: Elevated roadway above wet ground.

caustic: Having strong corrosive properties.

cavalier projection: Oblique drawing in which a true view of the front face of an object is shown with all receding lines drawn to the same scale as corresponding lines in the true view.

cavetto: Concave molding containing at least a quadrant of a circle.

cavil: Heavy hammer with one blunt end and one pointed end. Used for rough dressing of stone at a quarry.

cavitation: 1. Localized cavities of air or other gases in a stream of liquid. 2. Surface erosion caused by liquid moving over the surface at a high velocity.

cavitation damage: Pitting of freshly placed concrete resulting from collapse of vapor bubbles in flowing water.

cavity wall: Partition or masonry wall with at least a 2″ void between faces. Increases resistance to thermal and sound transmittance. HOLLOW WALL.

C/B ratio: Comparison of the weight of cold water and boiling water absorbed by a masonry unit. Used to indicate the resistance of a masonry unit to damage from freezing and thawing.

C-clamp: Steel C-shaped clamp used to apply pressure to materials being clamped. One end has a fixed jaw and the other end contains a threaded shaft with a swiveling jaw attached. See *clamp**.

cedar shake: Roof and siding material produced from western red cedar that is split into tapered members.

cedar shingle: Roof and siding material produced from western red cedar that is sawn into tapered members.

ceil: To provide a ceiling.

ceiling: Overhead finish of a room used to conceal the floor or roof above.

ceiling area lighting: General lighting system in which the entire ceiling is one large lumenaire.

ceiling diffuser: Air duct grill installed in the ceiling. Used to distribute the air horizontally around the room.

ceiling flange: Escutcheon around a pipe

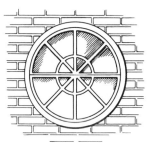

catherine-wheel window

caulking gun

caulking iron

projecting through a ceiling.

ceiling floor: Framework, including joists and covering, that supports a ceiling.

ceiling joist: Intermediate horizontal structural member used to support finished ceiling material.

ceiling light: Electrical light fixture attached to a ceiling.

ceiling outlet: Electrical receptacle installed in a ceiling.

ceiling suspension system: See *suspended ceiling*.

ceiling tile: Square or rectangular piece of fiberboard stapled to a ceiling or inserted in a suspended ceiling system.

cell: 1. Single, enclosed tubular space in a cellular metal floor. Axis is parallel with the axis of cellular metal floor members. 2. See *cellular cofferdam*.

cellar: See *basement*.

cellular cofferdam: Self-supporting cofferdam with separate interior and exterior walls constructed of interlocking sheet piling.

cellular concrete: Lightweight concrete consisting of portland cement, lightweight aggregate, and a gas-forming or foaming agent that produces voids in the final product.

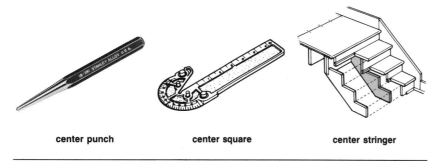

center punch   center square   center stringer

cellular-core door: See *hollow-core door.*

cellular metal floor: Floor constructed of prefabricated metal panels supported by a structural frame.

cellular metal floor raceway: Hollow space in a cellular metal floor that is acceptable as an enclosure for electrical conductors.

celluloid: Thermoplastic material formed in thin sheets. Inflammable, easily molded, and colored.

cellulose: Primary component of wood consisting of glucose units.

Celsius: Measure of temperature in which the difference between the freezing and boiling point of water at sea level is divided into 100 increments. Freezing point is 0°C and boiling point is 100°C.

cement: Material that sets and hardens due to chemical reaction with water. Used as a binding agent in concrete, plaster, and mortar.

cement-aggregate ratio: Comparison of the volume or weight of cement to aggregate in a mix.

cementation: Setting of a cementitious material.

cement-base waterproof coating: Compound that seals the pores in basement walls or floors, providing protection against water pressure and dampness in the ground from entering the structure.

cement-coated nail: Nail coated with cement to improve holding power and withdrawal resistance. See *nail*.*

cement content: Quantity of cement in a unit volume of concrete or mortar. Preferably expressed as weight.

cement gel: Mixture of cement and water that comprises the majority of hydrated cement paste.

cement gun: Pneumatic device used to spray fine concrete or mortar on a surface. See *shotcrete.*

cementing trowel: See *trowel* (1).

cementitious: Able to harden like cement.

cement joggle: Key formed between adjacent stones of a masonry wall by applying mortar into a squared channel that is recessed equally into adjoining faces. Used to prevent movement of the stones.

cement mason: Person who places and finishes concrete surfaces such as roadways, sidewalks, and stairways.

cement mixer: See *concrete mixer.*

cement mortar: See *mortar.*

cement paint: Paint composed of white portland cement, water, pigments, hydrated lime, water repellants, and salts.

cement paste: Concrete constituent consisting of water and cement.

cement plaster: Gypsum plaster with portland cement as a binder. Sand and hydrated lime are added at the job site.

cement rock: Natural limestone containing the approximate proportions required for the production of portland cement.

cement stucco: See *stucco.*

cement stucco dash brush: See *stucco brush.*

cement temper: Mixture of portland cement in lime plaster to increase durability and hardness.

center: Point exactly halfway between two other points or surfaces.

center bit: Wood-boring tool used with a brace consisting of a feed screw that pulls the bit into the wood and two sharpened spurs that score the circumference of the hole.

centering: 1. Frame used as a pattern and temporary support for a brick or stone arch; the falsework over which an arch is formed. **2.** Shoring or falsework used to support concrete forms.

center line: 1. Line used to indicate an axis of symmetry. Consists of a short dash between two long dashes.

center line of section: Surveying line connecting opposite quarter corners.

center-matched lumber: Tongue-and-groove lumber in which the tongue and groove are at the center of each piece rather than offset as in standard-matched lumber.

centerpiece: Ornament located in the middle of a ceiling.

centerplank: Quarter-sawn hardwood lumber cut from the center of a log.

center point: Point formed by the intersection of two center lines.

* center punch: Steel hand tool with one end formed to a conical point of approximately 90°. Used to make small indentations in hard surfaces, such as metal, by striking the surface with a hammer to establish a center point for drilling a hole.

center rail: See *meeting rail.*

* center square: Hand tool with a curved end used to locate the center of a circle.

* center stringer: Structural member used to support a flight of stairs at the midpoint of the treads.

center-to-center: Distance from the center of one structural member to the center of the adjacent member. ON CENTER.

centigrade: See *Celsius.*

centimeter: Metric measure of length equal to one-hundredth of a meter, or 0.3937″.

central air conditioning system: Air conditioning system in which the air is treated by equipment at one or more locations outside of the areas served. Air is conveyed to the conditioned areas by means of fans and ducts.

central angle of a curve: Angular measurement between two lines drawn at 90° from the tangent points of a curve.

central fan system: HVAC system in which air is treated and conveyed through ductwork with fans or blowers.

central heating system: Heating system in which heated air is supplied to the areas of a structure from one source.

central-mixed concrete: Concrete that is completely mixed in a stationary mixer and delivered to the job site.

central mixer: Stationary concrete mixer.

centrifugal fan: Direct- or belt-driven fan that receives air along its axis and discharges it radially.

centrifugal pump: Pump that draws fluid into the center of a rapidly spinning impeller and discharges it using pressure created by centrifugal force created by the impeller.

ceramic: Material made from clay or similar materials that are baked in an oven at high temperatures.

ceramic aluminum oxide: High-strength material with a uniform structure. Used as an abrasive for wood and metal.

ceramic bond: Bond between materials resulting from exposure to high temperatures close to the fusion point of the materials.

ceramic mosaic tile: Unglazed wall and floor tile manufactured in sheet form with several tiles properly spaced and mounted on backing sheets.

ceramic tile: Thin, flat piece of fired clay

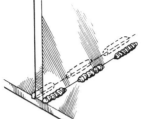

chain intermittent weld   chair   chair carrier

attached to a surface with cement, mastic, or other adhesive. May be formed using the plastic process (while clay is wet) or dust-pressed process (compressed clay powder).

ceramic tile cutter: Hand tool with a small carbon steel or carbide-tipped cutting wheel used for scoring tile and opposing jaws for breaking the tile at the scored point.

ceramic veneer: Terra cotta with a ceramic vitreous or glazed surface. Back is scored or ribbed to form a key with the mastic or adhesive.

cesspool: 1. See *dry well*. **2.** Small waterproof box used as a cistern in the gutter of a parapet where rainwater is discharged into a downspout.

chafe: To wear away.

chain: 1. Series of links of metal or other material. **2.** Surveyor's steel tape measure. **3.** Unit of length for surveying; 1 link = 7.92″, 100 links = 66′-0″.

chain binders: Lengthwise yarns in the back of woven carpet used to hold the exposed yarns together.

chain block: Grooved pulley or sheave designed to accept a chain. Commonly suspended from an overhead track.

chain bolt: Spring bolt attached to the top of a door and operated by pulling a chain.

chain bond: Masonry construction that is reinforced and bonded together with a chain or an iron bar.

chain door fastener: Security device used to limit the distance that a door may open. One end of the chain is secured to the door jamb and the other end is inserted in a slotted plate mounted on the face of the door along the edge.

chain fall: See *chain hoist*.

chain hoist: Pulley arrangement in which sheaves of two different sizes are fitted with spurred gears mounted on the same shaft. Large sheave takes in more chain than the smaller sheave, providing a mechanical advantage for lifting heavy loads. CHAIN FALL.

* **chain intermittent weld:** Evenly spaced welds on both sides of a weld joint in which the welds are approximately opposite those on the other side.

chain-link fence: Fence made of interwoven heavy-gauge steel wire fabric and supported by posts and rails. Commonly coated with zinc or durable plastic material to inhibit corrosion.

chain-pipe vise: Portable vise that uses a heavy chain to secure pipe in the jaw.

chain pull switch: See *pull switch*.

chain saw: Power saw with a projecting steel arm that is slotted around its pe-rimeter. An endless chain with cutting teeth is driven around the arm. Used for rough-cutting wood and lumber. See *saw**.

chain timber: Timber built into a masonry wall in a horizontal position to reinforce the wall and tie it together during construction.

chain tongs: See *chain wrench*.

chain vise: V block fitted with a chain that is tightened around a round or cylindrical object to secure it in the block.

chain wrench: Wrench used by a plumber to rotate pipe or conduit. Consists of a lever bar with sharp teeth at one end that engage the pipe and a chain that wraps around and secures the pipe in position. See *wrench**. CHAIN PIPE WRENCH, CHAIN TONGS.

* **chair:** Device used to support rebars in the proper position during concrete placement.

* **chair carrier:** Supporting device used to support wall-hung water closets, lavatories, and urinals.

chair rail: Horizontal trim molding attached to a wall 32″ above the floor as ornamentation and protection against damage from chairs being pushed against the wall. See *molding**.

chalet: Wood-framed structure characterized by exposed structural members, balconies, and stairs.

chalking: Formation of a powdery surface resulting from the disintegration of binder or elastomer in a painted surface. Caused by weathering.

chalk line: 1. String wound around a spool that is in a small box filled with powdered chalk. String is unwound from the case through a hole at one end, secured against the work, and snapped with the fingers to make a line. **2.** Colored line made with a chalked string.

chamber: Private room.

chamfer: 1. Oblique surface extending from the surface of a member to the adjacent surface.

chamfer bit: Router bit used to form a 45 ° angle. See *bit* (2)**.

chamfer plane: Hand plane with a V-shaped groove along the bottom to facilitate cutting chamfers.

chamfer strip: Narrow wood, metal, plastic, or rubber member placed in the outside corners of concrete formwork to produce a chamfer on the corner of the finished member. CANT STRIP.

chandelier: Ornate ceiling-mounted light fixture, commonly with multiple lights.

channel: 1. See *American standard channel*. **2.** Concave groove cut in a surface as a decorative feature. CHANNELING. **3.** Grooved molding used for ornamental purposes. **4.** Light-gauge metal member used for framing and supporting lath or drywall. **5.** Three-sided opening for the installation of a glass light or panel in a sash or frame. **6.** Enclosure in which the ballast, lampholders, starter, and conductors for a fluorescent lamp are contained.

channel block: Hollow concrete masonry unit with a recessed area for placement of reinforcing steel and mortar.

channel clip: 1. Metal device hung from a suspended ceiling channel; used to support a metal ceiling pan. **2.** Light-gauge sheet metal or wire fastener used to attach gypsum lath or drywall to steel channels.

channel cutter: Hand tool used to cut light-gauge metal framing members.

channel iron: See *American standard channel*.

channel pipe: Section of pipe with the upper quarter or half removed.

channel runner: Horizontal supporting member in a suspended ceiling system.

characteristic impedance: Electrical network in which the input and output impedance are equal.

charge: 1. Conversion of electrical energy to chemical energy in a battery. **2.** Explosives that are set in place and ready for activation. **3.** Amount of refrigerant in a refrigeration system.

charging: Loading material into a concrete or mortar mixer where they are further treated or refined.

charging connection: Fitting on a refrigeration system that allows for the introduction of refrigerant into the system.

* **chase: 1.** Enclosure in a structure that allows for the placement of electrical, plumbing, or mechanical wiring and piping. CHASEWAY. **2.** Groove or channel in the face of a masonry wall that allows for pipes, ducts, or conduit. **3.** To decorate the exterior surface of metalwork.

chase bonding: Joining new and existing masonry work by forming a vertical groove in the existing masonry.

chase wedge: Wedge-shaped tool used for bossing sheet lead.

chat: Stone mineral material used as a base for asphalt paving or as a road sur-

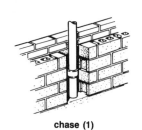

chase (1)

check valve

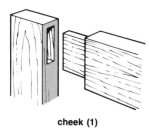

cheek (1)

cherry picker

face without finish paving.

chat saw: Stone-cutting device that uses small pieces of hardened steel as the cutting agent.

chatter: 1. Improper operation of a cutting tool resulting in intermittent skipping. **2.** Waviness on concrete flatwork surface caused by finishing the concrete with a trowel blade held at too great of an angle.

check: 1. Wood defect extending parallel to the grain of the wood; caused by separation of tissues during seasoning. **2.** Thin split occurring in finely figured crotches and burls; caused by strain in the seasoning of the wood. **3.** Door closing device. See *door check*. **4.** Partial or complete evaluation of an object or system. **5.** Small crack in steel caused by cooling too quickly.

checkerboard placement: Method of placing concrete in which adjacent concrete slabs are placed at different times. Minimizes shrinkage cracks between large slabs, creates control joints without sawing or jointing, and provides breaks when special colors or finishes are utilized.

checkered plate: Metal plate with a rough, waffle-like surface that provides better surface traction. CHECKER PLATE.

checkerwork: Pattern formed by laying masonry units so as to produce a checkerboard effect.

checking: Development of shallow cracks at regular intervals on the surface of concrete, plaster, or paint.

checking of wood: See *check*.

check nut: See *locknut*.

checkpoints: Marks made on a footing at 16″ intervals used as a guide when setting masonry units.

check rail: See *meeting rail*.

check stop: Molding used to restrain a sliding unit such as the bottom sash of a double-hung window.

check throat: See *drip*.

* **check valve:** Valve that allows the flow of liquid in one direction only. Prevents the backflow of water or other fluid.

* **cheek: 1.** Area on each side of a mortise. **2.** Area on either side of a tenon that has been removed. **3.** Beveled cut on the edge of a hip, jack, valley, or cripple rafter. CHEEK CUT. **4.** Flat side of the head of a claw hammer.

chemical foam insulation: Material sprayed into wall cavities that expands to fill voids and increase resistance to thermal transmittance. May also be sprayed onto exposed wall surfaces.

chemical membrane: Material sprayed on the surface of fresh concrete to pre-

vent evaporation of moisture.

chemical plaster: See *plaster*.

chemical stabilization: Injection of chemicals into soil to improve strength and reduce permeability of the soil.

chequer: Ornamental design used in furniture-making that consists of an ornamental arrangement of squares.

* **cherry picker:** Powered equipment used to lift personnel or materials. Consists of a boom with an enclosed platform. May be self-powered or mounted on a truck.

chert: Fine-grained dense rock composed of chalcedony, cryptocrystalline or microcrystalline quartz, or opal. May react with cement and therefore is undesirable for concrete aggregate.

chest pull: See *drawer pull*.

chevron: 1. L-shaped wood-fastening device used to join miter joints. **2.** Meeting of rafters at the ridge of a gable roof. **3.** Zigzag or V-shaped pattern used as ornamentation.

chiller: Refrigerating equipment that uses heat energy to cool water.

chimney: Noncombustible vertical structure containing one or more flues to convey smoke, flue gases, and fumes away from stoves, furnaces, fireplaces, or other sources of smoke and gas. Includes factory-built, low temperature (less than 600°F), masonry, medium temperature (between 600°F and 1000°F), and metal.

chimney arch: Curved header above a fireplace opening.

chimney back: Portion of a chimney wall at the rear of a fireplace.

chimney bar: Iron or steel lintel supporting a masonry arch above a fireplace opening.

chimney block: Concrete masonry unit used in laying up or surrounding a round flue.

chimney board: Low metal screen or frame in front of an open fireplace. FENDER.

chimney bond: Stretcher bond used to form the internal structure of a masonry chimney.

chimney bracket: Metal frame attached to corner poles for use as a masonry guide for constructing a brick chimney.

chimney breast: Section of a chimney that projects out into a room.

chimney cap: 1. Masonry member that forms the top of a chimney. **2.** Metal device used to improve the draft and prevent entry of rainwater into a chimney. See *chimney cowl*. BONNET.

chimney cowl: Revolving metal device at the top of a chimney used to improve the draft and prevent entry of rainwater into a chimney.

chimney effect: Rising of heated air or gas in a vertical passage or duct. Rising air or gas creates a draft that draws in cooler air or gases from below.

chimney head: See *chimney cap* (1).

chimney hood: Covering for the top of a chimney that prevents rainwater from entering.

chimney jamb: Vertical side of a fireplace opening.

chimney lining: See *flue lining*.

chimney piece: Ornamental construction placed over or around the opening of a fireplace; commonly a mantel or shelf.

chimney pot: Cylindrical brick, terra cotta, or metal pipe placed over a chimney flue to improve the draft.

chimney scaffold: Tubular steel framework set on top of a roof and adjusted to provide a temporary platform for the construction of a masonry chimney.

chimney shaft: Section of a chimney that extends above a roof.

chimney stack: 1. Tall, large diameter chimney used in industrial installations. **2.** Chimney that contains several flues.

* **chimney throat:** Area above the fireplace and below the smoke shelf of a chimney.

chimney tun: See *chimney stack* (2).

chimney wing: Side of a chimney above the opening that narrows to the chimney throat.

china marker: Marker with an opaque pigment used to mark glazed surfaces.

chink: Crack that is longer than it is wide.

chinking: Material used to fill a chink.

chip: 1. Small fragment of stone, wood, or other material. **2.** To cut with a cutting tool using quick blows.

chip ax: Small ax used to cut or shape structural stone or timber.

chip board: See *particleboard*.

chip seal: Spreading of fine crushed rock on asphalt oil and rolling flat with a heavy roller.

chisel: Hand tool with a cutting edge on one end used to shape, dress, or work wood, metal, or stone. Commonly driven with a mallet or hammer. See *butt chisel, cold chisel, firmer chisel, wood chisel.*

chock: Wedge-shaped block used to prevent movement of a member. Commonly used in pairs.

choker: 1. Self-tightening sling. **2.** Road shoulder that is higher than the level of the subgrade.

chop saw: Circular saw attached to a small table and that pivots at the rear. Saw is lowered into the workpiece and returned to original position by a spring. See *saw**.

* **chord: 1.** Horizontal member of a truss, commonly the bottom member. **2.** Straight line connecting two points on a curve. **3.** Span of an arch.

chromated copper arsenate (CCA): Preservative that is injected into wood. Reacts with cellulose in the wood.

chrome steel: Hard, wear-resistant steel with a high elastic limit. Contains 2% chromium and from 0.08% to 2% carbon.

chromium: Hard, brittle, corrosion-resistant metal that produces a highly reflective surface.

chromium steel: See *chrome steel*.

* **chuck:** Device with adjustable jaws used to hold a cutting bit or drill bit.

chuck key: Device with a central shaft and an enlarged, toothed ring used to tighten and loosen a chuck.

chute: 1. Inclined trough or tube used to convey concrete or other free-flowing material from a higher to a lower elevation. **2.** Temporary inclined trough used to remove debris from upper floors of a structure.

cimbia: Decorative band around a column shaft.

cincture: Encircling molding around the top or bottom of a column between the shaft and capital or the shaft and pedestal. GIRDLE.

cinder: 1. Residue that remains after material has been burned; e.g., clinkers that remain after coal has been burned. **2.** Slag from a furnace. **3.** Fragments of unburned lava from a volcano.

cinder block: Lightweight masonry unit composed of cement and cinder.

cinder concrete: Lightweight concrete composed of portland cement and clean, well-burned coal cinder used as coarse aggregate.

cinquefoil: Ornamental design with five arcs.

circle: 1. Plane geometric figure in which all points are equidistant from a center point. **2.** See *horizontal circle, vertical circle*.

circle cutter: 1. Hand tool used to cut circles in thin wood or drywall. Consists of a central pointed shaft mounted on a perpendicular bar with a cutter fixed to the end. Shaft is adjustable along the bar to allow cutting of circles up to 16″ diameter. **2.** Bit used with a drill press to cut large diameter circles in soft material such as wood. Consists of a central mandrel or pilot drill that is adjusted along a perpendicular bar with a cutter fixed to the end. See *bit* (1)*.

circle-head window: Semicircular window over a door or window.

* **circle sweep glass cutter:** Hand tool used to cut circles in glass. Consists of a central suction device mounted on a perpendicular bar with a cutter fixed to one end.

circline lamp: Circular fluorescent lamp.

circuit: 1. Conductor or series of conductors through which electric current flows. **2.** Consecutive series of surveying measurements of elevation from a predetermined benchmark elevation to another point and back to the original elevation. **3.** Piping loop for gas or liquid.

circuit breaker: Device that opens and closes a circuit by nonautomatic means, and automatically opens a circuit when a predetermined current overload is reached without damage to itself.

* **circuit tester:** Two- or three-pronged device connected to a neon lamp indicator used to verify the presence of current in an electrical circuit.

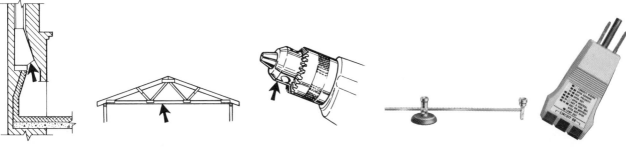

chimney throat   chord (1)   chuck   circle sweep glass cutter   circuit tester

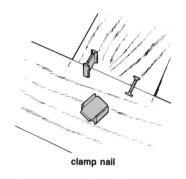

clamp nail

circuit vent: Branch vent that serves two or more traps. Extends from the last fixture connection of a horizontal branch to a vent stack. Used to prevent buildup of pressure within the system.

circular/angular measure: Standard measure of a circular or angular figure. Expressed in quadrants, degrees, minutes, and seconds; 60″ (seconds) = 1′ (minute); 60′ = 1° (degree); 90° = 1 quadrant.

circular arch: Curved arch with a radius equal to one-half the length of the span.

circular mil: Measure of the cross-sectional area of a circle or circular shape, such as a conductor. Equal to the square of the diameter of a circle; 1974 circular mils equal 1 square millimeter.

circular mil-foot: Measure of an electrical conductor with a cross-sectional area of 1 circular mil and a length of 1′-0″.

circular saw: Hand-held or table-mounted power saw with teeth around the circumference of a circular blade that is rotated at high speed on a central axis or shaft. Primarily used to cut wood. See *saw*.

circular stairway: Stairway with steps that radiate horizontally from a common center.

circulating line: System of piping designed to allow fluid to flow in a closed piping system.

circulating pump: Centrifugal-driven device that moves hot water through a heat piping system. CIRCULATOR.

circulation: 1. Flow of gas or liquid within a closed piping system. 2. Flow of air within ductwork and interior area of a structure. 3. Traffic pattern within a structure.

circumference: Perimeter of a circle.

circumscribe: To draw a line enclosing certain portions of an object, figure, or plane.

cissing: Small amount of shrinkage of a painted surface resulting in small surface cracks.

cistern: Artificial reservoir or tank, often underground, for storing water at atmospheric pressure.

civil engineer: Person who designs and assists in the construction of static structures such as roads, bridges, and other structures.

civil engineer's scale: See *engineer's scale.*

clack valve: See *check valve.*

cladding: 1. External covering of a building. 2. Metal coating bonded to another metal to improve corrosion or heat resistance. 3. Layer of material greater than 0.04″ thick used to improve corrosion resistance or other properties. SURFACING.

*** clamp:** Device used to secure workpieces together.

clamping plate: Metal connector used to reinforce a timber frame. See *timber connector.*

*** clamp nail:** Flat piece of metal with flanged edges used to join miter and butt joints. Installed by making a saw kerf in the adjoining members and driving the clamp nail in. Flanges draw the wood together, producing a tight joint.

clamshell: Mechanical bucket used with a crane or derrick for handling granular material. Two jaws close by means of a cable or hydraulic action.

clapboard: Horizontal wood siding used as an exterior covering on a wood-framed structure. Thickness is consistent throughout the width.

clapboard gauge: Device used to evenly space clapboards.

claw: Part of a tool used to remove a fastener; e.g., the claw of a hammer.

claw bar: See *wrecking bar.*

claw chisel: Hand tool with a serrated cutting edge used to shape stone.

claw hammer: Hammer with a flat head and a slotted claw used to remove nails.

claw hatchet: See *shingling hatchet.*

claw plate: Round timber connector with projections used to grip the wood.

clay: Natural mineral material that is compact and brittle when dry but plastic when wet. Compound of silica and alumina with small amounts of other ingredients. Occurs in three principal forms: surface clay, shale, and fire clay. Used in the manufacture of brick and tile.

clay mill: Location where clay is mixed and tempered. PUGMILL.

clay tile: Fired clay product used for roofing material, drainage pipe, and wall and floor covering.

*** cleanout:** 1. Pipe fitting with a removable plate or plug that provides access to plumbing or other drainage pipes for inspection or removal of material. 2. Opening at the foot of a chimney for

CLAMPS

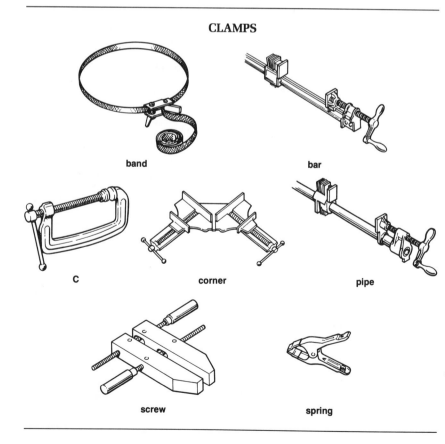

band

bar

c

corner

pipe

screw

spring

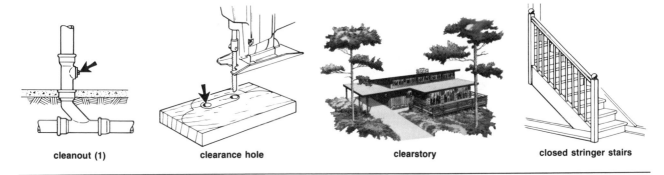

cleanout (1)   clearance hole   clearstory   closed stringer stairs

removal of soot and ashes. **3.** Temporary opening at the base of concrete forms for removal of debris. Opening is sealed before concrete is placed.

cleanout door: Hinged steel or cast iron plate and frame at the base of a chimney flue or ashpit for removal of ashes.

clean room: Room with a highly controlled environment that limits the amount of airborne particles and regulates air pressure, air motion, relative humidity, and temperature within close tolerances. Used for manufacture and assembly of precision products that may be affected by airborne particles.

clearance: 1. Open area between two objects. **2.** Amount of space between the cutting edge of a tool and surface of the workpiece.

*****clearance hole:** Opening in a surface for insertion of a saw blade to begin a cut.

clear and grub: Removal of vegetation and other obstacles from a building site in preparation for new construction. CLEARING.

clear lumber: Lumber that is free of knots or other defects.

clear span: Distance between the inside faces of structural supporting members.

*****clearstory:** Portion of a multistory structure that extends above a roof or other parts of the structure; has windows that admit light and provide ventilation. CLERESTORY.

cleat: 1. Strip of wood, metal, or other material attached to the surface of a member to restrain or support another member. **2.** Small board that is used to join formwork members or used as a brace.

cleavage: Splitting of rock or wood along closely spaced parallel planes.

cleavage plane: Natural plane of weakness in stone that allows for easy splitting. CLEAVAGE LINE.

cleft timber: Wood beam that is split parallel with the grain to an approximate specified dimension.

clench: See *clinch.*

clerestory: See *clearstory.*

clevis: Steel or iron U-shaped device with holes in the ends that receive a pin. Used for connecting loads. SHACKLE.

clevis hanger: Pipe hanger used to support horizontal pipe from a ceiling. Consists of a suspended U-shaped bracket with holes drilled in the upper ends to receive a pin and bracket. Bracket is threaded to receive a threaded hanging rod.

climbing form: Formwork raised vertically for successive lifts of concrete. See *gang form, slipform.*

climbing tower crane: Crane used in the erection of high-rises. Consists of a vertical mast fastened to structural members and moved up as construction reaches higher levels. Horizontal boom is rotated atop the mast to facilitate lifting.

clinch: To secure a fastener by bending over the projecting point.

clinch-on corner bead tool: See *corner bead clincher.*

clink: 1. 12″ long pointed steel rod used with a sledgehammer to break up concrete slabs or pavement. **2.** Sealed seam between pieces of flexible metal roofing.

clinker: 1. Vitrified stonelike material fused together and formed as a deposit in a furnace after coal containing impurities has been burned. **2.** Slag from a metal blast furnace. **3.** Unburned lava from a volcano. **4.** Point in cement manufacture where the ingredients are fused into small pieces.

clinometer: Surveying tool that measures angles of slope, elevation, or inclination.

clip: 1. Light-gauge sheet metal or wire fastener used to join various building materials. **2.** Brick that is cut to length.

clip bond: Masonry joint in which the corners of the stretchers are cut to form a V-notch to receive the corner of a diagonal header.

clip joint: Mortar joint that is thicker than normal to raise a masonry course to a given elevation.

clipped gable: See *Dutch hip roof.*

clo: Unit of clothing insulation required to keep a sitting person comfortable in a normally ventilated room with a temperature of 70°F and 50% relative humidity.

cloak rail: Horizontal wood member with hooks or pegs on which garments are hung.

cloakroom: Small room or closet near the main entrance of a structure where coats and other outer garments are temporarily stored.

clockwise: Movement in the same direction as the hands of a clock.

close: 1. Narrow entry; an alley. **2.** Enclosed area of land around a structure.

closed circuit: Continuous path for the flow of electrons.

closed-circuit grouting: Injection of grout into voids in concrete in a manner that the excess grout is returned to the pump for recirculation.

closed-coat abrasive: Abrasive material used for semifinish and finish sanding. Face of the material is completely covered with abrasive grains.

closed cornice: Trim on the underside of a roof overhang that is entirely enclosed by the roof, fascia, and soffit. BOXED CORNICE.

closed newel: Central column or shaft of a circular stairway or stairway that changes direction. Constructed as a continuous enclosing wall or shaft.

closed-panel system: Construction of prefabricated building components that contain electrical conductors, plumbing, insulation, and interior and exterior finish material.

closed stairway: Stairway enclosed by walls on both sides.

*****closed stringer stairs:** Stairway with finish stringers on both sides, enclosing treads and risers on both ends.

closed system: Piping for heating or cooling that is completely enclosed and not affected by outside sources. System is not exposed to the atmosphere except for an expansion tank at the highest elevation.

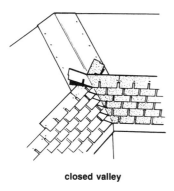

closed valley

coaxial

* **closed valley:** Intersection of two roof surfaces where courses of shingles meet and completely cover the flashing at the intersection.

close-grained wood: Wood with annual rings growing close together.

close nipple: Short piece of pipe externally threaded on both ends with no shoulder between the threads.

closer: 1. Last masonry unit laid in a course. May be a partial or whole brick. 2. See *door closer.*

closet: Small enclosed storage area.

closet bend: 90° soil pipe fitting installed directly under a water closet.

closet bolt: Threaded fastener with a T-shaped head; used to secure a floor-mounted water closet to the floor and soil pipe flange.

closet pole: Horizontal wood or metal rod in a closet on which clothes hangers are placed. CLOSET ROD.

closet valve: Valve that regulates the flushing cycle of a water closet.

closing device: See *door closer.*

closing stile: See *lock stile.*

closure: Short masonry units used at corners or jambs to maintain bond patterns.

closure bar: Flat metal bar connected to a stair stringer; used to close the opening between the stringer and the wall.

closure brick: See *closer* (1).

clothes chute: Inclined or vertical trough or duct through which soiled clothes are dropped from upper floors to a laundry area. Commonly installed in a hallway wall or closet.

clouding: Loss of luster in paint resulting from a porous undercoat.

clout: Metal plate attached to wood member to protect it from abrasion.

clout nail: Fastener with a large flat head, rounded shank, and blunt point; designed to be clinched on the back side of the workpiece.

clove hitch: Knot for quick fastening of rope around a post or stake. See *knot*.*

clustered columns: Several columns or shafts grouped and fastened together to form a single structural unit.

clutch: Mechanical device that allows the drive train of equipment to be connected and disconnected from the power source.

clutch-head recess: Screw head recess with two tapered wings projecting from a central point. See *screw*.*

coach bolt: See *carriage bolt.*

coak: 1. See *tenon.* 2. Hardwood or metal dowel used to connect overlapping timbers.

coalescence: 1. Growing together or joining of welded base metal components. 2. Formation of resinous or polymeric film as water evaporates from a latex mixture.

coaming: Curbing around a roof or floor opening used to prevent water from flowing into the opening.

coarse aggregate: Aggregate retained on a U.S. Standard No. 4 sieve.

coarse-grained wood: Wood with wide and conspicuous annual rings in which there is a significant difference between springwood and summerwood.

coarse-textured wood: Wood with large pores.

coarse thread: Design of screw, bolt, and nut threads that allows for rapid assembly and disassembly. UNIFIED NATIONAL COARSE (UNC) THREADS.

coat: Layer of any substance or material that covers another layer; e.g., coat of paint.

coat and hat hook: Hardware with two or more projecting prongs; the longest for holding hats, the shorter for coats or other garments.

coated electrode: See *electrode.*

coated nail: See *cement-coated nail.*

* **coaxial:** Transmission line consisting of two or more conductors placed so that one path surrounds the other throughout its entire length.

coaxial conductor: Electrical conductor composed of an outgoing and incoming current path with a common axis; one of the paths surrounds the other throughout its entire length.

cob: Mixture of unburned clay, aggregate, and straw used as a binder in forming walls.

cobble: Rock fragment ranging from approximately 3″ to 10″ in diameter.

cobblestone: Rounded or semirounded rock fragment ranging from approximately 3″ to 12″ in diameter.

cob wall: Wall constructed of blocks made of unburnt clay, aggregate, and straw.

cobwork: Timber or logs placed horizontally with the ends joined; commonly overlapped to form an enclosure.

cochlea: 1. Circular tower for a winding stairway. 2. Winding stairway.

cock: 1. Valve consisting of a tapered plug with a slot or other opening that is fitted into a body to control the flow of fluids. 2. See *faucet.*

cock bead: Circular or semicircular molding that projects beyond the surface of the adjoining member.

cocking: See *cogging.*

cockle stairs: See *winding stairway.*

cocoa mat: Wood fiber fabric that evenly distributes water over a flat surface.

code: Regulations and requirements devised to maintain uniformity of work and proper standards of procedure; e.g., building code. Designed to protect the health and welfare of the public.

coefficient of beam utilization: Comparison of amount of light reaching a specified area to total light emitted by the source.

coefficient of conductivity: Measure of the rate at which a material conducts thermal energy. Equal to the product of the uniform thickness of a material (in inches) and the heat flow rate (in Btu/hour) divided by the product of the surface area (in square feet) and the difference in temperature across the surface (in °F). Expressed as k. COEFFICIENT OF THERMAL TRANSMISSION.

coefficient of discharge: Comparison of the effective area of an air diffuser to the free area.

coefficient of expansion: Measure of the change in length or volume of a material per unit length. Equal to the product of the length or volume of a material (in feet) and the difference in temperature (in °F) divided by unit length or volume.

coefficient of friction: Comparison of the amount of force required to slide an object across a surface to the normal force exerted acting perpendicular to the surface that presses the surfaces together.

coefficient of performance: 1. Comparison of the heat produced by a heat pump to the energy expended. 2. Comparison of the heat removed in a refrigeration system to the energy expended.

coefficient of thermal transmission: See

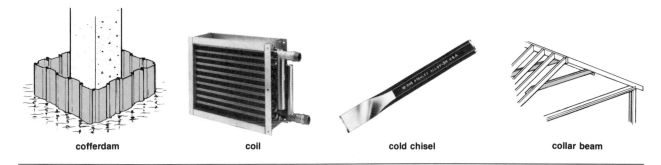

| cofferdam | coil | cold chisel | collar beam |

coefficient of conductivity.

coefficient of variation: See *standard deviation.*

coffer: Ornamental recessed panel in a ceiling or soffit.

* **cofferdam:** Watertight enclosure used to allow construction or repairs to be performed below the surface of water. May be formed with sheet piling, as in bridge pier construction, or attached to the side of a ship, as in repairs to a ship.

cog: Solid portion remaining in a structural timber after it has been notched.

cogged joint: Wood joint in which the two mating surfaces have been notched to receive one another.

cogging: Joining two members that have been notched.

cohesion: Attraction of particles, such as soil, in which the molecular attraction is greater than the force required to separate the particles.

cohesive soil: Soil primarily composed of fine particles of silt and clay.

coign: Projecting wedge or corner.

* **coil:** Cooling or heating element consisting of a spiral- or helical-shaped pipe or tube. Perimeter of pipe or tube may have fins or other means of increasing the surface area to increase radiation of heat energy.

coil depth: Number of rows of coils in the direction of air flow through a cooling or heating system.

coil expansion loop: Continuous pipe or tube formed into a loop to absorb the effects of thermal expansion and contraction.

coil face area: Measure of the surface area of a heating or cooling coil. Calculated as the length of a coil multiplied by the width.

coil heating: Radiant heating method in which hot water pipes or electrical conductors are embedded in concrete or plaster.

coil length: Dimension of an exposed coil face parallel to the air flow in a heating or cooling system.

coil tie: Concrete formwork hardware consisting of two threaded inserts welded to ends of small diameter steel rods. Bolts are inserted through the forms and threaded into ends. Allows adjustment of forms for various widths of concrete placement. See *tie* (2) *.

coil width: Dimension of an exposed coil face perpendicular to the air flow in a heating or cooling system.

coil without support: Filler metal package used for welding; consists of a continuous section of electrode in coil form without an internal support.

coil with support: Filler metal package used for welding; consists of a continuous section of electrode wound on a cylindrical support without flanges.

coin: See *quoin.*

coincidence: Arrangement of a prism in a leveling instrument in which the opposite ends of the leveling bubble are observed as a single image.

cold-air return: Ductwork that conveys cold air back to the furnace for reheating in a heating system.

* **cold chisel:** Chisel used to cut and chip metal, concrete, and other hard material. Cutting edge is tempered to maintain durability.

cold cracking: Underbead separation of base metal in the heated zone of an electric arc weld.

cold-drawn: Process used in the manufacture of steel and other metals in which the metal is drawn through a set of dies without being heated. Commonly used in fabrication of wire and tubing.

cold flow: 1. Permanent distortion in the shape of a member as a result of sustained application of force. 2. Spread and distortion of asphalt unless it is very cold.

cold glue: Unheated liquid adhesive.

cold joint: Joint or discontinuity in concrete resulting from delay in the placement of successive lifts.

cold mix: Slow-curing asphalt mixture applied without heat.

cold molding: Process in which plastics are formed at room temperature and hardened by baking.

cold-rolled channel: C-shaped 16-gauge steel channel that is ³/₄″ or 1 ¹/₂″ wide. Used in furred walls and suspended ceilings.

cold-rolled process: Process used in the manufacture of steel and other metals in which the metal is rolled without being heated. Produces members with high tensile strength and reduced ductility.

cold-shrink fitting: Process in which two members are joined by cooling the inner member to reduce its size so that it fits inside the adjoining member. As the temperature of the members equalize, a tight joint is formed.

cold soldered joint: Incompletely formed solder joint caused by insufficient application of heat in the base metal.

cold weather concreting: Placement of concrete in temperatures less than 40°F. Precautions are taken in mixing, placing, and curing concrete.

cold welding: Solid-state welding process in which pressure is applied at room temperature to join metal with extreme deformation at the weld.

collar: 1. Encircling band. 2. Molding encircling the leg of a piece of furniture. 3. Two short timbers with through bolts that are tightened to secure the timbers around a concrete pile. 4. Reinforcing metal in a nonpressure thermit weld.

* **collar beam:** Horizontal member that ties rafters together above the wall plate. COLLAR TIE.

collar joint: Vertical joint between two wythes of masonry.

collar tie: See *collar beam.*

collection line: Section of a drainage system that receives the discharge of all soil and waste stacks in a structure and conveys it to a sewer.

collector: See *solar collector.*

collet: Clamping ring or device that secures another member in place by tightening around it.

collet chuck: Split tapered cylinder inside a nut that is tightened on a hollow shaft. Tightening of the nut compresses the cylinder to secure a cutting bit in position such as in a router.

collimator: 1. Vertical telescope with a horizontal eyepiece. Used to establish an optical plumb line. **2.** Surveying accessory with a graduation in the lens that is used to align lines along a parallel plane.

colloid: Substance in a state of division with many small particles that prevent passage through a semipermeable membrane.

colloidal grout: Grout that has artificially induced cohesiveness and can retain dispersed solid particles and maintain suspension.

Cologne earth: Lignite that yields a deep-brown transparent pigment.

colombage: See *half-timbered construction*.

colonette: Small decorative column.

colonial architecture: Style of architecture used in America from approximately 1770 through 1840. Characterized by a centrally located front door flanked on either side by double-hung windows, gable roof with a small overhang, and clapboard siding. Modification of the English Georgian style.

colonnade: Series of columns spaced at regular intervals, usually supporting an entablature.

colorimetric test: Method of determining amount of inorganic matter in aggregate used for concrete.

colorimetric value: Number used to indicate the amount of inorganic impurities in fine aggregate.

coloring pigment: See *pigment*.

column: Vertical structural member used to support axial compressive loads. Height is at least three times its largest diameter.

column cap: Horizontal concrete beam that ties several concrete columns together, such as in bridge construction.

column capital: See *capital*.

* **column clamp:** Two perforated steel angles, approximately 36″ long, that are hinged with a steel pin at one end to form an L-shape. Used in pairs and fastened with steel wedges to encompass a square column form.

columniation: See *intercolumniation*.

column schedule: Print schedule that lists the amount, size, and placement of steel, wood, and concrete columns in a structure. Columns on the print are identified with a letter and number code that relates to the schedule.

column strip: Flat panel over a column consisting of two adjacent quarter panels on each side of a center line.

comb: 1. Ridge of a roof. COMB BOARD. **2.** Tool used to finish the face of stone. DRAG. **3.** Tool used to paint a grain finish. **4.** To scrape damp plaster with a scarifier.

combination blade: Circular saw blade used with a power saws. Used for both crosscutting and ripping material. See *blade**.

combination column: Column composed of structural steel and concrete.

combination door: Exterior door with an interchangeable screen and glass panel that allows for ventilation in warm weather and thermal resistance in cold weather.

combination faucet: Faucet attached to hot and cold water supply lines and that discharges water from a central spout.

combination fixture: Fixture with a kitchen sink and laundry tray in one unit.

combination lock: Lock in which the tumblers are controlled by means of a rotating dial inscribed with letters or numbers. A U-shaped bolt is released to open the lock after the appropriate combination of letters or numbers is dialed.

combination plane: Wood smoothing plane with interchangeable plane irons used to form various shapes.

combination rasp: Rasp with both coarse and fine teeth.

combination sewer: Waste disposal piping that conveys sanitary waste and storm water.

* **combination square:** Hand tool with a head that slides along a 12″ metal rule with a central groove on one side. Head

may be locked into any position along the rule.

combination trap: P-trap manufactured in two pieces to allow for adjustment during installation.

* **combination window:** Window with an interchangeable screen and glass panel.

combination wrench: Hand tool with an open end wrench on one end and a closed end wrench on the other end. See *wrench**.

combined aggregate: Mixture of fine and coarse aggregate for use in concrete.

combined escutcheon plate: Flat metal trim for a door with both a knob socket and a keyhole.

combined footing: Structural unit used to support more than one column.

combined load: Total load placed on a structure. Includes live and dead loads.

combined sewer: See *combination sewer*.

combined stress: Total of axial and bending stress placed on a structural member.

combined stress ratio (CSR): Combined stress divided by the allowable stress on a structure. CSR should be less than 1.00 for a structural member.

comb plate: Toothed portion of a threshold plate at each end of an escalator.

comb roof: See *gable roof*.

combustible: Capable of burning.

combustible liquid: Liquid with a flash point equal to or greater than 100°F.

combustion chamber: Area in a boiler or heater where fuel and air are mixed and burned.

come-along: 1. Hand tool with a blade approximately 4″ high and 20″ wide used to spread concrete. **2.** Device used to tighten chains for securing loads on a truck bed.

comfort zone: Range of temperature over which a majority of people feel comfortable. COMFORT LINE.

commercial bronze: Alloy containing 90% copper and 10% zinc.

comminution: Screening and processing sewage into particles of a sufficient fineness to pass through a secondary screening process.

commode: See *water closet*.

common area: Space designed for use by all occupants of a structure or group of structures but not for use by the general public.

common bond: Brickwork pattern in which every sixth course is a header course and the intervening courses are stretcher courses. See *bond* (1)*.

common brass: Alloy containing 65% copper and 35% zinc.

common brick: See *building brick*.

common catenary: See *catenary*.

common ground: Section of property

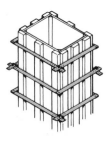

column clamp

combination square

combination window

used by all owners in a building development but not by the general public; governed and maintained by a committee of owners.

common lap: Offset of one-half the width of a shingle when applying roof shingles.

common lime: See *quicklime*.

common nail: Nail with a smooth cylindrical shaft and a flat head. Available in lengths ranging from 2d (1″) to 60d (6″). Used to join wood framing members and in general carpentry work. See *nail**.

* **common rafter:** Structural member that extends from the top wall plate to the ridge in a direction perpendicular to the wall plate and ridge.

common vent: Vent connecting two fixture drains and serving as a vent for both. UNIT VENT.

common wall: Vertical divider jointly used by two parties, one or both of whom are entitled to it under provisions of a lease.

communicating door lock: Hardware used on doors between connecting rooms. Commonly equipped with a knob latch that has a thumb bolt.

communicating frame: Double-rabbeted door frame used with two single-acting doors, each swinging in opposite directions.

compact: To increase the firmness or bearing capacity of soil, subgrade material, or other porous material such as freshly placed concrete by compressing, vibrating, or tamping to eliminate voids.

compacting factor: Comparison of the weight of concrete falling into and filling a standard container under standard conditions to the weight of fully compacted concrete filling the same container.

compaction: Process of reducing the volume of soil, subgrade material, or other porous material such as freshly placed concrete by compressing, vibrating, or tamping to eliminate voids.

compaction pile: Pile driven vertically into the ground to compact an area of loose soil and increase its bearing capacity.

compactor: Mechanical or manual device used to improve the bearing capacity of soil by tamping or vibrating.

compass: 1. Instrument used to determine the direction of a body relative to meridians. Locates the magnetic pole of the earth. 2. Layout instrument with two legs joined at the top; one leg is pointed and the other leg is fitted with a marking device. Used to lay out circles, curves, and other circular plane figures.

compass bearing: Directional indication on plot and plat plans expressed in degrees, minutes, and seconds; refers to the deviation of a line from the true north/south direction.

compass brick: Curved or tapering masonry unit used in curved masonry work, such as an arch.

* **compass plane:** Plane used to smooth concave or convex wood surfaces. Radius of the sole is adjustable.

compass rafter: Rafter that is curved on one or both sides.

compass roof: 1. Roof with curved rafters on one or both sides. 2. Roof used in timber structure in which the rafters, collar beams, and braces form an arch.

compass saw: Hand saw with a 12″ to 14″ narrow tapered blade. Used to cut small curves or circles or detailed designs in wood. See *saw**.

compass window: See *bow window*.

complete fusion: Fusion occurring over the entire surface of the base metal to be welded and between layers and weld beads.

component: 1. Part of a whole. 2. Part of a structure assembled before delivery to the job site.

composite arch: Arch formed with four centers. MIXED ARCH, LANCET ARCH.

composite board: Lumber panel product composed of a particleboard core and veneered faces. Used for thermal insulation.

composite column: Concrete column composed of a structural steel member, pipe, or tubing reinforced longitudinally.

composite joist: Intermediate horizontal structural member composed of different types of building materials; e.g., concrete with reinforcing steel. See *joist**.

composite pile: 1. Pile formed of two different materials, such as wood and concrete. 2. Steel pile formed by splicing two or more prefabricated lengths together end to end.

composite wall: Bonded masonry wall made of several types of masonry units.

composition board: Manufactured panel that consists of several types of materials pressed together with a binder. Used as sheathing or thermal or acoustical insulation.

composition joint: Bell and spigot pipe joint packed with oakum and rosin or cement.

composition roofing: Roofing consisting of asphalt-saturated felt building paper with a bituminous covering. May be covered with small coarse aggregate. Commonly applied to flat roofs or roofs

common rafter

compass plane

compound action snips

with a small pitch. ROLL ROOFING.

composition shingles: See *asphalt shingle*.

compound: 1. Plaster base coat to which sand is added at the job site. See *neat plaster*. 2. Material used for insulating electrical conductors. Consists of fillers, softeners, plasticizers, catalysts, pigments, and/or dyes.

* **compound action snips:** Hand tool used to cut sheet metal. Force applied to the handles is increased through mechanical advantage.

compound arch: Arch formed of several concentric arches placed within one another.

compound beam: Rectangular beam composed of small timbers with boards nailed over them for reinforcement.

compound leverage pipe wrench: Pipe wrench consisting of a jaw that exerts pressure in a clockwise direction along one section of the circumference of a pipe and an offset trunnion that exerts pressure in a clockwise direction to another part of the circumference. COMPOUND WRENCH.

compregnated wood: Wood impregnated with a thermosetting resin and bonded under intense heat and pressure. Increased density, strength, and decay resistance.

**computer-aided design
and drafting (CADD)**

concrete mixer

compression: 1. Inward or crushing force. **2.** Change in dimension resulting from an inward force.

compression bearing joint: Joint between two structural members in compression. One transmits compressive stress to the other.

compression connector: Sleeve installed between two adjoining conduit members by tightening a nut that secures the conduit, providing a rigid joint. See *conduit fitting**.

compression failure: Deformation of a member resulting from excessive compression.

compression faucet: Faucet in which the flow of liquid is controlled with a washer that is forced down onto a seat.

compression flange: Widened portion of a beam that is shortened by bending under a load.

compression joint: 1. Wood joint in which one member is drilled or mortised to receive a mating member that is slightly larger than the hole or mortise. End of tenon is rounded on the end and temporarily pressed to allow it to fit within the mortise. End expands when in position, locking the mortise and tenon together. **2.** Bell and spigot pipe joint formed by inserting a compression gasket into the bell and pressing the spigot into position. Gasket forms around the spigot to form a watertight joint.

compression lead: Guide for the pile and pile hammer used to withstand the load of pile extraction.

compression roll: Drive wheel on a steel-wheeled roller for compaction of earth or other materials.

compression set: Permanent distortion of a member after a compressive load has been applied.

compression stop: Non-rated globe valve.

compression tank: Tank or reservoir that absorbs water pressure exerted in a hot water heating system.

compression test: 1. Method of verifying compressive strength of a material or joint by subjecting it to a compressive load. **2.** Method used to determine the compressive strength of mortar or concrete by subjecting the material to a compressive load. Mortar samples are 2″ cubes and concrete samples are 6″ cylinders that are 12″ high.

compression wood: Abnormal lumber formed on the lower side of branches and leaning and crooked softwood trees. Hard and brittle and shrinks lengthwise more than normal wood.

compressive strength: 1. Maximum amount of compression a material withstands without failure. Expressed as pounds per square inch (psi). **2.** Maximum resistance of mortar or concrete to axial loading. Expressed as force per unit cross-sectional area.

compressor: 1. Component in the cooling cycle of an air conditioning system that condenses a refrigerant gas into a liquid coolant **2.** Equipment that retains air at a higher pressure than atmospheric pressure. AIR COMPRESSOR. **3.** Mechanical device used to increase gas pressure.

*computer-aided design and drafting (CADD): Design and drafting of a component, member, or structure using a computer. Increases productivity by producing drawings with consistency that can be easily reproduced. Uses hardware (physical components) and software (computer program).

computer-aided drafting (CAD): See *computer-aided design and drafting (CADD)*.

computer-aided engineering (CAE): Mechanical, structural, civil, and electrical engineering using a computer. Allows engineer to evaluate and revise systems as required.

concameration: 1. Arch or vault. **2.** Apartment.

concave: Curved recess.

concave joint: Mortar joint between masonry units formed with a slightly rounded recess, producing a weather-resistant joint. See *mortar joint**.

concealed casework hinge: Hinge installed on a cabinet without face frame. Completely concealed when door is closed. See *hinge**. EUROPEAN HINGE.

concealed fouling surface: Surface of a plumbing fixture not visible and not cleaned or flushed with each operation.

concealed gutter: Gutter that is not visible from the exterior of a structure. Removes rainwater from a roof. HIDDEN GUTTER, CONCEALED BOX GUTTER.

concealed heating: Heating system in which the heating units or elements are not plainly visible or integrated into the architectural features of the room. PANEL HEATING.

concealed hinge: Hinge used on swinging doors; mortised into both the jamb and edge of the door. Completely concealed when the door is closed. See *hinge**.

concentrated load: Weight localized on and supported by one structural member or a small area of a structure.

concentric: With a common center.

concentric arch: Arch composed of several arches that have a common center.

concentric tendons: Tendons that are aligned with the gravity axis of the prestressed concrete member.

concha: Smooth concave surface of a vault.

concordant tendon: Concrete reinforcing member that is aligned with the pressure line produced by the tendon.

concrete: Composition of a binding medium and aggregate. Commonly consists of a mixture of cement, aggregate, and water in varying proportions. Mixture is worked in a plastic state and gains hardness through the hydration of water with the cement.

concrete admixture: See *admixture*.

concrete block: See *concrete masonry unit**.

concrete containment structure: Reinforced concrete structure used to protect against the release of radioactive or hazardous materials into the atmosphere.

concrete cover: Distance from the face of the concrete to the reinforcing steel. CLEARANCE, CONCRETE PROTECTION.

concrete cylinder: Cylindrical mold 6″ in diameter and 12″ high in which three separate placements of concrete are placed and rodded. Cylinder and concrete are subjected to a compression test after curing to determine compressive strength of the concrete mix.

concrete form: See *form*.

*concrete masonry unit: Precast hollow or solid masonry unit made of portland cement and fine aggregate, with or without admixtures or pigments. Formed into modular or non-modular dimensions to be laid with other similar units.

*concrete mixer: Portable or stationary

equipment with a rotating drum or paddles; used to mix concrete ingredients. Ingredients are introduced into the open end and discharged by tilting the drum. CEMENT MIXER.

concrete nail: Hardened steel nail with a flat head and diamond point. Used for nailing into concrete and masonry surfaces. See *nail**.

concrete pump: Device that forces concrete through a pipe or hose to the location of placement.

concrete saw: Power saw used with an abrasive rotating blade to score and cut hardened concrete. See *saw**.

concrete vibrator: See *vibrator*.

concurrent heating: Application of additional heat during arc welding or cutting.

concussion: See *water hammer*.

condensate: Liquid formed by condensation of vapor.

condensation: Conversion of a substance from a vapor to a liquid state by removing heat. Results in formation of condensate.

condensation point: Temperature at which vapor condenses if latent heat is removed.

condensation pump: Device used to remove condensate from air conditioners and steam returns.

condenser: 1. Heat exchanger that removes heat from a refrigerant and converts hot pressurized gas into a cool liquid. 2. See *capacitor*.

condensing unit: Component of a refrigeration system consisting of a compressor, condenser, condenser fan motor, receiver, and controls.

condominium: Multifamily dwelling in which each unit and a share of the common areas, such as hallways or elevators, are owned by individual owners.

conductance: See *thermal conductance*.

conduction: Transfer of thermal or electrical energy from one member to another through direct contact.

conductive flooring: Floor surface designed to eliminate or minimize electrostatic accumulation and sparking.

conductivity: See *thermal conductivity*.

conductor: 1. Material that allows electrical current to flow through. May be bare, covered, or insulated. 2. Pipe that is used to convey rainwater from a roof gutter to the drainpipe. LEADER, DOWNSPOUT. 3. See *lightning arrester*. 4. Material that readily transmits heat.

conduit: 1. Tube or pipe used to support and protect electrical conductors. Includes rigid metal conduit, intermediate metal conduit (IMC), electrical metallic tubing (EMT), and flexible metal conduit. 2. Natural or artificial channel such as a pipe or canal used to convey fluids.

conduit bender: See *hickey*.

conduit body: Separate unit of a conduit or tubing system with a removable cover that provides access to the interior of the electrical system. See *conduit fitting**.

conduit box: See *junction box*.

conduit bushing: Short, internally threaded sleeve fastened to the end of conduit inside an outlet box. Edges of the sleeve are rounded to prevent damaging the conductors.

conduit coupling: Short sleeve into which the ends of two adjoining pieces of conduit are inserted and secured with set screws. See *conduit fitting**.

conduit elbow: Short conduit fitting formed to a standard angle such as 45° or 90°.

CONCRETE MASONRY UNITS

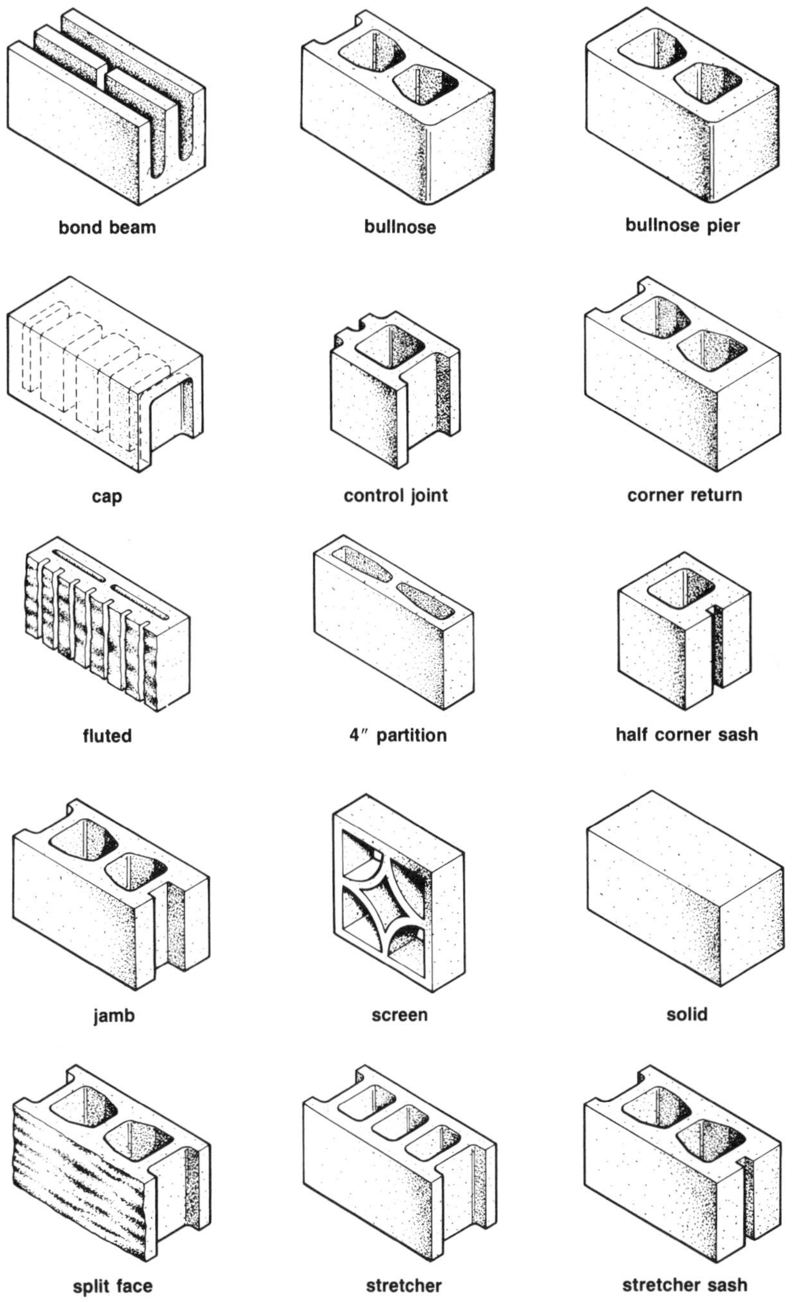

bond beam

bullnose

bullnose pier

cap

control joint

corner return

fluted

4″ partition

half corner sash

jamb

screen

solid

split face

stretcher

stretcher sash

*conduit fitting: Conduit accessory used to connect sections of conduit.

cone: 1. Three-dimensional geometric figure with a circular base and sides that taper to a point. 2. Part of an oxy-fuel gas flame adjacent to the welding tip orifice. 3. Plastic tapered device attached at the end of a wire snap tie used to spread and space concrete forms. Depression remaining from it is filled after forms are stripped.

cone bolt: Concrete hardware device with conical ends at each end inside the forms that spread and space the forms.

configurated glass: Glass with an irregular surface texture used to obscure vision.

confined concrete: Concrete with closely spaced transverse reinforcement designed to restrain the concrete in a direction perpendicular to applied stress.

confluent vent: Vent pipe that serves two or more fixture vents.

congé: Base molding that is flush with the wall above and has a small fillet at the intersection of the floor and wall.

conifer: See softwood.

connected load: Sum of the continuous ratings of all fixtures and equipment attached to an electrical system.

connecting bar: Round or hexagonal profile tool ranging from 24″ to 36″ long and from $^{3}/_{4}$″ to $^{7}/_{8}$″ in diameter. Used to position structural steel members and align bolt holes. See driftpin.

CONDUIT FITTINGS

Rigid Metal and Intermediate Metal Conduit

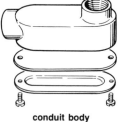
conduit body

bushing

locknut

compression connector

setscrew coupling

threaded coupling

45° elbow

90° elbow

nipple

Electrical Metallic Conduit

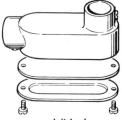
conduit body

compression connector

compression coupling

setscrew connector

setscrew coupling

45° elbow

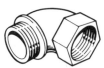
90° elbow

Flexible Metal Conduit

coupling

squeeze connector

45° squeeze connector

90° squeeze connector

connector: Device attached to the ends of two or more conductors or cables to allow connection of electric current without permanent splices.

conoid: See *cone* (1).

consistency: Degree of firmness or stiffness of a material such as concrete or mortar. Measured by the slump test for concrete and flow test for mortar, plaster, and grout.

console: 1. Decorative supporting bracket. **2.** Control panel for equipment.

consolidation: Process of creating a close arrangement of solid particles in concrete or mortar during placement by eliminating voids other than entrained air. Commonly accomplished by vibration or tamping.

constant-voltage transformer: Transformer that maintains a constant voltage ratio over the range of 0 to rated output.

construction: 1. Work done on a job site that ultimately produces an altered or new structure. Includes excavation, assembly, and erection of components. **2.** Product of the building process.

construction-grade lumber: Lumber that is generally free of waste that would result from defects.

* construction joint: Surface where two successive placements of concrete meet across which a bond is maintained between the placements. Reinforcement may or may not be interrupted. ISO-LATION JOINT.

contact cement: Adhesive that bonds two materials together upon contact. Adhesive is applied to both surfaces, allowed to dry to the touch, then joined to ensure a good bond.

contact closure input: Electrical signal sent to a relay, which changes the position of the contacts.

contact pressure: Pressure acting on the area between a structural footing and the supporting soil. Includes pressure resulting from all structural loads and stresses.

contact print: Process used to reproduce prints in which the original drawing is held against light-sensitive paper and exposed to light. Light passes through the blank areas of the original and exposes the light-sensitive paper. Light does not pass through the printed areas of the original, such as lines, producing an image on the light-sensitive paper.

contact resistance: Resistance to electric current flowing between two pieces of metal or the electrode and base metal in the arc welding process.

contact splice: Connection of rebars by overlapping the ends that are in contact with each other.

contemporary architecture: Style of architecture characterized by simplicity and straight lines.

continuity tester: Battery-powered test instrument used to determine the presence and location of open or broken connections and grounded circuits.

continuous beam: Beam supported by three or more supporting members.

continuous duty: Operation of equipment or system at a constant load for an indefinite period of time.

* continuous high chair: Reinforcement support consisting of a continuous top wire member welded to the top of equally spaced wire legs.

continuous hinge: Hinge that extends the entire length of the moving component. PIANO HINGE.

continuous load: Load with the maximum electric current passing through the circuit for a minimum of three hours.

continuous mixer: Concrete mixer in which ingredients are introduced without interruption and the finished mix is discharged in a continuous stream.

continuous rating: Maximum constant load carried continuously by electrical equipment without exceeding predetermined temperature rise limitations.

continuous ridge vent: Vent that extends the full length of a roof ridge. Commonly has a covering of metal and screen.

continuous sequence: Welding technique in which each pass is made from one end of a joint to the other.

continuous slab: Slab that extends over three or more supports in a given direction.

continuous vent: Extension of a vertical vent that is a continuation of the drain it serves.

continuous waste pipe: Waste pipe from two or more fixtures or compartments of the same fixture connected to a single trap.

contour: Profile of a figure.

* contour gauge: Hand tool consisting of many narrow steel pins enclosed in a frame that allows them to slide. Pins are pressed against an irregularly shaped object, causing them to conform to this shape. Gauge is then used to transfer the shape to another member.

contour interval: Vertical distance between adjacent contour lines.

* contour line: Dashed or solid line on a plot plan used to show elevations of the surface. Dashed line indicates existing elevation; solid line indicates finished

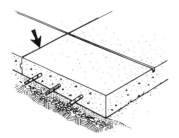

construction joint

continuous high chair

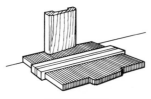

contour gauge

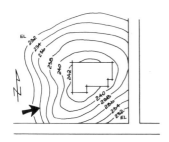

contour line

elevation. Actual elevation may be written on each contour line.

contraction: Shrinkage of material, often caused in building material by loss of moisture or lowering of temperature.

contraction joint: Groove in a vertical or horizontal concrete surface to create a weakened plane and control the location of cracking due to imposed loads.

contractor: Individual or company responsible for performance of construction work, including labor and materials, according to plans and specifications.

control factor: Ratio of the minimum compressive strength of concrete compared to the average compressive strength.

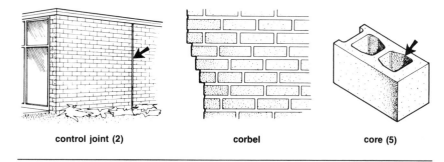

control joint (2)   corbel   core (5)

---

*control joint: 1. See *contraction joint*. 2. Continuous vertical joint in a masonry wall without mortar in which rubber, plastic, or caulking is installed. Used in long masonry walls where thermal expansion and contraction may cause cracking. 3. Thin strip of perforated metal applied on lath and plaster surfaces to relieve stress resulting from expansion and contraction in large ceiling and wall surfaces.

control joint block: Concrete masonry unit with a solid projecting end that fits within a corresponding key in an adjoining block. See *concrete masonry unit**.

controlled fill: Fill material placed in layers, compacted, and tested after each compaction for moisture content, depth of lift, and other bearing capabilities before additional layers are placed.

controller: Device or group of devices that regulates amount of electricity delivered to the equipment to which it is connected.

control station: Surveying reference point whose horizontal and/or vertical position is used as a reference for other surveying operations.

convection: Heat transmission produced by fluid, such as an air current resulting from difference in density of heated air. As fluid is heated it rises, displacing the cooler air and causing transmission.

convector: 1. Enclosed heating unit used to transfer heat through convection. Fan may be incorporated to facilitate air movement. 2. Surface designed to transfer heat to a surrounding fluid through convection.

convector radiator: See *radiator*.

convenience outlet: See *receptacle*.

conversion: 1. Change in the designed use of a structure to another use that has different code requirements. 2. Process of sawing lumber into smaller members.

conversion burner: Gas or oil burner used to replace a boiler or furnace that originally used wood or coal as a fuel source.

convex: Outward curve.

conveyor: Horizontal, vertical, or inclined mechanical device used to move bulk materials or prepackaged components along a predetermined path.

cooked glue: Adhesive that requires heating before use. See *hot melt adhesive*.

cooling leg: Section of uninsulated pipe that allows condensate to cool to prevent flashing when a trap opens.

cooling rate: Decrease in temperature in a given amount of time.

cooling tower: Structure over which water is circulated to decrease the temperature through evaporation to the surrounding air. Commonly installed on the roof of a structure or as a separate cooling structure.

coopered joint: Wood joint used in furniture construction that resembles joinery used in wood barrel construction.

coordinate: Linear and/or angular quantity that designates the location of a point as it relates to a given frame of reference.

coordinator: Hardware on a pair of double doors that allow the doors to close in the proper sequence. Used on doors with an overlapping astragal.

copal: Natural resin that provides hardness and gloss to varnish and lacquer.

cope: 1. To cut or trim the end of a molding at an inside corner so as to match the profile of the mating member. 2. To remove part of the flange at the end of a beam to avoid interference with adjoining members. 3. To cover the top of a wall with a masonry cap.

coped joint: Joint in which one member is cut on the end to match the profile of an adjoining member. See *joinery**.

coping: 1. Cap or other protective member used to cover the top of a vertical structure such as a parapet or chimney. Commonly beveled or curved and projects from the face of the structure for additional protection. 2. Process of splitting stone by drilling a row of holes and driving wedges in them.

coping block: Concrete masonry unit with a solid upper surface. Used for coping at the top of a vertical structure.

coping brick: Masonry unit specifically manufactured for coping at the top of a vertical structure.

coping saw: Hand saw with a fine-toothed narrow blade secured in a U-shaped frame attached to a short straight handle. Used to cut curves and coped joints. See *saw**.

copper: Metal element with high ductility, malleability, and tensile strength. Used as an electrical and a thermal conductor.

copper-clad aluminum conductor: Electrical conductor formed from an aluminum rod with copper bonded to it. Copper forms a minimum of 10% of the cross-sectional area of a solid conductor or each strand of a stranded conductor.

copperlight glazing: Fire-retardant glazing consisting of several individual lights separated by strips of copper.

copperplating: Depositing copper coating on the surface of a base metal by dipping or using the electrolytic method.

copper tubing: See *tubing*.

coquina: Soft, porous limestone formed from marine shells and coral cemented together by calcite. Used as a raw material in the manufacture of portland cement.

*corbel: Short projection from the face of a wall formed with successive courses of masonry.

corbel table: Projecting course of masonry supported by corbels.

corbie gable: Gable with a stepped upper edge. CORBIESTEPS, CROWSTEPS.

cord: 1. Measure of firewood equal to a stack 4'-0" wide × 4'-0" high × 8'-0" long, or 128 cubic feet. 2. Flexible insulated electrical conductor enclosed in a flexible insulating covering and equipped with terminals.

cordless drill: Electric power tool in which a bit is inserted and used for boring holes. Electric power is provided by rechargeable batteries in the handle of the tool. See *drill**.

*core: 1. Central portion of a material or component; e.g., solid-core door. 2. Material removed from a mortise joint. 3. Central portion of a plywood panel to which veneers are glued. 4. Cylindrical piece of material cut by a hollow rotary drill; e.g., portion of earth removed for test drilling is a core sample. 5. Open area in a concrete masonry unit. 6. See *mandrel*. 7. Metal member used as a support for a handrail. 8. Fill material between a lintel and the relieving arches.

coreboard: Panel product consisting of a gypsum core encased with strong

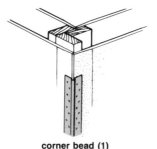

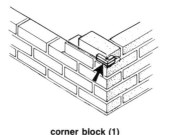

| core cock | corner bead (1) | corner block (1) | corner tape creaser |

liner paper, forming a 1″ thick panel. Covered with additional layers of gypsum panels for use in elevator shafts and laminated gypsum partitions.

core box bit: Router bit used to form semicircular grooves or flutes in wood and other soft material. See *bit* (2)*.

* **core cock:** Valve in which the flow of liquid is controlled by a cylindrical plug that fits closely in a machined seat. Part of the plug is bored to allow passage of liquid. PLUG VALVE.

core drill: Power equipment used to drill holes into hard material such as rock or concrete to obtain core samples. Sample is retained in the barrel of the equipment during operations.

cored solder: Solder wire or bar with a flux center.

core grease: Non-water soluble lubricant used for the interior mating surfaces of valves, cocks, and other plumbing control devices.

core sample: Test sample taken from rock, concrete, or other hard material in which the sample remains in its original sturcture.

core test: Compression test on a sample cylinder of concrete that has been cut from a section of hardened concrete.

coring: Removal of a core sample from a hardened material by cutting out a cylindrical portion.

Corinthian: Greek order of architecture characterized by ornate bell-shaped capitals adorned with rows of acanthus leaves.

cork: Soft material made by the outer bark of a cork oak tree. Used for acoustical and thermal insulation, vibration control, and as an interior wall finish.

corkboard: Granulated and compressed cork formed into sheets or rectangular blocks. Used as acoustical and thermal insulation or fabricated for use as bulletin boards.

corner: 1. Intersection of two adjacent faces not in the same plane. **2.** Point determined by surveying, which indicates a boundary of property.

corner angle: L-shaped metal device attached to outside or inside edge of a 90°

joint for reinforcement. Holes in the device are countersunk to provide for a flush fit with flathead screws. CORNER BRACE.

* **corner bead: 1.** Light-gauge, L-shaped galvanized metal device used to cover outside joints of gypsum drywall. Small perforations in the flanges allow for the adhesion of joint compound. Lengths range from 6′-8″ to 10′-0″ and flanges range from 1″ to $1\frac{1}{4}$″ in width. **2.** Light-gauge, L-shaped galvanized metal device used to cover outside joints in plaster walls. Available in 8′-0″ and 10′-0″ lengths with flanges ranging in width from $1\frac{1}{2}$″ to $3\frac{1}{4}$″.

corner bead clincher: Tool used to fasten drywall corner beads without using screws or nails. L-shaped steel member is fitted with two jaws that are activated by striking and crimping the edges of the corner bead into the drywall.

corner bit brace: Bit brace used to bore holes in tight locations. Handle is offset to facilitate turning the brace.

* **corner block: 1.** Small wood, metal, or plastic block used to secure a string to facilitate level placement of masonry units. Blocks are set on masonry corners and aligned with top of masonry course with string stretched between them. See *concrete masonry unit*. **2.** Triangular wood member along the inside corner of a cabinet or other furniture that reinforces the joint. **3.** See *corner return block*.

corner brace: 1. Diagonal brace let-in to framing members to reinforce wood-frame construction. **2.** See *corner angle*.

corner chisel: Hand tool with two straight cutting edges joined at a 90° angle for cutting inside corners of wood members.

corner clamp: Clamp used to secure members of a 90° joint during assembly. Consists of two fixed and two movable jaws. See *clamp*.

corner finisher: Hand tool used to distribute excess joint compound evenly across reinforcing tape and to feather the edges.

corner iron: L-shaped metal device attached to the face of a 90° joint for reinforcement. Holes in the device are countersunk to provide a flush fit with flathead screws.

Cornerite®: Narrow L-shaped strip of painted or galvanized expanded metal lath used to reinforce inside plaster corners where metal lath is not lapped.

corner joint: Joining of two pieces of metal at approximately 90° to each other. See *weldment*.

corner marker: Permanently fixed surveyor's reference point at the intersection of quarter sections according to township and section divisions.

corner molding: Wood, metal, plastic, or vinyl trim member used as a decorative and/or protective covering at the intersection of adjacent angular surfaces.

corner pole: Vertical rod with graduations used as a guide for masonry walls.

corner post: 1. Assembly in western framing consisting of two studs with blocks in between or three studs fastened together to provide additional structural support at wall intersections. **2.** Glazing mullion that holds two pieces of glass at an angle to form a corner.

corner return block: Concrete masonry unit with solid sides and a solid face at one end. See *concrete masonry unit*.

corner roller: Hand tool used to embed reinforcing tape along an inside drywall corner and force the excess joint compound out.

corner round bit: Router bit used to remove the square corner from a piece of stock. Produces a slight convex radius along the edge. See *bit* (2)*.

cornerstone: Large masonry unit permanently set in a location near the base of a structure. Contains information about the structure and may be hollow to accept a vault containing memorabilia.

* **corner tape creaser:** Hand tool used to crease and apply reinforcing tape to inside drywall corners before being covered with joint compound.

corner tool: Hand tool that simultaneously embeds reinforcing tape and applies joint compound to both sides of an inside corner in drywall construction.

* **corner trowel:** Hand tool with a V-shaped blade used to finish corners of plaster or concrete. Available for inside and outside corners. CORNER PLOW.

cornice: 1. Exterior structural trim along the intersection of the roof and top of a wall. Includes all framing and trim members. **2.** Wood or plaster ornamental molding extending around the walls of a room just below the ceiling. **3.** Uppermost section of an entablature.

cornice return: Continuation of the overhang trim members around the end of a gable roof.

corona: 1. Vertical portion of a cornice projecting beyond the bed molding. Used to divert rainwater away from the structure. **2.** Area around the nugget of a spot weld.

coronet: See *pediment*.

* **corporation cock:** Valve placed near the junction of the public water or gas main. Used to control water or gas supply to a structure. CORPORATION STOP.

corridor: Interior hallway that provides access to rooms or apartments.

corrosion: Deterioration of a material, such as metal or concrete, caused by a chemical or electrochemical reaction resulting from prolonged exposure to moisture, chemicals, or other corrosive substances.

corrugated: Formed with alternating parallel ridges and depressions.

corrugated anchor: Masonry anchor made of a thin strip of sheet metal formed with alternating parallel ridges and depressions. See *anchor**.

* **corrugated fastener:** Steel fastener with small parallel ridges and depressions used for fastening flush wood joints. One end is sharpened for penetration into the wood.

corrugated flashing: Sheet metal formed with alternating ridges and depressions used for moisture protection. See *flashing**.

cotter: Wedge-shaped wood or steel device used for fastening.

cotter pin: Split metal pin inserted into a hole; ends are bent to secure the pin.

coulisse: Timber with a longitudinal groove for use with a sliding member. CULLIS.

count: Number of openings per lineal inch in wire cloth.

counter: Horizontal surface used for display or as a work surface as in a kitchen.

* **counterbore:** To increase diameter of a hole through part of its length to receive the head of a bolt or nut.

counterbrace: Brace that counteracts the force imposed by another brace, such as a web in a truss.

counterclockwise: Movement in the direction opposite the hands of a clock.

counterflashing: Sheet metal installed in masonry joints and bent down over base flashing to prevent water from entering the joints and protecting the exposed ends of the base flashing.

counterflow: Two fluids in a heat exchanger that flow in opposite directions.

counterfort: Buttress, pilaster, or pier extending upward from the foundation of a masonry wall to provide additional resistance to lateral loads.

countersink: 1. To make a conical depression in a surface to receive the head of screw or bolt with a similar shape. Upper surface of the head is flush with the surface when installed. **2.** Bit used to make a conical depression in a surface. Produces tapered sides with an 82° angle. See *bit* (1)**.

counterweight: Weight that balances another weight; e.g., a sash weight that counterbalances the weight of a window sash.

coupled: Two equal and opposite parallel forces that produce a rotational motion. Their moment equals the product of one of the forces and the perpendicular distance between them.

coupled column: One of a pair of columns set close together with wider intercolumniation betweeen them.

coupler: 1. Tubular metal device used for quick connection and disconnection of a hose without screwing or turning the device. **2.** Tubular device inserted in the end of a scaffold section to provide a means for alignment and reinforcement of scaffold panels. **3.** Device used to connect steel reinforcing rods or prestressing tendons by joining them end to end.

coupling: Short collar or pipe with female threads at each end. See *pipe fitting**.

coupling pin: Short metal rod inserted in a scaffold coupler to secure the scaffold panels in position.

course: 1. Continuous horizontal layer of masonry units bonded with mortar. **2.** Horizontal row of roof shingles. **3.** Horizontal layer of concrete. Several courses comprise a lift. **4.** Bearing and length of a surveyed line. **5.** Azimuth and length of a surveyed line.

coursed ashlar: Stone construction in which the stones are of equal height within each course. All courses are not the same height. See *bond* (1)**. RANGEWORK, REGULAR COURSED RUBBLE.

coursed masonry: Masonry construction in which the individual units are laid in regular courses. COURSE WORK.

coursed rubble: Masonry construction consisting of roughly shaped stones of various sizes and small stones used to fill the voids.

court: Open area partially or entirely surrounded by buildings or walls. COURTYARD.

cove base: Concave member installed at the intersection of floor and walls.

cove bit: Router bit that cuts a concave shape along the edge of a wood member. See *bit* (2)**.

cove bracketing: Series of wood brackets or framing used to receive lathing for a plaster cove.

cove molding: Concave interior corner molding. See *molding**.

cover: 1. Smallest distance between the surface of reinforcement and the finish surface of concrete. **2.** Portion of a shingle or tile that is lapped under the

corner trowel

corporation cock

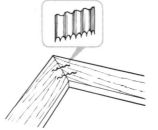

corrugated fastener

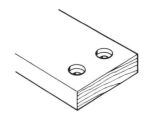

counterbore

succeeding course.

coverage: Measure of the total area to be covered by a liquid, such as paint. Expressed as square feet per gallon when spread to a 1 mil thickness.

covered conductor: Electrical conductor encased in non-conductive material of sufficient thickness and composition.

covered electrode: Composite filler metal welding electrode consisting of a bare electrode or metal-cored electrode with a coating that provides a layer of slag on the base metal.

covered joint: See *shiplap*.

* **cover plate: 1.** Flat member fastened to the flanges of a girder to increase the cross-sectional area. **2.** See *wall plate*. **3.** Removable piece of colorless glass, plastic-coated glass, or plastic that covers the filter plate of a welding helmet, hood, or goggles and protects it from weld spatter or pitting. **4.** Transparent material covering the solar absorber in a flat plate solar collector.

coverstone: Flat stone set on a steel beam to be used as a foundation for masonry to be placed on top.

cowl: 1. Protective hood on a vent pipe that prevents water and snow from entering. **2.** See *chimney cowl*.

cracking: Deep checks occurring in a surface, such as paint or lacquer, resulting from incorrect mixing of materials.

cracking load: Load that causes tensile stress in a member to exceed tensile strength.

cradle: 1. Structure that provides support to a pile hammer placed in front of the lead rails and travels on the rails of the leads. **2.** Footing designed to fit the conduit it supports.

cradling: Framing used to support the lath and plaster or masonry of a vaulted ceiling.

cramp: 1. U-shaped device used to secure adjacent masonry units together. **2.** Rectangular frame with a screw mechanism used to secure members in position during gluing operations.

crandall: Hand tool consisting of several pointed bars fixed to the end of a handle. Used to dress stone.

crane: 1. Hoisting equipment consisting of a boom, load cables, and a power source. Used to move a load vertically and horizontally. See *tower crane*. **2.** Siphon or bent pipe used to draw liquid from a large container.

crank brace: See *corner bit brace*.

crash bar: Horizontal bar of a panic bar system. Used to activate the opener.

crater: Depression at the end of a weld bead or in the molten weld pool.

crawler belt: Assembled tread shoes and

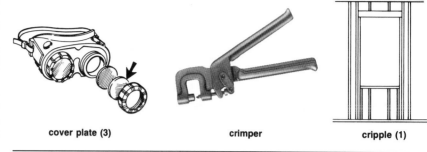

cover plate (3)   crimper   cripple (1)

connecting pins extending around rollers and drive sprockets. Contacts the ground and provides a means for movement of the equipment.

crawling: Painting defect in which the paint film breaks, separates, or raises. Results from the application of paint onto a slick surface, surface tension as a result of a heavy coat of paint, or the application of an elastic film over a hard, brittle surface.

crawl space: Unfinished accessible area below the first floor of a structure. Height ranges from approximately 1'-6" to 4'-0". Commonly used for components such as ductwork and piping.

crazing: 1. Small cracks in a finish paint coat caused by non-uniform shrinkage of the paint. **2.** Small cracks in a concrete surface caused by uneven contraction during hydration.

creaser: Hand tool used to fold reinforcing tape in half lengthwise. Tape is then used to conceal drywall corners.

creasing: 1. Corbeled tile or brick at the top of a masonry wall or chimney to shed water. Commonly projects 1" to 2" in each course. **2.** Slate or metal flashing.

creep: 1. Tendency of hardened concrete to deflect a greater amount under a sustained load over an extended period of time. **2.** Slow longitudinal movement caused by external forces. **3.** Slow movement by a piece of machinery. **4.** Elongation of a roof membrane resulting from thermal and moisture changes.

creeper: Brick that has been cut to fit the extrados of an arch.

cremorne bolt: Hardware used to lock French or casement windows in which a sliding rod is fastened to the edge of the window and actuated by a rotating handle. CREMONE BOLT.

creosote: Coal tar distillate used to impregnate wood as a wood preservative. Heavier than water and consists of liquid and solid aromatic hydrocarbons containing tar acids and bases.

crescent truss: Truss in which both the

top and bottom chords are curved upward or downward at different radii and intersect at the ends to form a crescent shape. See *truss**.

Crescent® wrench: See *adjustable wrench*.

crest: 1. Ornamental ridge covering of a roof. **2.** Ornamental finish at the top of a wall. CRESTING. **3.** Top of a dam or spillway.

cribbing: 1. Framing members or timbers used as permanent or temporary support for a structure above. CRIB. **2.** Wooden lining on the inside of a shaft. **3.** Framework of horizontal timbers, steel beams, or concrete used as a retaining wall. CRIB, CRIBWORK.

cribbled surface: Surface covered with raised or recessed dots.

cricket: See *saddle* (1).

crimp: 1. Sharp bend in sheet metal. **2.** To apply pressure to a flexible material or member to sharply deform it. May be used to join dissimilar materials. **3.** Indenting ends of pipe or tubing.

crimped wire: Wire that is shaped into a serpentine pattern to increase its bonding capacity with concrete and provide reinforcement.

* **crimper:** Hand tool with jaws that pierce a metal stud and runner and interlock the members.

* **cripple: 1.** Vertical component in wood-frame construction that is shorter than a full-length stud. Commonly installed over headers and under sills. CRIPPLE STUD. **2.** Change of direction in a chimney.

cripple jack rafter: Roof framing component installed between a hip and valley rafter. Does not bear on the ridge or the wall plate.

critical angle: Maximum angle of a stairway or ramp before it becomes unsafe and uncomfortable. Critical angle for stairways is 50° and for ramps is 20°.

critical density: Unit weight of a saturated granular material above which strength is increased and below which strength is decreased.

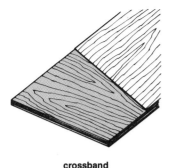

crossband

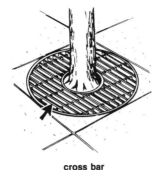

cross bar

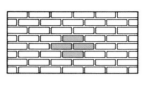

cross bond

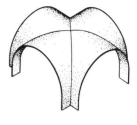

cross vault

critical height: Maximum height that unsupported soil will remain in position.

critical level: Mark on a backflow preventer or vacuum breaker that indicates minimum distance above the flood level rim of a plumbing fixture that the device should be installed.

critical pressure: Pressure at which the corresponding properties of gas and liquid are equal.

critical saturation: Degree to which freezable water fills pore space in cement paste and affects the response of the cement paste to freezing. Assumed to be 91.7% because 8.3% of the volume of water changes from liquid to a solid state.

critical slope: Maximum horizontal angle that unsupported soil will remain in position.

critical speed: Maximum angular speed of rotating equipment above which excessive vibration is produced.

critical state: State of a material at which the corresponding properties of gas and liquid are equal.

critical temperature: 1. Temperature at which the corresponding properties of gas and liquid are equal. **2.** Temperature below which a substance becomes superconductive.

critical velocity: Maximum velocity of fluid flow in a pipe above which the fluid does not flow in a direction parallel with the pipe.

crocus cloth: Fine abrasive material used to polish soft nonferrous metal.

crook: Longitudinal warpage in wood resulting in a convex or concave edge. See *defect**.

cross: 1. Pipe fitting consisting of four branches at right angles to one another. See *pipe fitting**. **2.** Design consisting of two lines intersecting at a right angle.

cross ambient: Thermal control system that operates with a bulb temperature that is different from the surrounding air temperature.

* **crossband:** Layer of veneer in plywood adjacent to the face veneer. Grain is perpendicular to the grain of the face veneer. CROSSBANDING.

* **cross bar:** Connecting bar that extends across the bearing bars of a grate.

cross-bedding: Inclined lamination in sedimentary rock that gives stone texture and pattern.

* **cross bond:** Brickwork pattern in which the joints between stretchers are aligned with centers of stretchers of adjacent courses.

cross brace: Diagonal braces used in pairs that cross each other to transfer or resist imposed loads.

cross break: Separation in a wood member that is perpendicular to the grain.

cross bridging: Diagonal braces set in pairs between floor joists or timbers. Used to transfer or resist imposed loads. See *bridging**.

crossbuck door: Door with intersecting angular rails installed in an X design in the lower portion of the door. Upper portion of the door is commonly fitted with glass. See *door**.

cross connection: Pipe connection between two separate piping systems, one of which is potable water and the other containing waste, sewage, or other sources of contamination.

crosscut: To saw wood across the grain.

crosscut blade: Circular saw blade with fine teeth used to cut wood across the grain. See *blade**.

crosscut saw: Hand saw with sharpened, pointed teeth designed to cut across the grain of wood. See *saw**.

crossette: Projection or ear at the corner of the architrave of a door or window. ANCON, ELBOW.

cross-garnet hinge: T-shaped decorative hinge.

cross grain: Wood grain that is not parallel to the longitudinal fibers caused by twisting of the wood fibers or cells.

crosshair: Horizontal and vertical marks in the field of view in the telescope of a surveying instrument. Perpendicular

crosshairs intersect in the middle of the field of view to facilitate locating objects.

crosshatching: Series of closely spaced parallel lines used to indicate a cross section. Various materials are identified using different symbols.

crosslap: Wood joint formed by removing half the thickness of the mating sections of adjoining members. Total thickness of the joint is equal to the thickness of the stock.

cross light: Illumination of an object with two or more light sources from two or more opposite angles.

cross nogging: See *cross bridging*.

crossover: 1. U-shaped pipe fitting with the ends turned outward. Used where one pipe passes around another pipe in the same plane. CROSSOVER FITTING. **2.** Connection between two water supply pipes in the same system.

cross peen hammer: Heavy striking tool with one flat face and one wedge-shaped face used for breaking and chipping hard surfaces. Wedge-shaped face is perpendicular to the handle.

cross rail: See *lock rail*.

cross section: View of an object showing the internal construction of a part or an assembly. Made by taking an imaginary cut through an object at a right angle to the longitudinal axis.

cross springer: 1. Diagonal rib of a groined roof. **2.** Diagonal arch of a ribbed groin vault.

cross stile: Intermediate vertical member in a door.

cross tee: Light-gauge metal suspended ceiling component placed between main runners or main tees.

* **cross vault:** Vault formed by the intersection of two barrel vaults at right angles.

cross ventilation: Circulation of fresh air in an area in which open windows, doors, or vents are in opposite sides of the area.

crosswalk: Section of roadway designated for pedestrian traffic.

cross welt: Seam between sections of

flexible metal sheet roofing usually parallel to the ridge or gutter.

crotch: Area of a tree where the branches join the trunk.

crotch veneer: Veneer from lumber produced from the crotch of a tree. Unusual decorative grain pattern.

crotchwood: See *crotch veneer.*

crowbar: Rounded or hexagonal profile steel prying tool with one end flattened and offset and the opposite end curved and formed with claws for pulling nails. Used for prying and as a lever when lifting heavy objects.

* **crown: 1.** Uppermost member or highest point of an object or structure. **2.** Highest point in the center of a road; used to divert water to the edges. **3.** Slight convex deformation in a board along the bearing edge. **4.** Upper section of a tree that contains leaves and twigs. **5.** Section of a plumbing trap where it changes direction from upward to horizontal.

crown knot: Knot tied in the end of a rope to keep the strands from unlaying and unraveling.

crown molding: 1. Trim member at the top of the cornice and immediately beneath the roof. **2.** Trim member along the intersection of interior walls and ceilings. See *molding*.*

crown of an arch: See *crown* (1).

crown of a trap: See *crown* (5).

crown post: See *kingpost.*

crown saw: See *hole saw.*

* **crown weir:** Highest point along the bottom interior surface of the crown of a trap.

crow's foot: 1. Small electrical fitting to which fixtures are attached; fastened in an outlet box. **2.** V-shaped mark used to indicate locations and lay out points. **3.** Mark on a stake that indicates finish grade.

crowstep: See *corbie gable.*

crumb: Very fine soil texture.

crumbing shoe: Attachment on a wheel trencher; retains loose dirt in the digging buckets. CRUMBER.

crushed gravel: Product resulting from artificial crushing of gravel with a minimum percentage of fragments having one or more fractured faces. See *coarse aggregate.*

crushed stone: Product resulting from artificial crushing of stone, rocks, boulders, or large cobblestone in which all edges are well-defined.

crusher-run aggregate: Aggregate that has been artificially crushed but not screened for size.

crushing strain: Compressive pressure that causes failure of a member. CRUSHING STRENGTH.

crush plate: Piece of wood fastened to the edge of a concrete form to protect the form from damage during stripping.

cryotemperature: Temperature below 120°K.

cryptoporticus: Enclosed gallery with walls consisting of openings or windows instead of columns.

cube: 1. Solid geometric figure consisting of six square sides. **2.** Number multiplied to the third power; e.g., $N \times N \times N = N^3$.

cube strength: Load per unit area at which a concrete test cube fails.

cubic content: Volume of a structure or area.

cubic feet per second: Measure of liquid flow calculated as flow (cubic feet) passing through a given point in one second.

cubic foot of gas: Volume of natural gas that occupies 1 cubic foot at 60°F and under pressure of 30″ of mercury.

cubic measure: Measure of volume in cubic units; e.g., 1728 cubic inches = 1 cubic foot, 27 cubic feet = 1 cubic yard, 231 cubic inches = 1 gallon, 128 cubic feet = 1 cord.

cul-de-sac: Street or alley with one outlet and a large circular area for turning around.

cull: Building material, such as lumber or brick, that has been rejected because of its low quality.

culvert: 1. Wood, metal, or concrete passageway under a roadway or sidewalk that allows for drainage. **2.** Small bridge with a minimum length of 20′-0″.

cumulative batching: Measuring two or more concrete ingredients in the same container by bringing the batcher scale into balance as each ingredient is introduced into the container.

cuneiform pile: Tapered and/or step tapered pile.

cup: Convex or concave deformation across the grain of lumber. See *defect*.*

cup base: Device used to secure the base of a Lally column in position.

cupboard: Enclosed storage area, especially one in a kitchen.

cup brush: Coarse-bristled metal wheel used with an electric drill to clean metal surfaces.

cup chuck: Lathe accessory with a deep recessed face; used to secure lumber in the lathe.

cup esutcheon: Plate with a recessed area that provides a fingerhold. May contain a ring that is flush with the surface of the door.

cup joint: Straight joint between two sections of copper tubing. One end of a section is flared to accept the end of the other section. Molten solder is used to fill the void between the two sections.

cupola: Small ornamental structure or dome projecting above the ridge of a roof.

* **cup point:** Indentation in the point of a setscrew.

cup shake: Wood defect in which annual rings separate from each other to create a semicircular flaw. WINDSHAKE.

curb: 1. Raised rim along the edge of a roadway. **2.** Wall plate that supports a dome at the springings.

* **curb box:** Vertical section of pipe extending from a curb cock to grade level and fitted with a removable cover. Provides access to the curb cock. BUFFALO BOX.

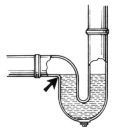

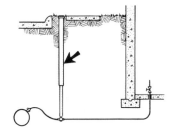

| crown (5) | crown weir | cup point | curb box |

curb edger

cusp

curb cock: Control valve in a water service pipe installed between a curb and the structure. Regulated by means of a long key that is inserted through the curb box. Used to turn off water supply. CURB STOP.

** **curb edger**: Hand tool with a curved edge used to impart the desired shape to the edge of curbs, sidewalks, and driveways.

curb level: Elevation of a street curb.

curb plate: See *curb* (2).

curb roof: See *mansard roof*.

curb shoe: Grader blade attachment that allows the blade to form a shape similar to a street curb.

curb stop: See *curb cock*.

curb-stop box: See *curb box*.

curb stringer stairway: See *closed stringer stairs*.

curdling: Thickening of varnish in the container.

cure: 1. Change in the properties of an adhesive or sealant as a result of chemical reaction. 2. See *curing*.

curing: 1. Maintenance of the proper moisture content and temperature in concrete during its early stages to obtain desired properties. 2. Hardening and final adhesion of glue.

curing blanket: Built-up covering placed over freshly placed concrete to maintain proper moisture content and temperature during hydration.

curing compound: Liquid chemical membrane applied to the surface of concrete to retard loss of water and improve the quality and durability of hardened concrete.

curl: Spiral or curved woodgrain.

current: Flow of electrons through an electrical circuit; measured in amperes. See *alternating current, direct current*.

current-carrying capacity: Maximum electric current a device is designed to carry continuously or over a predetermined amount of time.

curstable: Stonework course with cut moldings on the face.

curtail: Spiral termination of an architectural member, such as a volute.

curtail step: Lowest tread in a stairway that is rounded in a spiral or scroll shape on one or both ends and that projects beyond the newel post.

curtain grouting: Injection of grout or slurry into a subsurface formation to create an area of grouted material along a line in the same direction as the anticipated flow of water.

curtain wall: 1. Non-bearing exterior panel suspended on or supported by the primary structure. 2. Partition between chambers of an incinerator.

curtilage: Ground adjacent to a dwelling, such as a yard.

curved-claw hammer: Hammer with claws curved toward the handle and a tapered slot between the claws. Curvature of claw is used to facilitate pulling of nails. See *hammer**.

curved jaw pliers: Pliers with the tips of the jaws bent at an angle of 60° to 70° for reaching in tight locations. See *pliers**.

curved-tooth file: Hand tool with a serrated face with serrations in a curved pattern across the face. Used to smooth metal or hardwood surfaces. See *file**.

curvilinear: Consisting of curved lines.

cushion: 1. Cap that protects the pile hammer and head of a pile from damage resulting from the driving process. 2. Padding used to reduce the effects of vibration. 3. Stone that receives and distributes a vertical load.

** **cusp**: Intersection of small arcs or curves in tracery.

cut: 1. To sever with an sharp-edged instrument. 2. Depression resulting from the excavation of an area. 3. Excavated material.

cut and cover: Pipe-laying operation in which a trench is cut into the surface, pipe set in place, and the trench refilled with the excavated material.

cut and fill: Excavating operation in which material is removed from one area and placed in another area.

cutback asphalt: Bituminous material thinned with solvent and applied without using heat. Used for sealing and dampproofing roofs, concrete, and masonry.

cut nail: Wedge-shaped nail with a blunted end used in heavy timber or masonry construction. Nail is cut from sheet steel. See *nail**.

cutoff blade: See *crosscut blade*.

cutout: Manual or automatic electrical device used to interrupt flow of current through equipment or apparatus.

cutout box: Surface-mounted electrical enclosure that houses fuseholder blocks, fuses, and circuit breakers. Swinging door or cover is used to enclose the box.

cut stone: Building stone that is shaped to a specific dimension for installation in a specific location.

cutter: Soft brick that is cut to desired shape with a trowel and rubbed smooth.

cutter mattock: Tool used to loosen hard surface materials. Has a long handle with two flat wedge-shaped edges on the head that are perpendicular to each other.

cutting attachment: Device used to convert an oxyfuel gas welding torch to an oxygen cutting torch.

cutting diagram: Layout used in cabinetmaking to show how several small pieces will be cut from one large piece of lumber with minimal waste.

cutting gauge: Hand tool with an adjustable stop similar to a marking gauge. Used to cut thin stock.

cutting in: To carefully paint around the edges of an object or area.

cutting list: See *cutting diagram*.

cutting nozzle: See *cutting tip*.

cutting oil: Lubricant applied to metal during drilling or threading to reduce the ambient temperature.

cutting plane line: Heavy broken line on a drawing with perpendicular arrows and letters or numbers at each end. Used to indicate the plane at which a cross section is taken and the direction of viewing the object.

cutting pliers: Pliers with flat jaws having sharp edges.

cutting screed: Sharp-edged concrete tool used to trim excess shotcrete to the proper shape.

cutting shoe: Metal reinforcement placed at the tip of an open-end pile or caisson.

cutting tip: Point at which gas is released from an oxygen cutting torch.

cutting torch: 1. Device used to direct the preheating flame produced by the combustion of fuel gases and control cutting oxygen. 2. Device used for plasma arc cutting, which controls the position of the electrode, transfers current to the arc, and directs the flow of plasma and shielding gases.

cut washer: Flat steel disc with a hole in the center. Used to distribute pressure of a fastener head or nut to the surrounding surface. See *washer*.

* **cycle:** Interval of time in which one set of events occurs; e.g., the flow of electric current in an alternating current circuit.

cycle rate: Frequency at which certain events are repeated.

cyclic loading: Method of testing piles in which loads are imposed and removed at regular intervals. Used to determine bearing capacity of the pile.

cycling life: Number of times a device can perform a similar function before it fails.

cyclone cellar: See *basement*.

cyclone collector: Conical container that collects and separates particles in the air by centrifugal action. Commonly used in industrial exhaust systems.

cyclopean concrete: Mass concrete in which large stones, each exceeding 100 pounds, are embedded in concrete as it is placed.

cyclostyle: Structure built of a circular design of columns without a core.

cylinder: **1.** Solid geometric figure formed with two circular ends and joined with a parallel plane. **2.** Assembly in a lock that contains the tumbler and keyway. **3.** See *test cylinder*. **4.** See *gas cylinder*.

cylinder lock: Door lock consisting of a cylinder containing the locking mechanism separate from the lock case. See *lock*.

cylinder ring: Washer installed under the head of a cylinder lock that allows the installation of a longer cylinder in a thin door.

* **cyma:** Molding with a double-curvature profile.

cymatium: Crowning molding of an entablature in the form of a cyma.

cyrostyle: Curved projecting portico supported by columns.

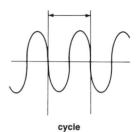

cycle

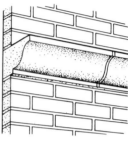

cyma

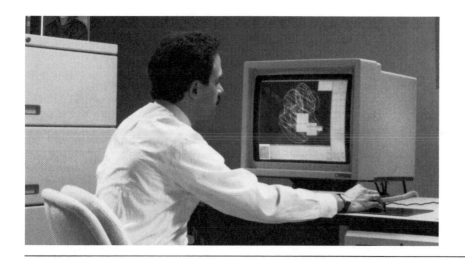

dab: See *daub*.

dabber: Soft bristle brush used to apply varnish.

dado: 1. Groove across the grain of a member that receives another piece. Commonly used at a 90° intersection of two members. See *joinery**. 2. Vertical face of a pedestal between the base and surbase or cornice.

* dado head: Power saw blade used to cut dados or rabbets. Commonly consists of two identical saw blades separated by at least one chipper.

dado rail: See *chair rail*.

dais: Raised speaking platform; head table.

dallage: Pavement consisting of marble, stone, or tile.

damp: Moderate absorption or covering of moisture.

damp course: Horizontal layer of waterproof material used to prevent moisture from the ground or lower course of masonry from entering the structure. DAMP CHECK.

* damper: 1. Movable assembly used to regulate flow of air in a flue or duct. 2. Device used to reduce vibration of equipment.

damper actuator: Automatic control device used to adjust the position of a damper.

damp location: Partially protected exterior location subjected to moderate amounts of moisture, such as under canopies or roofed open porches. Also includes interior locations such as basements or cold-storage facilities.

dampproof course brick: Masonry unit with water absorption less than 4.5% by weight. DPC BRICK.

dampproofing: Treatment of mortar or concrete to prevent passage of water or water vapor. Available as a coating for exposed surfaces, admixture to concrete or mortar, or plastic film.

dancing stair: See *circular stairway, elliptical stairway*.

dap: Notch cut in a structural timber to receive another member.

darby: 1. Tool used to smooth concrete surfaces immediately after screeding. DERBY, SLICKER. 2. Tool used to smooth or float plaster brown coat. Also used to apply a finish coat to obtain level surfaces.

dash-bond coat: Thick slurry of portland cement, sand, and water that is splashed onto a surface to provide a base coat for plaster. May also be used as a finish coat.

data plate: Information plate required on all safety valves. Includes manufacturer's name or trademark, manufacturer's design or type number, seat diameter, popping pressure setting, blowdown pressure, capacity, lift of the valve, year built, ASME symbol, and serial number.

datum: Reference point to which other elevations, angles, or measurements are related.

datum point: Established, permanent elevation reference. Indicates the height above mean sea level.

daub: 1. To create a texture by applying an irregular, rough coat of plaster.

2. To dress the surface of a stone with a scaling hammer that finishes the surface with small holes.

day: One division of a large window.

daylight glass: Blue glass used with an incandescent lamp to simulate characteristics of daylight.

D-cracking: Series of cracks in concrete along the edges or joints; structural cracks.

dead: Electrical conductor that is not connected to a power source. DEENERGIZED.

dead-air space: Unventilated space between structural members and/or finish materials.

dead blow hammer: Striking tool with the head made of steel pellets encased in a plastic coating. This design prevents rebounding of the head. See *hammer**.

deadbolt lock: Door bolt that is operated from both sides by a key or turn piece. See *lock**.

* dead end: 1. Section of pipe extending from soil, waste, or vent pipe and is closed at the end with a cap, plug, or other closed fitting. Extends at least 2'-0" from the nearest connection. 2. Termination of a rope. 3. End of a prestressing tendon opposite the jacking end.

dado head

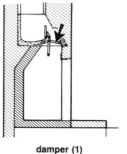

damper (1)

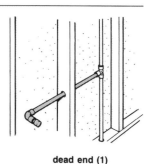

dead end (1)

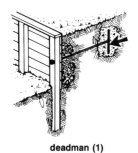

deadman (1)   deep seal trap   deformed bar

deadening: Use of insulating and damping materials to restrict transmission of sound.

dead front: Electrical equipment with no live exposed parts on the operating side.

dead knot: Knot that has lost its fibrous connection with the surrounding wood. Can be easily removed.

dead latch: Door lock with a spring bolt that operates independently of the primary lock. Operated from the outside by a key and from the inside with a knob or handle. NIGHT LATCH.

dead load: Permanent, stationary load composed of all building material and fixtures and equipment permanently attached to the structure.

dead lock: Door lock incorporating a deadbolt.

* **deadman: 1.** Anchorage for a guy line, retaining wall, or other vertical member. Consists of a log, rod, or concrete block that is buried in the soil and uses its own weight and resistance of the soil to secure the vertical member in position. **2.** Anchorage installed during assembly of sections of underground piping to prevent the pipe from separating. Remains in position when backfilling.

dead sand: Unwashed and ungraded fill sand.

dead shore: Temporary vertical timber used to support a dead load during structural alterations to a building.

deal: Fir or pine cut to minimum dimensions of 3″ wide and 9″ wide.

decarburization: Loss of carbon from the surface of carbon steel when it is heated.

decay: Disintegration of wood fiber caused by fungi.

decenter: To lower or remove shoring or centering.

decibel: Measure of sound intensity.

deciduous: Shedding leaves annually. Includes most hardwood trees and a few softwood trees.

decimal: Fraction or mixed number in which the denominator is a power of 10; e.g., $^1/_{10}$ = .1, $^1/_{100}$ = .01, $^1/_{1000}$ = .001.

decimal equivalent: Fraction or mixed number expressed as a decimal; e.g., $^1/_4$ = .25, $3^5/_8$ = 3.625.

deck: 1. Surface to which a roof covering is applied. **2.** Flat, open platform, commonly wood. **3.** Flooring of a structure. **4.** Horizontal form on which a concrete floor or roof is placed. **5.** Concrete floor or roof.

deck clip: See *panel clip.*

decking: 1. Light-gauge metal sheets used to construct a floor or deck form, or used as floor or deck members. **2.** Separation of explosive charges with an inert material. **3.** Wood members used as finish surface of a deck.

deck paint: Enamel paint with high abrasion resistance. Used in high wear areas.

deck roof: See *flat roof.*

declination: Downward bend.

declination arc: Graduated arc attached to a surveyor's transit with high power of magnification. Provides a correction factor for the natural downward arc of the earth.

decoupling: Leaving voids between building materials or members to limit sound and thermal transmission.

deenergize: To disconnect from a power source.

* **deep seal trap:** P-trap with a minimum water seal depth of 4″. Provides additional protection against sewer gas backup. ANTISIPHON TRAP.

* **defect:** Irregularity in a material that reduces its strength, durability, and/or appearance.

deflagration: Rapid combustion of a material.

deflection: Deformation or deviation of a member as a result of its own weight, an applied force, or other stress.

deflector: Fixed or adjustable device used to change the direction of air flow.

deformation: Modification in the shape or form of a member without rupture or break in continuity of the member.

* **deformed bar:** Rebar with surface ridges that improve the bonding capabilities with the surrounding concrete.

deformed reinforcement: Metal bars, wire, or fabric with surface ridges that improve the bonding capabilities with the surrounding concrete.

degradation: Deterioration of paint film as a result of weathering.

degrade: To reinspect and reduce the grade of a lumber product from its original classification.

degree: 1. Angular measure equal to $^1/_{360}$ of the circumference of a circle. Denoted with °; e.g., 45°. **2.** Measure of thermal energy. Measured in Fahrenheit or Celcius. Denoted with °; e.g., 70°F, 25°C.

degree-day: Unit of measure used to estimate fuel consumption and specify the nominal heating or cooling load of a building. Equal to the difference between the mean temperature and 65°F in a 24-hour period.

degree-hour: Measure of concrete strength gain expressed as the product of the temperature and a specific time interval.

degree of curve: Measure of the curvature in roadwork. Equal to the angular measurement at the center of a circle enclosed by a 100′-0″ chord length. Greater angular measurement results in a longer, smoother curve.

degree of saturation: Ratio of the volume of water in soil to the volume of the voids. Expressed as a percentage.

dehumidifier: 1. Air conditioner that cools and dehumidifies by passing air over cooling coils in which the surface temperature is below dew point. **2.** Equipment used to extract excess moisture from the atmosphere.

dehydrate: To remove moisture from the atmosphere by means of absorption or adsorption.

deicer: Chemical used to melt ice and snow. Common deicers are sodium and calcium chloride.

delamination: Separation of plies or pieces of reconstituted wood due to adhesive bond failure.

deliquescence: Absorption of moisture from the air by certain areas in plaster and brick. Results in damp, dark spots on the finish.

deluge sprinkler system: Automatic dry-pipe sprinkler system that has open sprinklers connected to a water supply valve. Valve is activated by smoke- or heat-sensing devices.

delustered yarn: Low luster yarn used in carpet with a low sheen.

demand: Electrical load over a specified interval of time. Expressed as kilowatts, kilovolt-amperes, kilovars, and/or amperes.

demand factor: Ratio of the maximum

demand of an electrical system, or part of the system, to the total connected load or the part of the system under consideration.

demolding: Removal of molds from concrete test samples and precast products.

demolition: Organized destruction of a structure.

demountable partition: Movable non-bearing wall; may extend from floor to ceiling or partial height. Walls are moved without structural alterations to the building. Commonly used in office areas.

den: Indoor recreational or work area.

dense-graded aggregate: Graded aggregate that produces few voids and maximum weight when compacted.

density: Quantity per unit volume; e.g., pounds per cubic foot, persons per acre.

denticular: Decorated with an alternating pattern of voids and small rectangular blocks. DENTICULATED.

dentil: One section of a dentil mold.

dentil mold: Trim member with an alternating pattern of voids and small rectangular blocks. See molding*. DENTIL BAND.

depolished glass: Glass with a diffused surface; commonly formed by blasting or etching.

deposited metal: Filler metal added to the base metal during a welding operation.

depot: 1. Storehouse. **2.** Shelter for passengers and freight; railroad depot.

depressed ceiling panel: Recessed ceiling design produced by furring down the perimeter of a room. COFFERED CEILING.

depreter: Stucco finish formed by embedding pebbles in the plaster.

depth gauge: Tool used to measure the depth of recessed areas, such as the depth of a hole or groove. Consists of a narrow graduated rule that slides through a crosspiece.

deriliction: Reclamation of a land from a body of water.

derrick: Hoisting device consisting of a mast with a boom hinged at the bottom. Boom is raised and lowered with a block and tackle or winch. See A-frame.

desiccant: Absorbent or adsorbent material that removes moisture from a material. Used in refrigeration systems.

desiccate: To remove moisture from a material.

design load: 1. Maximum load that structural members are designed to support. FACTORED LOAD. **2.** Maximum heat load an HVAC system is designed to handle.

design strength: Assumed load-bearing capacity of a structural member. Determined by multiplying the nominal strength of the member by a strength reduction factor.

design working pressure: Maximum allowable working pressure for a component of a pressurized system. Expressed in pounds per square inch.

detached: Not connected.

detail drawing: Drawing showing a small part of a plan, elevation, or section view at an enlarged scale. Used to provide information not clearly shown in another part of the drawing. See print*.

detention door: Heavy steel door with fixed lights protected by steel bars. Used in prisons and mental institutions.

detention screen: Heavy-gauge stainless steel wire woven into a mesh and reinforced with a steel frame. Used as a window screen in prisons and mental institutions.

detention window: Narrow metal awning window approximately 6″ to 8″ wide. Used in prisons and mental institutions.

deterioration: Decomposition and failure of a material, resulting in diminished quality, appearance, or strength. Includes pitting, flaking, scaling, cracking, and spalling. Caused by environmental or internal characteristics of the material.

detonating cord: Flexible cord with an explosive core. Used to activate a detonator.

detonator: Device such as a blasting cap used to activate explosives.

detritus: Deposit of loose material produced by the disintegration of rock.

detrusion: Splitting of wood parallel with the grain.

developed distance: Shortest lineal distance between two points through which free air travels.

developed length: Length of a pipe measured along the center line of the pipe and fittings.

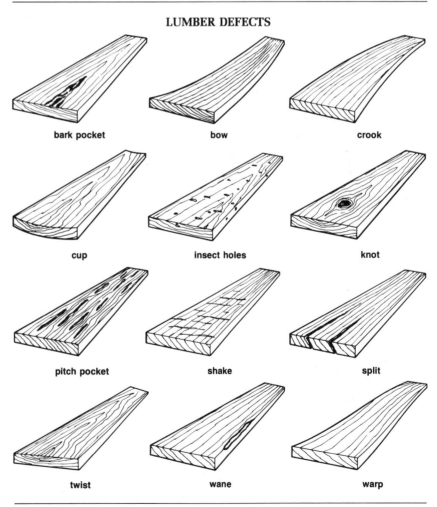

LUMBER DEFECTS

bark pocket   bow   crook

cup   insect holes   knot

pitch pocket   shake   split

twist   wane   warp

development length: Length of embedded concrete reinforcement required to develop the design strength of the reinforcement in a critical area.

device: Unit in an electrical system designed to conduct, but not use, electrical energy.

devil float: Wood float with two nails projecting from the base. Used to roughen a plaster surface to provide a key for the next coat. DEVIL, NAIL FLOAT.

dew: Water condensed on cold surfaces.

dewater: To remove water from a job site with pumps or drainage systems.

dew point: Temperature at which air is saturated with water vapor and at which the vapor begins to condense. Relative humidity and atmospheric pressure also affect dew point. DEW POINT TEMPERATURE.

diagonal: Inclined straight line or member extending across a surface.

diagonal bond: Brickwork pattern in which every sixth course is a header, with the bricks laid diagonally to the face of the wall. Used for thick masonry walls.

diagonal slating

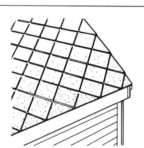

die (1)

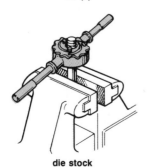

die stock

diagonal brace: Reinforcing member attached diagonally to a rectangular frame to give additional lateral strength.

diagonal crack: Inclined crack in concrete caused by shear stress.

diagonal cutting pliers: Pliers used to cut wire and light-gauge nails and bolts. See *pliers**. DIAGONAL PLIERS.

diagonal rib: Structural member that transversely crosses a bay of a Gothic vault.

* **diagonal slating:** Method of laying flat slate tiles with the diagonals of the tiles placed horizontally. DROP-POINT SLATING.

diagrammatic index: Surveying or plat index that indicates the arrangement of sheets or maps.

dial: Rotating plate or disk used as a control device.

dial lock: See *combination lock.*

dial saw: Adjustable tool used with an electric drill to cut holes of various diameters.

diameter: Distance across the circumference of a circular figure passing through the center point.

diamicton: Hollow masonry wall filled with rubble.

diamond blade: Circular steel blade with small diamond chips bonded to the surface. Used to cut tile, brick, concrete, concrete block, and stone. See *blade**.

diamond drill: Drill with small diamond chips bonded to the cutting edge. Used to bore holes in rock to create blasting holes.

diamond file: File with a diamond-shaped cross section. Used in tight locations such as corners.

diamond fret: Decorative molding composed of intersecting fillets, forming a series of diamonds or rhombuses.

diamond-matched veneer: Veneer that is cut diagonally and fit together to form a diamond pattern.

diamond mesh metal lath: Expanded metal plaster base used for flat, ornamental, and contoured shapes.

diamond tool: See *V tool.*

diaphragm: 1. Flexible divider between two areas. 2. Structural sheathing material applied over framing to create a load-bearing unit. 3. Steel plate between two structural steel members that increases rigidity of the members.

diaphragm wall: Underground concrete retaining wall.

diastyle: Space between two adjacent columns that measures three times the diameter of the column. See *intercolumniation.*

diathermanous: Substance or space that

allows passage of radiant heat.

diatomaceous earth: Fine siliceous material primarily composed of fossil remains of small marine life. Used as a paint extender, concrete aggregate, and abrasive material.

diazo-dry print: Print with blue line work on a light-colored background made by passing light through a transparent or translucent medium to a light-sensitive paper. Opaque images on the medium do not allow the light to pass, creating an image on the paper. The paper is fed through a developer that exposes it to ammonia vapors.

diazo-moist print: Print with black line work on a white background made by passing light through a transparent or translucent medium to a light-sensitive paper. Opaque images on the medium do not allow the light to pass, creating an image on the paper. The paper is fed through a developer that dampens the sensitized side with a developing solution.

dichroic reflector lamp: Incandescent light with an integral filter used to simulate natural light.

* **die:** 1. Tool used to cut external threads. 2. Portion of a pedestal between the base and cap.

die cast: Casting formed by forcing molten metal into a mold.

dielectric: Nonconductive of electrical current.

dielectric strength: Maximum amount of voltage a material can withstand prior to failure.

* **die stock:** Tool with two handles at 180° to each other and a central opening for the insertion of a die. Used to secure the die in position and apply pressure to cut external threads.

differential hammer: Pile hammer that uses steam, compressed air, or high-pressure hydraulic fluid to raise the ram and accelerate the fall.

differential hoist: Chain hoist with two sheaves at the top and one sheave at the load.

differential leveling: Method of determining the difference in elevation between two points using a level and rod.

differential pressure control: Method of maintaining a specified pressure difference between two separate pipelines or areas.

differential thermostat: Device used to control solar and water heating systems by determining the water temperature at the collector outlet and comparing it to the water temperature in the storage tank.

diffused lighting: Lighting obtained from many sources and/or directions.

diffuser: 1. Grill at the end of ductwork used to disperse flow of air. **2.** Device used to redirect light from a source. **3.** Perforated steel plate through which air is pumped into sewage to introduce small air bubbles.

diffusion brazing: Brazing process in which metal is joined by heating the metal to a suitable temperature and using a filler metal. Filler metal is distributed by capillary action or placed and formed along the faying surfaces. Pressure may or may not be applied.

diffusion welding: Solid-state welding process in which metal is joined by applying pressure at an elevated temperature. Filler material may or may not be used.

digitizing tablet: Electronic board that a drawing is placed on and converted to electronic data. Drawing is digitized (traced) to identify reference points for the computer-aided design and drafting system.

dike: Barrier that retains water. May be earthen or stone.

dilation: Expansion of concrete during cooling or freezing, calculated as maximum deviation from normal thermal contraction.

diluent: Liquid or solid substance mixed with active ingredients of a formulation to increase the bulk or lower the concentration.

dimension: Measurement; the distance between two points.

dimensional lumber: Lumber that has not been surfaced at the mill. Dimensions of the lumber are actual size; e.g., a 2 × 4 measures 2″ × 4″. DIMENSION LUMBER.

dimensional stability: Degree to which a building material retains its original shape despite changes in ambient and loading conditions.

dimension line: Line used on a print or drawing to indicate the dimension of a part or member. Dimension is noted above the line or at a break in the line. Extension lines are placed at both ends of the dimension line.

dimension shingles: Shingles cut to a uniform size for use as a wall or roof covering.

dimension stone: Stone cut and shaped according to specifications for a particular use, such as building stone. CUT STONE.

diminished arch: Arch with less than a semicircular rise or height. SKEEN ARCH, SKENE ARCH.

dimmer: Electrical control device used to vary the intensity of a light source by regulating the electric current.

* dimple: Slight depression made by the face of a hammer when used to drive nails into gypsum drywall. Later filled with joint compound to conceal nail heads.

dinging: Single rough stucco coat, commonly scored to imitate a mortar joint design.

dining room: Area primarily used for eating meals.

diorama: Large three-dimensional painting or series of paintings.

dip brazing: Brazing process in which heat required to join metal is produced by a molten chemical or metal bath.

* dip of a trap: Lowest point on the inner top surface of a trap.

dipper: Digging bucket attached to the arm of excavation equipment, such as a backhoe or power shovel.

direct-acting thermostat: Control device that activates a control circuit when a predetermined temperature is reached.

direct cross-connection: Continuous cross-connection that allows water to flow between two different systems with a slight pressure differential.

direct current (DC): Electrical circuit with unidirectional current.

direct current electrode negative: Arrangement of direct current arc welding leads in which the workpiece is the positive pole and the electrode is the negative pole.

direct current electrode positive: Arrangement of direct current arc welding leads in which the workpiece is the negative pole and the electrode is the positive pole.

direct drive: Method of transferring rotational motion in which the motor is attached directly to the device utilizing the motion.

direct dumping: Discharging concrete directly from a crane bucket or mixer.

direct heating: See *direct system.*

direction instrument: High precision theodolite in which the graduated horizontal circle is fixed to correspond to magnetic north during the surveying operation.

direct lighting: Lighting in which luminaires distribute 90% to 100% of the illumination toward the area to be illuminated. Commonly refers to light emitted in a downward direction.

direct load method: Method of testing the bearing capacity of piles by applying a load directly to the pile.

direct nailing: See *face nail.*

directory board: Informational sign with changeable letters, numbers, and symbols. Commonly installed in commercial and office structures to identify office locations.

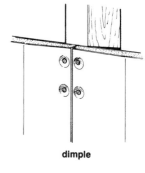

dimple

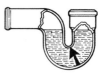

dip of a trap

direct return system: HVAC or refrigeration system in which the heating or cooling medium is passed through a radiator or convector and conveyed directly back to the source using the shortest possible path.

direct system: HVAC or refrigeration system in which heat is directly exchanged with the surrounding material or area.

disappearing stair: See *folding stair.*

disc: 1. Equipment with thin steel wheels that is pulled across the ground to cut into soil and roll it over. Used to aereate and mix soil. **2.** Flat circular design.

discharging arch: Arch constructed above the lintel of a door or window to distribute the imposed load of the wall above to both sides of the opening. RELIEVING ARCH, SAFETY ARCH.

discoloration: Variation from original color.

disconnecting means: Device or group of devices used to disconnect conductors of an electrical circuit from the energy source.

disconnecting switch: Switch used to isolate equipment or a circuit from the energy source. DISCONNECTOR, ISOLATOR.

discontinuous: Without connection; disjointed.

disc rasp: Circular rasp attached to a central arbor that is mounted in an electric drill. Used to smooth wood surfaces.

disc sander: Portable electric sander with a rotating, circular, abrasive disc. Used for smoothing and polishing surfaces of various materials.

disc seat: Section of a valve which, along with the disc, forms a watertight seal to prevent passage of fluid.

divided light

dog iron

dishwasher: Appliance used for washing dishes.

disk sander: See *disc sander*.

dispersant: Admixture used to keep finely ground materials in suspension. Used as a slurry thinner or grinding compound.

dispersing agent: Admixture that increases fluidity of mortar or concrete by reducing the attraction between particles.

displacement pile: Vertical foundation support driven into the ground and that displaces soil by compaction or vertical and/or horizontal shifting of the soil.

display model: Three-dimensional scaled representation of a structure in its finished form. Commonly includes landscaping and other details.

disposal: 1. Plumbing appliance used to grind and reduce size of food waste before it is discharged into the sanitary waste system. DISPOSAL UNIT, GARBAGE DISPOSAL. **2.** Removal and placement of undesirable material in a waste containment area.

disposal field: See *absorption field*.

disposal point: Location at which a building drain enters a public waste system.

dissolve: To liquefy or become fluid.

distance separation: Distance specified by fire prevention codes between an exterior wall and an adjacent building, a property line, or center line of an adjacent street.

distemper: Interior paint in which the pigments are mixed with a glutinous base such as sizing or glue.

distress: Cracking and distortion of a structure caused by stress and/or chemical action.

distressed finish: Wood finish with small scratches that imitate wear.

distributed load: Weight or stress acting equally across a bearing member.

distribution box: Concrete container installed in a septic system from which effluent is channeled into the absorption field.

distribution center: Point in an electrical circuit used for the control and protection of feeders or branch circuits. DISTRIBUTION BOX.

distribution cutout: Electrical fuse or disconnecting device that protects a circuit against damage from excess current. Primarily installed on distribution feeders and circuits.

distribution field: Portion of an absorption field that conveys liquid from the septic tank.

distribution panelboard: Electrical panel or group of electrical panels assembled into a single unit that is installed in a cabinet or cutout box accessible from the front only. Includes buses with or without switches and/or automatic overcurrent protection devices. DISTRIBUTION PANEL.

distribution switchboard: Switchboard used to distribute electrical energy at a common voltage for use within a structure. Fuses, air circuit breakers, or knife switches are used for circuit interruption.

distribution system: 1. Part of a cooling tower that distributes circulating hot water to air contact points. **2.** Section of an electrical system that delivers electrical energy from transmission points.

distribution tile: Tile or pipe laid in an absorption field to distribute effluent from a septic tank. Includes cylindrical concrete or clay tile, and perforated or permeable plastic pipe.

distyle: Having two columns in front.

ditch: Long, open trench or channel used for drainage, irrigation, or installation of underground utilities.

ditching machine: Excavation equipment used to cut a ditch or trench. Removes soil by means of buckets or teeth mounted on a rotating continuous chain or cable. DITCHER.

diverter: Raised piece of metal installed on a roof to channel precipitation to a desired location.

diverter fitting: Pipe fitting that allows water to flow to a heating unit while providing hot water to another heating unit in the same system.

diverting valve: Three-way pipe fitting used to supply a single source of fluid to either of two outlets.

* divided light: Window composed of several panes of glass separated by a mullion and/or muntins.

dividers: Device with two hinged and pointed legs used to lay out equal distances, transfer measurements, or scribe a line.

divider strip: Narrow piece of plastic or noncorrosive metal used as screeds in terrazzo works and to separate terrazzo panels.

division: One of the 16 major divisions used in specifying and categorizing construction data. Developed by the Construction Specifications Institute.

division wall: See *fire division wall*.

dock: Raised platform used to facilitate the loading and unloading of trucks.

dock bumper: Resilient device attached to a dock to absorb the shock of truck impact and prevent damage to the dock.

docking saw: Heavy-duty saw used to cut and shape timber. PILE BUTT SAW.

dog: 1. See *dog iron*. **2.** Concrete formwork member used to secure a snap tie in position.

dog-eared fold: External sheet metal corner formed by folding rather than cutting.

* dog iron: Iron bar with the pointed ends bent at 90° to the body of the bar. Used to secure two wood members in position. DOG, DOG ANCHOR.

dog-legged stairway: Stairway with two or more flights and a 90° or 180° turn at each landing. Rails and balusters of successive flights are aligned vertically.

dog nail: Large nail or spike with a head that projects over one side of the shaft.

dog's tooth: Brick laid with corners projecting from the surface of the wall.

dogtooth: See *dog's tooth*.

dolly: Small cart or truck used to move heavy loads.

dolomite: Mineral consisting of calcium carbonate and magnesium carbonate. Commonly a constituent of limestone used for structural stone.

dolphin: Cluster of driven piles used to protect piers.

dome: 1. Vaulted roof structure spanning an area. Includes circular, hexagonal, octagonal, and other geometric patterns. **2.** See *pan*.

dome damper: Curved damper used to regulate air flow in a flue.

domestic sewer: See *sanitary sewer*.

dook: Wood insert embedded in a wall for attachment of finish members.

* door: Hinged, folding, or sliding assembly used to close an opening in a wall or cabinet.

DOORS

Types

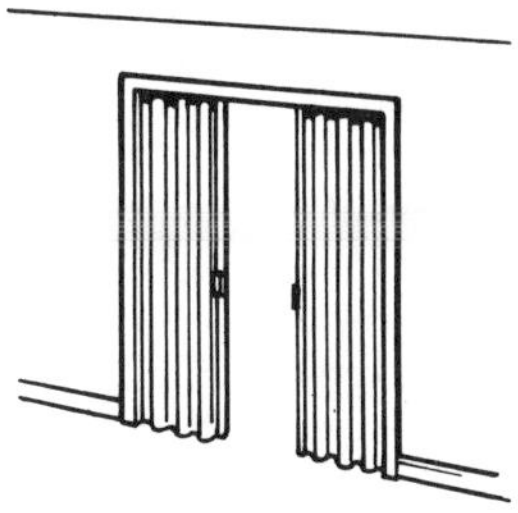

accordion

bi-fold

overhead

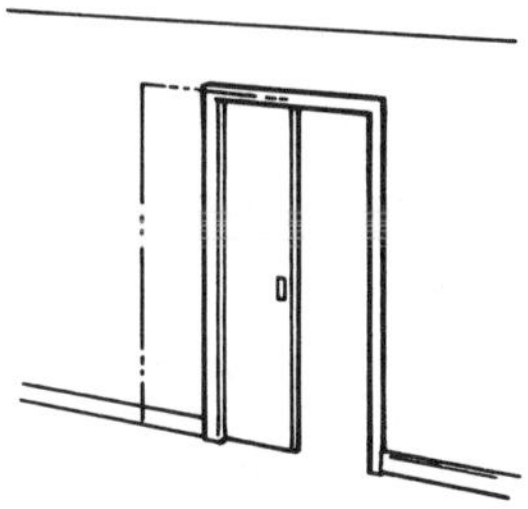

pocket

revolving

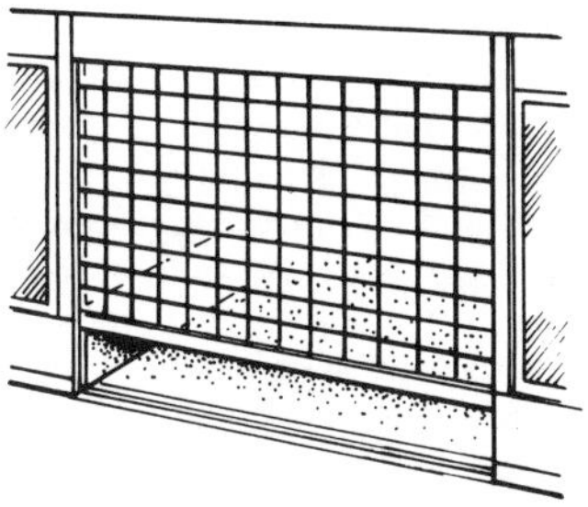

roll-up

sliding

swinging

Designs

crossbuck

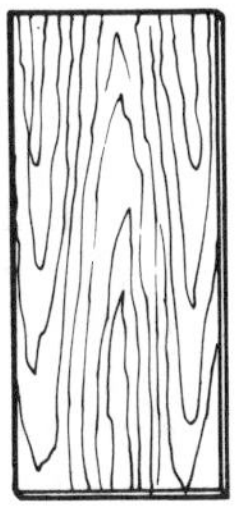

flush

French

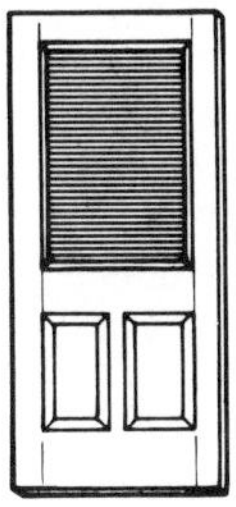

louvered

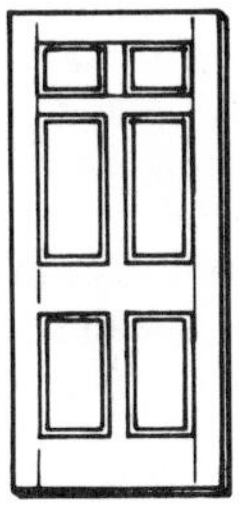

panel

Swing

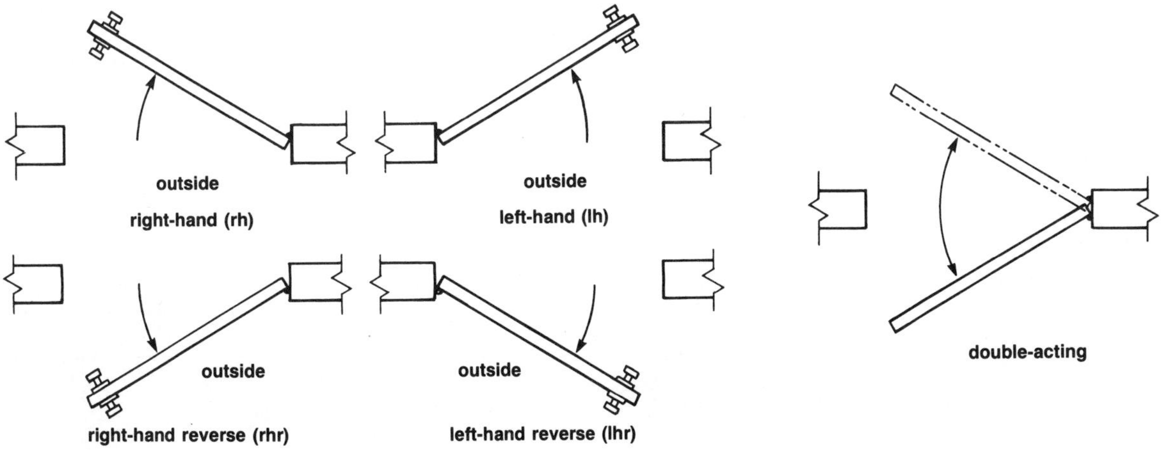

outside

right-hand (rh)

outside

left-hand (lh)

double-acting

outside

right-hand reverse (rhr)

outside

left-hand reverse (lhr)

doorbell: Signaling device in a structure that is activated by a control near an entrance.

door bolt: Manually operated sliding steel bar or rod in a frame attached to the inside of a door for security.

door buck: Wood or metal jamb set in a wall to receive a finished door frame.

door case: Finish frame around a door opening in a wall. Includes trim members and jamb.

door casing: Trim members concealing the space between the jamb and wall.

door check: See *door closer*.

door chime: See *doorbell*.

door clearance: Distance between a finished floor and the bottom of the door or between double doors.

* **door closer:** Device that closes a door and controls the speed and closing action of the door.

door contact: See *door switch*.

door dike: Resilient gasket along bottom edge of a door of a refrigeration compartment. Used as a barrier against heat flow into the compartment.

door frame: Framing and trim members that surround a door opening, including side and head jambs.

* **door guide:** Hardware fastened to the floor to keep sliding doors plumb.

door header: Horizontal structural member over the top of a door opening.

door holder: See *door stop* (2).

door jack: Device used to hold a wood door while it is being fitted with hinges or planed.

door jamb: Assembly of trim members along the sides and top of door opening.

doornail: 1. Large-headed nail that a door knocker strikes against. 2. Large-headed nail used as decoration or to strengthen a door.

door plate: Ornamental nameplate on exterior side of a door indicating occupant's name and/or address.

doorpost: See *door jamb*.

door pull: Handle used for opening a door.

door rail: See *rail*.

door schedule: Table that is part of the prints or specifications for a structure that gives size, type, location, and miscellaneous information about doors in the structure. Corresponding numbers or letters on the prints are cross-referenced to the schedule. See *schedule*.

door sill: See *threshold*.

doorstep: Riser and tread on the outside of an exterior door.

door stone: Piece of masonry used as a threshold.

doorstop: 1. Trim member that projects from the inside face of a jamb. Prevents a door from swinging too far. 2. Device used to hold a door open. 3. Hardware device used to prevent a door from swinging into a wall.

door strip: See *weatherstrip*.

door switch: Electrical contact mounted in or on a door frame that activates a circuit when a door is opened or closed.

* **door trim:** Casing around a door frame that conceals the joint between the jamb and wall.

door unit: 1. Door and jamb assembly. 2. Clear opening of a door for a required fire exit.

doorway: Opening in a wall with a door. Provides passage from one room to another.

dope: 1. Material applied to pipe threads to ensure an airtight joint. 2. Compound used to retard or accelerate setting time of mortar or plaster.

Doric frieze: Section of the entablature between the cornice and the architrave in Doric architecture.

Doric order: Oldest and simplest form of Greek architecture. Distinguishing feature is a column without a base.

dormant: Large supporting beam for smaller beams.

* **dormer:** Projection from a sloping roof, commonly including a window or louver. Provides additional interior area between two sloping roofs.

dormer cheek: Side wall of a dormer.

dormitory: Residential structure with sleeping facilities. Commonly on the

grounds of an institution such as a university.

dosing tank: Watertight tank installed between the septic tank and distribution box in a septic system. Pump or siphon automatically discharges sewage when a predetermined quantity has accumulated.

dot: Small spot of plaster applied at a thickness equal to the projection of the grounds to help the plasterer maintain a consistent thickness of plaster.

dote: Rotting or decay in wood. DOAT, DOZE.

dotted line: See *hidden line*.

doty timber: Timber that is beginning to decay. DOSY TIMBER.

double-acting butt hinge: Hinge mounted on a double-acting door that enables it to swing in both directions.

double-acting door: Door that swings in two directions. See *door**.

double-acting hammer: Pile hammer powered by steam or compressed air; used to raise the ram and accelerate its fall. Commonly has light rams and operates at high speeds.

double-acting spring hinge: Hinge mounted in the floor that acts as a pivot for a double-acting door. Two sets of springs are in tension in opposite directions to allow the door to move in two directions. Vertical pin is used as a pivot point at the top of the door.

double angle: Two L-shaped metal members joined back to back.

double back: To apply a second layer of plaster before the initial layer is dry.

double-bend fitting: S-shaped pipe fitting.

double-bitted ax: Tool with two cutting edges.

double-bitted key: Key that is toothed along both edges. Either edge can be used to move the tumblers.

double break: 1. Circuit breaker arrangement in which two circuit breakers are installed in series to reduce possibility of extended power outage caused by circuit breaker problems. 2. Condition caused by an electrical switch pro-

door closer

door guide

door trim

dormer

viding two air gaps in a single supply line.

double brick: Masonry unit that measures 4″ × 5$^1/_3$″ × 8″.

double bridging: Bridging that breaks the joist span into three sections.

* double corner block: Concrete masonry unit with solid rectangular ends and faces.

double cut: To cut two adjacent pieces of finish material by overlapping the pieces and cutting through the top and bottom pieces. Ensures a tight joint.

double-cut file: File with two sets of diagonal cutting ridges on the face of the file. Used for fast removal of surface material. Primarily used for metal. See *file**.

double-cut saw: Saw with teeth that cut on both the forward and backward strokes.

double door: Two double-acting doors hung in the same frame.

double door bolt: Security device with a vertical sliding bar at the top and bottom of a door. Bars move simultaneously in opposite directions to lock and unlock the door.

double egress frame: Door frame that receives two single-acting doors that swing in opposite directions. Commonly used in commercial food service establishments.

double-end tenoner: Production woodworking equipment with a fixed and a movable cutter, each containing a series of blades. Used to produce wood window and door components.

double-end-trimmed: Cut squarely on both ends.

double extra-strong pipe: Steel pipe with increased wall thickness to provide twice the strength of standard pipe.

double-faced hammer: Hammer that has a striking face on each end of the head.

double-Flemish bond: Brickwork pattern in which the inner and outer faces of a wall consist of headers and stretchers alternating in each course.

double framing: Using twice the standard number of framing members to provide additional structural strength.

double glazing: Two panes of glass with an air space between adjacent faces. Edges are sealed to provide thermal and sound insulation.

double-headed nail: Nail with two heads, one above the other. Lower head bears on the surface of the material into which it is driven, allowing access to the upper head for removal. Commonly used in concrete formwork and other temporary construction. See *nail**.

DUPLEX NAIL, FORM NAIL, SCAFFOLD NAIL, STAGING NAIL.

double header: Header over an opening consisting of two lumber members fastened together.

* double hem: Sheet metal edge formed by folding the edge over twice. Provides rigidity and reinforcement to the sheet metal.

double-hung window: Window with two vertically sliding sashes mounted one above the other in different tracks. Both sashes can be opened at the same time. See *window**.

double jack rafter: Roof framing member that extends from a valley to a hip. CRIPPLE JACK RAFTER.

double jack sledge: Heavy sledgehammer, weighing from 16 to 20 pounds. Used in excavation work.

double jointing: Application of mortar to both ends of a closure brick and to the ends of the opening provided for the brick.

double-layer application: Application of a face layer of gypsum drywall over a base layer of gypsum drywall that is directly attached to the framing members. Provides additional strength and better fire and sound resistance.

double-lock seam: Locking seam used in ductwork and metal roofing in which the adjacent edges of sheet metal are folded twice and dressed flat.

double-pitched: Having two pitches or slopes.

double-pitched roof: See *gambrel roof*.

double-pole switch: Switch with two blades and contacts for opening or closing both sides of an electrical circuit simultaneously.

double return stairway: Stairway with one flight extending from the main floor to a landing and two flights extending from the landing to the upper floor.

double-strength glass: Sheet glass with the thickness ranging between 0.113″ and 0.118″.

double-swinging door: See *double-acting door*.

double T-beam: Precast concrete member consisting of two vertical structural members with a top slab that projects beyond the vertical portions of the beams.

double-throw switch: Switch that can change circuit connections by moving the switch blade into one of two operating positions.

double vault: Vault consisting of an inner and outer shell.

double-welded joint: Weld joint produced by welding on both sides of the members.

double window: See *double glazing*.

doughnut: **1.** Large washer used to increase bearing area of bolts, nuts, and other hardware. **2.** Large washer used to position rebars a desired distance from forms.

dovetail: Interlocking joint consisting of angular teeth that are wider at the ends than at the base. Commonly used in 90° corners. See *joinery**.

dovetail anchor: Prefabricated metal tie with angular sides. Used to tie masonry to concrete walls. See *anchor**.

dovetail bit: Router bit used to cut a flat-bottomed recess with sloping sides (dovetail). Used for drawer construction and decorative work. See *bit (2)**.

dovetail cramp: Dovetail-shaped metal device used to lift masonry units.

dovetail dado: Recess perpendicular to the woodgrain in which base dimension of the recess is wider than the face. Strengthens the joint against perpendicular forces.

* dovetail half-lap joint: Joint formed by two members of equal thickness in which the end of one member is dovetailed and fitted into a corresponding mortise of a mating member. Dovetail and mortise are one-half the thickness of the members.

dovetail saw: Hand saw with fine teeth, a reinforced back, and a straight handle. Used to make narrow, straight cuts. See *saw**.

dovetail seam: Sheet metal seam used to join round or elliptical collars to flanges. Three types are plain, beaded, and flanged.

double corner block   double hem   dovetail half-lap joint

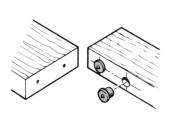

| dowel center (2) | doweling jig | downspout | drain tile |

dowel: **1.** Cylindrical wood or metal rod inserted into holes in adjacent members to align and strengthen a joint. **2.** Plain or deformed cylindrical steel rod that extends into adjacent sections of concrete construction. Used to transfer shear loads.

dowel bit: Bit with a semicylindrical section. Used with a brace to drill dowel holes.

* dowel center: **1.** Small circular button with small projecting points centered in each side. Used to lay out locations for dowel placement in two members simultaneously. **2.** Small circular button with a small projecting point centered in one side. Used to lay out dowel holes in one member by inserting the button into a predrilled hole in the other member. Manufactured in common dowel sizes.

* doweling jig: Device used to locate the center of a hole to be drilled and guide the drill bit into the workpiece. Available as self-centering or adjustable.

dowel joint: Wood joint that uses dowels as a means of reinforcement.

dowel lubricant: Material applied to a metal dowel to reduce the amount of surface adhesion and permit axial movement.

dowel pin: Deformed or barbed pointed metal pin used to fasten and reinforce mortise-and-tenon joints.

dowel screw: Metal dowel threaded on both ends for connecting wood members.

downdraft: Downward current of air.

down-feed system: HVAC or refrigeration piping system in which the supply mains are above the level of the individual heating, cooling, or refrigeration units.

downpass: Passage of gaseous substances between two incinerator chambers to carry combustion products downward.

downpipe: See *downspout.*

* downspout: Vertical pipe or sheet metal channel that conveys rainwater from the roof of a structure to the ground or into a sewer. CONDUCTOR, LEADER.

dozer: See *bulldozer.*

draft: **1.** Current of air flow resulting from imbalanced atmospheric pressure conditions. Includes balanced, forced, induced, and natural. **2.** Slight bevel produced when cutting a material. **3.** Narrow, smooth border around the face of stone. **4.** Force required to pull equipment.

draft hood: **1.** Device installed into or on top of a flue to prevent downdrafts. **2.** Enclosure over a gas-fired appliance that allows for the escape of flue gases and prevents a back draft.

drafting chisel: Chisel used to cut a narrow, smooth border around the face of a stone.

drafting machine: Drafting instrument that combines the features of a T-square, triangle, protractor, and scale. Commonly fastened to a drafting table.

drafting scale: See *architect's scale, engineer's scale.*

draft regulator: Device that admits air into breeching or flue connection to maintain a constant draft.

draft stop: Material installed in concealed open areas, such as in a frame wall, to prevent air movement. See *fireblock.* FIRE STOP.

drag: **1.** Hand tool with fine steel teeth used to dress the surface of stone. COMB. **2.** Amount of deflection in a jet of cutting gas as it passes through the base metal. **3.** Long-handled tool with teeth used to roughen the surface of plaster to produce a key for the next coat.

dragline: Bucket attachment for a crane that digs by pulling the bucket back toward itself by means of cables. Commonly used in soft material, such as marine or marsh work.

drag shovel: See *backhoe.*

drain: Pipe, channel, or trench used to convey waste water or waterborne waste.

drainage: **1.** See *drainage system.* **2.** Material conveyed by a pipe, channel, or trench.

drainage fill: **1.** Granular base course

placed under a concrete floor slab to prevent capillary rise of moisture. **2.** Lightweight concrete placed on a floor or roof to improve drainage of waste material.

drainage system: Pipes, channels, or trenches used to convey waste water or waterborne waste.

drainage system air break: See *air break.*

drainage system air gap: See *air gap.*

drainboard: Work surface adjacent to a sink with a slight pitch so as to drain into the sink.

drain cock: Valve or faucet installed at the lowest point in a tank or piping system to allow for removal of liquid. DRAIN VALVE.

drain field: See *absorption field.*

drainpipe: See *downspout.*

* drain tile: Pipe installed below ground to drain water from saturated soil or to convey and disperse fluids, such as in an absorption field. Includes perforated and non-perforated clay tile, porous concrete tile, and perforated and non-perforated plastic tubing.

drain trap: See *trap* (1).

drain valve: See *drain cock.*

drapery track: System of channels or tracks that supports draperies and allows them to be open and closed.

drawbar: Steel bar attached to the back of a tractor for attachment of equipment.

draw bolt: Sliding locking device used for securing a door. BARREL BOLT.

drawbore: Hole drilled in the tenon of a mortise-and-tenon which is not aligned with a hole in the mortise. When a pin is inserted and driven through the holes, the joint tightens.

drawbridge: Bridge that is raised or rotated to allow passage below it.

drawdown: Distance that groundwater elevation is lowered as a result of pumping, excavation, or dewatering.

drawer guide: Strip of wood, plastic, or metal on which a drawer slides when it is opened.

drawer pull: Handle or grip used to open and close cabinet drawers.

drawer roller: Prefabricated hardware used to facilitate opening and closing a drawer. Commonly consists of a metal or fiber wheel that moves in a metal track.

drawer stop: Device that prevents a drawer from falling out of the cabinet opening if pulled too far.

draw filing: Process in which a file is placed perpendicularly to the longest dimension of the workpiece and moved across the surface. Pressure is applied only on the forward stroke. Produces a smooth flat surface.

drawing knife: Hand tool with a hooked cutting blade used to cut thin sheet lead.

drawing up: Tightening a knot.

draw-in system: Electrical wiring system in which conductors are installed in conduits, raceways, and boxes, allowing for future removal without disturbing other structural components.

drawknife: Woodworking hand tool consisting of a curved or straight blade with handles that are perpendicular to the blade at each end. Used to smooth wood by pulling it toward the user. DRAWSHAVE.

drawn: Pulled through a single die or succession of dies to achieve a desired shape.

draw pin: Pin used to attach equipment to a drawbar.

draw tongue: See *drawbar*.

dredge: 1. To excavate underwater material. **2.** Equipment used for underwater excavation such as clamshell or power shovel connected to a suction line.

dress: To smooth the surface of a material, such as stone or lumber, by cutting, planing, or sanding.

dressed-and-matched: See *tongue-and-groove*.

dressed lumber: Wood in which one or more surfaces have been surfaced at a mill.

dressed masonry: Stone or brick with a smooth finish.

dressed size: Dimensions of lumber after sawing and planing; e.g., the dressed size of a 2 × 4 is $1^1/_2'' × 3^1/_2''$.

dresser: 1. Tool used to flatten sheet lead and straighten lead pipe. **2.** See *wheel dresser*.

dressing: 1. Projecting decorative molding. **2.** Smoothing and squaring lumber or stone. **3.** Liquid bituminous material applied as a protective covering over roofing felt. DRESSING COMPOUND.

dried strength: Compressive or tensile strength of refractory concrete within three hours after initial drying in an

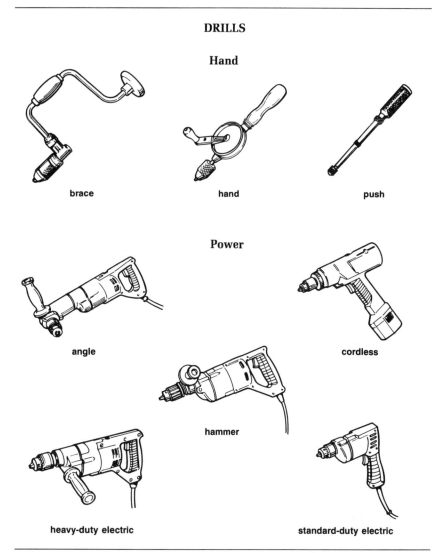

DRILLS

Hand

brace   hand   push

Power

angle   cordless

hammer

heavy-duty electric   standard-duty electric

oven at a temperature between 220°F and 230°F.

drier: 1. Substance that accelerates the drying of oil, paint, or varnish. **2.** Sifted cement and sand placed on the surface of fresh concrete flatwork to absorb surface water.

drift: 1. Deposit of loose material such as sand or rock. **2.** To enlarge a hole using a tapered shaft. **3.** Change in electrical characteristics of synchronous equipment caused by temperature changes or old components. **4.** Undesired change in the output of an electrical device occurring over a period of time. Expressed as a percentage of the maximum rated value of the component evaluated. **5.** Lateral deflection of a building. **6.** Entrained unevaporated liquid carried out of a spraying device by air movement. **7.** To move a suspended load horizontally with at least two pieces of hoisting equipment.

driftbolt: Rod or square bolt approximately 1′-0″ to 2′-0″ long driven into predrilled holes in timber to secure it to adjacent members.

driftpin: Tapered cylindrical steel dowel used to align and temporarily secure steel beams or other structural steel members.

drift plug: Hardwood plug driven into a lead pipe to straighten a bend.

drift punch: See *driftpin*.

* **drill: 1.** Manual or power-driven rotary tool used with a bit to bore holes in material. **2.** To bore a hole in a material.

drill bit: Accessory used with a drill to bore holes in wood, metal, or other material. Removes material from hole using a spiral flute.

drilled pile: Pile formed by boring into the ground to firm, bearing soil and placing concrete in the void. DRILLED-IN CAISSON.

drill press: Stationary equipment used to bore holes in material. Drill bit or other drilling accessory is inserted into the chuck and rotated using an integral electric motor. Workpiece is secured on a movable table mounted on a vertical column.

drinking fountain: Plumbing fixture that delivers a stream of potable water through a nozzle. Stream of water is pumped at an upward angle to facilitate drinking. Manufactured from vitreous china, enameled cast iron, and stainless steel.

drip: 1. Building member attached to the outside of a structure and designed to cause rainwater to run off. **2.** Groove along bottom surface of a sill or other horizontal exterior member. Used to divert rainwater away from the structure. **3.** Pipe or integral pipe and steam trap that conveys condensation from the steam side of a piping system to the water or return side of the system.

drip cap: Exterior molding with a sloped upper surface and a groove along the bottom surface that diverts rainwater away from the structure. Commonly installed above the exterior of a door or window. DRIP STRIP, WATER TABLE.

drip mold: See *drip cap.*

dripstone: Masonry member over a door or window that diverts rainwater away from the structure.

* **drive band:** Steel band secured around the top of a timber pile to prevent damage while being driven.

drive cap: Cushion between a pile hammer and a pile. Used to prevent damage to the pile while being driven. BONNET.

drive center: Piece inserted in the spindle of a lathe to hold the work in place. LIVE CENTER.

drive clip seam: Sheet metal joint in which the adjacent edges of two members are folded and secured with a C-shaped channel.

driven pile: Pile that is driven into its final position at the job site.

drive screw: See *screw nail.*

* **drive shoe:** Protective member placed on the tip of a pile to prevent damage

while being driven.

drive-through: Retail business, such as a bank, in which transactions occur while the customer remains in the vehicle.

driveway: Private access for vehicles.

driving band: See *drive band.*

driving home: Installing a fastener in its final position.

drop: 1. Vertical distance that an air stream falls from its initial elevation when leaving ductwork. Measured at the end of its throw. **2.** Vertical dimension from the finished edge of a cabinet lock to the center of the cylinder. **3.** See *gutta.*

drop apron: Metal strip bent downward around the edge of a roof.

drop box: Electrical outlet box that is suspended from above. Commonly used for temporary installations such as exhibit halls.

drop ceiling: See *suspended ceiling.*

drop chute: Device used to direct the fall of fresh concrete. Includes tapered metal cylinders and flexible canvas or plastic tubing.

drop cloth: Large sheet of plastic or cloth that temporarily covers and protects finished material or fixtures during construction.

drop cord: 1. Flexible electrical cable suspended from a ceiling. Commonly used in temporary construction, such as an exhibit hall. **2.** Flexible cable coiled around a spring-retracted drum mounted at ceiling level. Receptacle or other electrical device is attached to the lower end of the cable.

drop curtain: Curtain that moves vertically to conceal a theatrical stage from the audience.

drop door: Door on a piece of furniture that is hinged at the bottom and opens downward.

drop drawer pull: See *drop handle.*

* **drop elbow:** 90° pipe fitting with one or two ears on the side that allow the fitting to be secured to a wall or framing member. DROP ELL.

drop ell: See *drop elbow.*

drop escutcheon: Escutcheon with a piv-

oting metal plate that conceals a keyhole.

drop hammer: Heavy weight that moves freely between two vertical guide rails to exert force to the top of a pile. Used to drive piles for small projects and in remote locations.

drop handle: Drawer pull that pivots along an axis extending from the ends of the pull. Pull lays flat against the surface of the drawer when not used.

drop key plate: See *drop escutcheon.*

drop light: Suspended electric lampholder.

* **drop panel:** Thickened structural portion of a flat concrete slab in the area around a column or bracket. Used to distribute the imposed load.

dropped ceiling: See *suspended ceiling.*

drop ring: Ring used as a handle to operate a lock. Ring is flush with the surface of the door when not in use and raised and pivoted to operate the lock.

drop siding: See *horizontal lap siding.*

drop table: Hinged tabletop supported by brackets when lowered to its horizontal position.

drop tee: T-shaped pipe fitting with one or two ears on the side that allow the fitting to be secured to a wall or framing member.

drop-thru: Undesirable depression or irregularity in the surface of metal when welding or brazing. Caused by overheating or alloying between the filler material and base metal.

drop vent: Individual vent that connects to a drain or vent pipe at a point lower than the fixture.

drop window: Vertically sliding window that can be lowered into a pocket below the sill. Allows the entire window to be opened for ventilation.

drop wire: Electrical cable or conductor that extends from a utility pole or cable terminal to a structure.

drove: Chisel with a blunted edge approximately 2″ to 4″ wide used to face stone. BOASTER.

drum: 1. Portion of a ready-mixed concrete truck that contains, transports, and agitates concrete. **2.** Welding

drive band

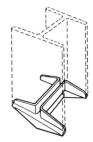

drive shoe

drop elbow

drop panel

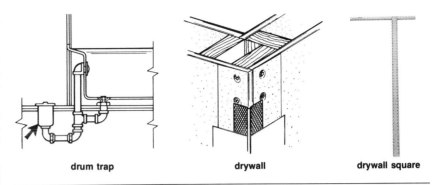

drum trap   drywall   drywall square

filler metal consisting of a continuous length of electrode coiled within a cylinder. **3.** Rotating cylinder that cable is wound around. **4.** Cylindrical section of a Composite or Corinthian column. **5.** Wall that supports a dome or cupola.

drum rasp: Cylindrical device with a rough, open circumference used with a drill to shape and smooth wood, plastic, and other soft material.

drum sander: Cylindrical device with an abrasive sleeve used to shape and smooth wood, plastic, or other soft material.

* **drum trap:** Cylindrical plumbing trap in which the body is in a vertical axis. Trap is closed on the bottom with a cleanout on the top. Commonly used as a bathtub trap.

dry-batch weight: Weight of all material in a batch of concrete, excluding water.

dry-bond adhesive: Adhesive material that is dry to the touch but adheres on contact; e.g., contact cement. CONTACT ADHESIVE.

dry bonding: Process of laying out masonry units without mortar to establish the bond pattern.

dry-bulb temperature: Air temperature as measured by a standard thermometer with an unmoistened bulb.

dry-bulb thermometer: 1. Standard thermometer, commonly filled with mercury. **2.** Thermometer in a psychrometer; thermometer has an unmoistened bulb.

dry cell: Cell of a battery in which the electrolyte is immobilized.

dry-cell battery: Type of battery that has a zinc cylinder, paste electrolyte, and central carbon rod.

dry clamping: Process of assembling all members of a cabinet or other assembly without using fasteners or adhesives to ensure that all joints fit together before final assembly.

dry construction: Construction using drywall, wood, or other materials that do not involve the use of water. Expedites construction and allows the structure to be occupied sooner.

dryer: See *drier*.

dry filter: Filter that removes undesirable material from the air by passing air through a series of screens.

dry glazing: Installation of glass using preformed material or gaskets without using putty or glazing compound. PATENT GLAZING.

drying oil: Paint component that absorbs oxygen and forms a hard, dry surface.

drying shrinkage: Contraction of a material resulting from loss of moisture. Commonly occurs in plaster and mortar.

dry joint: Masonry joint without mortar.

dry kiln: Chamber in which wood is seasoned and dried by applying heat and removing moisture.

dry lumber: Seasoned lumber with a maximum moisture content of 19%. May be dried using mechanical or natural means.

dry masonry: Masonry units laid without mortar.

dry mix: Concrete, mortar, or plaster mix that contains all ingredients except water. Commonly available in bags. **2.** Concrete with approximately zero slump.

dry mortar: Mortar that has enough moisture to set sufficiently but not enough to cause it to adhere to brick, block, or stone.

dryout: Condition of plaster in which moisture evaporates from it before setting. Produces a soft, powdery surface.

dry packing: Process of filling a void with zero slump or approximately zero slump concrete, mortar, or plaster and compacting it.

dry-pipe sprinkler system: Fire protection sprinkler system in which there is no water in the pipes until the system is activated. Used in areas where low temperatures produce a potential freezing hazard.

dry-pipe valve: Control valve that activates a dry-pipe sprinkler system. Must be protected from damage and freezing.

dry return: Pipe in a steam heating system that carries both air and condensation.

dry-rodded volume: Volume of dry aggregate that is compacted with a steel rod in a cylindrical container. Used to determine unit weight of aggregate.

dry-rodded weight: Weight per unit volume of dry aggregate that is compacted with a steel rod in a cylindrical container. Used to determine unit weight of aggregate.

dry rot: Decay of wood caused by fungi, which reduces the wood to a fine powder.

dry sampling: Method of obtaining an undisturbed test sample of earth by boring a hole using a boring tool that has a collection device at the top.

dry-shake: Mixture of hydraulic cement and fine aggregate that is worked into the top of concrete flatwork to impart desired surface qualities. Mixture may contain pigment to color the surface.

dry shrinkage: See *drying shrinkage*.

dry steam: Saturated steam that does not contain moisture in suspension.

dry strength: Strength of an adhesive joint immediately after drying under specified conditions.

dry-type transformer: Electrical transformer in which the core and coils are immersed in a gaseous or dry insulating medium.

* **drywall:** Interior surfacing material applied to framing members using dry construction. Fireproof gypsum core is encased with heavy paper on one side and liner paper on the other side. Sheets are 48″ wide with a variety of lengths in 24″ increments. Long edges are tapered for the application of joint compound and reinforcing tape. GYPSUM DRYWALL, SHEETROCK, WALLBOARD.

drywall compound: See *joint compound*.

drywall hammer: Striking tool with a symmetrical, serrated face used to drive fasteners into drywall and leave a dimple. Blade is used to pry and wedge drywall.

drywall mud masher: See *potato masher*.

drywall router: Portable electric power tool with interchangeable cutting heads used to make cutouts in drywall.

drywall saw: Short, straight-bladed hand saw with coarse teeth for piercing and cutting drywall. See *saw*.

* **drywall square:** T-shaped tool with a head approximately 22″ long and a blade 48″ long. Blade is used as a guide when cutting drywall.

drywall trowel: Hand tool with a handle attached to the back of a flexible steel blade. Used to apply joint compound and reinforcing tape to seams between sheets of drywall.

duckbill trowel   duplex lock   duplex receptacle

dry well: Covered pit in the earth that is filled or lined with coarse aggregate or other open-jointed material. Allows rainwater to accumulate and leach into the surrounding soil.

dual duct: Duct that is divided into two raceways for electrical conductors. Used to house two separate wiring systems, such as an intercom system and electrical power.

dual-duct system: 1. HVAC system in which two ducts are used to convey air; one conveys cool air and the other conveys warm air. Air from the two ducts converge at a mixing box and is then dispersed into the conditioned space.

dual-effect control: Control device that responds to thermostats in two different zones.

dual-element fuse: Fuse in which two current-responsive elements are connected in series.

dual-fuel system: Heating system in which two types of fuel can be used. Commonly fuel oil and natural gas.

dual thermostat: Thermostat with two predetermined settings: one for maintaining a daytime temperature and the other for maintaining a nighttime temperature.

dual vent: Vent pipe that connects two fixture branches and vents. COMMON VENT, UNIT VENT.

dub: To strike, rub, or dress to obtain a smooth surface finish.

* **duckbill trowel:** Masonry tool with a narrow blade and a rounded point.

duct: 1. Pipe or conduit used to distribute or return air in an HVAC system. Extends from the source to the register or other outlet or from the conditioned area to the source. **2.** Enclosed raceway for electrical cables or conductors. **3.** Enclosed channel in a concrete member for accepting a post-tensioning tendon.

ductile: Capable of being deformed without rupturing.

duct riser: Duct extending at least one story vertically.

duct tape: Heavy fabric tape commonly used to seal air ducts.

* **ductwork:** System of ducts for air distribution.

dumbwaiter: Small elevator housed within a structure, exclusively used to transport material.

dummy cylinder: Inoperative cylinder in a door lock.

dummy joint: 1. See *control joint*. **2.** Scoring in the surface of concrete flatwork for ornamental purposes.

dumpy level: Surveying instrument with a fixed horizontal telescope. Used to obtain relative elevation differences.

Dunagan analyis: Method of separating freshly mixed concrete or mortar ingredients to determine the proportions of the mix.

dunnage: 1. Wood strips or crating between material that provides air circulation and lifting space. **2.** Waste material. **3.** Members that support a structure, such as a cooling tower, but are not an integral part of the structure.

duplex: 1. Residential structure with two separate living areas. **2.** See *duplex cable*.

duplex apartment: See *duplex*.

duplex cable: Cable composed of two insulated conductors that are twisted together. May or may not have a common insulating covering.

duplex-headed nail: See *double-headed nail*.

duplex house: See *duplex*.

* **duplex lock:** Lockset with two cylinders that both operate the same bolt. One cylinder is operated with a change key, and the other cylinder is operated with a master key that operates several locks.

* **duplex receptacle:** Electrical contact device containing two receptacles.

durability: Ability of a material to resist deterioration caused by weather, chemicals, abrasion, or other wear factors.

duraluminum: Alloy primarily composed of aluminum, 4% copper, 0.2% to 0.75% magnesium, and 0.4% to 1% manganese. Used in extrusions and rolled sheets. Good corrosion resis-

DUCT FITTINGS

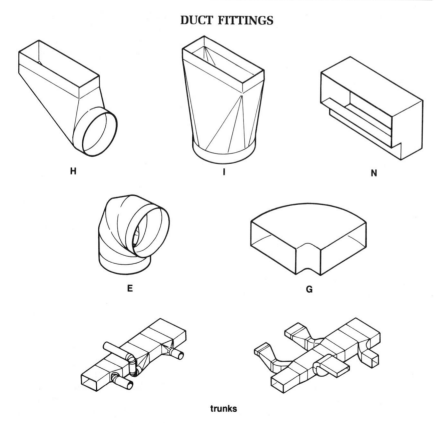

H   I   N

E   G

trunks

tance.

Durham system: Soil and waste piping system in which pipes and fittings are connected by threaded joints.

duriron: Alloy with a silicon content of approximately $14^1/_2\%$. Resistant to corrosive waste.

durometer: Instrument used to determine the hardness of a material.

dust-free time: Time required for freshly applied paint or varnish to form a skin that resists the adhesion of dust particles.

dusting: Development of powder on the surface of hardened concrete. Results from excess water in the mix, improper mix proportions, or premature troweling.

dust mask: Protective mask worn to cover mouth and nose from harmful airborne material.

* **Dutch arch:** Flat brick arch in which the bricks are pitched at an outward angle from the middle of the arch. FRENCH ARCH.

Dutch block: Y-shaped rigging device used to support a diesel-powered pile hammer and lead.

Dutch bond: See *English cross bond*.

Dutch colonial architecture: Architectural style characterized by gambrel roofs and overhanging eaves.

* **Dutch door:** Door that is divided horizontally into two separately hinged sections that swing independently of each other or as one unit.

Dutch door bolt: Door bolt that fastens the upper and lower sections of a Dutch door together so they swing as one unit.

Dutch hip roof: Style of roof framing in which the gable ends of a roof are built with a short hip section at the top. Hip section is often built approximately on the top quarter of each gable end.

Dutch lap: Style of roof shingle application in which each shingle is overlapped at the bottom and on one side.

duty cycle: 1. Percentage of time that electric arc welding equipment is loaded; e.g., welding equipment supplied by a 60 Hz system producing two 15 cycle periods per second has a duty cycle of 50% $[(2 \times 15) \div 60] \times 100$. **2.** Regularly timed schedule for an HVAC system. Used to reduce or limit energy consumption.

dwarf partition: Partition that does not extend the full ceiling height.

dwarf rafter: See *jack rafter*.

dwarf wall: Short wall less than one story tall, such as a wall for a crawl space. TOE WALL.

dwelling: Structure designed for use as living accommodations for one or more people. Includes living, eating, and sleeping areas.

dynamic analysis: Determination of stress in a structural system as a function of displacement under transient loading conditions.

dynamic head: Water pressure required to produce a specified flow velocity.

dynamic load: Variable, nonstatic load.

dynamic pile formula: Formula used to determine the bearing capacity of a

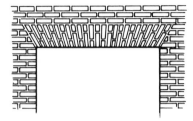

Dutch arch

Dutch door

driven pile. Based on the energy of the pile hammer and displacement of the pile resulting from repeated blows.

dynamic pressure: Pressure resulting from stoppage of a flowing fluid as compared to pressure of stagnant fluid.

dynamic resistance: Resistance of a pile during driving. Expressed as the number of blows per unit depth of penetration.

*ear: 1. Projection on a fitting that allows for attachment to a flat surface. 2. Projection on the end of a concrete stretcher block.

early English period: Period during 13th century when architectural style was refined. Characterized by inconspicuous masonry joints, pointed arches, and long narrow windows.

early stiffening: Early development of stiffening in portland cement. FALSE SET, QUICK SET.

early strength: Strength of concrete or mortar during a 72 hour period after placement.

early wood: See springwood.

earth: 1. Ground side of an electrical circuit. See ground. 2. Soil.

earth electrode: Metal conductor embedded in the earth to provide a ground for an electrical system. May be a metal rod, pipe, plate, or ring.

earthenware: Nonvitreous white ceramic with a maximum absorption rate of 3%. Used for plumbing fixtures.

earth pressure: Lateral force exerted against a soil-retaining structure such as a retaining wall.

earthwork: Digging and excavating operations.

eased edge: Edge of member that is slightly rounded over to eliminate a sharp corner.

easement: 1. Strip of land, commonly along the perimeter of property lines, that is used for placement of utilities. 2. Member cut at an angle that joins an inside stringer to the baseboard at the bottom of a stairway. 3. Curve formed at the intersection of two angular members. Used to provide a smooth transition.

eaves: Part of a roof, including the framing and trim members, that extends beyond the exterior wall line. See soffit, overhang.

eaves board: Tapered piece of wood along the lower edge of a roof to support the first course of shingles, tile, or slate. Also installed under valley shingles and along the edges of a gable to prevent water penetration. TILTING FILLET.

eaves trough: See gutter.

*eccentric fitting: Pipe fitting with an offset center line.

eccentric load: Weight that bears on a vertical support at an angle; nonsymmetrical load.

eccentric tendon: Prestressing tendon that extends in a direction that does not coincide with the gravity axis of the concrete member.

economizer: 1. Device used to preheat feedwater before it enters a boiler. Pipes are installed in a flue and water is circulated through them. 2. Control system used to reduce the load on a heating/cooling system by circulating fresh cooler air.

economy brick: Masonry unit with nominal dimensions of 4″ × 4″ × 8″. See brick*.

economy grade lumber: Lumber with a grade classification of No. 4 or 5. Has many knots.

economy 12 brick: Masonry unit with nominal dimensions of 4″ × 4″ × 12″.

economy wall: 4″ thick brick wall that is back-mortared and reinforced with pilasters. Corbels are used to support the floors and roof.

ecphora: Projection of an upper member beyond the face of the member directly below it.

edge: 1. Portion of a board that is perpendicular to the face and end. 2. To round the perimeter of concrete flatwork to prevent chipping.

edge band: Thin strip of wood or plastic laminate along the edge of a countertop or cabinet. Applied manually or automatically using an edge bander.

edge clamp: Clamp with two jaws at 180° to each other and one jaw at 90° to the first two. Used to apply pressure to edge or side of a member.

edge clearance: Distance between the edge of a glass pane and the surrounding frame.

*edge-flange weld: Weld with the two weld members flanged at the location of the weld.

edge form: Low form constructed around perimeter of an area to contain concrete being placed for flat surfaces such as floors and sidewalks.

ear (1)

eccentric fitting

edge-flange weld

edge grain: Lumber grain that forms at least a 45° angle to the face. VERTICAL GRAIN.

* **edge guide:** Router attachment that guides the cutting bit parallel to the edge of the work.

edge hem: Single fold along the edge of sheet metal. Used to reinforce the piece.

edge joint: Joint formed between two adjoining edge-to-edge surfaces.

* **edger: 1.** Concrete finishing tool used to round the edges of fresh concrete. **2.** Equipment in a saw mill that removes bark from rough-sawn lumber.

edge toenailing: See *toenail*.

edge guide

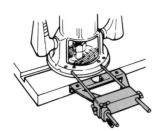

edger (1)

eggcrate diffuser

elbow catch

edge void: Defect along end or edge of a plywood panel in which an inner ply has split or broken during manufacture.

edge weld: Weld between the edges of two or more parallel or approximately parallel members. See *weldment**.

edging: 1. Strip of wood along perimeter of a veneered member that protects the edge of the veneer. **2.** Thin strip of material designed to cover and/or protect perimeter of a surface.

edging board: First piece of lumber cut from a log after the slab cut.

edifice: 1. Large structure or building. **2.** Front face of a building.

E elbow: Sheet metal fitting that makes a 90° transition between two round ducts. See *ductwork**.

effective area: Open space in an outlet or inlet device through which air or liquid can pass.

effective area of concrete: Portion of a concrete structure that is active in resisting applied stresses.

effective bond: Brickwork pattern in which each course is finished at the end with a $2^{1}/_{2}''$ closer.

effective depth: Depth of a concrete beam or slab measured from the compressive face to the central point of the tensile reinforcement members.

effective length: Distance between inflection points of a column.

effective opening: Minimum cross-sectional area of a pipe at the point of discharge. Expressed as a diameter.

effective prestress: Stress remaining in a concrete member, due to prestressing, after all losses have occurred.

effective span: Distance between supporting members or the distance between supporting members plus the effective depth of a concrete beam or slab, whichever is less.

effective temperature: Index that combines temperature, humidity, and air movement into a numerical equivalent of still, saturated air.

effective voltage: Working voltage in alternating current equal to .707 times the peak voltage.

efficiency: Comparison of the energy output of a device or equipment to the energy input.

efficiency apartment: Single-room apartment with a combination living room/bedroom with an adjoining kitchen alcove and bathroom.

efflorescence: White, crystallized deposit of soluble salts that forms on concrete or masonry walls. Results from calcium carbide in the mortar. Removed with builder's acid. BLOOM.

effluent: Liquid sewage discharged from a septic system or sewage disposal plant.

egg-and-anchor: Decorative design having an egg-shaped form alternating with an anchor-shaped form.

egg-and-dart: Decorative design having an egg-shaped form alternating with a dart-shaped form.

egg-and-tongue: Decorative design having an egg-shaped form alternating with a tongue-shaped form.

* **eggcrate diffuser:** Metal or plastic grill used to diffuse light from above.

eggshell gloss: Low-gloss finish that is between flat and semigloss.

egress: Exit or means of exiting an area.

ejector: 1. Pump used to raise liquid to a drain located at a higher elevation. **2.** Device used to increase fluid velocity and induce lower static pressure in an area of a hydraulic system. Used to draw another fluid into the high pressure line.

elastic deformation: Alteration in the length of a structural member as a result of tension or compression.

elasticity: Property of a material giving it the ability to return to its original shape and size after deformation.

elastic knife: Plastering tool, approximately 6″ wide, made of flexible steel. Used to trim the plaster brown coat at beads, grounds, etc.

elastic limit: Stress limit beyond which a member may not fully recover.

elastic shortening: Shortening of a concrete member during prestressing.

elastomer: Synthetic polymer that returns to the original shape and size after being deformed.

elastomeric: Made of a pliable synthetic polymer.

elbow: 1. Fitting used to join two pipes at a 45° or 90° angle. See *pipe fitting**. ELL. **2.** Fitting used to join two pieces of conduit at a 45° or 90° angle. See *conduit fitting**.

elbow board: See *apron*.

* **elbow catch:** Spring-loaded locking device commonly used to secure inactive leaf of a pair of cabinet doors. Hook on one end engages a catch and the other end (bent to a 90° angle) releases the latch.

electrical circuit: Path traveled by electrical current.

electrical insulator: See *insulator*.

electrical metallic tubing: See *conduit*.

electrical outlet: See *outlet*.

electrical plan: Print showing placement of electrical fixtures, appliances, and circuits. Exact placement of conductors is not specified. Commonly noted with a capital E preceding the sheet number. See *print**.

electrical resistance: See *resistance*.

electrical tape: Lightweight plastic with an adhesive backing. Covers and insulates electrical connections.

electric arc welding: See *arc welding*.

electric block plane: Hand-held power tool in which a rotating cutter projects through an opening in a short, flat bed to smooth wood and plastic. Bed is composed of two small adjustable pieces that are used to set the depth of cut.

electric demand: See *demand*.

electric discharge lamp: Lamp that produces light by passing electrical current through a vapor or gas.

electric eye: See *photoelectric cell*.

electric field: Area in which electric intensity can be detected.

electric heater: Appliance that produces heat by passing electrical current through resistive wires. The heat is dispersed by means of a fan or transferred to a sealed water supply.

* **electrician's chisel:** Tool used to cut rough openings for passage of conductors. Blade is approximately $2^3/_4''$ wide with overall length of approximately $8''$.

electrician's hammer: Hammer with extended striking face for reaching fasteners in tight locations, such as the inside back of an electrical outlet box.

electric meter: Device used to measure and indicate the amount of electricity used in a given amount of time.

electric plane: Hand-held power tool in which a rotating cutter projects through an opening in a flat bed to smooth wood or plastic. Bed is composed of two adjustable pieces that are used to set the depth and angle of cut.

electric receptacle: See *receptacle*.

electric resistance welding: See *resistance welding*.

electric sign: Fixed or stationary, self-contained, electrically illuminated utilization equipment displaying words or symbols.

electric squib: Blasting charge ignition device activated by an electrical charge.

electrocoating: Process used to stand abrasive material on end during application to a backing sheet in the manufacture of abrasives. Used to improve cutting ability.

* **electrode: 1.** Consumable or nonconsumable component of a welding circuit that conducts current to the arc, molten slag, or base metal. **2.** Conductors in a battery that produce a difference of potential electrical energy.

electrogalvanizing: See *electroplate*.

electrolier: Hanging electrical fixture.

electroluminescent lamp: Lamp with a rigid or flexible sheet that produces light by means of a phosphor that is excited by electromagnetic energy.

electrolysis: 1. Chemical decomposition produced by passing electrical current through an electrolyte. **2.** Conduction of electricity through fluid.

electrolyte: Electricity-conducting fluid.

electromagnet: Device that produces a magnetic field by passing current through coils around a metal core.

electromotive force (EMF): See *voltage*.

electronic distance measuring system: Instrument used to determine distance by measuring the time between emitting and receiving electromagnetic waves.

electronic distance meter (EDM): Surveying instrument used to determine distance electronically by emitting electromagnetic waves toward a target and measuring the time required for them to reflect. Actual distance is displayed on a monitor. See *surveying**.

electroplate: To apply a coating of metal to a surface. Metal is submerged in an electrolyte and electrons from another oppositely charged material are deposited on the surface.

electropneumatic: 1. Device that adjusts its pneumatic output when activated by an electrical control. **2.** Device that converts electric input into pneumatic output.

electroslag welding: Welding process used to join metal with molten slag that melts the electrode and base metal.

electrostatic precipitator: Smoke and dust filtering device in which particles are electrically charged as they pass through a screen. A second screen is charged with the opposite electrical charge and attracts and traps the particles.

elements: 1. Basic principles of a system. **2.** Source of heat in an electric water heater. **3.** Material consisting of only one type of atom such as copper or hydrogen.

elephant trunk: See *tremie*.

elevation: 1. Orthographic view of a vertical surface without allowance for perspective. See *print**. **2.** Height of a point or plane in relation to another point or plane. **3.** Height above mean sea level.

elevator: Enclosed platform used to transport people or material vertically from one level to another.

* **elevator bolt:** Fastener with a threaded end and a large flat head that finishes almost flush with the work.

elevator landing: Floor area directly outside elevator doors that allows for movement of passengers.

elevator shaft: Vertical fireproof chute used as a hoistway for an elevator.

ell: 1. Addition or extension at a 90° angle to the main structure. **2.** See *elbow*.

* **ellipse:** Plane figure in which the sum of the distances from two focus points is constant and equal to the major diameter.

elliptical: Shaped like an ellipse.

elliptical stairway: Stairway that is constructed within an elliptical stairwell.

emarginated: Having a notched perimeter.

embankment: Raised surface of rock, fill material, or earth used to retain water or support a roadway.

electrician's chisel

electrode (1)

elevator bolt

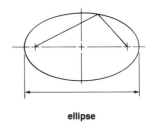

ellipse

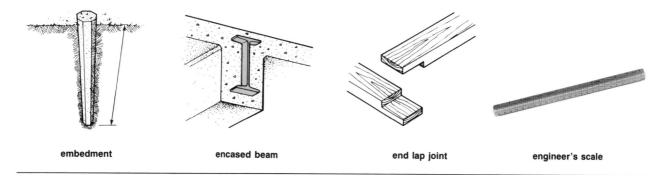

| embedment | encased beam | end lap joint | engineer's scale |

*embedment: Depth of a pile measured from the tip to ground level.

embellishment: Ornamentation or decoration.

emblazon: To decorate with heraldic or armorial designs, such as a coat of arms.

emboss: To decorate with a raised design.

embrasure: Opening with flared sides at an outward angle. CRENEL, CRENELLE.

emergency control station: Approved location at a job site where signals from emergency equipment are received.

emergency door stop: Spring-loaded pivoting latch installed in a door jamb to convert a double-acting door to a single-acting door.

emergency exit lighting: Lighting system that is activated when a power failure occurs and provides adequate lighting for exit from a building.

emery: Crushed stone used as an abrasive material.

emery cloth: Fine abrasive material consisting of a cloth backing coated with emery. Used wet or dry for polishing metal.

emery wheel: Abrasive disk primarily composed of emery; rotated at high speed for grinding and polishing metal.

emission: 1. Discharge of gaseous material into the atmosphere. 2. Blowback of water or gas into an occupied area caused by positive pressure.

emissivity: Rate at which electrical or thermal particles radiate from an object.

emittance: Percentage of solar energy absorbed by a solar collector.

emplection: Masonry wall in which the faces are constructed with alternating courses of ashlar set as headers and stretchers and a core filled with rubblestone.

EMT: Electric metallic tubing. See conduit.

emulsifier: Substance that modifies the surface tension of droplets and keeps them suspended.

emulsion: 1. Mixture of liquids that are insoluble. 2. Solution in which insoluble particles are suspended.

enamel: Paint with a large amount of varnish; has a hard finish when dry.

enameled brick: Masonry unit with a glazed surface.

encarpus: Ornamental design having fruit, leaves, and flowers draped between two points.

*encased beam: Steel beam surrounded by concrete.

encased knot: Knot in which the growth rings are surrounded, but not intergrown, with the growth rings of the wood member.

encaustic: Process in which color is applied to a material and fixed in position by heating.

enchased: Metalwork in which a relief pattern is formed by hammering and depressing the background.

enclosed: Surrounded by a structure, such as a wall or fence, that prevents accidental contact with electrically energized components.

enclosed knot: Concealed wood defect not visible from the face, edge, or end of the board.

enclosure: Equipment housing or structure surrounding equipment; provides protection to a person or equipment.

enclosure wall: Non-bearing partition.

end-bearing pile: Pile primarily supported by the tip driven into firm bearing stratum.

end channel: Steel reinforcement welded into the top and bottom of a hollow metal door to provide strength and rigidity.

end-cutting pliers: Hand tool used to cut wire or nails flush with a surface. Jaws are 90° to the handle and are sharpened to produce a knife edge. See pliers*. NIPPERS, PINCERS.

end grain: Portion of a piece of wood exposed when the fibers are cut across the grain.

end joint: Joint formed when the ends of two members are butted.

*end lap joint: Angular joint formed by removing one-half the thickness of two members and overlapping them. Faces are flush when joined.

end-matched lumber: Lumber with a tongue along one edge and a groove along the other edge.

endothermic reaction: Chemical reaction that absorbs heat.

end thrust: Pressure exerted in a direction parallel with the long axis of a structural member.

endurance limit: Maximum stress a member can withstand without failure.

end view: See side view.

energize: To connect to an electrical power source.

energy efficiency rate (EER): Comparison of the cooling capacity of an air conditioning system to the electrical input; expressed in watts. ENERGY EFFICIENCY RATIO.

enfilade: Alignment of a series of doors through several rooms.

engaged: Attached; in contact with.

engineered brick: Solid masonry unit with nominal dimensions of $4'' \times 3\frac{1}{5}'' \times 8''$. See brick*. ENGINEER BRICK.

engineered grade: Plywood panel designed to be used for structural purpose.

engineering drawing size: Standard-sized drawing medium denoted as A ($8\frac{1}{2}'' \times 11''$), B ($11'' \times 17''$), C ($17'' \times 22''$), D ($22'' \times 34''$), and E ($34'' \times 44''$).

engineer's chain: Surveying chain having 100 links, each 1' long.

engineer's rod: $6'\text{-}0''$ to $12'\text{-}0''$ surveying instrument graduated in hundredths of a foot. Used to determine elevation.

*engineer's scale: Measuring and scaling device graduated in the decimal system. Common scales include $1'' = 10'\text{-}0''$, $1'' = 20'\text{-}0''$, $1'' = 30'\text{-}0''$, $1'' = 40'\text{-}0''$, $1'' = 50'\text{-}0''$, and $1'' = 60'\text{-}0''$.

engineer's telescope: Surveying instrument used to determine direction and magnify an object.

English bond: Brickwork pattern com-

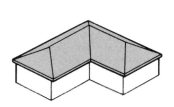

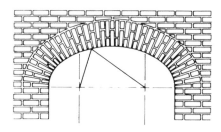

| equal pitch roof | equilateral arch | equilateral roof | escutcheon |

posed of alternating courses of headers and stretchers. Headers are centered on the stretchers and joints between stretchers in all courses are aligned vertically. See *bond* (1)*.

English cross bond: Brickwork pattern composed of alternating rows of headers and stretchers. Joints between the stretchers are displaced by half the length of a stretcher.

English garden wall bond: Brickwork pattern composed primarily of stretchers with each fourth course constructed with headers.

engrailed: Cut with a series of concave curves.

enrichment: Adornment; ornamental work fastened to flat surfaces.

entablature: Decorative and structural beam across two or more columns. Divided horizontally into the cornice, frieze, and architrave. Height can be divided into 100 equal sections to establish a scale that is used to determine the proportions of other parts of the architectural order or design.

entasis: Small amount of convex curvature in the vertical design of a column or pilaster.

entrained air: Small air bubbles incorporated into concrete to improve freeze- and thaw-resistance.

entrance: Entryway into a structure.

entrapped air: Undesirable voids in concrete resulting from improper consolidation. HONEYCOMB.

entry: Area immediately inside an entrance door; vestibule.

entry loss: Drop in pressure resulting from a fluid rapidly flowing into a pipe or other vessel.

envelope: See *exterior envelope*.

epistomium: Spout of a pipe.

epistyle: Lowest entablature member. ARCHITRAVE.

epoxy: Synthetic resin that dries to a hard finish and is chemical- and corrosion-resistant. Also used as an adhesive.

epoxy cement: Resin consisting of at least two parts, that when mixed, produces a strong adhesive.

epoxy concrete: Concrete mix composed of a catalyst, epoxy resin, and fine and coarse aggregate.

epoxy mortar: Mortar mix composed of a catalyst, epoxy resin, and fine aggregate.

epoxy plastic: Thermosetting plastic with good adhesive qualities and chemical- and corrosion-resistance. Slow burning rate. Used for adhesives and laminates.

equalizer: Internal or external pipes that maintain an even level in liquids in two or more separate chambers.

equalizing bed: Layer of sand, gravel, or concrete placed in the bottom of a trench to support underground pipes.

* **equal pitch roof:** Roof in which all surfaces are at the same angle.

* **equilateral arch:** Two-centered arch in which the length of the chords equals the span.

* **equilateral roof:** Roof in which the surfaces are sloped at a 60° angle, forming an equilateral triangle.

equilibrium moisture content (EMC): Level of moisture content of wood or soil at which it does not gain moisture from or lose moisture to the air.

equipment grounding conductor: Electrical conductor used to connect non-current-carrying metal parts of equipment or enclosures to the system grounded conductor or grounding electrode.

equivalent length: Measure of pressure drop in valves and fittings caused by resistance. Expressed as the equivalent length of straight pipe required to produce an equivalent amount of resistance.

erection: To build with structural steel.

erection wrench: Hand tool with a fixed jaw at one end and a tapered shaft at the other end. Used to turn nuts or bolts and align bolt holes in adjoining structural steel members. See *wrench*. SPUD WRENCH.

ergonomics: Design of appliances and furniture to fit the shape of the human body with minimal discomfort and fatigue.

erisma: Buttress for a wall.

erosion: 1. Removal of soil or other materials caused by water or wind. 2. Reduction in thickness of the base metal caused by abrasive action of gases, fluids, or solids in motion.

error of closure: Difference between a theoretical angular value and the actual angular measurement. Used in laying out several adjacent angles.

escalator: Inclined movable steps joined on an endless belt. Used to transport people from one level to another. MOVING STAIRCASE.

esconson: Open area at the intersection of two members in a window jamb.

* **escutcheon:** Protective cover placed around a door knob, wall switch, or vertical pipe.

escutcheon pin: Decorative nail with a round head; used to attach ornamental and protective covers.

espagnolette: 1. Lock for a casement window consisting of a long rod with hooks at the ends that engage the sash. Handle is used to rotate the rod and secure the hooks. 2. Decorative corners on Louis XIV furniture.

esplanade: Open landscaped area with paths.

Essex board measure: Table on a framing square used to calculate board feet of lumber.

estate: 1. Property, usually land. 2. Property of a deceased party at time of death.

estimate: To calculate amount of material, labor, and other costs involved in a construction project.

estimated design load: Total of potential heat transfer in a heating and air conditioning system.

estimated maximum load: Largest amount of heat transfer a heating and air conditioning system is required to provide.

* E-stud: Lightweight structural metal stud used to cap drywall separation wall panels at intersecting walls.

etch: 1. To engrave a design in glass, metal, or a hard surface by removing part of the surface. **2.** To remove surface concrete to create an exposed aggregate finish.

European hinge: See *concealed casework hinge.*

eustyle: Design of column placement in which the space between column shafts is equal to 2.25 times the column diameter.

evaporation rate: Rate at which water is absorbed into the surrounding air.

evaporation test: Boiler test used to verify operation of the low water fuel cutoff.

evaporative cooler: Air conditioner that cools an area by drawing air through a wet filter and dispersing a fine mist. Increases humidity of the atmosphere.

evaporator: Portion of a cooling system used to vaporize liquid refrigerant.

evener: Beam that distributes a load over a large area of overhead structural support.

even pitch roof: See *equal pitch roof.*

excavation: Removal of earth from its natural position.

excavator: 1. Mechanical equipment used to remove and transport earth to another location. **2.** Operator of excavating equipment.

excess air: Additional air in a combustion chamber of a furnace that is above the level required for combustion.

excess chalking: Undesirable finish of a painted surface. Results from applying too many coats of paint or lacquer to a heavy, porous undercoat or lack of the proper amount of binder in the paint.

excitation: Electric current and voltage that induce magnetic flux in a motor.

exedra: Raised, semicircular or elliptical platform with seats facing the center.

exfiltration: Outward air flow.

exfoliation: Deterioration of the surface of a material by delaminating or peeling in layers. Caused by chemical or physical weathering.

exhaust fan: Device used to remove air from an enclosed area.

exhaust system: Assembly of ducts, fittings, and registers used to remove air from an enclosed area and discharge it into the atmosphere.

exit: Opening that provides a safe and protected escape route from a structure.

exit access: Hallway or door that leads to an exit.

exit device: See *panic bar.*

exit light: Illuminated sign indicating an exit. Commonly placed above eye level.

exothermic: Chemical reaction that releases heat.

expanded: Enlarged by thermal, mechanical, or chemical means.

expanded metal lath: Open mesh plaster reinforcement formed by stamping or slitting sheet metal in one direction and stretching it perpendicular to the slots. Open mesh interlocks with the plaster.

expanded polystyrene: Foamed styrene plastic with high thermal insulation properties and strength-to-weight ratio.

expanded shale: Lightweight aggregate used in lightweight structural concrete. See *haydite.*

expanded slag: Lightweight porous aggregate used in concrete mix. Formed by processing blast furnace slag with water. EXPANDED BLAST FURNACE SLAG.

expansion: Swelling; increasing length or volume of a material.

expansion bead: Strip of metal used in plastering to allow movement between two adjoining materials.

* expansion bend: Horseshoe bend in a pipe run that allows for expansion or contraction of the pipe resulting from changes in thermal conditions.

expansion bit: Adjustable wood-boring tool with a pilot tip and one or more cutting edges. Used to bore holes of various diameters. See *bit* (1) *. EXPAN-

SIVE BIT.

expansion bolt: See *expansion shield.*

expansion end: End of a structural member that has an allowance for movement.

* expansion joint: 1. Separation between adjoining sections of a concrete slab to allow movement caused by expansion and contraction. **2.** Device that counteracts movement in pipes as a result of temperature changes. Available as sliding-sleeve or siphon bellows design.

expansion pipe: Pipe extending from the relief valve of a water heater to a drain. Hot water or steam discharged from the water heater is diverted to the drain.

expansion shield: Concrete or masonry fastener with a split casing that allows it to expand and wedge into position as a bolt is inserted. EXPANSION BOLT.

expansion sleeve: 1. Collar placed around pipes to provide clearance for movement through concrete or masonry walls and floors. **2.** Metal tube used to house a dowel bar and allow longitudinal movement. See *anchor*.*

expansion strip: Soft, resilient material used to fill an expansion joint.

expansion tank: Container or area above a hot water heating system that allows for expansion of heated water. See *expansion pipe.*

expansion valve: Control device used to regulate refrigerant flow to a cooling element.

expansive bit: See *expansion bit.*

expansive cement: Cement that increases in volume when mixed with water. Available as Type K, Type M, and Type S. EXPANSION CEMENT.

expletive: Stone used to fill a hole in a masonry surface.

* exploded drawing: Pictorial drawing that shows an object disassembled and in relation to its assembled position. EXPLODED VIEW.

explosionproof apparatus: Apparatus enclosed in a case that will withstand a vapor or gas explosion within it.

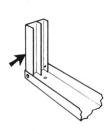

E-stud

expansion bend

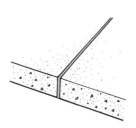

expansion joint (1)

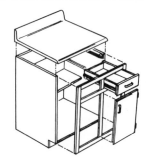

exploded drawing

explosion welding: Solid-state welding process in which the members are joined by a controlled detonation that causes high-speed movement between the pieces.

explosive: Material used to produce an explosion. Three classes are Class A: explosives that present a detonating hazard; Class B: explosives that present a flammability hazard; and Class C: explosives that contain controlled quantities of Class A and B explosives.

explosive rivet: Rivet with an explosive-filled shank. Rivet is inserted into pre-drilled hole and upset by striking the head with a hammer.

exposed: 1. Able to be seen when construction is complete. 2. Capable of being accidentally touched or approached at an unsafe distance. 3. On or attached to the surface to allow access.

exposed-aggregate finish: Concrete finish in which the outer layer of concrete or mortar is removed to expose decorative stone at the surface.

exposed joint: Mortar joint on the face of a masonry wall above ground level.

expressway: Highway with full or partial access.

expulsion: Forced ejection of molten metal from a weld.

extended heating surface: Fins or ribs that are heated by conduction from another surface.

extender: Inert material used to add bulk to paint, resins, or adhesives.

extensibility: 1. Maximum tensile stress that concrete can withstand without cracking. 2. Ability of a material to stretch.

extension: Projection or structure added to an existing building.

extension bit: 6″ to 18″ shaft with one end designed to be inserted into a drill and the other end designed to hold a drill bit.

extension cord: Electrical conductor with a plug on one end and a receptacle on the other.

extension handle: Tubular rod with a threaded fitting at one end. Attached to handles of various hand tools, such as paint rollers or concrete floats, to provide a greater reach.

* **extension ladder:** Ladder having two or more sections, with each section sliding within and locking with the other to provide greater length.

extension line: Line on a drawing or print that extends from the object line to the dimension line.

* **extension rule:** Rule with a sliding, graduated metal strip fitted into the last wood section. Sliding strip allows for

measuring inside dimensions.

exterior: 1. Outer surface of a structure or object. 2. Material grade that denotes ability to withstand weathering without deterioration or failure.

exterior envelope: Element within a structure that encloses a conditioned space.

exterior finish: Protective and/or decorative finish on the exterior of a structure.

exterior-protected construction: Building with noncombustible exterior and firewalls.

exterior trim: Finish material applied to the exterior of a structure.

exterior-type plywood: Plywood used for marine or outdoor applications with the plies bonded with 100% waterproof adhesive.

external interrupt: Device that disconnects a load from a power source but is not an integral part of the equipment.

externally operable: Device that can be operated without exposing the operator to live parts.

external tooth washer: Flat washer with an interior cylindrical opening and a toothed perimeter. See *washer**.

external vibrator: Device attached to the outside of concrete forms and causes high-frequency consolidation of the concrete. Commonly used for precasting or tunnel-lining forms.

extract: To pull previously driven piles out of the ground.

extractor: 1. Tool with tapered spiral shaft that is driven into the top of a threaded fastener that has broken off. Turned counterclockwise to remove the fastener. 2. Device used to remove driven piles from the ground. Equipment that delivers heavy upward blows or vibrates the pile is utilized. 3. Device in a forced air system that directs air from a supply duct to a branch line.

* **extrados:** Exterior curve of an arch or vault.

extra-fine threads: Close, narrow threads on a fastener that allow for small accurate adjustments. Used for thin-walled tubing and thin nuts where maximum thread engagement is required. Designated as UNEF.

extra-heavy pipe: Pipe with a thicker wall than standard pipe.

extruded: Formed to shape by forcing through a shaped opening.

extruded mortar joint: See *skintled joint*.

extruded polystyrene insulation: Rigid foam sheet material used to increase resistance to thermal transmission.

extruder: Equipment that pumps caulking or other semifluid material out of bulk containers and through a hose to

extension ladder

extension rule

extrados

an applicator.

extrusion: Forming into shape by forcing material through a die.

exude: To force fluid or semifluid material through an opening.

eye: 1. Opening on top of a cupola. 2. Center of a circular member. 3. Opening in the head of a tool that receives a handle; e.g., eye of a hammer. 4. Hardware device with the end bent into a semicircular shape for receiving a hook. SCREW EYE.

eyebolt: Threaded fastener bent on one end to form a circular opening. See *bolt**.

eyebrow: Window or ventilation opening that extends through a roof but does not form an acute angle with the roof.

eye of dome: Opening at the top of a dome.

eyepiece: Part of a surveying instrument that magnifies an image and rotates to focus an object.

eye splice: Loop made at the end of a rope by unraveling the strands at the end, making a bend, and weaving the strands into the standing part. See *knot**.

fabricate: 1. To form and assemble components of a structure. **2.** To cut, bundle, and tag rebars at a location away from the job site.

facade: Exterior face of a structure often distinguished by ornamentation.

* face: 1. Widest side of a piece of lumber, perpendicular to the edge and end. **2.** Finished side of a piece of rough lumber. **3.** Exposed vertical plane of an arch. **4.** Side of a framing square bearing the name of the manufacturer. **5.** Striking surface of a hammer. **6.** See *face of weld.* **7.** Highest graded side of a sheet of plywood. **8.** Vertical surface of an exposed rock wall. **9.** To add a layer of a finish material to a base material. **10.** Exposed surface of a masonry unit.

face block: See *faced block.*

face brick: See *facing brick.*

face bushing: Pipe fitting with external and internal threads and at least two projecting lugs to provide a gripping surface when turning.

face clearance: Distance between the face of a panel or pane of glass and the face of its frame.

faced block: Concrete masonry unit with a special glazed, ceramic, plastic, or polished face.

faced insulation: Batt insulation with a paper, plastic, or metallic facing material on one or both sides that acts as a moisture barrier.

faced wall: Masonry wall in which one or both sides are finished with material different from the body of the wall. Exterior surfaces and body are tied together to form a single load-bearing unit.

face edge: 1. Edge of a board with the best appearance. **2.** Edge of a board from which the ends, faces, and other edge are laid out and formed.

face feed: Application of filler metal to a brazed or soldered joint. Usually applied manually.

* face frame: Casework on the front of a cabinet that conceals the edges of the cabinet sides and bottom. FACEPLATE.

face glazing: Glazing compound applied after glass has been set and fastened in a rabbeted sash without stops.

face hammer: Hammer with one blunt face and one wedge-shaped end for rough dressing masonry units.

face joint: Exposed joint in the face of a masonry wall. Commonly struck or pointed with a tool.

face mark: Mark made on the surface of a piece of lumber that denotes it as the work face.

face mix: Concrete mixture applied to an exterior face of a cast stone wall. Used to enhance the appearance and durability of the wall.

face mold: 1. Template used to lay out an ornamental design on wood. **2.** Template used to examine the shape of wood or stone surfaces.

face nail: To drive a nail perpendicular to the face of the work.

face of weld: Exposed surface of a weld joint on which the weld is performed.

face plate: See *face frame.*

* face putty: 1. Pliable glazing material applied on the exposed side of glass after the glass is set in place. **2.** See *face glazing.*

face seam: Carpet seams that are made without turning the carpet to expose its backing.

face shell bedding: Setting concrete masonry units with mortar applied to the outside webs of the units only.

face shield: Protective device that covers the face area.

face side: See *face* (7).

* face stringer: Finished lumber that conceals the rough stairway stringer and the joint between the wall and treads and risers.

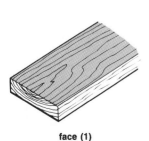

face (1)

face frame

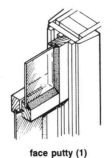

face putty (1)

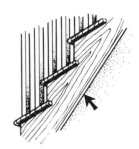

face stringer

facet: Flat surface between two flutes, as in a column. FACETTE.

face tier: See *face brick.*

face velocity: Rate at which air passes through a register in a forced-air system.

face veneer: Outer veneer of plywood with the higher grade.

facing: Finish material applied over a rough, unfinished surface.

facing brick: Brick commonly used for an exposed surface. Composed of selected clay and often treated to produce a desired surface color or texture. FACE BRICK.

facing hammer: Hammer with two wedge-shaped serrated faces. Used to dress stone and concrete.

factor of safety: 1. Ratio of the maximum stress or yield point of a material to the designed working load. **2.** Ratio of the maximum load or shear of a structural member to the designed working load or shear.

factory edge: Edge of a building material as it is received at the job site.

factory lumber: Broad category of lumber used to produce sashes, stiles, rails, and trim members. SHOP LUMBER.

factory select: Factory lumber grade that is at least 70% No. 1 material.

fade: To lose intensity of color as a result of age or weathering.

Fahrenheit: Measure of temperature in which the freezing point of water is 32° and the boiling point is 212° under standard atmospheric conditions.

faïence: Earthenware with a transparent glaze.

faïence mosaic: Glazed ceramic tile approximately $3/8''$ thick and less than 6 square inches in area.

failure: Inability of a structural material or structure to support an imposed load for which it was designed.

fair face concrete: Concrete surface that does not require finishing after forms are removed.

fairlead: Block, ring, or pulley used to guide a load line and ensure that the line winds smoothly around a drum.

fair raking cutting: Cutting exposed brick at an angle to the horizontal plane, such as along a gable.

fall: Slope of a channel, conduit, or pipe, expressed in inches per foot or as a percentage; e.g., a pipe with 3'-0" of vertical fall per 100'-0" of horizontal length, or a 3% fall.

false ceiling: See *suspended ceiling.*

false header: 1. One-half of a brick in a wall that gives the appearance of a header. **2.** Wood members installed between floor joists along the inside of a foundation wall.

false set: Rapid hardening of freshly mixed concrete, mortar, or portland cement paste without the production of excessive heat. Plasticity is regained by remixing the material without additional water. HESITATION SET, PREMATURE STIFFENING, RUBBER SET.

falsework: Temporary shoring or support used to support work under construction.

fan: 1. Device with rotating blades or vanes that move air. **2.** Building or architectural design having several lines radiating from a common point.

fan-coil unit: Air conditioning unit located in the area to be conditioned. Consists of heating and/or cooling coils, air filter, and a fan.

fanlight: See *fan window.*

fan shroud: Protective enclosure surrounding the moving parts of a fan that also directs air flow.

fan tracery: Plaster ceiling design having ribs radiating from a central point, producing a radiating pattern.

* **fan window:** Semicircular or elliptical window with muntins radiating from a center point. Commonly installed over a door or rectangular window. FANLIGHT.

* **fascia: 1.** Horizontal trim member at the lower end of roof rafters. FACIA. **2.** Complete assembly of exterior trim members at the lower end of an overhang. **3.** Flat vertical member in an entablature.

fascine: Woven wood mattress laid on a earthen bank to inhibit erosion.

fastener: Mechanical device used to secure two or more members in position or join two or more members. Common types include nails, screws, and bolts.

fast-pin hinge: Hinge with a nonremovable pin. FAST-JOINT BUTT.

fat concrete: Concrete containing a large amount of cohesive mortar.

fathom: Lineal measure equal to 6'-0".

fatigue: Progressive weakening of a material resulting from repeated or alternating application of stresses that are below the maximum tensile strength of the material. May result in cracks or complete failure of the material.

fat lime: Plastering and masonry material obtained by burning limestone. RICH LIME.

fat mix: Mortar or concrete containing a high lime or cement content. Facilitates spreading and working the mortar or concrete. See *fat mortar.*

fat mortar: Mortar containing a high lime or cement content, producing a mixture which adheres to a trowel.

fat spot: Thickened area in bituminous paving.

faucet: Device used to control the flow of liquid at the outlet end of a pipe. Consists of a valve and spout. SPIGOT.

fault: 1. Partial or total failure of insulation or conductivity of an electrical conductor. **2.** Physical condition that causes a component to fail to operate as designed. **3.** Vertical displacement on each side of a slab or member that is adjacent to a crack.

* **favus:** Tile or marble slab cut into a hexagonal shape and assembled to form a honeycomb pattern.

faying surface: Mating surface of a wood or metal member that contacts or is in close proximity to another member to which it is to be joined.

* **feather: 1.** Thin piece of wood inserted in a groove to create a joint between two members. **2.** To blend the edge or surface finish of a material with an older or previously placed material. **3.** To flatten a sheet metal seam.

fan window

fascia (1)

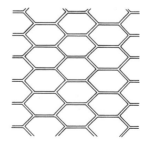

favus

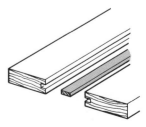

feather (1)

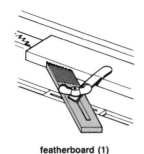

featherboard (1)

female thread

fence pliers

---

* **featherboard:** 1. Piece of lumber with closely spaced slots approximately 5″ to 6″ long cut into the end. Used as a guide and anti-kickback device on a table saw. 2. Horizontal lap siding with a small amount of overlap. CLAPBOARD.

featheredge brick: See *arch brick*.

featheredged coping: Stone surface that slopes in one direction only. SPLAYED COPING, WEDGE COPING.

featheredge rule: Straightedge used to smooth a material to a fine, tapered edge. FEATHEREDGE.

feather joint: See *spline*.

feature strip: Contrasting stripe, line, or trim member applied to a surface.

feeder: 1. Circuit conductors between the service equipment and the last branch-circuit overcurrent protection device. 2. Device that supplies material to a conveyor system.

feed rate: Rate at which material is introduced into equipment, such as a planer, without inefficient operation of the equipment.

feedwater: Water supplied to a boiler at the proper pressure and temperature.

feedwater check valve: Automatic valve that prevents backflow of water from a boiler to the feedwater line.

feedwater heater: Heating unit used to heat feedwater and vent air and other noncondensable gases to the atmosphere.

feedwater line: Piping extending from the boiler to the feedwater pump.

feedwater pump: Pump used to convey water from the feedwater heater to the boiler at the proper pressure.

feeler gauge: Precision instrument used to measure the thickness of a gap. Consists of a series of blades with different thicknesses, each marked with the thickness in thousandths of an inch. THICKNESS GAUGE.

feet of head: See *head*.

feint: Edge of flashing that is slightly bent to form a capillary break.

felt paper: See *building paper*.

female plug: See *receptacle*.

* **female thread:** Threads on the inside of pipes or fittings.

femerell: Ventilator used to exhaust smoke when a chimney is not provided.

fence: 1. Device fastened parallel to a saw blade or other cutting tool used to guide material during cutting or shaping operations. 2. Enclosing framework for exterior areas, such as yards or gardens.

* **fence pliers:** Flat-nosed pliers that includes wire cutters, staple and wire grips, and claw used to remove fasteners.

fender: Protective curb or device surrounding a structure, such as a bridge pier, to prevent damage to the structure.

fenestral: Small window.

fenestration: Arrangement of windows in a structure.

ferriferous: Broad category of raw material component of cement that supplies iron to the mix.

ferrous metal: Metal with iron as its primary element.

ferrule: 1. Metal sleeve fitted with a screwed cap; attached to the side of a pipe to provide access for inspection or maintenance. 2. Metal sleeve on the end of a screwdriver, chisel, or similar tool into which the handle or tang is inserted. 3. Short metal sleeve crimped around a cable to form a loop at the end. 4. Short piece of unthreaded pipe placed over another pipe or tube. 5. Metal cap or sleeve on a wood post or furniture leg.

fertilizer: Organic or inorganic nutrient.

fiberboard: Broad category of sheet material consisting of wood or other vegetable fiber that is added to a binder and compressed. Used for structural or decorative applications. See *particleboard, hardboard*.

fiberglass: Thin glass filaments or strands formed by spinning or pulling molten glass into random lengths.

Used as thermal and acoustical insulation, reinforcing material, and in glass fabric. FIBROUS GLASS, GLASS FIBER.

fiber pipe and fittings: Plumbing pipe and fittings composed of interwoven fibrous threads impregnated with a bituminous material.

fiber reinforcement: See *fibrous concrete*.

fiber saturation point: Point in drying or wetting wood at which wood fibers are saturated with water, but the cell cavities are empty.

fiber stress: Longitudinal compressive or tensile stress in a wood structural member, such as a beam or girder.

fibrous concrete: Concrete in which plastic fibers have been mixed. Reduces unit weight of the concrete and improves tensile strength.

fibrous plaster: Cast plaster reinforced with canvas.

fiducial mark: Index line or point used as a surveying reference point.

field: Continuous expanse of brickwork between openings and/or corners.

field conversion: Conversion of fractional parts of a foot to decimal parts of a foot by using a memorized standard table. Based on five accuracy points: 0″ = .00′, 3″ = .25′, 6″ = .50′, 9″ = .75′, and 12″ = 1.00′, with 1/8″ increments equal to approximately .01′.

field-cured cylinder: Concrete test cylinder cured in similar conditions as the concrete in the structure. Used to measure the strength of concrete and indicate when additional loads can be imposed.

fielded panel: Surface bordered with a decorative design or divided into several areas.

fieldhouse: Large auditorium used for athletic events.

fieldstone: Loose stone in soil or on the surface.

field tile: See *drain tile*.

fifth wheel: Swiveling, weight-bearing connection between a tractor and semitrailer.

figure: Natural grain pattern in wood formed by an unusual arrangement of wood fibers or color.

figured glass: Translucent sheet glass with a relief pattern on one side, creating a limited degree of obscurity.

figure-eight knot: Knot used to prevent the end of a rope from sliding out of a tackle or pulley. Formed by making an underhand loop, bringing the end over the standing part, and passing it under and through the loop. See *knot**.

filament: Thin wire in an electrical fixture, such as an incandescent lamp that produces light and/or heat.

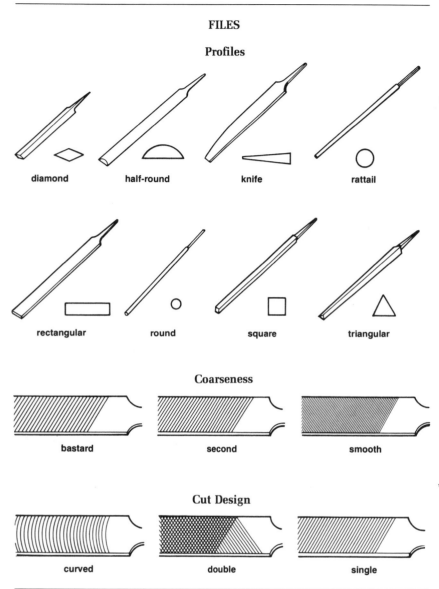

FILES

Profiles

diamond   half-round   knife   rattail

rectangular   round   square   triangular

Coarseness

bastard   second   smooth

Cut Design

curved   double   single

* **file:** Abrasive tool with single or double rows of fine teeth cut into the surface. Used to form and/or smooth material such as wood or metal.
* **file brush:** Hand tool used to clean files; one side has fiber bristles and the other side has steel bristles.

fill: Inert material, such as soil or rock, used to raise an existing grade.

filler: 1. Fine inert material, such as limestone or silica, used to reduce shrinkage and improve workability of portland cement, paint, or other material. **2.** Material used to fill a void in formwork.

filler coat: Initial coat of paint or varnish. PRIMER COAT.

filler metal: Metal added to a welded, brazed, or soldered joint.

* **fillet: 1.** Narrow, flat strip between two angular surfaces; e.g., between the lower end of stairway balusters. LISTEL. **2.** Concave joint at the intersection of two surfaces.

fillet chisel: Mason's chisel used to shape and apply detailing to stone.

fillet weld: Triangular-shaped weld at the intersection of two surfaces at approximately a 90° angle to one another, such as a T-, corner, or lap joint.

fillister: 1. Rabbet along the outside edge of a muntin for the glass and putty. **2.** Small plane used to groove timber.

fillister head: Screw head with a straight-shanked body and a slightly domed upper surface. See *screw**.

film glue: Thin sheet of paper impregnated with thermosetting adhesive. Used to eliminate liquid adhesive from bleeding through veneers.

filter: 1. Device used to separate solid material from liquid or gas. **2.** Capacitor or inductor that eliminates the effects of an alternating current signal across a direct current circuit. **3.** Device used to change the transmission or reflection of light.

* **fin: 1.** Sheet metal projection used to increase the heat transfer surface area of a pipe, such as in a baseboard heater. **2.** Narrow projection on a formed concrete surface resulting from mortar flowing into narrow voids in formwork. **3.** Blade in a concrete mixer drum.

final set: Degree of stiffening of concrete mix. Expressed as the amount of time required to obtain a specified degree of hardness.

final stress: Stress in a prestressed concrete member that exists after all stress losses have occurred.

fine aggregate: Aggregate passing through a $3/8''$ sieve and retained on a No. 200 sieve.

fineness: Measure of particle size.

fines: Soil which passes through a No. 200 sieve.

fine solder: Solder with a low melting point of approximately 370°F.

fine stuff: Plaster made of pure lime and water. Used as a finish coat.

finger cup: Concave trim member recessed in the face of a sliding door; provides a grip for opening and closing the door.

finger joint: Wood joint consisting of

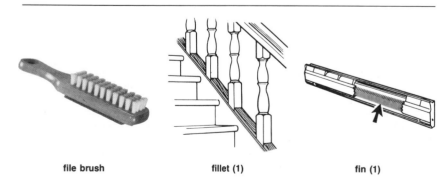

file brush   fillet (1)   fin (1)

interlaced finger-like projections. Used to join wood members end to end. See *joinery**.

finial: Ornamental member, such as a spire, at the top of a vertical member.

finish: 1. Material, hardware, or surface coating that is exposed at the completion of construction. **2.** Texture of an exposed concrete surface.

finish coat: Final coat of a material such as plaster, joint compound, shotcrete, or paint.

finished stringer: See *face stringer*.

finisher: See *cement mason*.

finish floor: Wearing surface of a floor.

finish flooring: Material used as a finish floor surface; commonly tile or hardwood.

finish grade: Elevation of soil, walks, drives, and other improved surfaces after final grading operations.

finish hardware: Exposed hardware, such as hinges and knobs, that is functional and also adds aesthetic value. ARCHITECTURAL HARDWARE, BUILDER'S FINISH HARDWARE, BUILDER'S HARDWARE.

finish nail: Nail with a smooth shank and a small barrel-shaped head. Head creates a small void in the surface of the workpiece that is filled after setting the nail head. Available in sizes from 2d to 20d. See *nail**.

finish size: Final dimensions of a completed component or assembly. FINISHED SIZE.

Fink truss: Symmetrical truss used to support large sloping roofs. Consists of three isosceles triangles with the center triangle having its base along the bottom chord. See *truss**. BELGIAN TRUSS, FRENCH TRUSS.

finned length: Distance between the two outside fins on a pipe or tube.

fin pitch: Number of fins in a unit length of a tube.

fire alarm system: Electrical system installed in a residential or commercial structure to sound an alarm when heat and/or smoke is detected. Includes automatically and manually activated devices.

fire area: Area in a structure enclosed by fire division walls and/or exterior walls that restrict the spread of fire.

fire assembly: Assembly of a fire door, window, or damper, including all hardware.

fireback: Back wall of a fireplace constructed of heat-resistant material which radiates heat into the living area. CHIMNEY BACK.

* fire block: Wood member nailed between studs or joists to restrict air movement. See *fire stop*.

firebox: Combustion chamber in heating equipment.

fire break: 1. Space between structures or groups of structures in a city to prevent fire from spreading to adjacent areas. **2.** Fire-resistant doors, walls, and floors that prevent the spread of fire to adjacent areas.

firebrick: Brick manufactured from refractory ceramic material that resists high temperatures.

fire clay: Clay that is highly resistant to heat and does not deform. Used in the manufacture of brick.

fire cock: See *fire hydrant*.

fire cut: Angular cut along the width of a joist supported by a masonry wall. Wall remains intact in case the joist burns through and falls.

fire damper: Device that automatically seals an air duct in the event of a fire. Restricts the flow of air and passage of fire and smoke. FIRE CONTROL DAMPER.

fire division wall: Fire-resistant wall between fire areas that prevents the spread of fire. FIRE SEPARATION WALL, FIREWALL.

fire door: 1. Fire-resistant door assembly, including the frame and hardware, commonly equipped with an automatic door closer. Rated according to the amount of time it will prevent the spread of fire; e.g., 2-hour fire door will resist the spread of fire a minimum of 2 hours. The designations of fire doors are *Class A* (3-hour door)—openings in fire division walls and walls that divide a single structure into fire areas; *Class B* (1- or 1$\frac{1}{2}$-hour door)—openings in enclosures of vertical communications through structures and in 2-hour partitions providing horizontal fire separation; *Class C* ($\frac{3}{4}$-hour door)—openings in walls or partitions with a fire-spread rating of 1 hour or less; *Class D* (1$\frac{1}{2}$-hour door)—openings in exterior walls subject to severe flame exposure from the exterior of the building; *Class E* ($\frac{3}{4}$-hour door)—openings in exterior walls subject to moderate to light flame exposure from the exterior of the structure. **2.** Opening in a boiler or furnace through which fuel is added for combustion.

fired strength: Compressive and tensile strength of refractory concrete after its first firing and cooling.

fire endurance: Length of time an assembly or component resists failure resulting from exposure to fire.

fire escape: Continuous, unobstructed path of escape from a structure in case of a fire. Constructed of fire-resistant or noncombustible material.

fire extinguisher: Portable device de-

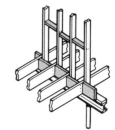

fire block

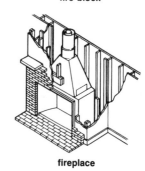

fireplace

signed to be used to suppress fire immediately. Designated as *Class A* (used for ordinary combustible material; e.g., wood, cloth, paper), *Class B* (used for flammable liquids/gases and greases), *Class C* (used for energized electrical equipment where the electrical nonconductivity of the extinguishing medium is important), and *Class D* (used for combustible metals; e.g., magnesium, sodium, potassium).

fire grading: See *fire-hazard classification*.

fire hazard: Potential that fire will ignite and spread, smoke and gases will be generated, or an explosion will occur in a structure, possibly endangering safety of occupants.

fire-hazard classification: Classification by the American Society for Testing and Materials of a fire hazard as high, medium, or low. Based on contents and type of operations being performed within the structure, or on the flame-spread rating.

fire hydrant: Supply outlet from a water main that provides a means for attachment for emergency equipment. FIRE PLUG.

fire load: Measure of the combustible contents of a structure per unit of floor area. Expressed as pounds per square foot or Btu per square foot.

fire partition: Interior partition with a fire rating of not less than 2 hours, but does not qualify as a fire division wall.

* fireplace: Opening at the base of a chimney in which combustible material, such as wood, can be safely burned to produce heat.

fireplace throat: See *throat* (1).

firepoint: See *flash point*.

fireproof: 1. Noncombustible; highly fire-resistant. **2.** To apply a material to a structural member to make it fire-resistant.

fireproofing: Material applied to a structural member to increase fire resistance. May be a chemical treatment or a coating, such as concrete.

fire protection: Practice of preventing fire or minimizing potential loss of life or property resulting from fire. Includes use of detection and/or suppression equipment, and training of occupants in a structure in fire safety and evacuation procedures.

fire rating: See *fire-resistance rating*.

fire-resistance rating: Measure of resistance of a material or component to failure when exposed to fire. Expressed as the number of hours a material or component will retain its integrity.

fire-resistive construction: Structure in which the members, including walls, floors, ceilings, partitions, columns, and roof, are of noncombustible material.

fire retardant: Chemical applied to combustible material to inhibit combustion and spread of fire.

fire separation: Wall, floor, or ceiling with a fire-resistance rating required by authorities having jurisdiction. Used to prevent spread of fire within a structure.

fire separation distance: Distance from a structure face to the closest lot line, adjacent building line, or center line of a street.

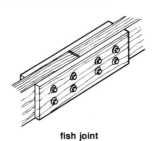

fish joint

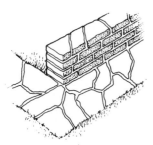

flagstone

fire stop: Material or member that seals open construction to inhibit spread of fire. DRAFT STOP.

fire tower: Vertical enclosure with a stairway having a high fire-endurance rating.

fire tube boiler: Boiler in which heat and gases of combustion pass through tubes surrounded by water. Includes scotch marine boiler, firebox boiler, and vertical fire tube boiler.

firewall: See *fire division wall*.

fire window: Window assembly with a fire-endurance rating required by code for the specific application.

firmer chisel: Long, narrow chisel used for mortising.

firsts and seconds (FAS): Highest grade of hardwood lumber, commonly used for long, wide cuttings. Must be 6″ or wider and 8′-0″ or longer, and produce a minimum of 83.33% clear cuttings.

fish beam: Composite timber beam consisting of two timbers placed end-to-end and joined with fish plates.

fish-bellied member: Structural member in which the bottom chord or flange is convex.

fisheye: Small, round void in an electric arc weld, or concrete or plaster finish.

* **fish joint:** Heavy timber joint in which two members are placed end to end and reinforced with fish plates on the sides.

fish plate: Reinforcement on the side of a butt joint. Commonly a steel plate.

fish scale: Wood shingle pattern in which ends of the individual shingles have semicircular ends.

fishtail: Wedge-shaped wood member inserted between tapered pans in concrete joist construction.

fish tape: Flexible cable that is pushed in one end of a conduit system until it comes out the other end. Flexible wires are then attached to the trailing end of the semi-rigid cable and pulled through the conduit. SNAKE.

fissure: Crack.

fitch: 1. Thin, long-handled paintbrush with bristles that are protected on the side. Used to paint in small recessed areas. **2.** See *flitch*. **3.** Member forming part of a flitch beam.

fitting: 1. Standard plumbing component used to join two or more pieces of pipe. **2.** Accessory, such as a bushing or other similar part of an electrical system, designed to perform a mechanical function rather than an electrical function.

fitting brush: Short brush with steel bristles arranged in a cylindrical design. Used to clean the interior of pipe and fittings.

fixed: Permanently attached, not movable.

fixed beam: Structural beam with fixed ends.

fixed glass: See *fixed light*.

fixed light: Window or section of a window that does not open. FIXED GLASS, FIXED WINDOW.

fixed window: Window unit with immobile pane of glass. Commonly installed adjacent to movable windows or in high-rise structures. See *window**.

fixing: Installing glass in a ceiling, wall, or partition.

fixture: 1. Device that holds an electric lamp and is secured to a wall or ceiling. **2.** Part of a plumbing system that provides access for water or waste disposal; e.g., lavatories, water closets, sinks. **3.** Device that secures components or members while they are machined, welded, or drilled. Does not guide the tool during the operation.

fixture branch: Water supply pipe that extends between the fixture supply pipe and distribution pipe.

fixture drain: Drain that extends from a fixture trap to a junction with a drain pipe.

fixture fitting: Device used to control or guide flow of water into or around a plumbing fixture.

fixture flow rate: Measure of the amount of water that can pass through a fixture per unit of time. Determined by dividing the number of gallons of water that passes through a fixture in one minute by 7.5. FIXTURE UNIT FLOW RATE.

fixture joint: Connection of two electrical conductors formed by crossing the bare ends, wrapping one conductor around the other, and folding them over.

fixture stud: Threaded fitting used to fasten a light fixture to an outlet box.

fixture supply: Water supply pipe extending from the fixture to the branch pipe.

fixture unit: Measure of the amount of water that is delivered to or discharged from a fixture per unit time. One fixture unit is equal to $7\frac{1}{2}$ gallons of water per minute. See *water supply fixture unit*.

fixture vent: Vent pipe which extends from a drainage pipe to the atmosphere or another vent.

flagging tape: Brightly colored strands of plastic used to mark surveying reference points or the edges of dangerous areas.

* **flagstone:** Slab of limestone or sandstone ranging from 1″ to 4″ thick used for paving, decorative retaining walls, and walkways.

flakeboard: See *particleboard, oriented strand board.*

flaking: Separation of a paint coating from the surface.

flame impingement: Combustion flame striking the interior surface of the furnace to complete the consumption of fuel.

flameproof: Able to resist spread of flames and not easily ignited. Less fire-resistant than a fire retardant material.

flame retardant: Able to resist spread of flame.

flame safeguard: Device that automatically shuts off fuel supply to a burner when burner ignition is not functioning properly.

flame-spread rating: Numerical designation, created by the American Society for Testing and Materials, given to a structural material to indicate its resistance to combustion. Ranges from red oak flooring with a designation of 100 to cement-asbestos board with a designation of 0.

flammable: **1.** Easily ignited and rapidly combustible. **2.** Liquid with a flash point below 100°F.

flanch: To widen and slope the top of a chimney stack to divert water away from the flue. FLAUNCH.

flange: **1.** One of the parallel faces of a structural steel member that is perpendicular to the web. **2.** Excess cloth or fiber material along the edge of batt insulation that allows fastening of the insulation into position. **3.** Collar that is perpendicular to the axis of a pipe, fitting, or valve with holes that allow for attachment to another pipe, fitting, or valve.

flange dovetail: Sheet metal joint used to join cylindrical ductwork to a flat surface. Flange is formed on the cylindrical duct, opening is cut in the flat member to the inside dimensions of the duct, sleeve with a series of slots cut into the perimeter along one end is inserted into the duct, and the flanges formed by the slots on the sleeve are bent over. Flat member is secured between the flanges on the sleeve and the flange on the cylindrical duct. Holes are drilled through duct and sleeve and secured with rivets. See *seam* (1)*.

* **flange union:** Method of joining two sections of threaded pipe by screwing flanges onto the ends and bolting the flanges together.

flank: Side of an arch.

flanking path: Path by which sound is transmitted around a member that is designed to inhibit the transmission.

flap valve: Plumbing device used to prevent backflow in a pipe. Consists of a hinged disc that obstructs the pipe when backflow occurs. See *check valve.*

flap-wheel sander: Disc with short strips of abrasive material that rotates in an electric drill to sand flat and contoured surfaces.

flare: To widen the diameter of a pipe with a gradual taper.

flare fitting: Mechanical fitting for soft metal tubing. One end of the tubing is flared and sealed with a mechanical device such as a coupling or nut.

* **flaring tool:** Tool used to flare the end of soft metal pipe. Available in various sizes for different size tubing.

flash: **1.** Color variation on the surface of a masonry unit, such as a brick, resulting from the surface fusion. **2.** To install flashing. **3.** Material forced out a weld joint and formed around the weld.

flashback: Recession of a flame into or in back of the mixing chamber of a torch.

flash coat: **1.** Thin coating of metal less than .002″ thick. **2.** Thin coating of shotcrete applied to a concrete surface to conceal imperfections.

* **flashing:** Pieces of sheet metal or plastic installed in conjunction with exterior finish materials to prevent water from penetrating a structure. Commonly in-

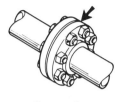

flange union

flaring tool

stalled around wall openings and intersections of walls and roofs.

flash point: Minimum temperature of liquid at which sufficient vapor is produced to form a flammable mixture with air.

flash set: Rapid development of hardness in freshly mixed portland cement concrete, mortar, or paste with the production of a considerable heat. Water must be added to the mix to regain plasticity. GRAB SET, QUICK SET.

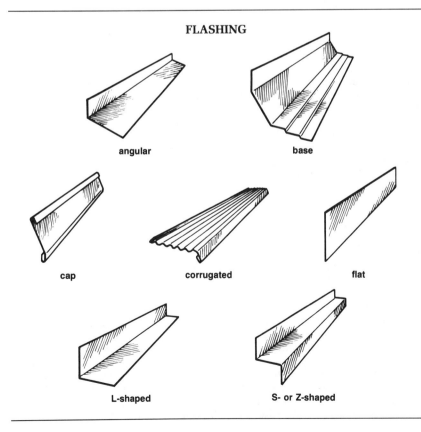

FLASHING

angular

base

cap

corrugated

flat

L-shaped

S- or Z-shaped

flat position

flatted

flexible connector (1)

flash-type cove base: Cove base formed with a continuous piece of resilient flooring that covers the floor and extends onto a wall or partition. Creates a small radius at the intersection of the floor and wall.

flash welding: Resistance welding process that joins metals simultaneously over the entire area of abutting surfaces using heat obtained from resistance to electric current between the surfaces. Pressure is applied to the weld joint after heating is completed.

flat: 1. Having little or no slope. 2. One floor of a multilevel structure or an apartment occupying one floor. 3. Low-gloss paint finish.

flat arch: See *jack arch*.

flat cable cutter: Hand tool with long parallel jaws used to cut wide, flat electrical cable without distorting the cable.

flat coat: See *filler coat*.

flat flashing: Sheet metal member manufactured in a continuous flat piece. Commonly cut and formed at the job site to custom shapes. Installed around

wall openings to prevent water leakage. See *flashing**.

flat-grained: See *plain sawn*.

flat head screw: Screw or bolt with a flat top and a conical bearing surface. Upper surface of top is flush with surface of workpiece. See *screw**.

flat Howe truss: Prefabricated roof or floor unit with a horizontal top chord and reinforced with alternating vertical and angular struts. Angular struts are inclined toward the center of the truss. See *truss**.

flat nose pliers: Pliers with a long blunt nose and knurled jaws. Used in tight locations. See *pliers**.

flat paint: See *flat*.

flat plate: Concrete slab without column capitals or drop panels.

flat plate collector: Solar collector panel that converts sunlight into heat.

flat pointing: Process of finishing mortar joints in which the joints are flush with the surface of the work.

* **flat position:** Welding position in which the workpiece is approximately horizontal and is welded from the upper side of the joint.

flat Pratt truss: Prefabricated roof or floor unit with a horizontal top chord and reinforced with alternating vertical and angular struts. Angular struts are inclined away from the center of the truss. See *truss**.

flat-rolled steel: Steel members that are manufactured by rolling steel between two flat rollers.

flat roof: Roof with a nearly horizontal surface. Slight amount of pitch is required to facilitate drainage of rainwater and other precipitation. See *roof**.

flat seam: Sheet metal seam between two adjacent pieces. Formed by turning up the adjoining edges, folding them over, flattening them, and soldering the joint. See *seam* (1)*.

flat slab: Concrete slab reinforced in two or more directions with beams and/or girders to transfer loads to supporting members.

flat spot: Dull finished area on an otherwise reflective surface.

flat stock: Metal machined to a precision flatness and thickness. Used for fabricating machine parts, shims, and gutters.

* **flatted:** Surface of a log that has been straightened and smoothed.

flat tie: Concrete formwork hardware made of a flat piece of metal with perforations at each end. Used to secure wall forms a predetermined distance apart. See *tie* (1)*.

flatting: 1. Straightening wood veneer

that is buckled. 2. Removing the gloss from paint.

flat truss: Prefabricated roof or floor unit with a horizontal top chord reinforced with vertical struts. See *truss**.

flatwork: Concrete slabs such as sidewalks, driveways, and floors that require finishing.

fleche: Narrow spire above the ridge of a roof.

fleck: Small mark or spot on a wood member. Caused by irregular wood rays or growth characteristics.

fleet angle: Angle between the center line of a sheave to the location at which a rope winds onto a hoisting drum.

Flemish bond: Brickwork pattern in which headers and stretchers alternate in each course. Headers are aligned to the center of the stretchers of the adjoining courses. See *bond* (1)*.

Flemish cross bond: Brickwork pattern in which headers and stretchers alternate in each course with two additional headers replacing stretchers at each course.

Flemish diagonal bond: Brickwork pattern in which one course consists of alternating headers and stretchers and the adjacent course consists entirely of stretchers.

Flemish garden bond: Brickwork pattern in which three stretchers are set between each header. Headers are aligned to the center of the stretchers of adjacent courses.

fleuron: Small flower-like ornament.

* **flexible connector:** 1. Airtight connection between two ducts or a duct and a fan. Used to isolate noise and vibration. 2. Pipe connector composed of a nonmetallic material or metallic mesh. Used to minimize vibration transmitted by the piping. 3. Electrical connection that permits movement caused by expansion, contraction, vibration, or rotation.

flexible corner tool: Hand tool with a wide blade used to form plaster corners.

flexible metal conduit: Flexible electrical tubing with a circular cross section. Used for protecting electrical conductors.

flexible pavement: Pavement that distributes loads and remains in contact with the subgrade and generally uses a bituminous binder for stability.

flexural bond: Stress occurring between a prestressed tendon and concrete member resulting from application of an external load.

flexural rigidity: Measure of stiffness of a structural member. Calculated as the product of the modulus of elasticity

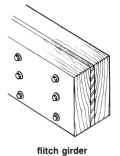
flitch girder

float (1)

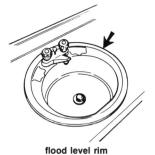
flood level rim

floor framing

and the moment of inertia divided by the length of the member.

flexural strength: Ability of a material to resist failure as a result of bending.

flier: Stair tread in a straight flight of stairs with parallel sides. FLYER.

flight: Continuous series of steps in a stairway without landings.

flint: Dense, fine-grained stone used in the manufacture of abrasive paper.

flint paper: Abrasive paper used for manual sanding and removal of paint. Has poor durability.

flitch: 1. Large timber flatted on two sides. **2.** Wood veneer stacked in the sequence in which it was cut from a tree. FITCH.

* **flitch girder:** Horizontal support composed of a steel plate bolted between two timbers. FLITCH BEAM, SANDWICH BEAM.

flitch plate: Steel plate bolted between two timbers to form a girder.

* **float: 1.** Hand tool used to finish concrete or plaster surfaces by providing an even texture to the material before it sets. Commonly made of wood, aluminum, or magnesium. **2.** Tool used to polish marble. **3.** To allow the blade of earth-moving equipment to rest on its own weight. **4.** Device that rests on the surface of a liquid and actuates a valve in a plumbing system. **5.** Flat trailer with a tilting bed and ramp used to transport heavy earth-moving equipment.

float coat: Layer of plaster applied to a surface with a float. Commonly applied between the scratch and finish coats.

floated coat: See *float coat.*

float finish: Rough concrete surface finish obtained by finishing the surface with a float.

float glass: High-quality, smooth-surfaced sheet glass. Manufactured by applying molten glass to a bed of molten metal.

floating: 1. Process of spreading plaster to an even thickness with a float. **2.** Process of finishing a mortar or concrete surface using a float. Commonly precedes troweling. **3.** Supported by, but not rigidly attached, to another structure.

floating foundation: Structure support designed to move with the surface of the ground without a loss of structural integrity. See *mat foundation, raft foundation.*

float rod: Threaded rod extending between a float and a flush valve in a water closet.

floatstone: Stone used to rub and remove rough marks from brickwork.

float trap: Valve in a steam heating system that is activated by a float that inhibits the flow of steam and allows air and condensation to flow through.

float valve: See *float* (4).

flocking: Short soft fibers attached to a fabric base for use in wall and floor covering.

flood coat: 1. Top layer of bituminous material of a built-up roof. Usually poured on and allowed to flow. **2.** See *flow coat.*

flooding: Method of curing concrete flatwork in which a layer of water completely covers the surface.

* **flood level rim:** Upper edge of a plumbing fixture from which water overflows.

floodlight: High-intensity lamp used to illuminate an object or area to a brightness considerably brighter than its surroundings. Usually can be pointed in various directions.

floodplain: Land that is subject to water coverage; commonly located along a river or stream.

floor: 1. Surface within a structure upon which a person walks. **2.** One level of a structure. STORY.

floor chisel: Steel chisel with a 2″ to 3″ wide blade. Used to remove flooring boards.

floor clearance: Distance between the finish floor and bottom of a door.

floor clip: See *sleeper.*

floor drain: Waste receptacle located in a floor that receives drainage from the floor and conveys it to the drainage sys-tem. Drain is covered with a grate; floor is usually sloped toward the opening.

floor flange: Circular metal fitting used to secure pipe to a flat surface such as a floor. Fitting is tapped in the center with standard pipe threads, which allows for attachment of the pipe.

* **floor framing:** Components, such as sills, joists, bridging, and headers, used to construct wood floors.

floor hinge: See *double-acting spring hinge.*

floor hole: Opening in a floor with a minimum dimension of 1″ and a maximum dimension of 12″.

flooring: Material used to construct the walking surface of a floor.

flooring nail: See *screw nail, ring-shank nail.*

floor joist: Horizontal wood or light-gauge metal member that supports a floor. Wood members are commonly 2 × 8s, 2 × 10s, or larger.

floor load: Weight placed on a floor, including live and dead loads.

floor opening: Opening in a floor with a minimum dimension of 12″.

floor plan: Plan view of a structure that shows arrangement of walls and partitions as they appear in an imaginary section taken horizontally approximately 5′-0″ above floor level. Electrical, plumbing, HVAC, and finishing information are also noted. See *print*.

floor plug: See *floor receptacle.*

floor receptacle: Electrical receptacle installed in a floor in such a way that the face of the receptacle is flush with the finished floor. FLOOR PLUG.

floor register: Grate or opening installed in a floor for use as an outlet in an HVAC system.

floor sander: Electrical finishing equipment used to smooth hardwood floors. Large abrasive belts or abrasive sheets are rotated to produce the sanding action.

floor tile: Modular unit of resilient material, such as vinyl, rubber, and linoleum, used as a finish floor covering.

floor underlayment: See *underlayment.*

floriated: Decorated with floral ornamentation or design.

florid: Very ornately decorated.

floury: Very fine soil texture.

flow: 1. Continuous motion of fluid in pipes, ducts, and other openings. **2.** Measure of the consistency of freshly mixed mortar, concrete, or cement paste. **3.** Characteristic of paint that allows it to spread and fill brush marks.

flowability: 1. Ability of a liquid or plastic material to spread or flow. **2.** Ability of molten filler metal to spread over a metal surface.

flow brazing: Brazing process in which metal is joined by heating the metal by pouring molten nonferrous filler metal over it to obtain brazing temperature. Filler metal is deposited in the joint by capillary action.

flow coat: See *flood coat.*

flow cone: 1. Device used to measure the consistency of concrete grout in which a predetermined volume of grout is allowed to escape through a hole of a specified size. **2.** Mold used to prepare a specimen for a flow test.

flow control valve: Device that prevents natural circulation of water in a hot water system when hot water is not being conveyed through the system.

flow gradient: Downward slope of a drain.

flow pressure: Water pressure in a water supply pipe near a faucet or other outlet while the faucet or outlet is fully open and water is flowing.

flow rate: Volume of water utilized by a plumbing fixture within a specified amount of time. Expressed as gallons per minute (GPM).

flow valve: Valve that closes when liquid velocity or pressure reaches a predetermined level.

flow welding: Welding process in which the metal is heated by pouring molten filler metal over it to obtain welding temperature. Filler metal is deposited in the joint by capillary action.

flue: Heat-resistant passage in a chimney that conveys smoke or other gases of combustion from a fireplace or furnace.

flue connection: Passage used to convey gases of combustion from a fuel-fired appliance to a vent pipe.

flue lining: See *flue pipe.*

flue loss: Undesirable loss of heat from an appliance into a flue.

* **flue pipe:** Pipe that transports smoke and gases of combustion from a furnace or fireplace to the atmosphere. Made from fire clay, terra cotta, or sheet metal and has a round, rectangular, or square cross section. FLUE LINING.

fluidifier: Admixture that decreases the flow factor of grout without affecting the water content. PLASTICIZER.

fluing: Expanding or splaying; e.g., window jamb that is wider at the bottom than at the top.

fluing arch: Arch that is wider at one end than at the other end.

flume: Inclined chute used to convey liquid.

fluorescent lamp: Electric discharge lamp in which a phosphor coating transforms ultraviolet energy, which is produced by the discharge, into light.

flush: 1. In the same plane as the surrounding surface. **2.** To expel waste by washing with a large volume of liquid.

* **flush bolt:** Door bolt with a shaft that slides in a sleeve mounted on the face of a door or mortised into the door. Hole in the jamb or keeper receives the sliding shaft to secure the door.

flush door: Door with two flush faces; stiles and rails are concealed within the door. See *door*.*

flush joint: Mortar joint in which the mortar is finished even with the face of the work. See *flat pointing.* See *mortar joint*.*

flushometer valve: Plumbing valve actuated by direct water pressure to supply a fixed quantity of water for flushing purposes.

flush paneled door: Door in which the surface of the panels on one or both sides are flush with the stiles and rails.

flush plate: 1. Door ornament that is mortised into the door and flush with the face of the door. **2.** Cover on an electrical box designed to finish flush with the surrounding finish material.

flush rim: Portion of a plumbing fixture, such as water closet or urinal, through which water is released to flush the inner surfaces of the fixture.

flush tank: Tank that retains a supply of water used to flush one or more plumbing fixtures.

flush valve: Plumbing valve located at the bottom of a flush tank that allows a discharge passage to flush waste from a plumbing fixture.

* **flute:** Concave groove formed in a member, such as a column.

fluted block: Concrete masonry unit with a series of parallel grooves on the face. Used as an ornamental finish block. See *concrete masonry unit*.*

flute reamer: See *reamer.*

flux: 1. Material used in soldering to prevent formation of oxides or facilitate removal of oxides and other undesirable superficial substances. **2.** Petroleum by-product used to soften asphalt. **3.** Magnetic lines of force extending between the north and south poles in a magnet.

fly ash: 1. Fine residue resulting from combustion of ground or powdered coal. Used as an ingredient in hydraulic cement.

flyer: See *flier.*

* **flying bond:** Brickwork pattern in which a header is set at intervals of four to seven stretchers. MONK BOND, YORKSHIRE BOND.

flying buttress: Characteristic feature of Gothic architecture in which a straight, sloping section of masonry is supported by an arch and a solid pier or buttress.

flying form: Large prefabricated concrete forms designed to be moved as a single unit. See *gang form.*

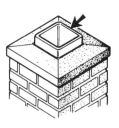

flue pipe

flush bolt

flute

flying bond

flying scaffold: Platform suspended from cantilevered beams using ropes or cables. SUSPENDED SCAFFOLD, HANGING SCAFFOLD.

fly rafter: Common rafter that is parallel with a gable and projects from the face of the gable to create an overhang along the gable end.

flywheel: Heavy wheel attached to a rotating shaft to reduce fluctuation in rotation.

foam: Lightweight cellular material that contains a large amount of air or gas.

foamed concrete: Lightweight cellular concrete formed by adding prepared foam or by generating gas in the un-hardened mix.

foamed polystyrene: Lightweight non-porous material used as thermal and sound insulation.

fogging: Process of spraying freshly placed concrete with a fine mist of water to increase hydration in the final set. FOG CURING.

foil: 1. Metal formed into a thin sheet by rolling; e.g., thin sheet of metal on batt insulation. **2.** See *trefoil, quatrefoil*.

foil-backed gypsum board: Panel product made by laminating aluminum foil to the back of fire-rated gypsum panels. Foil is used as a vapor retardant when applied with the foil next to the framing members.

folded-plate construction: Construction in which flat steel, concrete, or timber members are rigidly connected to each other at an angle to form an accordion-like design. Used to support a load over a long span. HIPPED-PLATE CONSTRUCTION.

folding door: Door formed with panel sections joined along the vertical edges with hinges and supported by rollers in a horizontal upper track. See *bi-fold door, accordion door*.

* **folding rule:** Measuring tool made of short graduated sections fastened together to allow them to pivot. Sections are unfolded for measuring and folded back together for convenient carrying. Various graduations include inch and fractional measurements, foot and decimal measurements, and brick layout. ZIGZAG RULE.

* **folding stair:** Series of ladder sections that are hinged and attached to an overhead door with spring hinges. Entire assembly is fastened in a ceiling scuttle to provide access to an attic or loft.

foldstir mixer: T-shaped device with several projecting wings. Device is rotated to consolidate and mix drywall joint compound or other semifluid material in a bucket.

foliated: Ornamented with designs in the shape of leaves, flowers, and branches.

follower: 1. Structural member that is an extension of a pile. Used to drive the pile below ground or beyond the reach of the pile hammer. FOLLOW BLOCK. **2.** Portion of a tap or die that keeps the thread straight.

foot: 1. Measure of length equal to 12″. **2.** Base of a column. **3.** Lowest step in a stairway. **4.** Lowest supporting member of a piece of furniture. **5.** Projection on a cylindrical roller that is used to compact fill or other porous material.

* **foot bolt:** Door bolt consisting of a shaft that slides in a sleeve mounted on the face of a door along the bottom edge. Keeper or hole in the floor receives the sliding shaft to secure the door.

footcandle: Measure of illuminance. Equal to 1 lumen per square foot.

foot cut: See *seat cut*.

footing: Section of a foundation that supports and distributes structural loads directly to the soil. FOOTER, SPREAD FOOTING.

* **foot lift:** Pivoting device used to lift and position gypsum drywall. One end is placed under the lower edge of the drywall; pressure is applied downward to the other end to lift the drywall.

foot-meter rod: Surveying rod graduated in feet and tenths of a foot on one side and meters and centimeters on the opposite side.

footpiece: Section of ductwork in an HVAC system used to change the direction of air flow.

foot plate: See *sole*.

foot-pound: Measure of mechanical work. Equal to the distance that the object is displaced (measured in feet) multiplied by the applied force (measured in pounds).

foot screw: See *leveling screws*.

foot valve: 1. Check valve located at the lower end of a suction pipe. **2.** Foot-operated valve.

forced-air heating/air conditioning: Method of heating or cooling an area by means of a blower that conveys heated or cooled air through a series of ducts and air returns to heat or cool a structure.

forced circulation: Circulation of air, water, or other medium by mechanical means such as a fan or pump.

forced convection: Transfer of heat resulting from the forced circulation of air, water, or other medium by fans or pumps.

forced draft: Air flow produced by mechanical means such as a fan.

forced drying: Accelerating the rate of drying in paint by introducing moderate heat up to 150°F.

forced fit: Joining of two members by forcing them together without using fasteners.

forecourt: Entrance plaza.

forehand welding: Welding technique in which the welding gun or torch is directed toward the weld.

fore plane: Wood smoothing hand tool approximately 18″ long. See *plane*(1)*.

forepole: Support member used to secure the roof or walls of a tunnel during excavation.

foresight: Point to which a surveying instrument is sighted to establish elevation and/or horizontal position.

folding rule

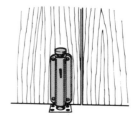

folding stair

foot bolt

foot lift

FOUNDATIONS

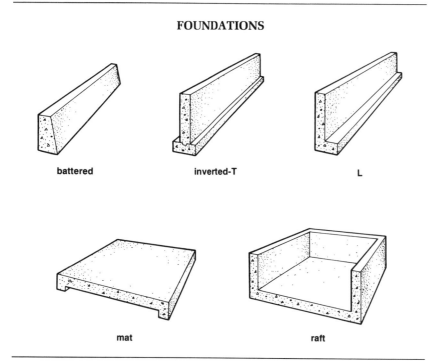

battered   inverted-T   L

mat   raft

forge welding: Solid-state welding process in which metal is joined by heating the workpieces in a forge and applying pressure or blows to permanently deform the intersection of the members.

forklift: Hoisting equipment with horizontal arms that project from the front and are moved vertically to transport and position heavy objects.

form: Temporary structure or mold used to retain and support concrete while it is setting and hardening.

Formica®: See *plastic laminate.*

form liner: Material used to line the face of concrete forms. Used to impart a smooth or patterned finish to the concrete or absorb moisture from the concrete. FORM LINING.

form lines: Contour lines drawn from visual observation and without accurate elevation information.

form oil: Non-staining oil applied to the interior surface of formwork to facilitate removal of the forms. FORM COATING.

form stop: See *bulkhead.*

form tie: Device made of high tensile strength steel used to prevent forms from spreading caused by the pressure of freshly placed concrete. See *tie* (2).

formwork: Entire system of support for freshly placed concrete. Includes forms, hardware, and bracing.

Forstner bit: Wood-boring tool that creates a flat-bottomed hole. See *bit* (1)*.

45° elbow: Pipe or conduit fitting that forms a 45° angle between two adjoining pipes or conduit. See *pipe fitting*.

* **foundation:** Primary support for a structure through which the imposed load is transmitted to the footing or earth.

foundation plan: Print that shows materials used for and placement of piers, footings, foundation walls, and basement columns and beams. See *print*.

four-way switch: Electrical switch used in conjunction with two three-way switches to provide control of a circuit from three or more locations.

foxtail saw: See *dovetail saw.*

* **fox wedge:** Wedge driven into the end of a tenon to spread the tenon and secure it in the mortise. FOXTAIL WEDGE, FOX TENON.

foyer: Entrance hall or lobby.

fractable: Coping on the gable of a structure where the gable projects above the roof line.

fraction: Numerical expression of part of a whole; e.g., $1/4$ denotes one part of a whole that is divided into four equal parts.

frame: 1. Skeleton of a structure or component that encloses and supports other structural components. **2.** To construct the skeleton of a structure or component. **3.** Decorative member that encompasses a flat area or feature for the purpose of highlighting the area. **4.** Prefabricated section of scaffolding.

frame construction: Structure constructed primarily of wood structural members.

framework: See *frame* (1).

framing: See *frame* (2).

framing anchor: Sheet metal device used to join and reinforce joints between wood framing members, such as studs, joists, and rafters.

framing square: L-shaped layout tool used to lay out angles and calculate lengths when laying out and erecting framing members. One leg is 24" long and the other leg is 16" long. Often embossed or etched with tables used for roof framing, board foot calculation, octagonal layout, and brace length calculation.

framing square gauge: See *stair gauge clamp.*

Franki pile: See *pressure-injected footing.*

* **Franklin stove:** Freestanding wood-burning stove.

free area: Total open space in a section view of a duct or pipe through which air or material flows.

freeboard: Distance between the normal highest fluid level in a container and the top of the container.

free end: Cantilevered section of a beam or joist.

* **free fall: 1.** Dropping freshly mixed concrete into formwork without drop-chutes, tremies, or other means of confinement. **2.** Distance that freshly mixed concrete is dropped. Measured from the bottom of the means of con-

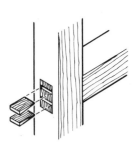

fox wedge

Franklin stove

free fall (2)

finement, such as a dropchute, to the final position of the concrete.

free-hanging hammer: Device suspended by cables without the support of leads. Used to drive piles.

free moisture: Moisture that is not absorbed by aggregate and has the same properties as pure water in bulk.

freestanding: Not fastened or attached to a support.

freestone: Fine-grained stone used for ornamental work. Commonly limestone or sandstone.

freeze-thaw cycle: Process of repeated cooling below 32°F and warming above 32°F. Temperature change results in expansion and contraction of materials caused by change of composition of water molecules.

freezing point: Temperature at which a material solidifies as a result of absence of heat. Freezing point of water at sea level is 32°F.

freight elevator: Elevator, usually larger than a pedestrian elevator, used to move heavy loads on which the operator and persons required for unloading are permitted to ride.

French arch: See *Dutch arch.*

French curve: See *irregular curve.*

French door: Swinging door with panes of glass that constitute nearly all the surface area. Commonly used as an exterior door. See *door*.*

French drain: Trench filled with rock and covered with soil. Used to collect excess water before it is absorbed into the soil. BOULDER DITCH, RUBBLE DRAIN.

French roof: See *mansard roof.*

French window: 1. Casement window that extends from the floor to the ceiling. **2.** See *French door.*

French window lock: See *cremorne bolt.*

Freon®: Nontoxic and nonflammable fluorocarbon refrigerant.

frequency: Number of periods or oscillations per unit of time.

fresco: Mural painted on fresh lime plaster.

fresh air inlet: See *vent.*

fresno trowel: Concrete finishing tool with a long extension handle.

fret: Ornamental design in which raised or depressed bands form a continuous rectangular shape.

fret saw: See *compass saw.*

friable: Easily powderized or crumbled.

friction: Resistance to movement between two mating surfaces.

friction catch: Device that, when engaged with a catch, is secured into position with friction. Commonly used on cabinet doors. FRICTION LATCH, BULLET CATCH.

friction loss: Stress loss in a prestressing tendon resulting from friction between the tendon and duct.

friction pile: Load-bearing pile that receives support from friction between the pile and the surrounding soil. Used when firm bearing soil is too deep to provide support for a bearing pile.

friction ring: Shaft seal component that provides a gastight seal.

friction tape: See *electrical tape.*

friction welding: Solid-state welding process in which heat is produced by a mechanically induced sliding motion between mating surfaces. Workpieces are secured together under pressure.

* **frieze: 1.** Horizontal exterior trim member used as a transitional piece between the top of the siding on a wall and the soffit or roof sheathing. **2.** Middle horizontal member of an entablature between the architrave and the cornice. **3.** Carpet yarn that is tightly twisted and has a rough appearance.

frit: Partially fused mixture of sand, soda, and limestone used for the manufacture of glass.

* **frog: 1.** Recess in the bed surface of a masonry unit. Provides a means for the unit to interlock with the mortar joint. **2.** Plane component that secures the plane iron.

frontage: Length of a property line or building line along a street or body of water forming a boundary.

frontage road: Roadway along a highway that provides access to the highway and adjoining property.

frontal: See *pediment.*

front-end loader: Excavating equipment consisting of a bucket and assembly used to lift the bucket. Attached to the front of a tractor. Bucket is raised or lowered using hydraulic or pneumatic means.

front hearth: Fireproof platform projecting from a fireplace into a room at the floor level or slightly above floor level. OUTER HEARTH.

frontispiece: See *pediment.*

frosted: Finished with a fine granular surface to eliminate a glossy finish. MATTE.

frost heave: Lifting of a soil surface or pavement as a result of freeze-thaw cycle occurring in the underlying soil.

frost line: Measure of the maximum penetration of frost in the soil in a given geographic location. Depth of penetration varies with climatic conditions. Foundations below this depth are not affected by freeze-thaw cycle.

frostproof closet: Water closet that does not contain standing water, thereby unaffected by freezing temperatures.

fuel: Material used to produce energy.

fuel oil system: Heating system that burns fuel oil in a furnace to produce heat.

fugitive dye: Dye whose color fades in a few days with exposure to ultraviolet light. Used to color membrane-curing compounds so as to observe coverage of a surface.

full bath: Bathroom with a water closet, lavatory, and bathtub or shower.

full bond: Brickwork pattern in which all bricks are headers.

full frame: Frame in which all joints are mortised and tenoned.

full-load current: Electric current that rotating equipment draws from a power source while operating at the rated voltage, speed, and torque.

full mortise-and-tenon: Wood joint in which a channel is cut entirely through one member (full mortise) to receive a projecting portion of adjoining member (tenon). End of tenon is exposed. See *joinery*.*

* **full pitch:** Roof incline equal to two units of rise for each unit of run; e.g., 24 on 12 pitch. Total rise is equal to total span.

full-size detail: Detail drawing scaled to actual size.

full-surface hinge: Hinge installed on the surface of a door and jamb without mortising.

furlong: Lineal measure equal to ¼ mile.

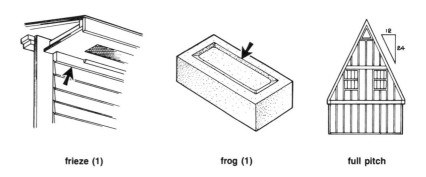

frieze (1)   frog (1)   full pitch

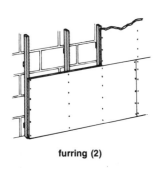

furring (2)

furnace: 1. Heating system in which heat is transferred from the point of combustion to the air supplied to the system. **2.** Section of a boiler in which combustion occurs.

furnish: Wood by-products such as sawdust and shavings used in the manufacture of panel products such as particleboard and fiberboard.

furred: 1. See *furring.* **2.** Encrusted with lime or salt, such as the interior of water heating pipes.

*** furring: 1.** Application of narrow strips of wood or metal to a structural surface to create a base for fastening finish materials. Air space is created between the finish material and structural surface. **2.** Wood or metal strips fastened to a structural surface to provide a base for fastening finish material. FURRING STRIPS, FURRING CHANNELS.

furrowing: 1. Striking off mortar joints to create a V shape. **2.** Process of making V-shaped grooves in a bed joint of a masonry wall with a trowel. Used to increase the speed of bricklaying.

fuse: Electrical overcurrent protection device with a fusible portion that is heated and broken by the passage of excessive current. Used to open a circuit and prevent overheating.

fuse box: Electrical service panel containing fuses for electrical circuits.

fuse clip: Current-carrying part of a fuse support device used to engage the fuse, fuseholder, or blade.

fuseholder: Device that mechanically supports a fuse and connects it to the circuit.

fuse link: 1. Replaceable assembly composed primarily of conductive material; used to open a circuit in the event of excessive heat or current. FUSIBLE LINK. **2.** Replaceable piece composed of metal with a low melting point. Used to close a fire-activated door closer in the event of excessive heat. FUSIBLE LINK.

fusestat: Electrical overcurrent protection device with a time delay that permits an electrical overload for only a short period of time.

fusetron: Electrical overcurrent protection device that carries an electrical overload for a specified interval of time before failure.

fusible metal: Alloy that melts at a low temperature, usually below 300°F.

fusion: Melting together of two members, commonly metal or plastic.

fusion zone: Area of the base metal that is melted, as determined in a cross section of the weld.

fust: Shaft of a column or pilaster.

gabion: Retaining wall or other bank stabilization constructed of rock randomly piled and encased in rectangular sections of heavy wire mesh.

* **gable:** Three- or four-sided section of an end wall that extends from the top wall plate to the ridge. GABLE END.

gableboard: See *bargeboard*.

gable roof: Roof with gabled ends that slopes in two opposite directions from the ridge. See *roof**.

gablet: Small triangular-shape ornament over a window or door.

gable window: 1. Gable-shaped window. **2.** Window built into a gable.

gad: Pointed chisel or rod used to break and loosen rock.

gaffs: Sharp spurs strapped to the inside of both ankles for climbing wood poles. Spurs stick into the pole to gain a foothold.

gage: See *gauge*.

gain: Mortise; notch to receive another member.

gallery: 1. Raised section of an auditorium or large theater used to increase seating capacity; a mezzanine. **2.** Tunnel that collects water in a rock or concrete dam. **3.** Long covered corridor.

gallet: Stone chip created by chiseling. SPALL.

galleting: Filling in joints of a rubble-stone surface with small stone chips set in the fresh mortar. GARRETING.

galley tile: Quarry tile with indented face.

gallon: Liquid measure equal to 8.33 pounds of water.

gallons per day (GPD): Measure of volume of fluid flow. Used to measure large volumes of continuously flowing fluids such as rivers.

gallons per minute (GPM): Measure of fluid flow. Used to measure small volumes of intermittently flowing fluids such as pump discharges.

galvanic action: Interchange of electrical charge between two metals immersed in an electrolytic solution.

galvanize: To coat iron or steel with a thin layer of zinc, which inhibits oxidation and corrosion.

gambrel roof: Roof sloped in two opposite directions from the ridge, with each surface divided into two sections with different pitches. Lower section of the roof slopes steeply to the outside walls. See *roof**. BARN ROOF.

* **gang form:** Prefabricated form panels used to construct a larger form to retain concrete during placement and until it sets. Used to facilitate erection and stripping of forms. GANGED FORM, GANGED PANEL FORM.

* **gang nail:** Steel plate with a series of projections pointed outward. Used to fasten wood truss and heavy timber components.

gang saw: Power tool with a series of parallel saw blades. Used to cut logs into boards or large stone into slabs.

gangway: Elevated temporary passageway.

ganister: Ground quartz and fire clay composition used as fireproofing for furnaces.

ganosis: Dulling on the face of polished marble.

gantry: Framework that supports a lifting device or equipment.

* **gantry crane:** Crane mounted on a pivoting platform that is suspended above the work area. Travels on a track.

gap-graded aggregate: Mixture of aggregate consisting of large and small particles. Intermediate-size particles are omitted.

garage: Enclosed structure for parking and protecting motor vehicles and general storage.

gable

gang form

gang nail

gantry crane

gas cock

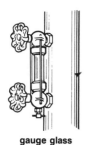

gate hook

gate valve

gauged arch

gauge glass

---

garbage disposal: Device installed in a sink drainage pipe that chops waste material into a size that can be removed by standard plumbing waste piping. WASTE DISPOSAL.

garden apartment: 1. One-, two-, or three-story multifamily dwelling with access to a common garden space.

2. Apartment with floor level below grade with access to an outdoor area.

garden wall bond: Masonry pattern in which three to five stretchers are placed between headers in each course. See *bond* (1)*.

garland: Ornamental design in the shape of draped flowers.

garnet paper: Abrasive paper coated with finely crushed garnet. Used for finishing and polishing.

garreting: See *galleting*.

garrett: See *attic*.

gas: Matter without shape or volume.

*ᐧ**gas cock:** Quick-closing valve for stopping flow of gas.

gas concrete: Lightweight concrete formed by introducing gas into the plastic mix to create voids. Gas is generated from the chemical reaction between cement and aluminum powder admixture.

gas cylinder: Metal container used to store and transport compressed gas.

gas-fired: Heated by the combustion of natural gas or gasoline.

gas/fuel system: System that burns gas to sustain combustion and create heat.

gasket: Soft, pliable material inserted between two mating surfaces to prevent liquid or gas leakage.

gaskin: Rope ring used as caulking in a stoneware pipe joint prior to sealing with mortar.

gas metal arc cutting: Process used to sever metal by melting it with an electric arc produced between a consumable electrode and the metal being cut.

gas metal arc welding: Process used to join metals by heating them with an electric arc produced between a consumable electrode and base metal.

gas meter riser: Pipe fitting that connects an underground gas line to a meter above ground.

gas outlet: Connection in a gas piping system to which equipment is attached.

gas pliers: Scissors-like tool with concave, toothed jaws for gripping small pipes.

gas pocket: See *porosity*.

gas pressure regulator: Control device that delivers gas to the point of combustion at a given pressure.

gas shielded arc welding: See *gas metal arc welding*.

gas tungsten arc welding: Process used to join metals by heating them with an electric arc produced between a nonconsumable tungsten electrode and base metal. Gas or gas vapor shielding is used.

gate: Passageway installed as part of a fence.

gate hinge: Ornamental hinge used for exterior applications. Active leaf is similar to strap hinge.

*ᐧ**gate hook:** L-shape metal device screwed or driven into the side of a fence opening with its shank projecting upward. Hinge is placed over shank to support the gate.

gatehouse: Small building adjoining an entrance gate.

*ᐧ**gate valve:** Device for regulating the flow of liquid in which a threaded stem raises and lowers an inner plate or wedge-shaped member. The position of this piece in the inner chamber restricts or opens the flow through the device.

gauge: 1. Brick or masonry units that are cut to produce a taper. Narrower end is placed along the interior of a curve to create mortar joints of equal thickness. **2.** To add plaster of paris to a mortar mix to decrease hardening time. **3.** Spacing of wheels or tracks on a piece of earth-moving equipment. **4.** To rub the face of a brick to create a contrasting surface with the surrounding work. **5.** Exposed portion of a shingle or other roofing material. **6.** To cut or shape a masonry unit. **7.** Measure of thickness for glass or metal. **8.** Measure of diameter for drill bits or wire. **9.** Horizontal distance between the inside of two railroad tracks. **10.** See *dividers*. GAGE.

gauge board: 1. Template used to lay out tread and riser locations on a housed stringer. **2.** Flat board on which mortar or plaster is mixed.

*ᐧ**gauged arch:** Arch in which each brick has been cut so as to produce a tapered brick. Narrower end is placed along the inside of the arch to create mortar joints of equal thickness.

gauged brick: Masonry unit cut or ground to an exact dimension or shape.

gauged mortar: Mixture of sand, cement, lime, water, and an additive of plaster of paris to speed hardening of the mix. GAUGED STUFF, GAUGING PLASTER.

*ᐧ**gauge glass:** Clear window or tube that allows for visual inspection of fluid levels in a device.

gauge glass cutter: Cutting tool used to cut glass tubing by scoring the glass either on the inside or outside.

gauge line: Layout mark made parallel with the length of the member for positioning fastener holes in structural steel.

gauge size: Width of the cutting edge of a drill bit.

gaul: Void in plaster or mortar resulting from improper troweling.

* **gazebo:** Open air structure with a decorative roof supported on posts or columns. Often a hexagonal or octagonal shape.

gear: Circular component with teeth or notches cut into the perimeter. Used to transmit rotating motion to another component.

geared chuck: Drill chuck in which a toothed key is fitted into a circular recess and rotated to loosen and tighten the jaws around the bit. Standard method used for electric drills.

G elbow: Sheet metal fitting that forms 90° angle in rectangular ducts. 90° angle is parallel to longest duct dimension.

gemel: Two building components considered a pair. CHYMOL, GIMER, GYMMER, JIMMER.

general diffuse lighting: Lighting in which 40% to 60% of the emitted light is distributed upward and the remainder downward.

general-purpose branch circuit: Electrical branch circuit that supplies a number of lighting and appliance outlets.

general-use snap switch: Electrical switch constructed to be installed in flush device boxes or outlet box covers. Capable of interrupting current in general-purpose branch circuits.

general-use switch: Electrical switch capable of interrupting current in general-purpose branch circuits. Rated in amperes.

generation capability: Maximum electrical demand a generator is capable of incurring from electrical equipment without exceeding temperature and stress limits.

generator: 1. Mechanical or electromechanical equipment that creates electrical power from mechanical energy. 2. Component of an absorption system that uses heat to separate refrigerant from a solution.

* **geodesic dome:** Dome-shape structure constructed of a series of straight lightweight members that form a triangular

gridwork.

geodetic survey: Survey in which the curvature of the earth is considered when determining reference points of large areas.

geometric: Consisting of straight and curved lines and surfaces.

geometrical stairway: Elliptical, curved, or spiral stairway.

Georgia buggy: See buggy.

Georgian architecture: Architectural style popular in England during the 18th century. Characterized by a rectangular-shape structure with windows and doors equally spaced.

* **geotextiles:** Sheets or rolls of material that stabilize and retain soil or earth in position on slopes or other unstable conditions.

gesso: Hard, plaster base coat used for painting of gilding.

gib: 1. Guide between a diesel pile hammer and the lead. 2. Steel strap used to connect two members.

gild: To apply a decorative finish of gold or brass.

gillet: Horizontal timber used to guide a cable or rope that raises and lowers a bucket in a shaft.

Gillmore needle: Measuring device used to determine the setting time of hydraulic cement.

gimlet: Small hand tool that holds a wood-boring bit at 90° to the handle. Used for boring small holes.

gin pole: Vertical post secured at the base with the top braced with at least three guy lines. Hoisting device is attached to the top of the pole. Pole is tilted in the direction of the object being hoisted.

* **girder:** Large horizontal structural member constructed of several steel, reinforced concrete, or timber members that support loads at isolated points along its length.

girt: 1. Horizontal supporting member in balloon framing used to support floor joists. LEDGER. 2. Horizontal stiffener placed at an intermediate level between studs, posts, or columns.

GIRTH. 3. Horizontal stiffener placed around the perimeter of a structure.

girth: Circumference of an object. GIRT.

girth hitch: Knot used to secure a mason's line to a nail.

glacial gravel: Non-graded aggregate without fines.

glacial till: Deposit of aggregate, sand, clay, and boulders. Good load-bearing capabilities.

gland: Soft pliable ring placed on a pipe or fitting and compressed to form a seal. Installed before a joining pipe. GLAND RING, GLAND SEAL.

glass: Hard, brittle, inorganic material that is normally transparent or translucent. Composed of silica, flux, and stabilizer, which are heated together.

glass block: Hollow opaque or transparent block made of glass. Used in non-bearing walls and partitions. GLASS BRICK.

* **glass cutter:** Hand tool with a small sharp wheel or diamond tip mounted on one end. Used to score glass.

glass fiber: See fiberglass.

glass stop: 1. Trim molding that secures glass in a door or window. GLAZING BEAD. 2. Component at the lower end of a patent glazing bar that prevents the glass from sliding down.

glass-wool insulation: Spun glass fibers used as thermal and acoustical insulation and as a filter in an air or a water filter.

glassy: Smooth and shiny.

glaze: 1. To install glass and trim members in a window or door. 2. To fill a rough grinding stone or wheel with fine metal particles during sharpening. 3. To smooth the surface of the lead in a cast iron pipe joint by heating the surface. 4. Ceramic, glossy surface coating.

glazed brick: Masonry unit with a smooth and shiny surface.

glazed tile: Ceramic or masonry tile with a smooth and shiny surface. FACE TILE.

glazement: Waterproof surface finishing material that creates a smooth, hard surface.

gazebo

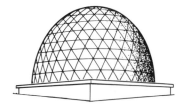

geodesic dome

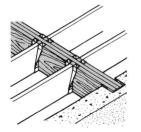

girder

glass cutter

glazier's point   glitter gun   globe valve

glazier: Tradesperson who cuts and installs glass. Duties include caulking, puttying, and installing rubber and plastic gaskets around the glass.

glazier's chisel: Tool with a sharp edge used to apply putty and glazing compound around glass.

* **glazier's point:** Small triangular- or diamond-shape piece of sheet metal used to secure glass in a wood frame. After glass is fitted in the opening, a sharp edge is pushed into the wood and the projecting part secures the glass. Installed before applying glazing compound. GLAZING BRAD, SPRIG.

glazier's point driver: Spring-loaded tool that drives glazier's points used to hold glass in position.

glazier's putty: Glazing compound consisting of plaster of paris and linseed oil. Used as pliable mastic when setting glass into a frame.

glazing bar: See *muntin*.

glazing bead: Wood trim member fastened in a window sash or door to retain glass in position.

glazing compound: Putty or caulking compound placed around the perimeter of glass to seal the joints between the glass and the sash or frame.

* **glitter gun:** Device used to spray and embed glitter flakes in fresh plaster walls.

* **globe valve:** Compression valve used to control fluid flow. Circular disk is compressed against a seat by means of a threaded spindle. Produces turbulence

in flow due to fluid changing direction within the valve.

globular transfer: To convey large drops of molten metal from a consumable electrode across an electric arc.

glue: General description of adhesive products.

glue block: Small wood member fastened with adhesive to the inside corners of cabinets and furniture. Used to strengthen the joints.

glue gun: Hand or power tool used to apply heated or room temperature adhesives.

glulam (glued-and-laminated): Description of a structural member constructed by bonding several layers of lumber with adhesive. Commonly used for large curved structural members.

glyph: Short ornamental groove.

go-devil: Burlap, paper, or fabricated device inserted into pump end of a pipeline and forced through the pipe with water pressure.

gooseneck: 1. Curved part at the upper end of a finish handrail. Used to connect the straight section of handrail to a newel post. **2.** Arched section between a truck tractor and trailer. **3.** Curved, flexible fitting or pipe.

gorgerin: Portion of a column where the shaft and the capital intersect; the neck.

Gothic: Architectural style popular in Europe from 12th to 15th century; characterized by pointed arches, flying buttresses, and relatively tall structures.

gouge: 1. Chisel with a rounded cutting edge that creates a groove or channel in the material, commonly wood or stone. GOUGE CHISEL. **2.** To remove metal from base metal by forming a groove or bevel.

government anchor: V-shape anchoring device used to reinforce the connection between masonry and structural steel members.

grab bar: Hand grip installed in an area where additional support is needed for standing or climbing, such as a shower or bathtub.

gradation: Distribution of granular particles among various standard sizes. Expressed as the percentages of materials that are larger or smaller than given sieve openings.

grade: 1. Level or elevation of the earth on a job site. See *finish grade, natural grade*. **2.** To change the surface of the earth by mechanical or manual means. **3.** Roadbed that has been prepared for the application of the base or subbase paving materials. **4.** Mixture of various size aggregate to ensure correct proportion of large and small particles. **5.** Measure of quality for lumber products or the durability of brick. **6.** Downward slope of a waste pipe.

* **grade beam:** Reinforced concrete beam placed at ground level. Member is supported by piles or piers at the end and intermediate positions.

graded aggregate: Aggregate in which the particles conform to a standard grading system.

graded sand: Aggregate less than $1/4$" in diameter in which the particles conform to a standard grading system.

grade line: Line usually marked with stakes to indicate the elevation relative to a common reference point.

grade mark: Impression in steel that identifies the quality of the member.

* **grader:** Earth-moving equipment used for leveling and grading. Single pivoting blade is mounted below the frame and may be elevated on one or both ends.

grade rod: 1. Rod graduated in feet and hundredths of a foot. Used for surveying purposes. See *engineer's scale*. GRADING ROD. **2.** Piece of wood or metal that is marked to assist in verifying elevations when surveying.

grade stake: Marker placed in the ground that has an elevation noted.

grade strip: Thin piece of wood nailed to the inside of concrete forms to act as a guide for the top of a concrete lift.

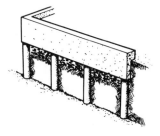

grade beam

grader

gradient: Change in elevation of a member or material in a given distance. Expressed as a percentage; e.g., a pipe pitched vertically 5′ for each 100′ of length has a 5% gradient.

gradual switch: Device that regulates compressed air or gas pressure.

graduate: 1. To mark a surface with incremental divisions. **2.** Progressive change in size from smaller to larger.

graffito: Plaster surface design applied over a colored base coat. Top layer is scratched to expose a portion of the lower layer for a decorative design.

grain: Pattern of fibers in wood. Strength of member is greatest in the direction of the grain.

graining comb: Painting tool with a number of teeth. Used to apply a surface coating simulating woodgrain.

grandstand: Structure designed to support a large group of spectators.

granite: Extremely hard stone that is cut and polished for use in building construction.

granolithic concrete: Concrete with granite or other hard aggregate. Used in high wear or high abrasion areas. GRANOLITH.

granular: Made of many non-cohesive particles of approximately the same size, such as sand or loose-fill insulation.

graph: Visual representation used to facilitate understanding of an idea; e.g., bar graph.

grappler: Wedge-shape device driven into a masonry joint to provide an attachment point for a scaffold.

grasscloth: Wallcovering made from woven vegetable fibers that are laminated to a paper backing.

* **grate:** Frame used to cover openings to drains or other disposal systems. Allows for passage of fluid and small solid material but restricts passage of larger objects. GRILL.

gravel: Granular material that varies in size from approximately $1/4''$ to 3″ in diameter and is retained on a No. 4 sieve. Results from natural disintegration of rock or crushing and processing of rock.

* **gravel strip:** Metal flange installed around perimeter of a built-up roof. Used to retain gravel and prevent it from falling off the roof. Also seals the outside roof edge to prevent leaks. GRAVEL STOP.

gravity convection: Movement of fluid resulting from warm fluid rising and cool fluid dropping.

gravity feed: Supply system in which material is conveyed by the gravitational pull of the earth. GRAVITY FLOW,

GRAVITY SUPPLY.

gravity hammer: Pile hammer that drives a pile into the ground only by the gravitational pull of the earth exerted on the hammer.

gray cast iron: Cast iron containing a large amount of free graphite.

grease cup: Cup-shape lubricating fitting into which grease is placed.

grease extractor: Exhaust equipment mounted above cooking equipment that traps grease particles in baffles in the exhaust system.

grease gun: Hand pump that accepts and distributes grease for lubricating. Hose on the end of the gun is attached to grease fitting and transfers grease to equipment.

grease trap: Plumbing device that separates fats and grease from waste water. Baffles slow flow of waste water and allow grease to cool, solidify, and float to the top of the trap where it is retained and removed. GREASE INTERCEPTOR.

Greek architecture: Architectual style developed in the Mediterranean region characterized by elliptical and parabolic shapes. Three orders are Doric (simplest), Ionic (curved and scrolled), and Corinthian (very decorated and flowered).

green: 1. Condition of building material that has not reached its final set; e.g., green concrete. **2.** Unseasoned, freshly cut; e.g., green lumber.

Greenfield: See *flexible metal conduit.*

greenheart: Hardwood species used for marine piles because of its immunity to marine borer attack.

greenhouse: Glass-enclosed structure with temperature and humidity control. Used to support plant growth.

grid: Arrangement of perpendicular lines at regular intervals.

grid ceiling: See *suspended ceiling.*

griffe: Ornamental design at the base of a column shaped like a tongue, leaf, or claw.

grill: Perforated or louvered cover installed at the outlet or intake of an air duct.

grillage: 1. Temporary shoring consisting of timber, steel, or reinforced concrete beams laid perpendicular to one another. Used to spread a heavy load over a large area. Installed in an existing building during repairs to the foundation. **2.** Suspended metal members with perpendicular metal channels attached to the bottom to form a suspended ceiling support.

grille: See *grill.*

grind: To shape an object or material by abrasive action.

grate

gravel strip

grinder

* **grinder:** Tool used to shape an object or material by abrasive action.

grindstone: Flat piece of sandstone used for sharpening, shaping, and polishing.

grip: 1. Thickness of the material or part a fastener is designed to secure in position. **2.** Thickness of material or part through which a rivet passes.

grip-handle lock: Door lockset in which the bolt is activated on the outside of the door by a thumbpiece. See *lock*.*

grip length: Minimum length of steel rebar required to be anchored in concrete. Expressed as the number of bar diameters. BOND LENGTH.

grit: Granular abrasive material applied to cloth, paper, or wheels used for sanding, grinding, polishing, and for non-skid surfaces.

groin: 1. Curved edge at the intersection of two vaults. **2.** Retaining wall that is built perpendicular to the shoreline to prevent erosion of sand and scouring.

groined ceiling: Ceiling with two vaults that intersect.

groined rib: Wood or stone member projecting from a ceiling to cover the intersection of two vaults.

groin point: Intersection line of two vaults.

* **grommet:** Circular eyelet used to reinforce a hole in fabric, canvas, or metal.

groove: 1. Indentation or depression in an otherwise flat surface. **2.** Channel

grommet

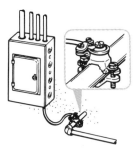

groover

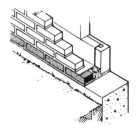

ground (1)

ground course

in wood surface that is parallel with the grain.

grooved seam: Sheet metal seam formed by folding the two adjoining edges, joining them, and interlocking the edges with a hand groover. Forms a flush interior surface and a ridge along the exterior surface.

groove joint pliers: See *tongue-and-groove pliers.*

* **groover:** Concrete finishing tool used to form control joints or grooves in a slab before it hardens. Constructed with a flat face with a projecting rib.

groove weld: Weld made in a gap between two pieces of metal.

gross: Total; entire amount; e.g., gross weight.

gross vehicle load: Total weight of a vehicle and load.

grotesque: Ornamentation involving distortions of animal and human forms.

grotto: Natural cavern or cave.

* **ground: 1.** Intentional or accidental electrical connection between a circuit or equipment and the earth or other conducting member. **2.** Guide for the application of plaster or other pliable finish that is used as a straightedge and thickness gauge. **3.** First coat of paint on a surface. GROUND COAT. **4.** Pulverized or crushed into smaller size pieces. **5.** Finished with an abrasive surface; e.g., ground glass.

ground beam: Horizontal support near the surface of the earth that ties wall and/or column footings together.

ground coat: Base coat; initial layer.

ground coupling: Closed pipe or duct system installed below grade for use as a heat exchanger. Fluid is circulated between the earth and a structure to maintain a consistent temperature within the structure.

* **ground course:** Horizontal row of masonry units closest to the earth.

grounded conductor: Intentionally grounded electrical system.

ground effect: Occurrence in electric arc welding at a point where the current changes direction in which flux builds up and causes the arc to move in a direction opposite the position of the ground.

ground-fault circuit-interrupter (GFCI): Device that automatically de-energizes a circuit or a portion of a circuit when the grounded current exceeds a predetermined value that is less than required to operate the overcurrent protection device. Used to protect personnel.

ground-fault protection of equipment: Electrical equipment protection provided by disconnecting the equipment

from the power source when a predetermined current-to-ground value is obtained.

ground floor: Level of a building that is closest to the level of the earth; first level above grade.

ground glass: Glass with light-diffusing characteristics produced by grinding the surface with an abrasive material.

grounding conductor: Conductor used to connect electrical equipment or a circuit with grounding electrode(s).

grounding electrode conductor: Conductor connecting the grounding electrode to the equipment grounding conductor and/or the grounded conductor of the circuit at the service equipment.

ground joint: 1. Tight-fitting masonry joint without mortar. **2.** Tight-fitting machined metal joint.

ground-key valve: Valve in which a slotted circular plug or key rotates in a cylindrical or conical seat to allow fluid to flow.

ground plane: Horizontal plane of projection in perspective drawings.

ground pressure: Weight of an object divided by the surface area in contact with the earth.

grounds: See *ground* (2).

groundwater: Water near the surface of the earth; absorbed through the subsoil.

ground wire: 1. Thin, high-strength steel wire used to establish a grade for shotcrete. ALIGNMENT WIRE, SCREED WIRE. **2.** See *grounding conductor.*

grouped columns: Three or more vertical posts placed on a common pedestal or spaced closely together.

group vent: Branch vent connected to at least two individual traps.

grout: Fluid mortar mixture consisting of cement and water with or without aggregate.

grout bag: Conical-shape container with a nozzle at the pointed end for the application of grout.

grouted-aggregate concrete: Concrete formed by placing coarse aggregate and injecting a grout into it.

grout saw: Hand tool with a blade designed to quickly remove hardened grout between ceramic tile.

growth ring: Annual cross-sectional growth layer of a tree. ANNUAL RING.

grozing iron: 1. Tool for finishing soldered pipe joints. **2.** Type of glass cutter.

grub: To remove roots, stumps, and undergrowth from an area.

grub axe: See *mattock.*

grub saw: Hand saw used to cut stone.

guard: Device on power equipment used to protect operator from possible dan-

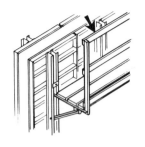

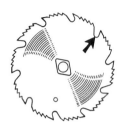

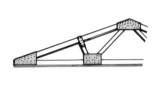

| guardrail (1) | gullet | gusset | gutter bracket |

---

ger during operation.

guarded: Covered, enclosed, or protected to prevent persons from being endangered.

* **guardrail: 1.** Protective railing fastened approximately 42″ to 45″ above a deck or plank. **2.** Horizontal divider between opposing lanes of traffic or along the road edge.

gudgeon: 1. Metal rod used to secure adjoining masonry units. **2.** See *gate hook.*

guided-bend test: Method of determining strength of a weld by bending a sample to a predetermined shape.

guild: Association of tradesworkers with common interests.

guilloche: Ornamental design made of curved lines that intersect and form waves with circular openings between each wave. Circular openings are filled with round ornaments.

guinea: Wood stake driven to grade and marked with blue paint.

* **gullet:** Low area between the teeth of a saw blade.

gumming: Process of deepening gullets of a saw blade after repeated sharpening of the blade.

gun: 1. Shotcrete delivery and application equipment consisting of double chambers under pressure. **2.** See *spray gun.*

gun finish: Shotcrete as it is applied to a surface without hand finishing.

gunite: See *shotcrete.*

* **gusset:** Plate or brace fastened to the face of two members to secure them together. GUSSET PLATE.

gutta: Pendant-shape ornamental design.

gutter: 1. Channel along lower edge of a roof used to conduct water to downspouts. **2.** Groove along edge of pavement for conducting water into catch basins. **3.** Channel used to supplement wiring space adjacent to distribution centers and switchboards. Encloses conductors and busbars, but not switches or overcurrent protection devices.

gutterboard: Vertical member attached to ends of rafters. Used as attachment surface for gutters. See *fascia.*

* **gutter bracket:** Connector used to secure a gutter along lower edge of a roof. GUTTER HOOK.

guttering: See *gutter.*

gutter tool: Trowel with a $4^1/_2$″ sidewall and a 6″ bottom wall. Used to finish concrete gutters.

guy: 1. Rope, wire, or cable used to stabilize a vertical member. **2.** To brace a vertical member with a line.

guy derrick: Lifting device with a mast (central pole) anchored at bottom and guyed in vertical position. Boom is hinged at bottom of the mast. Entire assembly swivels to lift and move objects and materials.

gymnasium: Large open area used for physical education or athletic events.

gypsum: Soft mineral (hydrous calcium sulfate) used as a core for drywall and other plaster products.

gypsum board: See *drywall.*

gypsum cement: See *Keene's cement.*

gypsum sheathing: Exterior wallboard consisting of a water-repellent gypsum core with a water-repellent surface paper.

habitable space: Area used for living, sleeping, eating or cooking. Excludes bathrooms and storage areas.

hack: 1. To divide one course of stone into two courses. Used when large stone is not available. **2.** To lay brick with bottom edge recessed from the surface of the wall. **3.** To roughen a surface by striking with a tool. **4.** To stack bricks in a kiln or on a kiln car.

hacking knife: Hand tool used to remove old glazing compound or putty from a window.

hacksaw: Metal-cutting handsaw with adjustable steel frame for holding various lengths and types of blades. Blades are inserted with the teeth pointing away from the handle. See *saw*.

haft: Handle for a cutting or marking tool such as an awl.

hairline crack: Thin cracks with random pattern that do not completely penetrate a surface.

hairpin: 1. U-shape concrete formwork hardware. **2.** Light gauge U-shape rebar. Used for shear reinforcement in beams and columns. **3.** Inverted U-shape, gravity-driven pile hammer. Used to initially drive sheet piles.

half bat: One-half of a brick.

half bath: Bathroom consisting of a water closet and lavatory.

half-corner sash block: Concrete masonry unit shaped on one end to accept a door or window sash. See *concrete masonry unit*.

half hatchet: Striking and cutting tool used for applying wood shingles. One end has a flat broad face and the other end is a straight sharpened blade.

half hitch: Knot for temporarily fastening a rope for a 90° pull. See *knot*.

half lap: Joint formed by two members of equal thickness in which half the thickness of adjoining faces is removed. Surface of the adjoining members is flush when fitted. See *joinery*.

* **halfpace landing:** Stair platform between two flights of stairs that make a 180° turn. HALF LANDING, HALF SPACE LANDING.

half pitch roof: Roof in which the rise is equal to one-half the span.

half round: Semicircular trim member.

half-round file: File that is flat on one face and convex on the opposite face. Used for shaping and smoothing metal and wood. See *file*.

half section: Drawing that shows the interior of one-half of the object while showing external features of the other half. Primarily used in showing symmetrical objects.

half story: Upper floor of a structure in which the walls are formed by vertical studs and the bottom of the sloping roof rafters.

* **half-timbered construction:** Construction style consisting of exposed wood members with the holes between the members filled with plaster, masonry, or other exterior finish material.

half turn: See *halfpace landing*.

* **halide torch:** Refrigerant testing device that indicates a leak by a change in color. Natural blue flame changes to green when refrigerant vapor is detected.

hall: 1. Large room or assembly area. **2.** Corridor connecting two or more rooms. HALLWAY.

halon fire extinguisher: Fire-suppression system that inhibits the spread of fire by covering the area with bromotriflourimethane, an odorless, colorless, and non-conductive gas. Used in areas that house electronic and computer equipment.

halfpace landing

half-timbered construction

halide torch

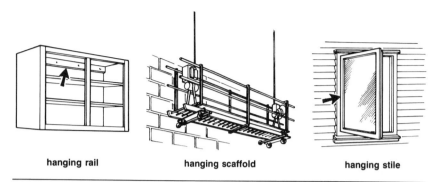

hanging rail   hanging scaffold   hanging stile

*hammer: 1. Striking or splitting tool with a hardened head fastened perpendicular to the handle. 2. Steam-, pneumatic-, or electric-powered device used to drive piles.

hammer drill: Electric or pneumatic tool used to bore holes in hard surfaces such as masonry and concrete. Hardened bit is rotated and moved in a reciprocating motion. See drill*.

hammer finish: Surface texture produced by applying powdered metal which gives the appearance of hammered metal. HAMMERED FINISH.

hance: 1. Horizontal trim member at the top of a window that conceals a light source. 2. Small arch that connects a larger arch or lintel to a jamb.

hand: 1. Direction of door swing. When determining door hand, the viewer stands on the outside of the door (hallway or street side of an entrance door). Left-hand door has hinges on the left and the door swings away. Left-hand reverse door has hinges on the left and swings toward the viewer. Right-hand door has hinges on the right and swings away. Right-hand reverse has hinges on the right and swings toward the viewer. 2. Direction of stair turn when descending a stairway. Right-hand refers to a stairway in which a person turns clockwise when descending. Left-hand refers to a stairway in which a person turns counterclockwise when descending.

hand brace: See bit brace.

hand drill: Hand tool used to drill into thin and soft material. Manually operated crank rotates the drill bit. See drill*.

hand float: Hand tool used to prepare concrete flatwork for troweling or to produce a textured surface. Flat face is constructed of aluminum, magnesium, or wood.

hand level: Hand-held surveying instrument used for approximate leveling over a small area. Consists of a small metal viewing tube in which a spirit level is viewed opposite the horizontal crosshair.

hand lever punch: Hand tool used to punch holes in sheet metal. WHITNEY PUNCH.

handrail: Horizontal or angular member along the edge of a stairway or platform. Used as a support or protective barrier to prevent falls.

handrail bolt: Straight threaded fastener that accepts a nut at each end. Inserted in a mortised portion of a butt joint to secure the members.

handsaw: Wood-cutting hand tool consisting of a straight, toothed blade attached to a handle. Blade is moved back and forth against wood to produce cutting action.

hand screw: Woodworking clamp consisting of two wood jaws connected by two parallel threaded rods. Jaws are adjusted to position by turning the rods. See clamp*. SCREW CLAMP.

hand signal: Hand and/or arm movement and position used to indicate specific operation to be performed. Commonly used in rigging or surveying where verbal commands cannot be used.

hand spike: Pole used to position a pile in the leads.

hand-split cedar shakes: Exterior finish material in which red cedar is split with the grain of the wood to create a rough-textured, irregular finish.

hanger: 1. Metal strap, wire, or rod that supports a suspended member such as a pipe or conduit. 2. Worker who installs gypsum drywall.

hanger bolt: Fastener with lag bolt thread on one end and machine bolt thread on the other end. Used in heavy timber construction.

hanging buttress: Freestanding buttress that projects from a wall. Decorative only; is not a structural support.

*hanging rail: Horizontal member fastened in the upper back of a wall cabinet to create a strong attachment point.

*hanging scaffold: Temporary work platform suspended from a structure with steel cables. Scaffold height is adjusted with manual cranks or electric motors.

*hanging stile: Vertical member of a door or casement window that accepts the hinges. BUTT STILE, HINGE STILE.

hardboard: Dense panel product contructed from wood fibers that are joined with heat and pressure. Available in standard, service-tempered, and underlayment grades. Used for applications where finished appearance is not important, such as cabinet backs and

HAMMERS

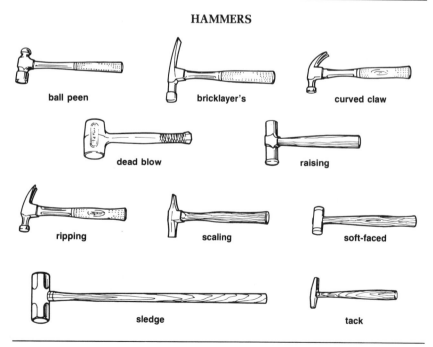

ball peen   bricklayer's   curved claw
dead blow   raising
ripping   scaling   soft-faced
sledge   tack

drawer bottoms.

hardened concrete: Concrete that has changed from a plastic to solid state through hydration.

hardener: Chemical compound applied to concrete to improve abrasion resistance.

hardfacing: Type of surfacing in which a coating is applied to base metal to reduce abrasion and erosion.

hardpan: Dense soil, clay, or gravel that is difficult to excavate.

hard solder: Solder with a melting point above that of alloyed lead and tin.

hardware: **1.** Fasteners, connectors, and trim members used in construction. Finish hardware includes knobs, latches, and hinges. Rough hardware includes nuts, bolts, and nails. **2.** Pins, wedges, or other devices used to secure concrete forms and form ties during concrete placement.

hardware cloth: Steel woven wire screen with $^1/_8''$ to $^3/_4''$ openings.

hard water: Water containing large amounts of minerals and salts.

hardwood: Lumber produced from deciduous and broadleaved trees. Does not refer to actual hardness of wood.

hardwood plywood: Plywood with various types of core material and one or two sides covered with hardwood veneer.

* **hasp:** Hinged metal device with a slot that fits around a U-shape piece that accepts a bolt or lock.

hatch: Opening in roof or floor of a building used for ventilation or passage of people or materials. Small door is used to cover the opening.

hatchet: Hand tool with straight, sharpened blade on one end and flat striking head on the other end.

hatchet iron: Plumber's soldering iron with a flat broad tip.

hatchway: See *hatch*.

haunch: **1.** Portion of an arch between the crown and point where it rests on a wall or column. **2.** Portion of a beam that has been deepened and increases in depth toward the support. **3.** Thickened portion of a concrete slab to support additional loads.

haunch board: Vertical formwork for the side of a concrete beam, stair, or slab.

haunched: Widened or thickened at the point of support.

* **hawk:** Square wood, aluminum, or magnesium plate used by plasterers to hold and carry plaster. Wood handle is attached to the bottom of the plate.

hawser: Rope that is 5″ to 10″ in diameter.

haydite: Expanded clay used as lightweight aggregate in concrete.

hazardous location: Area containing highly flammable or combustible products, vapors, or fumes.

haze: Dull finish.

H beam: See *HP shape*.

H boot: Sheet metal fitting used as a transition piece to connect a round duct to a rectangular duct. Longest dimension of rectangular duct is parallel with the run of the round duct. See *ductwork**.

head: **1.** Highest member of a member or structure. **2.** Column capital. **3.** Amount of fluid pressure between points at different elevations. Expressed in feet of fluid. **4.** Striking portion of a hand tool. **5.** Top of a pile.

head casing: Horizontal trim member at the top of a door or window jamb.

* **header:** **1.** Horizontal structural member over the top of a wall opening that distributes the load to either side of the opening. **2.** Floor or ceiling joist or roof rafter that is perpendicular to the common joists or rafters. Distributes loads to other joists or rafters around openings. **3.** Masonry member laid perpendicular to the face of the wall to tie the front and back portions of the wall together. See *brickwork**. **4.** Transverse raceway for electrical conductors in a cellular metal floor. Raceway has access points that permit the installation of conductors from a distribution box to different areas. **5.** Pipe with many outlets. Outlets are parallel and perpendicular to the center line of the pipe.

header block: Masonry unit with a recess for laying a brick in the header position. SHOE BLOCK.

header bond: Masonry pattern in which the masonry units are laid perpendicular to the face of the wall.

header course: Horizontal row of masonry in which the masonry units are laid perpendicular to the face of the wall.

head jamb: Uppermost member of a door or window jamb. YOKE.

head joint: Vertical mortar joint between the ends of masonry units. CROSS JOINT.

headlap: Upper part of a shingle covered by the shingle above it.

headless screw: Externally threaded fastener that is threaded over its entire length. Narrow slot is cut into the top of the fastener to allow turning with a straight screwdriver. See *screw**.

headroom: **1.** Vertical distance from front and upper surface of a stair tread to the ceiling above. **2.** Vertical distance from the floor to the bottom of a door jamb header.

headstone: Keystone or cornerstone.

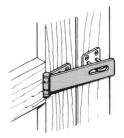

hasp

hawk

header (2)

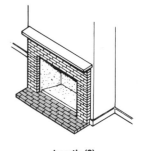

hearth (2)

headwall: Sidewall of a culvert.

heart bond: Masonry bond, used for walls too thick for stones extending through the wall, in which two headers meet in the middle of a wall and another header is set to cover the joint.

* **hearth:** **1.** Lowest horizontal portion of a fireplace. **2.** Fireproof floor immediately surrounding a fireplace.

heartshake: Defect in lumber in which splits radiate out from center of the log.

heartwood: Lumber from the core of a tree characterized by darker color and high resistance to decay. TRUE WOOD.

heat: Energy transferred resulting from temperature difference between two materials.

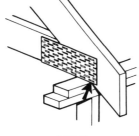

heat pump   heel (4)   helix (1)

heat-actuated door closer: Hardware device with fusible link that deteriorates at high temperatures and automatically closes a door.

heat-affected zone: Area of metal in welding or soldering in which the mechanical properties of the metal have been altered by high temperatures, but the metal is not melted.

heat duct: Ductwork designed to facilitate flow of heated air from a furnace or other heat source to desired area.

heated space: Area in a structure with a heat supply.

heater: Device designed to raise the temperature of a substance or an area. Includes stoves and other heat-producing appliances.

heat exchanger: Device that transfers heat between two fluids that are physically separated. Commonly composed of a cylinder with longitudinal tubes in which one fluid flows within the tubes and the other fluid flows through the cylinder.

heat gain: Net rise in temperature in a structure resulting from radiation from outside sources.

heating degree day: See *degree-day.*

heating element: Portion of electrical appliance that produces heat by electric current.

heating plant: System used to heat a building or group of buildings. Commonly includes a boiler, transfer piping, and radiators. HEATING SYSTEM.

heating surface: 1. Exposed portion of a heating unit. **2.** Portion of a boiler in which heat transfer occurs between water and a heating medium.

heating unit: Heat exchanger that transfers heat from hot water or steam to the air.

heating value: Heat released by combustion of a given volume of fuel. Expressed in Btu.

heating, ventilating, and air conditioning (HVAC): System and process of regulating temperature and surrounding atmosphere in a building. See *print*.*

heat load: Total amount of heat per unit time that must be supplied to a space, building, or group of buildings to maintain a desired temperature.

heat loss: Net decrease in temperature in a structure due to heat transmisssion to the outside.

*heat pump: Refrigeration system designed to carry heat to a transfer medium to provide cooling in the summer, and absorb heat for heating in the winter.

heat sink: Material or area where undesirable heat is discharged to.

heat transfer fluid: Liquid used to carry heat away from the source to be cooled.

heat transmission: Rate at which heat flows through material caused by conduction, convection, and radiation of heat.

heat trap: Device in a water heater outlet that equalizes water temperature. Prevents cool water from circulating in hot water pipes.

heave: Upward or lateral movement in a material resulting from freeze/thaw cycles, displacement, or other forces.

heavy concrete: See *heavyweight concrete.*

heavy-edge reinforcement: Welded wire fabric in which one to four edge wires have a larger diameter than the other longitudinal wires. Used for highway pavement slabs.

heavy joist: Timber between 4″ and 6″ thick and at least 8″ wide.

heavy timber construction: Structure in which fire resistance is achieved by using wood structural members, floors, and roofs of a specified dimension and composition.

heavyweight concrete: Concrete with greater density than normal concrete. Produced by using heavyweight aggregate. Used for radiation shielding. HEAVY CONCRETE.

*heel: 1. Portion of a beam, rafter, or joist that rests on the top of a wall or other support. **2.** Outside corner of a framing square. **3.** Trailing edge of an angled excavating blade. **4.** Bottom inside edge of a concrete footing.

heel bead: Bead formed with glazing compound that is applied after glass is set in place but before removable stop is placed.

heel cut: See *seat cut.*

heel plate: Gusset in a roof truss where the horizontal member joins the roof rafter.

height: Vertical distance between two points.

height of instrument (H.I.): Elevation of a line of sight when using a transit, builder's level, or other surveying instrument.

helical: In a spiral shape.

heliostat: Automatic adjusting mirror that follows the movement of the sun and reflects its rays toward a solar collector.

helium: Inert shielding gas used for gas tungsten arc welding and gas metal arc welding.

*helix: 1. Rebar bent to form a spiral shape. Used to reinforce concrete columns. **2.** Spiral shape used for ornamentation.

helm roof: Roof with four steeply pitched faces extending from four gables to form a spire.

helve: Handle for a striking tool such as a hammer or hatchet.

hermaphrodite caliper: Layout tool with two hinged legs. One leg is fitted with a marking point and the other leg hooked inward for layout reference.

hermetic compressor motor: Completely sealed electric motor with refrigerant and oil enclosed. Designed for single- and three-phase operation.

herringbone bond: Decorative masonry pattern in which rows of headers are set at 90° angle to each other to form zigzag design. See *bond (1)*.* HERRINGBONE, HERRINGBONE WORK.

hertz: Unit of frequency equal to 1 cycle per second.

hew: 1. To shape timber with an ax or hatchet. **2.** To roughly dress stone with a mallet and chisel. HEWE.

hexagon: Plane figure with six equal sides and six equal angles.

hexagonal asphalt shingle: Waterproof roof and wall finish material made by saturating felt paper with bituminous material. Cut into six-sided pieces measuring 12″ across and coated on the face with small aggregate. See *asphalt shingle*.*

hexagonal nut: Internally threaded fastener with a six-sided external profile. See *nut*.*

hexastyle: Structure with six columns across the front.

hex cap screw: Bolt with a six-sided

head and a shaft that is threaded its entire length. See *screw**.

hex head screw: Externally threaded fastener with a six-sided head designed to be turned by a wrench. See *screw**.

hex nut: Six-sided member with internal threads for mating to a bolt.

hex socket screw: Externally threaded fastener with a six-sided recess in the head. Designed to be turned with a hex key. See *screw**.

* **hex wrench:** Hexagonal-profile bar used to tighten and loosen hex socket screws.

H fitting: Pipe fitting designed to be used only as a vent.

H hinge: See *parliament hinge*.

hickey: 1. Device used to bend conduit, pipe, and rebar. Fitting at one end secures material and a long handle is used to gain leverage. **2.** Threaded electrical fitting used to connect a light fixture to an outlet box. HICKY.

hick joint: See *flush joint*.

hidden line: Dashed line on drawings or prints that indicates an object or part of an object that is obscured. Used on architectural drawings to indicate items such as cabinets, shelving, and interior soffits.

high-carbon steel: Steel with carbon content between 0.6% and 1.5%.

high centered: In the condition whereby tracks or wheels of excavating equipment become embedded in soft soil and allow the undercarriage to contact the ground. Further movement of the equipment is impossible.

* **high chair:** Device that supports rebars for concrete flatwork.

high-density overlay: Exterior grade plywood with a resin-impregnated fiber veneer. Smooth, hard surfaces that are highly resistant to abrasion and chemical damage. Used for concrete formwork and countertops.

high-discharge truck: Concrete mixer that dispenses concrete from upper end of the rotating drum.

high early-strength cement: Type III portland cement that hydrates faster

HINGES

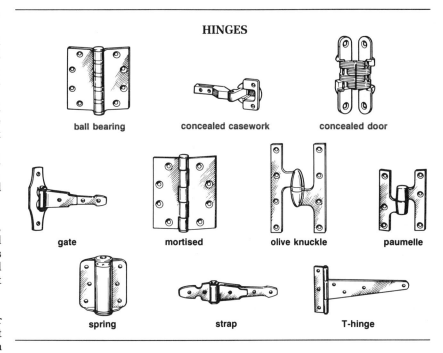

ball bearing

concealed casework

concealed door

gate

mortised

olive knuckle

paumelle

spring

strap

T-hinge

than normal cement and attains greater strength in shorter period of time.

high gloss: Shiny, reflective surface.

high-limit control: Temperature-sensitive switch that turns a heating system off when upper temperature limit is reached. Prevents overheating and damaging of components.

high line: Suspended electrical power cable.

high-pressure plastic laminate: Hard-surfaced veneers made of sheets of kraft paper impregnated with plastic resin and subjected to pressure between 1200 and 2000 pounds per square inch. Used for surfacing countertops and cabinetwork. Commonly called by the manufacuturer name ''Formica.''

high-pressure steam heating system: Heating system that uses steam at pressures exceeding 100 pounds per square inch to convey steam through pipes.

high relief: Sculptured relief in which figures project from the surface at least

one-half their thickness.

high rise: Building with many floors or levels above ground level.

high side: Sections of a refrigeration system that are subject to a pressure equal to condenser pressure.

highway: Public roadway including the right-of-way.

* **hinge:** Pivoting hardware that joins two surfaces or objects and allows them to swing around the pivot.

hinge strap: Ornamental plate on the face of a door that gives the appearance of a strap hinge.

* **hip:** External angle formed at the intersection of two adjacent sloped roof sections.

hip drop: Distance a hip rafter is lowered to allow for the thickness of the construction members and proper fitting of the jack rafters.

* **hip jack rafter:** Roof member that extends perpendicularly from the wall plate up to the hip rafter.

hex wrench

high chair

hip

hip jack rafter

hip knob: Ornament, such as a ball or finial, fastened to the apex of two opposing sloped roof sections.

hip rafter: Roof member that extends diagonally from the end of a ridge to the outside corner of the exterior walls.

hip roof: Roof formed by adjacent inclined planes that slope down toward all exterior walls. Intersecting corners of roof intersect at a hip rafter that extends diagonally from the ridge to the outside corners of the exterior walls. See *roof**.

hip truss: One of a series of prefabricated roof trusses that forms the end of a hip roof. Horizontal central portion along the top and sloped on each side. Height and length of horizontal portion varies with placement within series. See *truss**.

hitch: 1. Knot for temporarily fastening a rope to an object for a 90° pull. 2. Connection between two machines; e.g., trailer hitch. 3. Shelf in a rock wall used to support horizontal timbers.

H mold: See *H molding*.

H molding: H-shape metal or plastic trim member. Used to support and conceal a butt joint between two members of equal thickness. See *molding**. H MOLD.

* **hod:** V-shape trough with one enclosed end used to carry mortar on a job site. Attached to top of a pole and carried on a person's shoulder.

hod carrier: Person who mixes and transports brick, stone, and mortar for use by bricklayers and stone masons. Also helps erect scaffolding for brick-

hod

hoe: 1. Long-handled tool with a wide blade for leveling and moving fresh concrete. 2. See *backhoe*.

hog: Unlevel course in a masonry wall.

hoist: 1. Assembly of pulleys, sheaves, and/or mechanically driven motors for lifting materials or people. 2. To lift material or people.

hoist arm: Scaffold attachment that pivots and projects from the edge of a scaffold tower and supports a well wheel or other lifting device.

hoistway: Vertical shaft used to house an elevator or dumbwaiter. Extends from pit to underside of overhead machinery area.

* **hold-down clamp:** Device used to temporarily secure material to the top of a bench.

hole saw: Rotating saw blade with a mandrel mounted in the center. Used to cut larger holes than can be cut with drills. Primarily used to cut wood, fiberglass, ceramic, and soft metal such as copper. See *bit* (1)*.

hollow brick: Clay or shale masonry unit with a net cross-sectional area in a plane parallel to the bearing surface greater than 60% of the gross cross-sectional area.

hollow concrete block: See *hollow masonry unit*.

* **hollow-core door:** Door consisting of a corrugated center with veneer exterior. Edges of the door are banded with stiles and rails to allow for securing of hinges and fitting of the door.

hollow ground: Slight convex shape in cutting surface produced by grinding the edge with a grinding wheel.

hollow masonry unit: Masonry unit cast with voids in the center.

hollow metal: Light-gauge metal formed to produce jambs, window frames, and doors.

hollow-metal door: Door constructed with a light-gauge sheet steel exterior and reinforced with light-gauge steel channels around the edges. Insulating or fireproofing material may be incor-

porated in the door.

hollow newel stair: See *open newel stair*.

hollow punch: Hand tool with circular point designed to stamp holes and circles in sheet metal when struck with a hammer or mallet.

hollow wall anchor: Metal or plastic sleeve inserted in a hollow wall to secure a material or member. Sleeve is secured in or behind wall after a screw or other fastener is inserted. See *molly, toggle bolt*.

home run wiring: Electrical conductors that extend from the power source or central control point to the point of use.

homogeneous: 1. Made or formed of the same material. 2. Fitting together as one.

hone: 1. To sharpen a cutting tool to a very fine edge by drawing across a stone. 2. Stone used for sharpening cutting tools.

* **honeycomb:** 1. Voids in concrete caused by segregation and poor consolidation during concrete placement. 2. Cellular structure used to create voids in a material or member; e.g., hollow-core door.

honeycomb slating: Method of applying slate tile to a roof in which the diagonal of the tile is positioned horizontally. Bottom corners of the tile are removed.

honeysuckle: Flowered ornamentation.

hood: 1. Projecting protective cover or canopy. 2. Ductwork, connected to an exhaust fan, used to capture unwanted fumes or smoke. 3. See *welding hood*.

hood mold: Projecting trim over an arched door or window.

hook and butt joint: See *scarf joint*.

* **hook and eye:** Two-piece fastener consisting of a rod bent to form a circular opening that is threaded on the end and a rod that is curved at one end to fit into the opening. Two pieces are secured to adjacent surfaces.

hook bill knife: Hand tool with a curved cutting edge for trimming drywall and cutting detailed shapes. LINOLEUM KNIFE.

hold-down clamp

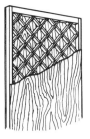

hollow-core door

honeycomb (1)

hook and eye

hopper: Vertical funnel-shaped chute used to temporarily store loose material, such as aggregate and sand.

* **hopper gun:** Spray gun and hopper combined to form one unit; used to apply drywall texture materials. Materials are gravity-fed through hopper and applied to the surface by using compressed air.

hopper window: Window in which the sash is hinged at the bottom of the frame and that opens inward at the top. See *window*. HOPPER FRAME, HOPPER LIGHT.

horizontal: Level; parallel with the horizon.

horizontal branch drain: Soil or waste pipe that extends from a soil or waste stack or building drain at less than a 45° angle. Receives discharge from fixture drains.

horizontal circle: Component of surveying instrument used to measure horizontal angles. Graduated in degrees, minutes, and seconds.

horizontal lap siding: Exterior wall-covering constructed of horizontal overlapping strips of wood or wood by-products. See *siding*.

horizontal pipe: Pipe that forms less than a 45° angle from horizontal.

* **horizontal position:** Welding position in which the axis of the weld is in a horizontal position and the face of the weld is in a vertical position.

horizontal sliding window: Window with sashes that travel horizontally on rollers in a track.

horse: 1. See *stringer, carriage.* 2. See *sawhorse.*

horsepower: 1. Mechanical unit of measure equal to the force required to lift 550 foot-pounds per second. 2. Electrical unit of measure equal to 746 watts.

* **hose bibb:** Faucet outlet threaded for attachment of a hose fitting. HOSE BIB.

hose clamp pliers: Hand tool with offset grooves formed in the jaws. Grooves are designed to fit around the projec-tions of a wire spring-tension hose clamp. See *pliers*.

hose station: Fire extinguishing system consisting of a valve, hose, nozzle, and storage rack.

hose stream test: Measure of fire resistance in which a material is subjected to one-half the indicated fire exposure and immediately sprayed with water of a predetermined pressure and distance.

hose thread: Standard screw thread having 12 threads per inch on a $3/4''$ pipe.

hospital arm pull: Handle with a hook that projects upward to enable a door to be opened by pulling with an arm.

hot-air heating: See *forced-air heating/ air conditioning.*

hot-dip galvanizing: Application of a protective coating for metal done by dipping the metal in molten zinc.

hot drawn: Shaped to a specific design while being heated, as applied to metal.

hot-melt adhesive: Glue that liquefies when heat is applied and solidifies when cooled. Provides excellent holding power.

hot mix: Heated paving material composed of aggregate and asphalt. ASPHALT, BLACKTOP.

hot-setting adhesive: Thermosetting glue requiring temperature of at least 212°F to set.

hot water: Water heated to at least 110°F.

hot water boiler: High- or low-pressure boiler that uses water to provide and transport heat.

hot water heating system: Method used to heat an area by conveying heated water from a boiler through pipes to heating units that radiate heat.

hot wires: Conductors carrying electrical current.

house: 1. To fit a member into another. 2. To enclose.

house drain: Main waste pipe in a residence. Connects to sewage disposal system.

* **housed stair:** 1. Stairway in which the stringers are mortised to accept risers and treads. Risers and treads are glued and wedged in position. 2. Stairway enclosed between two walls. CLOSED STAIRWAY.

house slant: T- or Y-shape sewer pipe fitting that connects the sewer to the house drainage system.

housing: Protective case or contained enclosure for moving parts.

Howe truss: Truss with vertical and diagonal webs. Vertical webs absorb tensile stresses and diagonal webs absorb compressive stresses.

HP shape: Hot-rolled steel member used for piles and other load-bearing applications. Webs and flanges approximately equal in thickness and length. See *pile (1)*. H BEAM, H PILE, STEEL H PILE.

hub: 1. Corner stake or reference point used to establish elevations or other reference points. 2. Enlarged end of a cast iron soil pipe. BELL. 3. Central reinforced section of a gear or wheel. 4. Portion of a lock through which the spindle passes.

huddling chamber: Portion of a safety valve that increases the disk area.

humidifier: Mechanical device used to increase moisture content of the air. May be integrated into the forced air system of a structure or stand alone.

humidistat: Automatic regulating device that maintains the relative humidity in an area.

humidity: Amount of water vapor suspended in the air.

humus: Fertile soil produced by decomposition of animal and vegetable matter.

hung ceiling: See *suspended ceiling.*

HVAC: See *heating, ventilating, and air conditioning.*

hybrid beam: Steel beam constructed of materials with different yield strength in the web and flanges.

hydrant: Connection for obtaining water from a water main supply pipe.

hydrate: Chemical combination of water with another chemical or material.

hopper gun

horizontal position

hose bibb

housed stair (1)

hydraulic jack

hypotenuse

---

hydrated lime: Material produced from chemical combination of calcium oxide and water. Used as an undercoat for finish plaster. QUICKLIME, SLAKED LIME.

hydration: Chemical reaction of water with another material, such as cement, that bonds molecules, resulting in hardening of the mixture.

hydraulic: 1. Operated by fluid pressure. 2. Hardening of a substance while in contact with water.

hydraulic cement: Cement capable of setting and hardening while submerged under water.

hydraulic fill: Earth or other material conveyed and placed by running water. Water is drained so that only fill remains.

hydraulic glue: Adhesive that is waterproof.

* **hydraulic jack:** Device that exerts pressure and moving force through the use of a fluid forced through a small internal pump.

hydraulic joint: Pipe joint formed by forcing a pliable material into belled end of a pipe by fluid pressure. Used for water main connections.

hydraulic test: Method of determining soundness of a piping system by filling the system with water and applying specified amount of pressure.

hydrocal: High-strength molding plaster.

hydrochloric acid: Strong cleaning fluid. See *muriatic acid.*

hydroelectric power: Electricity generated by turbines that are rotated by falling water.

hydronic system: Heating or cooling system that operates with forced water.

hydrostat: Electrical device that regulates the fluid level in a container.

hydrostatic pressure: Pressure equivalent to the pressure exerted on a surface by a column of water of a given height.

hydrostone: High-strength plaster used for casting and patternmaking.

hygrometer: Device for measuring humidity of the air.

hygrometric expansion: Measure of the tendency of materials to expand when humidity is increased.

hypalon roofing: Elastomeric roof finish material that is highly resistant to thermal movement and weathering.

* **hypotenuse:** Side of a right triangle opposite the 90° angle.

hypotrachelium: Beveled lower edge of the capital and the column shaft in Doric architecture.

HY steel: Steel with minimum yield strength of 80,000 pounds per square inch.

I beam: Structural steel member with cross-sectional area resembling capital letter I.

I boot: Sheet metal duct fitting with a vertical and horizontal 90 ° transition. See *ductwork**.

identified: Recognizable for a specific purpose, as applied to electrical equipment.

* idler: 1. Wheel or gear in a train that rotates in direction opposite a chain or belt. **2.** Wheel or gear used to reduce slack in a train.

igneous rock: Type of rock formed by solidified molten material.

ignition pilot: Pilot light that operates only during the lighting cycle. Remains closed during operation of the main burner.

ignition point: Temperature at which combustible material burns.

ignition transformer: Electrical device that supplies a spark for igniting a gas pilot light.

illumination: Luminous flux density on a surface. Expressed in lumens per square foot or foot candles.

imbow: To arch over.

imbricate: To overlap in a regular pattern; e.g., horizontal lap siding.

immersion vibrator: See *internal vibrator.*

impact insulation class: System for rating and comparing ability of various types of construction to insulate against noise. Higher ratings indicate better noise transmission resistance. IMPACT NOISE RATING.

impact resistance: Ability of a material to withstand shock.

impact wrench: Electric or pneumatic tool used to supply short, rapid impulses to sockets. Provides maximum torque for tightening and loosening power. See *wrench**.

impedance: Total resistance and reactance in an alternating current circuit. Expressed in ohms.

impeller: 1. Pump member that uses centrifugal force to discharge liquid. **2.** Rotating portion of a device.

impending slough: Consistency of shotcrete which contains maximum amount of water without sagging after placement.

impermeable: Not permitting passage of liquid; e.g., asphalt shingles. IMPERVIOUS.

impervious: Not permitting the passage of fluid or liquid.

imposed load: All loads on a structure except dead loads.

impost: 1. Highest portion of a column or other vertical member that supports the lower end of an arch. **2.** Mullion in a double window.

impoundment: Lake, pond, tank, or other storage area for water control.

impregnate: To thoroughly saturate a material, such as timber, under pressure.

incandescent lamp: Light source in which a tungsten filament is heated by an electrical current to the point of illumination.

* incertum: Rough stonework without a regular pattern of horizontal joints.

inch: Unit of lineal measure. Equal to one twelfth of a foot.

inch of water: Unit of pressure equal to the hydrostatic pressure of a 1 " high water column at 39.2 °F.

incident radiation: Solar energy as it is received on the plate of a solar collector.

incinerator: Type of furnace for burning waste material.

incise: To engrave or inscribe.

incline: Sloped surface.

inclined plane: Surface at an angle to the horizontal.

* inclined position: Position of a pipe joint with the pipe axis at approximately 45 ° angle from horizontal. The pipe is not rotated during the welding process.

inclusion: Presence of trapped slag or dirt in a finished surface.

incombustible: Unable to ignite into flame or support fire at 1200 °F over a period of 5 minutes. NON-COMBUSTIBLE.

idler (1)

incertum

inclined position

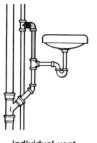

| individual vent | information stake | insert (1) | inside caliper |

increaser: Pipe coupling or vent with a diameter larger on one end than the other.

indent: Recessed area; surface depression.

indented wire: Concrete reinforcement or pretensioning tendons with series of surface indentations that increases bonding to the concrete.

index mark: Surveying line or point used as a sighting reference and base point from which measurements are taken.

indicator bolt: Door bolt with visible portion that indicates occupancy if locked and vacant if unlocked. Used primarily on bathroom and toilet doors.

indicator pile: Driven vertical foundation support used to evaluate required depth and length of piles.

indicator valve: Control device for fluid that externally indicates whether it is open or closed.

indirect lighting: Illumination produced by partially concealed light sources; 90% to 100% of the emitted light is directed upward. Commonly installed above a soffit around the perimeter of an area or above ceiling beams.

indirect vent: Vent pipe that vents a fixture trap. Connected to the venting system above the fixture being served.

indirect waste pipe: Waste pipe not directly connected to the drainage system. Discharges waste into fixture or receptacle that is properly trapped.

individual branch circuit: Electrical circuit that serves only one piece of equipment.

* **individual vent:** Vent pipe connected to a single fixture to allow gases to escape. BACK VENT.

indoor/outdoor carpet: Weatherproof floor covering used for both interior and exterior areas.

induced draft: Air flow produced mechanically by blower or fan.

inductance: Production of electromotive force by varying amount of current in an alternating current circuit.

induction brazing: Brazing process in which pieces of metal are joined by heat produced from the resistance of the metal to induced electrical current.

induction motor: Alternating current motor that has its primary winding connected to the electric source and secondary winding carrying induced current.

induction welding: Welding process in which pieces of metal are joined by heat produced from the resistance of the metal to induced electrical current, with or without pressure.

inelastic: Incapable of returning to original shape after deformation.

inert: Lacking properties to react chemically or biologically with other materials.

inert gas metal arc welding: See *gas metal arc welding*.

infill housing: Residential development on random vacant lots in an otherwise developed area.

infilling: Any material installed between structural members to increase insulation, vibration resistance, or strength.

inflammable: See *flammable*.

* **information stake:** Excavation marker that provides surveyor's measurements, such as elevations and angles, to an excavation crew.

infrared: Wavelength of electromagnetic energy just above the visible spectrum.

infrared lamp: Incandescent lamp that emits more radiant power than common incandescent lamps. Lamp is tinted to reduce visible light.

inhibitor: Compound added to paint to slow drying and reduce skinning and other undesirable defects.

initial graphics exchange specification: Nonproprietary computer-aided design and drafting file format defined by ANSI Y14.28. Permits operators to transfer CADD files from one system to another.

initial set: Hardening of concrete or mortar to a degree that it can resist known amount of penetration by testing devices. Expressed as the time in hours and minutes required to initially harden.

injector: 1. Device that forces a steam jet of water against boiler pressure. **2.** Device that lifts and pumps one fluid using another fluid at a higher pressure to create a partial vacuum.

inlaid: See *inlay*.

inlay: Pattern achieved by setting one or more materials into and flush with another surface. INTARSIA, MARQUETRY.

inlet: Surface connection to a drainage system.

input rating: Fuel-burning capacity of an appliance. Expressed in Btu/hr.

insect hole: Wood defect in the form of small hole made by wood-boring insects. See *defect**.

* **insert: 1.** Hardware installed in concrete or masonry for attachment of temporary or permanent braces or lifting devices. **2.** Non-structural and/or decorative surface patch such as an inlay.

insertion thermostat: Thermostat with sensing element designed to be placed directly in the air stream.

inside angle tool: Masonry float for shaping mortar along inside corners.

* **inside caliper:** Measuring tool used to determine inside dimensions of a cylinder or distance between shapes or features.

inside micrometer: Precision measuring tool used to measure inside diameter of a cylinder.

inside stop: Strip of wood, metal, or plastic on interior side of a window used to restrict movement of the sash to a vertical plane. BEAD STOP, INNER BEAD, STOP BEAD, WINDOW BEAD, WINDOW STOP.

inside thread: See *female thread*.

in-situ: In place, undisturbed; e.g., an in-situ pile is cast in place.

instrument: Surveying tool for measuring elevations and angles. See *builder's level, dumpy level, transit, and theodolite*.

insulated conductor: Electrical conductor encased in a material that does not carry electric current and is recognized by the National Electrical Code®.

insulating concrete: Concrete that decreases thermal and sound transmission.

* insulating glass: Assembly of two or more individual panes of glass with an air space between them and the perimeter of the panes sealed. Used to decrease thermal and sound transmission through the glass.

insulating resistance: Ability of an insulating material to inhibit flow of electric current.

insulation: 1. Material used as barrier to inhibit thermal and sound transmission. 2. Material used as barrier to inhibit flow of electric current.

insulation board: Wood fiber or expanded foam sheet or panel used as barrier against thermal or sound transmission.

insulator: 1. Electrical fitting or device with a large amount of resistance to electric current. Used to separate conductors from each other or conductive objects. 2. Person that installs insulation.

intaglio: Engraving or carving depressed below the surface.

intake: Opening or device that supplies air, water, or fuel to pipe or equipment.

intake belt course: Masonry course in which the brick faces have been shaped to span between walls of varying thicknesses.

intarsia: Ornamental inlaid woodwork.

integral: 1. Acting as one, formed into a single unit; e.g., monolithic concrete placement creates an integral unit. 2. Important, indispensable.

integrated system: Building monitoring system that combines systems, such as fire and security systems.

* intercepting drain: Trench or other depression filled with material that allows flow of fluids into a drain pipe.

intercepting sewer: Large sewer that receives flow of water from several smaller sewers. TRUNK SEWER.

* interceptor: Device that separates and retains undesirable waste matter from normal waste. Allows normal waste to discharge into drainage system.

intercolumniation: Open space between columns measured between the lower parts of the shafts.

intercom: Audio system that allows communication at various points within a building or complex. INTERCOMMUNICATION SYSTEM.

interference body bolt: Fastener with high bearing and shear strength. Slight deformation of the threads occurs during tightening to create a secure fit.

interior: 1. Material grade or description that denotes inability to withstand exposure to weather. 2. Inside of a

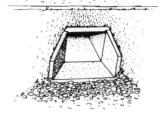

insulating glass   intercepting drain   interceptor

structure.

interior finish: Exposed interior surfaces of a building, including trim members. INTERIOR TRIM, INSIDE FINISH.

interior trim: Moldings and other decorative members installed inside a structure; finish materials not suitable for exterior application.

interlocking siding: Exterior finish material made of aluminum, steel, or vinyl with edges designed to interlock with the adjoining piece to provide a watertight seam. Available in widths ranging from 8″ to 12″ and lengths ranging from 10′-0″ to 12′-0″. See *siding**.

intermediate: Member placed between two other members; e.g., the intermediate rail of a handrail.

intermediate glue: Adhesive used in interior-grade plywood with a degree of moisture resistance. Not suitable for permanent exterior exposure.

intermittent duty: Operation of electrical circuit or device for alternate intervals of load and no load; load and rest; or load, no load, and rest.

* intermittent weld: Weld with irregular unwelded spaces along the joint.

internal dormer: Vertical window structure recessed in a sloping roof.

internal/external tooth washer: Circular washer with short projecting tabs along inside and outside diameter. Designed for light-duty applications between a bolt or cap screw and nut. See *washer**.

internal friction: Resistance to movement within soil.

internal mix: Spray gun in which mixing of air and paint takes place inside the cap of the gun.

internal partition trap: Trap that forms a seal by means of internal partitions. Considered undesirable because of possibility of holes forming in the partitions, which results in loss of water seal.

internal tooth washer: Circular washer with short projecting tabs along inside diameter. Designed for light-duty applications between a bolt or cap screw and nut. See *washer**.

* internal vibrator: Gasoline-powered device used to consolidate concrete by immersing the head in freshly placed concrete. Flexible steel spring extending from the motor to the head transmits vibrations to the concrete.

interpolate: 1. To calculate approximate location of a point based on measurements made between several other points. 2. To estimate a value between other known values.

interrupted: Noncontinuous or built as one member, as applied to a structure or material.

interrupting rating: Highest electrical current at a known voltage that a protective device can disconnect in testing conditions.

intersect: 1. To cut across or through. 2. To come in contact with.

* intersecting roof: Roof in which there are two or more ridges at different angles. Combination of basic roof styles.

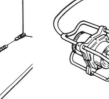

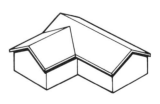

intermittent weld   internal vibrator   intersecting roof

intrados

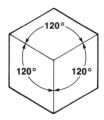

isolation joint

isometric drawing

intersection: Point or plane where two lines or objects come in contact or meet.

* **intrados:** Bottom side of the interior curve of an arch.

intrinsic barrier: Electric circuit that limits amount of voltage available to a hazardous location, thereby creating a situation incapable of ignition.

inverse: Reverse; opposite from normal position.

invert: 1. Lowest part of the inside of horizontal pipe or conduit. 2. To turn or move in an opposite position.

inverted arch: Masonry arch with the keystone placed at lowest point of the arch.

inverted ballast: Lamp ballast that operates on direct current. INVERTING BALLAST.

inverted-T foundation: Foundation formed with a concrete stem wall placed on a concrete spread footing. Commonly used for a structure with a basement. See *foundation**.

invisible hinge: Hinge completely mortised and concealed in door and jamb.

involute: Spiral shape in increasing diameter from a common center; e.g., clock spring.

Ionic: Order of Greek architecture characterized by column capitals with volutes.

iron: Metallic element mined from earth's crust as an ore in hermatite, lemonite, and magnetite. Used to produce metal such as steel.

ironworker: Tradesperson who works with various types of iron and steel and performs operations such as installing structural iron and steel, and assembling and installing reinforcing steel.

irregular coursed masonry: Rubblestone masonry formed with uneven and staggered mortar joints.

irregular curve: Drafting tool used to draw irregular, curved lines.

I-section: Cross section of a beam that has horizontal top and bottom flanges connected by a vertical web.

island cabinet: Stand-alone cabinet not connected to any other cabinets.

isolate: To make inaccessible unless special means of entry is provided.

* **isolation joint:** Separation between adjoining sections of concrete or plastered areas used to allow movement of the sections.

isolator: Device in an electrical circuit that is removed to break the circuit when no current is flowing.

* **isometric drawing:** Type of pictorial drawing in which all horizontal lines are drawn at a 30° angle from the horizontal and all vertical lines are drawn at a 90° angle from the horizontal. Isometric axis is at 120° angle.

Izod impact test: Impact resistance test of a material in which a single pendulum is dropped onto the surface.

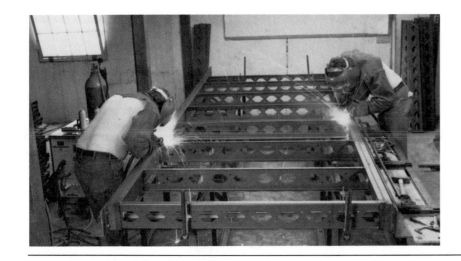

*jack: 1. Equipment used to lift or move loads a short distance by using mechanical advantage. **2.** Shortened vertical support member installed on the inside of window and door openings in wood-framed walls. Used to support window and door headers. TRIMMER. **3.** Female receptacle for temporary electrical connection. **4.** Mechanical device used to apply tension to prestressing tendons in prestressed and post-tensioned concrete embers. **5.** Device used to create tension on piles for movement resistance tests.

jack arch: Arch with flat intrados (highest point of arch). FRENCH ARCH.

jacked pile: Pile driven into the ground by applying force between the top of the pile and a stable object above.

*jacket: 1. Insulated covering for pipes or equipment. **2.** Sealed housing installed around a piece of equipment. Open area between equipment and jacket is used to circulate a fluid to maintain a specific temperature or heating or cooling.

jack hammer: Electric or pneumatic equipment with pointed or flat cutting edge. Used to penetrate hard surfaces.

jacking force: Temporary tensile stress exerted on prestressed tendons.

jack plane: Hand tool, approximately 14″ to 15″ long, used for smoothing and fitting mating surfaces. See *plane* (1)*.

jack rafter: Roof member that does not extend the entire distance from the ridge to the top plate of a wall. Commonly used in hip roof construction. See *cripple jack rafter, hip jack rafter, valley jack rafter.*

jackscrew: Jack that uses a threaded rod to lift loads.

Jacob's ladder: Ladder constructed with ropes or chains that support wood or metal rungs. Commonly used in ma-

rine construction.

Jacob's staff: Single vertical rod used to support surveyor's instrument. Used instead of a tripod.

jag: Notch or indentation.

jagger: Toothed chisel used to dress stone.

jalousie window: Movable window consisting of narrow glass louvers mounted in a frame. Louvers pivot outward at the bottom simultaneously to provide circulation of air. See *window**.

jamb: Wood or metal finish members installed along the inside of door or window openings. Includes the top, sides, and bottom of the opening.

jamb anchor: Light gauge metal device inserted in the back of hollow metal jambs. Used to secure jamb to framing or masonry.

jamb block: Concrete masonry unit with a recess formed along the end to receive a jamb. See *concrete masonry unit**.

jamb shaft: Decorative column placed along or forming a door or window jamb. ESCONSON.

jambstone: Masonry positioned along the edge of a wall opening. Used to form door or window jamb.

japan varnish: Dark, resinous varnish used as a drier in paint.

jaw: Clamp or tool component used to secure material during an operation.

jaw crusher: Machine used to crush rock or other hard material. Two inclined jaws, one or both actuated by reciprocating motion.

J bend: See *return bend.*

J bolt: J-shape anchor bolt with one threaded end. Serves as attachment between concrete or masonry foundation or footing and sill plate or other member. See *bolt**.

J channel: J-shape metal or plastic trim member used to support trim material such as a soffit or siding. Longer leg is fastened to a backing surface with the projecting lip providing the support.

jedding ax: Striking tool with one flat and one pointed face. Used to shape stone.

jerkin head: See *Dutch hip roof.*

jesting beam: Ornamental or false horizontal beam.

jetting: 1. Process of settling unstable ground or backfill. Hose is inserted below earth's surface and water is pumped to the end of the hose to consolidate the loose dirt. **2.** Drilling holes into the earth using high-pressure water or compressed air.

jetty: 1. Structure built along a waterfront to protect a harbor from erosion and improve the ability of a channel of water to remove silt from navigational areas. **2.** Portion of a structure that projects beyond the section immediately below.

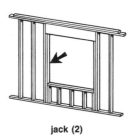

jack (2)

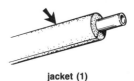

jacket (1)

Jiffy® mixer   Jitterbug® tamper

jib: Lifting arm of a crane.

jib boom: Hinged extension attached to upper end of a crane boom to provide additional boom length.

jib crane: Crane with a pivoting boom and jib.

jib door: Door that is flush with the surface of a wall and finished similar to the wall to create a concealed opening. Hardware is not visible on the exterior face.

Jiffler® mixer: Shaft with five perforated fins used to mix semiliquid materials such as drywall compound and plaster. Shaft is attached to an electric drill motor and rotated to distribute materials in opposite directions for complete consolidation.

* Jiffy® mixer: Shaft with two finned circular rims used to mix semiliquid materials such as drywall compound and plaster. Shaft is attached to an electric drill motor and rotated for complete consolidation.

jig: Device used to temporarily secure a material in position and guide a tool when shaping or forming the material; e.g., doweling jig.

jigsaw: 1. Electric hand tool with interchangeable, reciprocating blade. Used to cut curved or irregular designs in thin material. See *saw**. **2.** See *scroll saw*.

* Jitterbug® tamper: Concrete flatwork finishing tool used to bring fines to the surface and distribute large aggregate below the surface. Consists of a mesh screen fastened to a frame that is rolled over the surface.

jog: Change in direction or irregularity from a straight line.

joggle: 1. Projection from a wall used to support a brace. **2.** Stub tenon on the end of a mortised member for reinforcement and to prevent lateral movement. **3.** Notch or projection on one member that interlocks with an adjoining member. Used to prevent the member from slipping.

joggle joint: Joint between two masonry members in which a projection in one member fits into a recess of another member. Used to prevent lateral movement.

joggle post: Vertical support member in a truss with notches cut into the lower end to support the ends of struts. JOBBLE PIECE.

joiner: Carpenter who fabricates fitted members such as cabinet doors.

* joinery: Process of fabricating and fitting wood members together.

joining saddle tie: Wire tie that is doubled and wrapped around two members to secure them in position. See *tie**.

joint: 1. Intersection of two members.

WOOD JOINTS

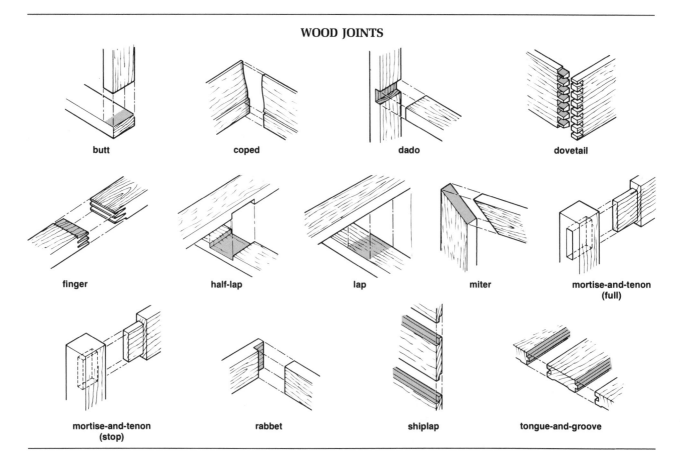

butt   coped   dado   dovetail

finger   half-lap   lap   miter   mortise-and-tenon (full)

mortise-and-tenon (stop)   rabbet   shiplap   tongue-and-groove

2. To finish masonry joints. **3.** To cut grooves partially through fresh concrete. **4.** To straighten the edge or face of a piece of wood. **5.** Initial operation when sharpening a cutting tool. Consists of grinding or filing the tips of all teeth or knives to the same height.

joint cement: Dry powder that is mixed with water to create a paste suitable for filling holes and cracks in gypsum drywall or plaster.

joint compound: Material that is applied to the joints and indentations in gypsum drywall. Available as dry powder, which is mixed with water, and ready-mixed. Three types are taping (designed for first fill coat and embedding reinforcing tape), topping (designed for second and third coats), and universal (designed to incorporate both taping and topping characteristics).

jointed core: Plywood core that has been machined squared. Gaps between pieces of core do not exceed $3/8''$ and the average of all gaps does not exceed $3/16''$.

joint efficiency: Ratio of the strength of a welded or brazed joint to the strength of the base metal. Expressed as a percent.

jointer: 1. Steel hand tool with a narrow blade used to shape mortar between masonry units. JOINT FORMER, JOINT ROD. **2.** Concrete flatwork finishing tool with a rib projecting from a flat bottom surface. Used to cut joints partially through fresh concrete. See *groover*. **3.** Woodworking machine used to smooth and plane rough stock. Cylindrical cutterhead that houses sharpened knives is rotated as the stock is fed.

jointer plane: Hand tool approximately 22″ to 24″ long used to true edges of lumber and smooth the face of long boards. See *plane* (1)*.

jointing material: Sheets of rubber or other flexible material that are cut to make gaskets.

joint knife: Hand tool with 4″ to 6″ flexible blade for applying and smoothing joint compound on drywall joints and indentations.

joint penetration: Minimum depth a flange or groove weld extends into the base metal. May include root penetration.

joint plane: Direction of breaks or seams in a rock layer.

joint reinforcing tape: See *reinforcing tape*.

joint rule: Steel plastering tool with 45° angle cut at one end for forming plaster miters.

*ᐧ**joint runner:** Incombustible material,

JOISTS

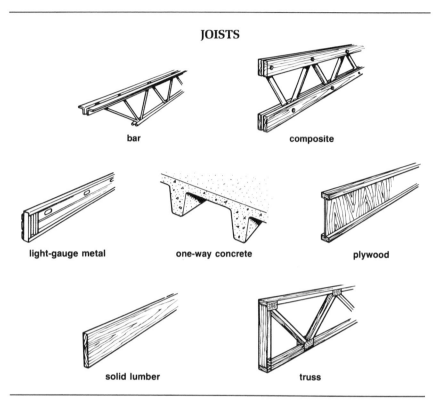

bar

composite

light-gauge metal

one-way concrete

plywood

solid lumber

truss

such as oakum, used to retain molten lead when forming a horizontal bell-and-spigot soil pipe joint.

*ᐧ**joist:** Horizontal support member to which finish floor and ceiling materials are fastened. Common joist material includes wood, steel, and concrete.

joist anchor: Beam or wall anchor or metal tie used to tie a masonry wall and wood structural members together. Fastened to a masonry wall and to the bottom of the joist or beam where it intersects with the wall.

joist hanger: Bent metal device used to support a joist where it intersects another horizontal member or wall.

journal: Portion of a shaft that rotates in a load-supporting bearing.

journeyman: Person who has completed apprenticeship training for a construction trade and is skilled in that field.

jumbo: 1. Multiple pieces of drilling equipment mounted on a traveling carriage. Used for drilling tunnels in rock or other hard material. **2.** Traveling support for concrete forms. Commonly used in tunnels.

jumbo brick: Masonry unit larger than a standard brick, commonly 4″ × 4″ × 12″ (including mortar joint). See *brick**.

jumper: 1. Short length of wire or cable used to make a temporary electrical connection. **2.** Stretcher stone that spans at least one cross joint.

*ᐧ**junction box:** Container that protects conductor splices and provides support for the electrical connection. Designed to allow conductors to enter through knockouts in the sides and back.

junction chamber: Section of sewer system where several sewers converge into a main sewer.

junior beam: I-beam made from lightweight structural steel. Used for shoring concrete forms.

jut: To project outward.

jute: Plant fiber used for carpet backing.

jutty: See *jetty*.

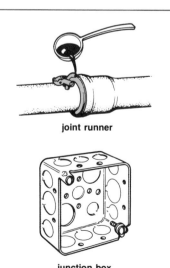

joint runner

junction box

kalamein door: Composite door made of a wood core covered with galvanized sheet metal.

Keene's cement: Cement product composed of finely ground calcined gypsum and other material. Used to form high-density plaster.

keeper: See *strike*.

kellastone: Stucco with a rough, crushed finish.

kelly: Square or fluted shaft that slides vertically through an opening in a rotating driving head. Used for rotating an earth auger. KELLY BAR.

kelly ball: Thirty pound, 6″ diameter weighted cylinder with rounded bottom used to determine consistency of fresh concrete. Cylinder is dropped onto a concrete surface and penetration measured. May be used as replacement for slump test.

* **kerf: 1.** Narrow groove in material produced by a saw blade. **2.** To cut a series of grooves partially through a member to allow for bending.

kern area: Area in which compressive force is applied without tensile stress resulting in surrounding areas.

kerosene value: Measure of the volume of kerosene absorbed by a piece of felt. Used to determine the ability of the felt to absorb asphalt.

kevel: 1. Hammer with flat face at one end and pointed peen at other end. Used for dressing stone. **2.** Piece of wood for securing the end of a rope.

* **key: 1.** Void or groove in a member designed to accept a corresponding projection from another member. Used to improve interlocking capabilities between members. **2.** Piece of metal inserted in a groove of a shaft and external member to prevent relative movement between interlocking parts. **3.** Wedge driven into the end of a tenon to secure it in a mortised joint. **4.** Plaster forced into the openings of backing lath. **5.** Roughened surface on the back of a material to increase bonding strength with an adjoining material. **6.** Slotted metal device that operates a lock. **7.** See *keyway*. **8.** See *chuck key*. **9.** See *keystone*.

key brick: Wedge shaped, tapered masonry unit; installed in arches.

key drop: Small hinged piece of metal used to cover a keyhole; escutcheon cover.

keyed: Shaped with a recessed area to provide better interlocking with adjoining member.

keyed alike: Operated with the same key, as applied to locks.

keyed joint: See *concave joint*.

keyhole: Welding technique in which a hole is melted completely through a workpiece and forms a hole along the front edge of the molten metal. Molten metal fills in behind the hole to form the weld bead.

keyhole saw: Handsaw with thin, tapered, interchangeable blades for cutting curved shapes and inside holes. Used for wood, metal, and drywall. See *compass saw*. See *saw**.

keyplate: Small metal plate on a door with a keyhole only.

* **keystone: 1.** Stone wedge at uppermost point of an arch that secures other arch members in position. **2.** Wedge-shape member installed above an arch for ornamentation.

key switch: Switch operated by a removable key.

* **keyway: 1.** Groove in a lift or placement of concrete that is filled with concrete of the next lift. Used to improve shear strength of the joint. **2.** See *key*. **3.** Opening in a lock cylinder that receives the key.

K factor: See *K value, thermal conductivity*.

kick: Slight indentation in a brick. FROG.

kickback: Forcing of material away from a rotating blade. Occurs because material being cut pinches the blade.

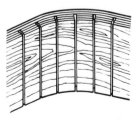

kerf (2)

key (2)

keystone (1)

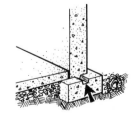

keyway (1)

kicker (1)

kick plate

knee boards

knee pads

knife-blade fuse

* **kicker: 1.** Wood block or board attached to concrete formwork member to stabilize the forms. Commonly placed at the bottom of forms to prevent spreading during concrete placement. **2.** Brace to prevent the bottom of an object from moving. **3.** Member installed above a cabinet drawer to prevent the drawer from tipping when extended.

kicker block: See *kicker* (2).

kick lift: See *lifter*.

kickout: Faulty, accidental movement of a concrete form or shoring, caused by inadequate or improper bracing.

* **kick plate:** Protective plate attached to the bottom of the face of a door to prevent marring.

kick rail: Short rail attached near the bottom of a door to allow the door to be opened by using a foot. Primarily used on institutional doors.

kiln: Heated chamber used for firing brick and tile or drying timber.

kiln-dried lumber: Lumber seasoned in a kiln to a moisture content of 6% to 12% by controlled heat and humidity.

kiln run: Masonry or other fired units that have been heated but not graded for burning, color, or size.

kiloampere: Measure of electrical current equal to 1000 amperes.

kilovolt: Measure of electromotive force equal to 1000 volts.

kilowatt: Measure of electric power equal to 1000 watts, or 1.34 horsepower.

kilowatt hour: Measure of electric power equal to 1000 watts utilized for 60 minutes.

king closer: Brick that is cut diagonally to have one 2″ end and one 4″ end. Used to fill a gap in a course where required. BEVELED CLOSER.

king pile: Central pile that extends higher than other piles in a group. Used for marine application.

kingpin: Vertical pivot or swivel mechanism that allows a piece of equipment to pivot. Supported both at top and bottom.

kingpost: Central vertical support that spans from the peak to the lower chord of a roof truss. See *truss**.

king stud: Vertical support installed along both sides of a framed opening for a window or door. Extends from the top to bottom plate.

king valve: Control device installed at the outlet of a receiver tank.

kink: 1. Sharp bend in an object. **2.** Sharp bend in a rope. Destroys the lay of the rope and weakens the strands.

kiosk: 1. Open garden pavilion. **2.** Semi-enclosed pavilion used for retail sales.

kip: Measurement of force equal to 1000 pounds.

kitchen: Room used for storage, preparation, and cooking of food.

kitchenette: Small kitchen.

kite winder: Triangular-shaped tread in a staircase winder with three steps.

knapping hammer: Steel hammer used for splitting cobbles and shaping paving stones or roughly sized stone. Either two square faces or a square face and a wedge-shape peen.

knee: 1. Short supplementary brace. **2.** 90° bend in a pipe. ELBOW.

* **knee boards:** Flat pads used by a concrete finisher to distribute weight when hand troweling fresh concrete. Reduces marring of the concrete.

knee brace: Angular member that stiffens and strengthens a building frame or formwork.

kneeler: Stone or brick that changes the direction of the masonry pattern, such as along an arch or window. KNEESTONE, SKEW.

* **knee pads:** Rubber or leather pads strapped onto a person's knees for protection. Commonly used by concrete finishers to provide a comfortable kneeling surface.

knee wall: 1. Wall that supports rafters at an intermediate point along their length. Used to shorten the span of rafters. **2.** Short interior or exterior wall.

knife: 1. Flat-bladed tool used to spread soft, pliable material. See *putty knife, broad knife*. **2.** Cutting tool with at least one sharpened edge; e.g., jointer knife. See *hook bill knife, utility knife*. **3.** Cutting edge of excavating equipment.

* **knife-blade fuse:** Cartridge fuse with projecting flat metal blade at each end.

knife file: Hand tool with a blade cross section tapering from a square to a pointed edge. Used to smooth areas enclosed by an acute angle. See *file**.

knob: Projecting handle for operating a lockset or door bolt.

knob bolt: Door lock in which the bolt is operated by a knob or thumb latch on one or both sides of the door.

knob latch: Door latch with a spring bolt operated by a knob on one or both sides.

knob lock: Door lock with a spring bolt operated by a knob on one or both sides and a dead bolt operated by a key.

knob rose: Round disk attached to a door face around a door knob spindle.

knob shank

knockout

knuckle

knurling

* **knob shank:** Projecting stem of a door knob that receives the spindle.

knocked down: Prefabricated components or structure delivered to a job site and then assembled or installed.

knocker: Hinged or pivoting metal device installed on exterior of an entrance door. Used to strike or knock on the door.

* **knockout:** Prestamped piece in an electrical box that is removed to create a passage for conductors.

* **knot: 1.** Hard, dense defect in lumber resulting from embedded branch or limb in the tree. See *defect**. **2.** Intertwining of two sections of rope to join or secure the two pieces. **3.** Ornamental design resembling intertwined ropes.

knothole: Void occurring when a knot drops out of lumber or veneer.

* **knuckle:** Projecting cylindrical portion of a hinge through which the hinge pin passes.

knuckle joint: Rafters that intersect at less than a 90° angle on each side of a gambrel roof.

knurl: Burl in wood.

* **knurling:** Rough surface finish consisting of small machined ridges or beads. Used to provide better grip.

kraft paper: Strong brown paper product made of a sulfate pulp. Provides moisture resistance when applied as a building paper.

K truss: Prefabricated truss in which the webs form a repeating pattern of the letter K.

K value: Measure of the thermal conductivity of a material. Equal to the Btu/hr. transferred through an area measuring 1″ thick × 1′-0″ × 1′-0″ for each 1°F difference in the faces of the material. Inverse of R value.

kyanize: To prevent decay in wood by treating it with a mercuric chloride solution.

KNOTS

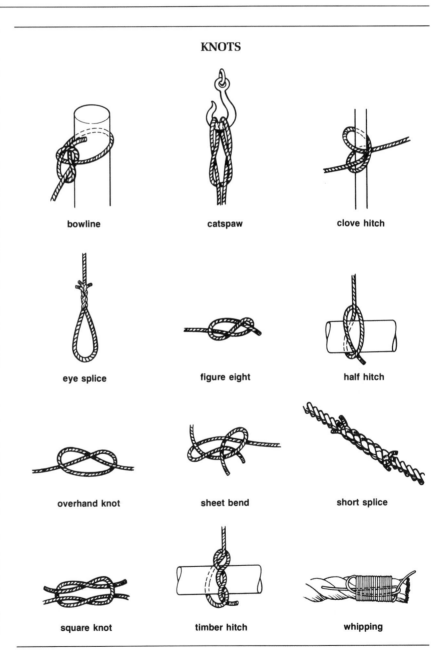

bowline

catspaw

clove hitch

eye splice

figure eight

half hitch

overhand knot

sheet bend

short splice

square knot

timber hitch

whipping

label: Stone trim member that extends over the top of a door or window and down the sides of the opening for a short distance. LABEL MOLDING.

labeled: Bearing a tag or other identifying mark that denotes inspection by an approved agency having jurisdiction.

label stop: Finish at lower ends of a decorative header.

laborer: Construction worker with various duties; e.g., placing concrete, carrying materials, and cleaning up.

labyrinth: Maze; series of twisting passageways.

lace: To fit interlocking or overlapping members together.

laced fall: Rope that is fed through several pairs of blocks and/or pulleys in hoisting equipment.

laced valley: Valley of a roof in which roof shingles are overlapped in alternating rows to form an interlocking pattern.

lacing: 1. Members used to connect components of a composite column or girder to enable them to act as one member. 2. Patches installed between lagging or shoring members to prevent soil from falling into the excavated area.

lacing course: Course of brick in a stone wall used to bond the wall together and level succeeding stonework.

lacquer: Glossy enamel composed of volatile solvents, resins, and plasticizers. Solvents evaporate quickly upon application.

lacunar: Recessed area in a finish panel; e.g., ceiling panel.

ladder: 1. Structure for climbing up and down, consisting of a series of rungs (horizontal steps) fastened between two stiles (upright members). 2. Boom assembly of a hydraulic dredger or mechanical ditcher.

ladder ditcher: Excavating machine used to cut trenches with a series of buckets attached to a rotating chain that is guided by a boom.

ladder jack: Metal device attached to the rungs of a ladder and used to support a scaffold plank. Used on either the front or back of a ladder.

ladder tie: Masonry reinforcement installed in bed joints. Consists of long parallel rods joined by perpendicular cross rods; similar to a ladder. See *tie* (1)*.

* **ladle:** Fire-resistant cup with a handle used to pour molten lead into a cast iron pipe joint.

lag: Time delay occurring between initial reponse of a sensing element in a control device and actual response of the component or equipment.

lag bolt: Fastener with a square or hexagonal head and a thread design for use in wood. See *bolt**. COACH BOLT, LAG SCREW.

* **lagging:** 1. Shoring used to retain earth on the side of a trench or excavation. Used to prevent cave-ins. 2. Contact area on a flat pulley. 3. Thermal pipe insulation.

lag hook: Fastener with a hook-shaped head and a thread designed for use in wood.

lag screw: See *lag bolt*.

laitance: Layer of weak material on freshly placed concrete flatwork consisting of the accumulation of fine particles of cement, aggregate, and impurities. Results from too much water in the mix, overworking concrete, and overvibration.

Lally column: Vertical metal pipe filled with concrete for additional support.

lambert: Measure of brightness of light. Equal to 1 lumen per square centimeter.

lamellar crack: Fracture in steel resulting from shrinkage in a weld joint. Occurs perpendicular to the direction in which the steel was rolled at the mill.

laminate: To bond layers of material to form a single unit.

laminate trimmer: Electric hand-held tool used to cut and finish edges on plastic laminate. See *bit* (2)*.

lamp: Mechanical light source. BULB.

lamp circuit: Branch circuit that supplies electricity only to lighting fixtures.

* **lampholder:** Electrical fixture that supports and makes electrical contact with a lamp.

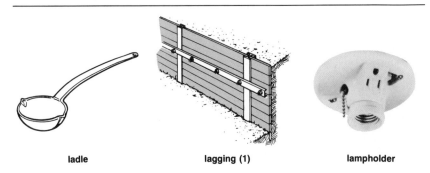

ladle   lagging (1)   lampholder

lancet

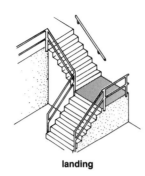

landing

lattice

laundry tray

lamppost: Freestanding post supporting an illumination source with internal wiring connections.

lancet: Narrow window with a steep pointed arch.

* **landing:** Horizontal platform in a stairway used to break a run of stairs. Also provides a location for a change of direction in a staircase. PACE, STAIR LANDING.

landing newel post: Vertical member that supports a handrail at a landing or change of direction in a stairway.

landing tread: Horizontal board directly above the highest riser of a stairway landing.

landmark: Permanent marker or monument used as a location device in surveying.

land tile: Clay pipe with open joints that allow for seepage through the pipe.

lane: Narrow road or passageway.

lang lay: Strands of rope woven in the same direction as the weaving in each strand.

lantern: See *cupola*.

lanyard: Piece of rope or cable with a snap hook on each end. Clipped to a safety belt on one end and a lifeline at the other end.

lap: 1. To position material so that one member extends over another. OVERLAP. **2.** Distance that a member overlaps another.

lap joint: 1. Wood joint in which two members are joined face-to-face without removing material from the faces. See *joinery**. **2.** Weld joint formed by overlapping two members and welding along the edge. Members should overlap a minimum of three times the thickness of the thinnest member. See *weldment**.

lap seam: Sheet metal seam in which two adjoining pieces are lapped and joined using mechanical means such as rivets or screws. See *seam* (1)*.

* **lap siding:** Horizontal siding with each succeeding row covering a section of the previously applied row.

lap splice: Connection of rebars made by overlapping the ends of the rebars.

large knot: Knot in a piece of wood with at least a $1^1/_2''$ diameter.

laser: Concentrated beam of light used for applications such as surveying or laying pipe. Acronym for light amplification by simulated emission of radiation.

laser beam cutting: Cutting wood or metal using heat generated by a concentrated beam of light. Produces finished cut with little waste. LASER CUTTING.

laser beam welding: Joining metal by melting the metal with heat produced by a concentrated light beam. LASER WELDING.

laser level: Leveling device in which a concentrated beam of light is projected horizontally or vertically from the source and used as a reference for leveling or verifying horizontal or vertical alignment. See *rotating laser, pipe laser.*

laser target: 1. Electronic surveying accessory that emits an audible and/or visual signal when aligned with a laser beam projecting from a laser level. See *surveying**. SENSOR. **2.** Graduated reflective device used to level suspended ceiling members. Device is secured to the members with a magnetic backing.

lash: To secure members with rope.

latch: Fastening device used to secure a door or window in position. Does not require a key.

latch bolt: Spring-loaded device in a lockset that is beveled so as to retract when pushed against the strike plate and project from the door when in a fully closed position. BOLT.

latent heat of evaporation: Measure of heat equal to the amount of heat required to convert 1 pound of liquid into vapor without raising the temperature. LATENT HEAT.

lateral: 1. Positioned at 90° to another line or object. **2.** Diagonal brace. **3.** Small irrigation ditch or pipe. **4.** Underground electrical conductor. **5.** See *lateral sewer.*

lateral sewer: Sewer that discharges into another sewer without any other sewer flowing into it.

lateral thrust: Force applied at 90° to the length of a structural member.

latex: Water emulsion of synthetic rubber or plastic.

latex paint: Water-base paint suitable for interior and exterior application.

lath: Wood, metal, or gypsum product fastened to structural members to create a support for plaster.

lathe: Stationary power tool that rotates a piece of wood or metal to allow for the shaping of the piece with sharpened or blunt tools.

lather: Tradesperson who fastens lath to structural members in preparation for plaster.

lath hammer: See *lathing hatchet.*

lathing: See *lath.*

lathing channel: Cold-rolled, 16-gauge steel member that is $^3/_4''$ or $1^1/_2''$ wide and used for furring, lathwork, partitions, and suspended ceilings.

lathing hatchet: Hand tool with a striking face on one end and a sharp edge on the other end. Used to cut and fasten lath to structural members.

latitude: Angular distance north or south of the equator. 1° latitude ranges from 68.704 to 69.407 miles.

latrine: Public toilet.

* **lattice:** Gridwork of narrow, parallel strips of wood or metal fastened at 90° to one another to produce interwoven appearance. Creates a pattern with open areas between the strips.

laundry chute: Vertical open duct used to transfer laundry from a collection area to the laundry cleaning area by gravity. CLOTHES CHUTE.

laundry room: Area with a clothes washer and dryer, and other equipment for cleaning laundry.

* **laundry tray:** Deep sink used for handwashing clothes. LAUNDRY TUB, SET TUB.

laurel: Leaf-shaped decorative design. Resembles a leaf from a laurel or number of laurel leaves in a group.

* **lavatory: 1.** Wash basin with a water

lavatory (1)

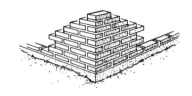

lead (1)

leaf (1)

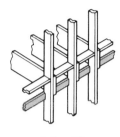

ledger (1)

supply and drainage system. **2.** Room equipped with a wash basin and water closet both with water supply and drainage system.

lay: Direction in which the strands of a rope are twisted. Most rope is right-lay, which means when the rope is held vertically the strands spiral up and to the right.

layer: 1. Single veneer in plywood. **2.** Two or more plies laminated together with parallel grain. **3.** Stratum of weld metal made up of one or more weld beads. **4.** Electronic overlay used in a CADD system that allows the operator to view various versions of the same drawing. One layer may include the floor plan; a second layer, the electrical plan; a third layer, the mechanical plan. All layers may be viewed simultaneously or separately.

lay-in ceiling: Suspended metal ceiling grid that supports ceiling tile that are placed on top of them.

laying line: Continuous marks on felt roofing paper used as a guide for overlapping the paper.

layout: Plan or design indicating the position of a component or object in relation to another.

lay out: To establish points and mark building material to indicate placement of other material or cutting.

layout dye: Dark liquid painted on a metal surface to emphasize layout marks. Removed from the surface with denatured alcohol.

lay up: To assemble veneers in plywood manufacturing.

lazy Susan: See *revolving shelf*.

L/D ratio: Relationship of the span (L) of a floor truss to its depth (D). Expressed in inches.

leaching: 1. Separating liquid from solid material by percolating them into soil. **2.** Movement of liquid through rock or porous soil while seeking the path of least resistance.

* **lead: 1.** Corner of a masonry wall laid in a vertical and plumb position. Used

as a guide when laying the central portion of a wall. **2.** Portion of a pile driver that secures the pile and hammer in position when driving. LEADS. **3.** Axial travel of a screw thread in one full turn. **4.** Short electrical conductor.

lead: Soft metal used for forming watertight joints in cast iron soil pipe. Also used as an alloy in solder.

lead angle: See *travel angle*.

leaded glass: Window or lighting fixture in which individual pieces of glass are secured in place with strips of H-shaped lead (cames). LEADED LIGHT.

leaded joint: Cast iron soil pipe joint in which the bell and spigot of two sections of pipe are joined. Joint is stuffed with oakum then filled with molten lead.

leader: 1. See *downspout*. **2.** Duct that carries air from the blower to the individual areas to be heated or cooled.

leader line: Inclined line on a print or drawing that extends from a written note to the object that it refers to. Commonly terminated by a horizontal shoulder at one end and an arrowhead at the other end.

leading edge: See *lock stile*.

lead plug: 1. Piece of lead inserted into a predrilled hole and used as a attachment point for a nail or screw. **2.** Piece of lead formed in a joint between adjacent stones. Formed by pouring molten lead into a void between the surfaces.

lead shield: Lead sleeve inserted into a predrilled hole and used as anchorage for a threaded fastener. Fastener forms a thread in the lead and expands the sleeve.

lead wool: Thin strands of lead that are tamped into an opening such as a pipe joint. Force applied to the material causes it to consolidate and form a single piece.

* **leaf: 1.** Flat portion of a hinge. **2.** See *wythe*.

leaf and dart: Decorative design having

a regular pattern of leaf shapes alternated with arrows.

leaf and square: Plastering tool used to form ornamental designs and shapes. One end has a tapered point and the other end is flat.

leakage current: 1. Nonreversible electric current that remains in a circuit after capacitive current and absorption current are dissipated. **2.** Electric current flowing through a device after the device is turned off.

lean concrete: Mixture with low cement content, resulting in a weak product. LEAN MORTAR.

lean-to: Small building sharing a common wall with a larger structure. Often constructed with a shed roof.

ledge: Shelf or horizontal projection.

* **ledger: 1.** Horizontal member that supports joists or other members placed on top of it; e.g., wood member used to support upper floor joists in balloon framing. LEDGER BOARD, LEDGER STRIP. **2.** Horizontal batterboard used to support reference lines.

leech line: Perforated waste pipe extending from a septic system to allow treated waste to seep into the soil.

left bank: Left side of a stream or river while facing downstream.

left-hand door: Door that swings inward with the swinging stile on the left when facing the outside of the door. See *door**.

left-hand reverse door: Door that swings outward with the swinging stile on the left when facing the outside of the door. See *door**.

left lay: Rope in which the strands spiral up and to the left when the rope is held vertically.

leg: 1. One side of an angle iron. **2.** One side of a compass or set of dividers. **3.** Conductor carrying electric current on one side of an electrical circuit.

legend: Description or explanation of symbols and other information on a drawing or print.

length: Distance; lineal measurement.

let in: Notched into another member.

level: 1. Aligned with the horizon; parallel to the horizontal. **2.** Hand tool that is used to indicate the horizontal or vertical position of an object. Slightly curved liquid-filled vials are marked to indicate position of the bubble. When the bubble is centered within the marks, it indicates that the surface is level or plumb. **3.** See *builder's level, dumpy level, instrument, transit.*

leveling instrument: See *builder's level, dumpy level, instrument, transit.*

leveling rod: Pole graduated in hundredths of a foot and used to indicate elevations. Used in conjunction with a leveling instrument. See *engineer's rod, Philadelphia rod.*

* **leveling screws:** Threaded feet with large thumb screws located between the base plate and frame of a surveying instrument. Allows for setting the telescope and frame in a horizontal position.

level loop: Carpet with the face yarn

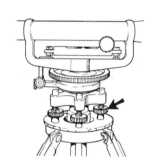

leveling screws

life belt

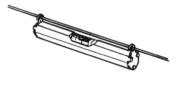

line level

woven or tufted to an equal height.

lever: Straight rod or bar that rests on a fulcrum (pivot point); used to increase mechanical advantage.

lever cap: Piece that holds the back iron and cutting iron in position on a plane.

lever handle: Style of finish hardware in which a horizontal handle acts as a control for a lock bolt.

lewis bolt: Dovetail-shaped device used to lift stone or other heavy masonry units. Lewis hole (tapered recess) made in the material to be lifted receives the bolt. LEWIS.

L foundation: Foundation having a spread footing with a wall extending vertically and flush with the outside edge. Used in an area with limited distance from the property line to the building line.

lierne rib: Short member in a Gothic vault that extends from one rib to another and creates a star-shaped pattern.

* **life belt:** Piece of leather or webbing fastened around the waist of a worker and secured to a lifeline. Used to prevent injury in case of a fall while working at heights.

lifeline: Rope or line with one end secured to a worker and the other end secured to a solid attachment point. Used to minimize injury in case of a fall.

lift: 1. Platform that moves vertically. Used to move objects from one elevation to another. **2.** Layer of concrete placed between two horizontal construction joints. **3.** Vertical step in an excavation. **4.** See *sash lift.*

lift angle: Angle between a vertical line extending from an object and the actual line between hoisting equipment and the object.

lifter: Device made of a short piece of metal that is tapered on one end and has a fulcrum on the bottom; used to position a sheet of gypsum drywall. Tapered end is placed under the bottom edge of the drywall and force is applied to the other end by stepping on it.

lift latch: Door latch that uses a flat rod that moves vertically and pivots in a hook attached to the door jamb. THUMB LATCH.

lift slab: Method of concrete construction in which horizontal slabs are cast on top of each other, jacked into position, and secured to columns at the desired elevation. Successive slabs are separated by a chemical-release agent.

light: 1. Pane of glass or translucent material in a door or window. **2.** Electrical fixture that illuminates. **3.** Illumination from a source.

light fixture: Electrical device that connects one or more lamps to a power source.

light-gauge metal joist: Light-gauge metal horizontal support member with a C-shaped cross section. Used in residential applications similar to those using solid wood joists.

lighting off: Initial fuel ignition.

lighting outlet: Electrical outlet designed for direct connection of a lampholder or light fixture.

lightning arrester: Protective device attached to a structure or electrical wiring system to prevent damage resulting from lightning or an electrical power surge. LIGHTNING ROD.

lightweight aggregate: Material that weighs between 40 and 70 pounds per cubic foot; mixed with concrete to reduce dead load of a structure or increase insulating properties. Includes expanded clay, natural pumice, volcanic cinders, and fly ash.

lime: Calcium oxide. Used as an ingredient in plaster and cement.

lime mixer: Drum-like container with a rotating shaft and several blades used to consolidate lime and other plaster products.

lime paste: Mixture of water and quicklime.

limit control: Switch that opens and closes an electrical circuit at a given temperature or pressure limit.

limit stop: Device that prevents adjustment of a controller beyond predetermined settings.

lineal: One-dimensional measurement.

linear: Straight narrow design.

line blocks: See *corner blocks.*

line drop: See *voltage drop.*

* **line level:** Slightly curved vial in a metal case that is suspended from a string to indicate relative position with the horizontal plane.

lineman's pliers: Hand tool used to cut and bend heavy-gauge wire. See *pliers*.*

linen pattern: Decorative design in the shape of folded or rolled fabric. LINEN FOLD, LINEN SCROLL.

line of sight: 1. Imaginary line extending from a surveying instrument to the object being sighted. **2.** Assumed position of a viewer in a perspective drawing.

line of travel: Area of a stair, approximately 12″ to 16″ from the handrail, in which the majority of traffic passes. Used when laying out winding stairs to ensure tread width at this point conforms to building code.

line oiler: Device installed in a compressed air system with a reservoir to lubricate and regulate amount of lubrication dispensed at a predetermined rate to pneumatic tools.

*line pin: Tapered steel shaft inserted in masonry work to secure a string in position, which acts as a guide for the masonry courses.

liner: 1. See *stiffback*. 2. See *lining*. 3. See *form liner*.

liner panels: 1″ thick gypsum sheets with fire-resistant core encased in moisture-resistant paper.

line stretcher: Device used to position a string a predetermined distance away from masonry work.

lining: Material used to cover, finish, or protect another surface.

link: 1. Unit of measure equal to 7.92″; used by surveyors. 100 links equal 66′-0″. 2. One segment of a chain.

linked switch: Two or more electrical switches physically joined to operate as a single unit.

linoleum: Resilient floor covering composed of ground cork and linseed oil; applied to fabric or canvas backing.

linoleum knife: See *hook bill knife*.

*linoleum roller: Roller made of a series of heavy cylindrical weights mounted on an axle; used to press flexible material, such as flooring material, in place. Attached to a handle, which allows operator to be in a standing position.

linseed oil: Drying agent used in paint. Derived from flaxseed.

lintel: Horizontal structural member above a door or window opening. Used to distribute the weight from the structure above to either side of the opening.

*lintel block: U-shaped concrete masonry unit designed to be filled with mortar and rebars to join it with adjoining similar units. Installed across tops of window and door openings.

lip: 1. Small projecting edge of an object. 2. Cutting edge of a tool or excavating bucket.

lip union: Pipe fitting with projecting edge around the inside diameter to prevent a gasket from being forced into the pipe and forming an obstruction.

liquation: Separation of metals that melt at a low temperature from alloys with a higher melting point.

liquefaction: Momentary loss of shearing resistance in saturated, fine-particled soil when exposed to vibrations.

liquid limit: Level of moisture absorption at which soil changes from solid to liquid.

liquid membrane forming compound: Compound sprayed on fresh concrete that forms a chemical barrier to prevent loss of moisture from the concrete.

liquid petroleum gas: See *LP gas*.

liquidus: Minimum temperature at which a metal becomes liquid.

live: Connected to an energized electrical source.

live load: Predetermined load a structure is capable of supporting. Includes moving and variable loads such as persons, furniture, and equipment.

live steam: Steam discharged directly from a boiler without pressure reduction.

living room: Room in a house designed for social activity.

load: 1. Weight or forces exerted on a structure resulting from environment, settlement, and dimensional changes. Includes dead, impact, internal, live, and wind loads. 2. Electric current flowing through a circuit. 3. Explosive blasting charge. 4. Amount of temperature differential per unit of time imposed on a heating or cooling system.

load-bearing: Supporting a load exerted from above.

*loader: Excavation equipment designed to lift or move loose material by means of a bucket mounted on hydraulic arms.

loader pump: Device used to convey bulk fluid from a large storage container to hand-held gun applicators.

load factor: 1. Measure of average weight moved by equipment. Expressed as a percentage of the equipment's maximum capacity. 2. Comparison of total volume of waste flowing from plumbing fixtures to total capacity of the drainage system at any given point in the system. 3. Comparison of average load on a cooling system to maximum load capacity of the system.

load indicating bolt: Threaded fastener with a head that has small projections on the underside. When torqued to the proper tension, the projections compress.

loading dock: Elevated platform for receiving deliveries. LOADING PLATFORM.

*loading dock leveler: Platform used to compensate for various truck bed heights by adjusting vertically.

loading dock seal: Resilient member around sides and top of a loading dock door that protects interior areas from rain and cold temperatures.

loading hopper: Chute into which concrete or free-flowing material is placed and loaded by gravity into transportation equipment.

loading ramp: Inclined ramp that spans an area between a loading surface and a truck.

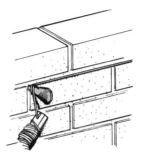

line pin

linoleum roller

lintel block

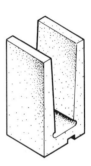

loader

loading dock leveler

LOCKS

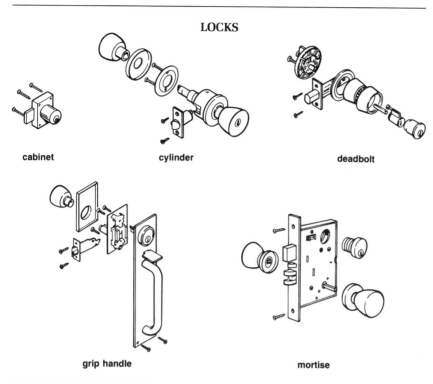

cabinet   cylinder   deadbolt

grip handle   mortise

load test: Method of determining ability of a structural member to support a design load. Pressure is applied to the member and evaluated before continuing with construction.

loam: Soil composed of clay, sand, and organic matter.

lobe: Rounded projection.

local vent stack: Vertical vent pipe installed on fixture side of a trap to remove noxious gases.

* **lock: 1.** Device used to secure a movable object from being moved or opened. **2.** Chamber in a compressed air system that is subjected to system pressure on one end and atmospheric pressure on the opposite end.

lock cap: Removable cover on a lockset.

locked rotor current: Amount of electric current a motor utilizes before the rotor begins rotating or while the rotor is unable to rotate.

lock hammer: Power tool with an anvil that rolls a lip in sheet metal and forms interlocking edge for ductwork.

locking pliers: Spring-loaded hand tool with jaws that clamp onto an object. Distance between the jaws is adjusted by turning a screw at the rear of the handle. Lever inside the handle is used to release the clamped jaws. See *pliers**. VISE GRIP.

lock joint: Pattern formed in two adjoining members that is used to secure them together with a minimum of fasteners or adhesive.

locknut: Secondary nut installed on a threaded rod or bolt to prevent the first nut from loosening. See *conduit fitting**.

* **lock rail:** Central horizontal member of a panel door used for installation of a lockset.

lock seam: Interlocking seam between two adjoining sheet metal members to form a tight, flat joint. LOCK JOINT.

lockset: Complete assembly of a bolt, knobs, escutcheon plates, and all mechanical components for securing a door and providing means for opening.

lockset bit: Wood-boring bit used to drill a 2$\frac{1}{8}$″ diameter hole for a lockset. Used with an electric drill. See *bit* (1)*.

* **lock stile:** Vertical member of a door used for installation of a lockset. CLOSING STILE, LOCKING STILE, STRIKING STILE.

lock washer: Round washer that is split and slightly bent; installed between a nut and the surface of the workpiece. Used to secure a nut in position and prevent loosening. See *washer**.

loess: Silty soil formed from wind-blown deposits.

loft: 1. Open raised platform above a main floor level. **2.** Attic. **3.** To lift piles into position for driving.

log cabin: Structure made of solid logs laid horizontally on one another to form exterior walls.

log cabin siding: Horizontal lap siding with a rabbeted edge and concave surface to give the appearance of logs. See *siding**.

long and short work: Stonework pattern around door or window openings in which stone is set alternately in a horizontal and vertical position.

long float: Long trowel for plasterwork requiring two people for operation.

longitude: Angular distance east or west of a meridian, such as the prime meridian in Greenwich, England.

longitudinal bond: Masonry pattern used in thick walls in which courses composed entirely of stretchers are randomly set.

longitudinal joint: Two members joined together parallel to their longest dimension.

long nipple: Section of pipe externally threaded on both ends with a longer blank center section than on a standard nipple. SPACE NIPPLE.

long nose pliers: Pliers with thin, extended, pointed jaws. Commonly used to reach objects in tight locations or to pick up small objects. See *pliers**.

long screw: 6″ long nipple with one end threaded longer than a standard nipple.

long splice: Splice used to join two ropes of equal size in which diameter of the splice is same as diameter of the rope. See *knot**.

long ton: Weight measure equal to 2240 pounds.

* **lookout:** Rafter, joist, or other framing member that is cantilevered to support

lock rail

lock stile

lookout

lucarne

lug (2)

lumber core

lute rake

---

a fly rafter or fascia. Projects beyond the exterior wall of a building. LOOKOUT RAFTER, TAIL, TAIL PIECE.

loop: Narrow window, vent, or opening in a wall. LOOPHOLE.

loop vent: Extension of a waste pipe that connects to a waste vent stack rather than a fixture vent stack. Also connected to a horizontal drain pipe and receives discharge from otherwise unvented fixtures. Similar to a circuit vent.

loose-fill insulation: Material manufactured as small granules or particles and poured or blown into voids in floors, walls, or roofs to reduce thermal or sound transmission.

loose knot: Defect in lumber caused by a limb growing perpendicular to the grain. Not tightly held in the lumber. May fall out, causing a serious defect in the lumber.

loose pin hinge: Hinge with a removable pin that rests in the knuckles. Pin may be removed for disassembly.

loose yard: Measure of soil or rock after it has been separated by excavation. Equal to a volume of 27 cubic feet.

lot: 1. Portion or parcel of land. **2.** Defined quantity.

lot line: See *property line*.

louver: 1. Vented covering for an opening that allows passage of air but prevents rain and pests from entering. Commonly made of wood, aluminum, or galvanized steel. **2.** Design of horizontal slats in doors and shutters. See *door**.

louver shielding angle: Horizontal angle above which objects are hidden when viewed through a louver.

low alloy steel: Steel with various metals added, such as nickel or chromium, to increase strength and corrosion resistance with minimal structural effect.

low boy: Trailer with a platform close to the ground for easy loading of heavy equipment. LOW BED.

low-carbon steel: Steel with less than 20% carbon content. Not commonly used for structural steel because of its ductility. MILD STEEL.

low-density concrete: Concrete with a unit weight of less than 50 pounds per cubic foot.

low-energy power circuit: Electrical circuit with a limited power supply. Used for devices such as doorbells and chimes. LOW VOLTAGE CIRCUIT.

low-pressure laminate: Plastic laminate material cured with maximum pressure of 400 pounds per square inch. Used in vertical and low-wear applications.

low-pressure steam boiler: Boiler with less than 15 pounds per square inch of steam pressure.

low side: Portion of a refrigeration system that is subject to evaporator pressure. LOW-PRESSURE SIDE.

low temperature brazed joint: Gas-tight pipe joint in which the metal alloys melt between 1000°F and 1500°F.

lozenge: Four-sided plane figure with all sides equal in length. Has two opposing acute angles and two opposing obtuse angles, forming a diamond shape.

LP gas: Liquefied petroleum gas containing propane, propylene, butane, and/or butylene. Acronym for liquid petroleum gas.

L-shaped flashing: Sheet metal bent into an L-shape and installed along the intersection of a roof and wall to prevent passage of water. See *flashing**.

* **lucarne:** Small window installed in a dormer.

Lucite: Manufacturer's name for various plastic sheet products.

* **lug: 1.** Projecting extension of a door or window stile beyond the rail. **2.** Terminal attached to electric cable that allows quick connecting and disconnecting. **3.** Projecting ridge or deformation on rebars. See *rib*.

lug hook: See *timber carrier*.

lug sill: Horizontal member at the bottom of a window opening in a masonry wall that extends beyond the sides of the window.

lumber: Wood that has been cut to size for building purposes.

* **lumber core:** Plywood with core constructed of strips of solid wood and outer plies that are veneer.

lumber crayon: Wax-type marker used to make water- and fade-resistant marks on lumber, steel, and concrete.

lumen: Measure of illumination.

luminaire: Complete light fixture.

luminaire efficiency: Ratio of the luminous flux emitted from a lamp or bulb to the total flux.

luminous: Lighted; transmitting light.

lump lime: See *quicklime*.

lunette: Crescent-shaped ceiling opening for a skylight.

* **lute rake:** Tool with a 36″ wide blade with one straight and one serrated edge. Used to move and smooth asphalt. LUTE.

luthern: See *lucarne*.

macadam: Pavement made of broken stone, slag, or gravel mixed with a grout, mortar, or bituminous filler. Base material is compacted to form an even surface and the filler is applied.

machine bolt: Square or hexagonal-headed bolt used in the assembly of members that do not require a precision fit.

machined: Shaped by cutting or grinding, as applied to metals.

machine rating: Measure of the power that an electric device produces without overheating.

machine screw: Fine- or coarse-threaded fastener used for general assembly of members. Head design varies, but slotted and Phillips head recesses are common. See screw*.

machine stress-rated lumber: Lumber evaluated by a piece of equipment that measures its structural properties, such as flexural strength and modulus of elasticity.

machine welding: Welding metal by using equipment that welds metal under the constant surveillance of a welder.

made ground: Land formed by filling a recess in the ground with fill such as rock, broken concrete, and other solid fill.

magazine: Storage facility for explosives.

* magnesium float: Tool used to finish concrete flatwork. Flat magnesium base has minimal amount of drag and creates a smooth finish.

magnetic base: See magnetic drill base.

* magnetic catch: Cabinet door catch in which the magnetic hardware is attached to the frame or faceplate and a steel plate is fastened to the door.

* magnetic drill base: Base of an electric drill in which electric current is used to magnetize and secure the base to a metallic surface. MAG BASE.

magnetic level: Short spirit level with a magnetic surface. Used to determine level and plumb position by attaching it to metal pipe or conduit.

magnetic overload relay: Electrical overcurrent protection device with contacts that are actuated by the load current.

magnetic screwdriver: Screwdriver with a magnetic tip that holds the screw when positioning the screw in tight locations.

magnetic switch: Electrical switch in which the contacts are controlled by an electromagnet. Commonly used in motors.

magnetic target: Graduated piece of plastic or metal attached to a magnet. Used to establish elevations with a laser level by attaching it to a suspended ceiling grid or other metal member.

magnetite: Mineral used as an aggregate in high-density concrete.

magnifier: See magnifying glass.

magnifying glass: Circular piece of glass or plastic used to magnify an object. Used by a surveyor to accurately read verniers of a surveying instrument. Commonly mounted in a case to prevent scratches. See surveying*.

mahlstick: Padded piece of wood used by painters to steady their hands while doing ornamental work.

mail slot: Small opening on an entrance door or sidelight with a hinged door. Used for delivery of mail.

main: 1. Principal supply pipe to which branches are connected. 2. Primary electrical circuit that feeds smaller circuits. 3. Primary; largest or most structurally important. 4. See main tee.

main runner: See main tee.

main sewer: Principal waste pipe to which branch sewers are connected.

* main tee: Inverted-T shape, light-gauge metal member used as the longest support member in a suspended ceiling. Member to which cross tees and ceiling tile are secured.

main tie: See bottom chord.

main vent: See vent stack.

major arch: Arch with a 6'-0" minimum span that can support loads over 1000 pounds per square foot.

magnesium float

magnetic catch

magnetic drill base

main tee

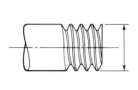

major diameter

male thread

mandrel (1)

* major diameter: Outside diameter of a screw thread.

makeup air: Air drawn into an enclosed area to compensate for exhausted air.

makeup water: Water added to a boiler to compensate for leaks, blowdowns, condensate, and evaporation.

male nipple: Short piece of pipe with external threads.

male plug: Electrical connector placed at the end of a power cord. This connector has two or more prongs that project outward for fitting into a receptacle or a female plug.

* male thread: Continuous spiral groove cut on the outside of a pipe or fitting. EXTERNAL THREAD.

mall: Public plaza.

malleable iron: Annealed cast iron.

malleability: Property of a metal that permits it to be hammered or formed without cracking.

mallet: Soft-faced striking tool. Used to strike tools such as chisels.

mall front: Glazed store front in a mall.

malm brick: Brick made from a mixture of clay and chalk.

* mandrel: 1. Solid core inserted in a hollow pile to prevent the shell from collapsing when being driven. Removed after placement. **2.** Cylindrical spindle used to guide a hole saw. Hole saw is fastened to the bottom section and the upper section is inserted in the chuck of an electric drill. Spindle may be replaced with a pilot bit.

manganese: Steel alloy used as a deoxidizer and hardener.

manifold: 1. Chamber with several inlets and one outlet, such as in an automobile exhaust system. **2.** Chamber with several outlets and one inlet, such as a pipe supplying gas to individual burners.

manila rope: High-strength fiber rope manufactured from organic material.

manipulative joint: Joint formed with soft metal pipe, such as copper, in which one end of the pipe is enlarged to receive the adjoining pipe. Pipes are forced together to form a tight joint.

manometer: U-shaped tube partially filled with water; used to measure gas and/or vapor pressure. One end of the tube is connected to the source of pressure; amount of water displacement is directly proportional to the pressure difference.

mansard roof: Roof with two different slopes on all sides of a structure. Lower edges of the roof are at a steeper angle. See roof*.

mansion: Large house or estate.

* mantel: Trim around or shelf above a fireplace opening.

mantle: Outer surface of a wall that differs from the structural members of the wall.

manufactured aggregate: Industrial by-products such as slag and cinders used as a concrete aggregate.

manufactured housing: See modular housing.

manufactured sand: Fine aggregate produced by crushing stone, gravel, or slag.

marble: Metamorphic rock primarily composed of calcium carbonate. Used for interior or exterior surfaces because of its high luster and durability.

marbling: Design on wood, metal, or other material that resembles polished marble.

margin: 1. Width of exposure of overlapped shingles or slate. **2.** Flat surface of stiles and rails that form a panel frame. **3.** Space between a door and the side and head jambs that allows easy opening. **4.** Perimeter of an object or area.

marginal wharf: Retaining wall built parallel with the shoreline used as a landing for boats and protection for the bank.

* margin light: Narrow window next to a larger window or door.

margin of safety: 1. Ratio of the ultimate load on a structure to the design load. **2.** Measure of the load-bearing capability of a member. Equal to the strength of a member divided by the unit stress on an area.

* margin trowel: 1. Small hand tool with flat, rectangular or triangular blade. Used to clean a larger trowel or small corner work. **2.** Plasterer's trowel with edges bent up at right angles to the surface. Used for corner work.

marigold window: See rose window.

marine glue: Adhesive that withstands exposure to water over an extended period of time without dissolving.

marine grade plywood: Plywood that is bonded with 100% waterproof adhesive and high veneer quality standards. Commonly used for marine applications such as boat building.

marking awl: See scratch awl.

marking gauge: Hand tool used to scribe a line parallel to the edge. Adjustable head slides along a graduated beam and is fixed in desired position. Marking pin projects from the edge of the beam to scribe the surface of the work.

marquee: Cantilevered, projecting structure over a building entrance. Not supported by columns.

marquetry: Inlaid wood or other material used as ornamental design.

marquise: See marquee.

mask: To tape a temporary covering over areas that are to be protected during painting or other finishing.

mason: Person who shapes and lays brick, block, and stone or finishes concrete. BRICKLAYER, CEMENT MASON, STONE MASON.

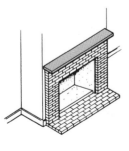

mantel

margin light

margin trowel (1)

Masonite®: Non-structural building material with one hard, smooth surface. Available tempered or untempered, with the tempered being more water-resistant and harder.

masonry: Construction involving assembly of a structure using individual units such as brick, block, stone, or tile bonded with mortar.

masonry anchor: Device embedded in a masonry wall that allows attachment of additional members, such as a door frame. See *anchor*.

masonry bit: Bit used in an electric drill to bore into hard surfaces such as concrete or masonry. Wide spiral flutes remove powdered material from bottom of the hole. See *bit* (1)*.

masonry blade: Circular abrasive blade made of silicon carbide grit and used to cut masonry material, such as block or brick.

masonry cement: Hydraulic cement used for masonry, consisting of portland cement, portland blast-furnace slag cement, portland-pozzolan cement, natural cement, and/or hydraulic lime.

masonry guide: Device, such as a corner pole, used as a gauge for accurately setting masonry units.

masonry nail: Hardened nail with fluted shank that can be driven into concrete or masonry surface without deforming. See *nail*. CONCRETE NAIL.

masonry unit: Hollow or solid building unit made of natural or manufactured material, such as clay or concrete.

mason's chisel: See *brick set*.

mason's hammer: See *brick hammer*.

* **mason's level:** Spirit level approximately 4′-0″ long with three separate vials used for plumbing and leveling masonry.

mason's lime: Material used to prepare plaster or mortar.

mason's putty: Mixture of water, hydrated lime, portland cement, and stone dust; used to fill voids in ashlar masonry.

* **mason's rule:** Rule graduated in increments equal to various mortar joint thicknesses. Used to determine layout for courses of brick.

mass haul curve: Line on a print or diagram that indicates amount of excavated material that can be used as fill.

mass profile: Sectional view of a road showing volume of cut and fill material. Expressed in cubic yards.

mast: 1. Vertical support for load lines. **2.** Long slender post used to support a light source.

master-keyed lock: Lock that can be opened by an individual key and a key that is common to other locks in the series.

master switch: Electrical device that controls two or more circuits.

mastic: Putty-like adhesive that maintains a degree of elasticity after setting.

mat: 1. Flat member that increases the bearing surface area of equipment. Commonly used where soft soil is encountered to distribute the load and prevent sinking of a concentrated load. **2.** Gridwork of rebars.

matched lumber: Lumber with fitted edges, commonly tongue-and-groove. MATCHED BOARDS.

matching lines: Aligning marks on prints of an area that is too large to be contained on one sheet. Two individual sheets can be laid next to one another and aligned to show a large area of the structure. MATCH LINES.

match marks: Marks on adjoining members used to indicate proper alignment for patterned work.

mat foundation: Continuous footing that supports a reinforced concrete slab covering a minimum of 75% of the total area within the exterior walls. Used to improve bearing capacity of the foundation. See *foundation*. MATT FOUNDATION.

matrix: 1. Cement paste in which fine aggregate is embedded. **2.** Mortar in which coarse aggregate is embedded.

matte finish: Non-glossy finish with low reflective characteristics.

* **mattock:** Digging tool used for loosening soil and other compact material. One end is pointed like a pick and the other end flattened like a chisel.

mattress: Concrete slab-on-grade used as a footing or supporting surface.

mat well: Depression in an entrance floor for placement of a rug or carpet. MAT SINKING.

maul: Long-handled, heavy striking hammer. MALL.

mausoleum: Structure used as a memorial or repository for tombs.

maximum demand: 1. Highest level of electric current in a system within a given time. **2.** Largest volume of flow in a water supply system or waste disposal system.

maximum density: Largest unit weight to which a material may be compressed.

maximum rated load: Largest amount of weight a structure can safely support.

meager lime: Building material with at least 15% impurities; used in plaster.

mean: Average.

mean horizontal candle power: Measure of illumination based on average amount of light radiated by a lamp in a horizontal direction.

mason's level

mason's rule

mattock

mean sea level: Average height of the surface of the sea; determined by hourly height measurements. SEA LEVEL DATUM.

means of egress: Continuous unobstructed path used as an exit from a structure, consisting of an exit access, exit, and exit discharge.

mean spherical candle power: Measure of illumination based on the average amount of light radiated by a lamp in all directions.

mean stress: Average stress incurred by a member during a single cycle of tensile and compressive loading.

measured drawing: Architectural drawing of an existing structure. Commonly drawn to scale.

measurement: 1. Process of determining dimension or quantity of an object or material. **2.** Dimension or quantity.

mechanical advantage: Comparison of the weight of a load lifted by a machine to the applied force.

mechanical analysis: Process used to determine particle size distribution in aggregate, soil, or sediment.

meeting rail

meeting stile

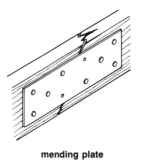

mending plate

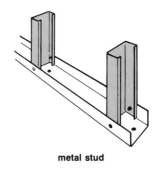

metal stud

mechanical application: The spraying or pumping of plaster or shotcrete rather than manual application.

mechanical bond: 1. Bond formed by interlocking or keying adjoining members; e.g., keyway in a concrete footing. **2.** Physical interlock between aggregate and cement paste or concrete and reinforcement. **3.** Keying a plaster coat with a base coat or partially embedding it in expanded metal lath.

mechanical contractor: Company or individual responsible for all phases of mechanical construction, including estimating, installation, and testing. Includes HVAC and piping systems.

mechanical equivalent of heat: Measure of mechanical energy equal to 1 Btu of heat; equal to 778 foot-pounds.

mechanical joint: Pipe or fitting connection that uses compression to form a watertight seal. Includes flanged, flared, and screwed joints.

mechanical plan: Print showing the placement of HVAC equipment and required piping or ductwork. Fire extinguishing systems may also be indicated. Commonly noted with a capital M preceding the sheet number. See *print**.

mechanical supply system: Forced-air system that uses mechanical means, such as fans and blowers, to move air.

medallion: Ornamental plaque.

median: Untraveled area between two roadways.

medicine cabinet: Storage cabinet for medicine and personal toiletries. Installed above a lavatory or vanity in a bathroom.

medium: Paint component that regulates the gloss, adhesion, durability, and overall appearance.

medium density overlay: Exterior-type plywood product finished with a fiber resin overlay. Heat and pressure compact the panels to a density ranging from 31 to 55 pounds per square foot. Used for furniture, signs, and displays.

medium-duty scaffold: Scaffold used to support a working load of more than 50 pounds per square foot.

medium solder: Solder composed of equal amounts of tin and lead.

M.E.E. (milled each end) pipe: Sewer pipe that is milled on each end and unfinished in the center.

* **meeting rail:** Horizontal member in a double-hung window that is the upper member in the lower sash and the lower member in the upper sash.

* **meeting stile:** Vertical abutting member in a pair of doors or windows.

melamine: Plastic compound used in the manufacture of high- and low-pressure plastic laminates.

melamine glue: Thermosetting adhesive cured with heat and pressure. Primarily used for commercial furniture construction.

melting point: Temperature at which a material changes from a solid to a liquid.

member: 1. Structural component. **2.** Single component of a series or set; e.g., three-member baseboard consists of three separate trim components.

membrane: 1. Moisture-impervious coating or film. **2.** Thin, flexible roofing material supported by a minimum air pressure equal to 1.5″ of water column.

membrane curing: Method of retaining moisture in freshly placed concrete by applying film to restrict evaporation. Membrane may be liquid form, such as sealers and emulsions, or non-liquid coating, such as plastic sheets or burlap.

membrane roofing: Roof covering that uses elastomeric sheets with sealed joints to provide a waterproof surface.

mende particleboard: Economy-grade wood panel product composed of wood fibers and resin and formed on a cylindrical press. Used for drawer bottoms and cabinet backs.

* **mending plate:** Flat piece of metal with staggered predrilled holes fastened across wood members to strengthen joints.

mensuration: 1. Process of measuring distances. **2.** Branch of mathematics involving computation of length, volume, and area.

merchant pipe: Pipe that is 5% to 8% lighter than standard pipe.

mercury switch: Electrical switch that uses mercury to make a silent contact. Commonly used in thermostats.

mercury vapor lamp: Electric discharge lamp that illuminates by creating an arc in an enclosed bulb or tube containing mercury vapor. MERCURY VAPOR LIGHT.

meridian: Surveying reference plane extending in a north-south direction.

meros: Flat surface of an object between grooves or channels.

mesh: 1. Gridwork of welded wire fabric members. See *welded-wire fabric*. **2.** To fit together, such as the teeth of two gears. **3.** Number of openings per square inch in wire screen.

mesh tape: See *reinforcing tape*.

metal: 1. Chemical element with the following properties: crystalline structure, high thermal and electrical conductivity, ability to be deformed when heated, and high reflectivity. **2.** Broken stones used for road construction.

metal arc cutting: Severing metal by melting through it using the heat of an electric arc between the metal and an electrode.

metal halide lamp: Electric discharge lamp that produces illumination from a mixture of metallic vapor, such as mercury, with sodium halides.

metal lath: See *expanded metal lath*.

metallic insulation: Thin sheets of reflective metal applied over insulating board or sheathing; used to resist heat transmission.

* **metal stud:** C-shaped, corrosion-resistant member used as a non-bearing structural component of interior partitions. Fastened between flanges or U-shaped metal track with screws or crimpers. Commonly prepunched for installation of electrical conductors and plumbing pipes.

metal ties: Rigid, corrosion-resistant steel rods embedded in a masonry wall to tie two wythes of a cavity wall together. Commonly formed into a rectangular or Z-shape.

metal track: U-shaped member fastened to floors, walls, and ceilings to be used as attachment points for metal studs.

meter: 1. Device used to measure and indicate flow of gas, liquid, or electric current; e.g., water meter. **2.** Metric unit of length equal to 39.37″.

meter stop: Valve in a water service that controls flow of water entering a water supply system from a municipal or other regulated source. METER VALVE.

metes and bounds: Property lines of a given area.

metric system: International measurement system based on a meter for length and gram for weight. Based on the base-ten system in which units are a multiple or divisor of ten; e.g., centimeter is $^1/_{100}$ of a meter.

mezzanine: Suspended gallery or balcony that covers less floor area than the level below.

mica: Natural material used to improve resistance to moisture penetration in roofing products and improve application properties of paint. Also used as a filler for thermal and electrical insulators.

microfarad: Measure of electrical capacitance equal to one-millionth of a farad.

microinch: Measure of length equal to one-millionth of an inch.

*****micrometer: 1.** Hand tool used to make very accurate measurements, often to the closest ten-thousandth of an inch. Various types are available, including inside micrometer, used to make inside measurements; outside micrometer, used to make outside measurements; and screw thread micrometer, used to measure root diameter of a screw thread. **2.** Measure of length equal to one-millionth of a meter. MICRON.

microsand: Fine sand that can pass through a No. 100 sieve.

middle lap: Wood joint formed by overlapping the end of one member across the middle of another member. Each member is reduced in thickness by one-half to form a flush, T-shaped joint.

middle rail: Intermediate horizontal member extending between two door stiles. May be exposed as in a panel door, or flush as in a flush door.

*****midrail:** Horizontal member approximately midway between the top handrail member and a floor or platform.

mid-wall column: Column that supports a wall that is thicker than the column diameter.

MIG welding: See *gas metal arc welding.*

mil: Measure of length equal to .001″. Commonly used for wire diameter and material thickness.

mildew: Fungus that grows in moist and unventilated conditions.

mild steel: Low-carbon steel with a carbon content of .15% to .20%.

mile: Measure of length equal to 5280′-0″.

mill: 1. Manufacturing plant that produces construction material, such as steel or wood products. **2.** To shape metal or wood with a rotary cutting tool.

mill bastard file: File with a thin flat face and coarse teeth. Used to sharpen saw blades. MILL FILE.

millimeter: Metric measure of length equal to one-thousandth of a meter.

mill scale: Oxide that forms on iron or steel when heated.

mill smooth file: File with a thin flat face and medium-coarse teeth.

*****mill stud:** Fastener that is threaded on both ends to accept nuts, and with a straight, unthreaded center section.

millwork: Wood construction products produced in a planing mill. Includes trim molding, window and door frames, and stairs.

millwright: Tradesperson who lays out, installs, and aligns heavy equipment.

mineral admixture: Inorganic substance added to concrete to reduce the amount of cement required in a mix and limit heat caused by hydration.

mineral aggregate: Aggregate consisting of inorganic nonmetallic material.

mineral fiber shingle: Flat, fireproof, and decay-resistant material composed of 30% mineral fibers and 70% portland cement or other binder and formed under pressure. Available in sizes ranging from 8″ to 12″ in width and 8″ to 36″ in length. Used as a roof covering.

mineral fiber siding: Flat, fireproof, and decay-resistant material composed of 30% mineral fibers and 70% portland cement or other binder and formed under pressure. Available in sizes ranging from 8″ to 12″ in width and 8″ to 36″ in length. Used as an exterior wall covering.

mineral spirits: Flammable liquid obtained from petroleum distillation; used as a paint thinner and solvent.

mineral wool: Fibrous insulating material formed from molten slag, glass, rock, or other inorganic material. Used as loose fill or formed into batts or blankets.

minor arch: Arch with less than a 6′-0″ maximum span and can support loads up to 1000 pounds per square foot.

*****minor diameter:** Smallest diameter of a screw thread.

minute: Measure of an angle equal to $^1/_{60}$ of a degree.

mirror: To create a copy of an original by reflecting the image around an axis.

mission roofing tile: Clay roofing tile with a semicircular cross-sectional shape and approximately 12″ long. Installed with concave surface alternately up and down.

mist coat: Thin spray coat of lacquer or paint.

miter: Intersecting of two members at an equally divided angle. Commonly a 90° angle formed with two members that have the ends beveled at a 45° angle. See *joinery*.

miter box: Device used to guide a saw when cutting bevels for miter joints. Boxes range from small, three-sided, wood boxes used to guide a hand saw to electric equipment that cut at variety of angles.

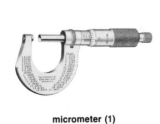

micrometer (1)

midrail

mill stud

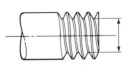

minor diameter

miter square

mixed garden wall bond

modular housing

miter clamp: Spring clamp in which the pointed jaws pinch the two members being joined. Commonly used on miters.

miter rod: Metal plate with one end cut at a 45° angle; used for finishing inside plaster corners.

miter saw: 1. Hand saw with a narrow blade having fine teeth and a reinforced back; ranges from 24″ to 30″ long. Used in conjunction with a miter box to make fine, accurate cuts. 2. Electric saw attached to a pivoting frame for positioning the saw at various angles. See *saw*.

miter square: Hand tool used for laying out 45° and 90° angles. Handle and blade are set at 90° to one another. The end of the handle is chamfered at a 45° angle for laying out 45° miters.

mitre: See *miter*.

mix design: Process of determining the amounts of water, cement, aggregate, and other materials needed to mix together to produce concrete with certain qualities.

mixed garden wall bond: Brickwork pattern in which every fourth course has alternate headers and stretchers. Other courses have stretchers only.

mixed-grained lumber: Laminated lumber with edge and face grain exposed.

mixer: Equipment used to consolidate and mix building material such as cement, aggregate, and water.

mixing box: See *dual-duct system*.

mixing speed: Rate of rotation of paddles or blades in a mixer. Expressed as revolutions per minute or feet per minute that a point along the perimeter of the blades rotates.

mixing valve: Valve that mixes fluids by manual or automatic means.

M.O.A. (machined over all) pipe: Sewer pipe milled from end to end.

mobile crane: Hoisting crane mounted on tracks or wheels.

mobile home: See *modular housing*.

model code: Building code developed by a regional committee of building officials. Recognized building codes in the United States are the *National Building Code*, published by Building Officials and Code Administrators International; *Standard Building Code*, published by the Southern Building Code Congress International; *Uniform Building Code*, published by the International Conference of Building Officials; and *CABO One and Two Family Dwelling Code*, published jointly by the three organizations.

modified portland cement: See *Type II cement*.

modillion: Ornamental bracket supporting a cornice.

modular: 1. Prefabricated as a unit and transported to the job site for installation. **2.** Based on a structural system that uses a 4″ measurement for laying out the placement of structural members.

modular brick: Brick in which the nominal dimensions (actual dimension plus the thickness of a mortar joint) are based on a 4″ unit.

modular housing: Housing unit that is prefabricated as a single unit or several large components and transported to the job site. Unit is attached to a permanent foundation or anchored to a concrete pad to prevent wind and storm damage.

modular ratio: Comparison of the modulus of elasticity of steel to the modulus of elasticity of concrete.

module: 1. Prefabricated structural unit. **2.** Standard unit of measure equal to 4″.

modulus of compression: Measure of the resistance of a material to inward pressure. Comparison of compressive stress to cubical compression.

modulus of elasticity: Measure of the ability of a material to resist deformation. Comparison of the unit stress to the unit strain on a member that has been subjected to strain below its lower limits.

modulus of rigidity: Measure of the ability of a material to resist shear. Comparison of the unit shearing stress to the unit shearing strain.

modulus of rupture: Measure of the maximum load-carrying capacity of a beam. Comparison of the bending moment of the rupture to the section modulus of the beam.

moellon: Fill, such as broken brick, placed inside a masonry wall.

Moh's scale: Hardness standard applied to nonmetallic elements and material. Ten degrees of hardness designated by a different element or material, with the hardness difference determined by which member in the series will scratch the preceding member: 1. talc, 2. gypsum, 3. calcite, 4. flourite, 5. apatite, 6. orthoclase, 7. quartz, 8. topaz, 9. corundum, and 10. diamond.

moist-air curing: Process of curing concrete in moist air with a minimum humidity level of 95% and at a temperature of approximately 72°F.

moisture barrier: See *vapor barrier*.

moisture content: 1. Amount of moisture in a given air space. Expressed as grains of moisture per pound of dry air. **2.** Amount of moisture in the cellular structure of a material such as aggregate. Expressed as a percentage of the dry weight of the material.

moisture gradient: Difference in moisture content between the exterior and interior of a material or object.

mold: Concave and/or convex pattern used to shape a semiliquid material such as plaster.

moldboard: Curved surface of a blade on excavating equipment. Moves material in a twisting motion.

molded anchor: Light-duty plastic sleeve with a cylindrical exterior shape and a tapered cavity to receive a threaded fastener. Inserted in a predrilled hole and expanded as the fastener is inserted.

molded belt course: See *belt course*.

molder: Piece of equipment with four rotating cutting heads for manufacturing wood molding and trim members.

molding: Decorative member used to conceal joints between various materials or to enhance the appearance of a structure.

molding plane: Small hand tool used to cut wood molding for furniture construction. Accepts various shape cutters to produce decorative strips of various designs.

molding plaster: Plaster used to form intricate, detailed designs. Made from finely ground calcined gypsum, lime putty, and water.

molly: Expansion shield in which four wings project from the perimeter as a threaded fastener is inserted. Wings wedge into the sides of a solid wall or spread out on the back of a hollow wall. Most commonly used in hollow material. See *anchor*.

molten alloy relay: Electrical device that opens contacts when electric current overheats a metal alloy.

moment: Load or stress that causes bending in a structural member.

momentary switch: Electrical switch that operates when held in the closed position. As switch is released, switch returns to its original position.

moment of inertia: Measure of bending resistance. Greater value denotes a greater bending resistance.

monel metal: Hard, corrosion-resistant alloy of nickel and copper. Used in the manufacture of marine equipment and components.

monial: See *mullion*.

monitor roof: Roof design in which a raised section of a roof straddles a ridge. Windows or louvers are installed along the perimeter for ventilation.

monolith: Structural member consisting of a large, single piece of stone or masonry.

monolithic concrete: Concrete formed as a single, continuous member. Only construction joints are used.

monolithic terrazzo: $5/8''$ layer of terrazzo topping applied directly to a concrete surface without an underbed.

monorail: Single structural member used to support material handling equipment. Equipment is transported along the member with a series of rollers or wheels.

Monotube® pile: Fluted pile constructed with 3 to 11 gauge steel and closed on the pile tip with a steel point. May be reinforced if needed.

monument: **1.** Commemorative structure; memorial. **2.** Permanent surveying marker used to establish boundaries of property.

mop sink: See *slop sink*.

mortar: Mixture of fine aggregate and cement paste used to fill voids between aggregate or masonry units and reinforce the structure.

mortar bed: See *bed joint*.

mortar board: Flat surface used to hold fresh mortar.

mortar box: Heavy-gauge, cold-rolled steel trough used for manual mixing of mortar.

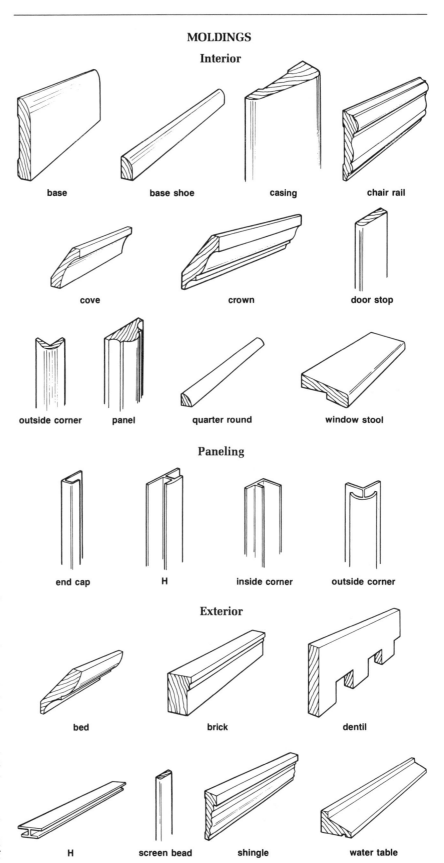

MOLDINGS

Interior

base   base shoe   casing   chair rail

cove   crown   door stop

outside corner   panel   quarter round   window stool

Paneling

end cap   H   inside corner   outside corner

Exterior

bed   brick   dentil

H   screen bead   shingle   water table

moving walk

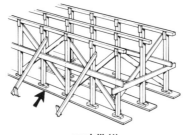

mudsill (1)

muntin (1)

*mortar joint: Vertical or horizontal joint between masonry units. Various designs are used to finish the joint.

mortar sand: Fine sand that can pass through a No. 8 sieve.

mortise: Recess or cavity in lumber or other material that receives a member such as a tenon or lock.

mortise-and-tenon: Joint in which a mortise (recess·or cavity) is formed in one member to receive a tenon (projecting portion) of an adjoining member. See joinery*.

mortise bolt: Door bolt housed in a recess or cavity in the edge of a door.

mortise chisel: Chisel with a long shank for reaching into deep cavities. Primarily used for cutting wood.

mortise hinge: Hinge in which the leaves are designed to fit into a recessed area of a jamb and edge of a door. Face of each leaf is flush with the surface after installation. See hinge*.

mortise lock: Door lock with a mechanism completely housed inside a door. A mortise is cut in the edge of the door and holes are drilled into the face for the mechanism and projecting knobs. See lock*.

mortising bit: Router bit used to cut a flat-bottomed recess. Commonly used to cut a recess for a mortise hinge. See bit (2)*.

mosaic: Large design formed by inlaying small pieces of glass, tile, or rock in a mortar or plaster underbed.

mosaic tile: Ceramic tile manufactured as small, individual squares or rectangles. Commonly fastened to a paper backing to form larger units.

motor: Machine that converts electrical energy to mechanical power.

motor grader: See grader.

mottle: Clouding or spotting that appears in stone or wood.

mottler: Heavy paintbrush used to make graining and marbling patterns.

mottling: Round discoloration in a spray painted surface.

moulding: See molding.

moulding plaster: See molding plaster.

mousing: Process of tying a rope or wire around the open portion of a hook to prevent a sling or choker from being detached when the line is slackened.

*moving walk: Continuously moving, passenger-carrying equipment that transports people on a horizontal plane or up a slight incline.

M shape beam: Structural steel member that cannot be classified as W, HP, or S shape. Availability of M shape beams is limited to a few manufacturers. See structural steel*.

mud: 1. Soil containing enough moisture to make it soft. 2. See joint compound.

mudjacking: Process of raising or filling voids under a foundation by injecting a slurry of soil, cement, and water under it and allowing it to set.

*mudsill: 1. Continuous timber placed on the ground that distributes a load and provides a level surface for scaffolding and shoring. 2. See sill.

mud slab: 2″ to 6″ layer of concrete below a structural concrete floor or footing to provide additional support on soft soil. MUD MAT.

mullion: Vertical dividing member between two window units.

multifoil: Ornamental design with at least five sections.

multioutlet assembly: Surface or flush electrical raceway that holds conductors and receptacles.

multiple lines: Single rope reeved around several sheaves to increase the mechanical advantage.

multiwire branch circuit: Branch circuit consisting of a minimum of two ungrounded conductors with a potential difference between them, and a grounded conductor with a potential difference between it and the ungrounded conductors; the ungrounded conductors are connected to the neutral conductor.

multizone system: HVAC system with individual control devices in at least two zones of the structure.

muntil: See mullion.

*muntin: 1. Vertical or horizontal member that divides a pane of glass into several smaller pieces. May be structural or decorative. DIVISION BAR, GLAZING BAR, SASH BAR, WINDOW BAR. 2. Intermediate vertical member that divides the panels of a door.

muriatic acid: Liquid used to clean masonry.

MORTAR JOINTS

| concave | flush | raked | scintled | struck | vee | weathered |

* **nail:** Fastener used to secure two or more pieces of material together, such as wood or masonry. Straight, slender piece of metal with one pointed end and one end that is struck with hammer or driven with pneumatic or powder-actuated gun.

nailable concrete: Lightweight concrete that is suitable having nails driven into it after it hardens.

nail claw: See *nail puller*.

nailer: Wood strip fastened to concrete, steel, or masonry surface. Used as an attachment surface for nailable materials.

nail float: See *devil float*.

nail gun: Mechanical device used to drive nails. See *powder-actuated tool*.

nail head: Flat enlarged end of a nail for striking.

nail pop: Circular irregularity occurring in the face of a finished surface caused by a nail backing out of its original position.

nail puller: Short bar that has a slotted curved head; used to remove nails.

nail set: Hand tool, approximately 4″ long, with a slightly cupped tip. Used to drive nail heads below the surface of the material. Tip is placed on a nail head and the other end is struck with a hammer.

narthex: Entrance vestibule in a church.

natatorium: Structure containing a swimming pool.

national coarse thread: Standard description of screw threads. Expressed as the number of threads per inch. Specific diameter threaded fasteners have a standard number of threads per inch.

National Electrical Code®: Nationally accepted guide regarding the safe installation of electrical conductors and equipment. Sponsored and controlled by the National Fire Protection Association.

national pipe thread straight: Pipe thread in which the diameter of the threads is constant throughout the entire length.

national pipe thread tapered: Pipe thread in which the diameter of the threads tapers $3/_4$″ per 1′-0″.

nattes: Decorative pattern that imitates a basket weave design. NATTE.

natural aggregate: Material, such as crushed stone, gravel, and sand, that is quarried or mined from the earth.

natural bed: Naturally formed stone surface.

natural draft: Movement of air or gas without mechanical blowers or fans.

natural finish: Transparent surface finish which does not obstruct natural color or grain of material.

natural gas: Combustible gas that occurs naturally. Used for residential and industrial power and heating.

natural grade: Original, existing elevation of a piece of property before excavation begins. EXISTING GRADE.

natural seasoning: Process of allowing wood to lose some of its moisture content by exposure to the air.

NAILS

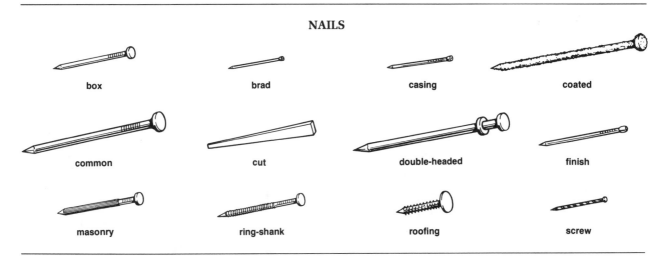

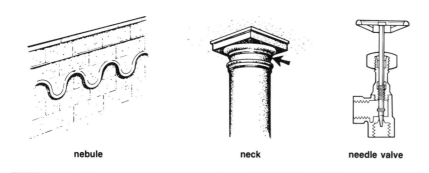

nebule   neck   needle valve

natural ventilation: Natural air movement through an area caused by cross-ventilation or non-powered ventilators.

N boot: Sheet metal fitting that makes a 90° transition in a rectangular duct. See *ductwork**.

neat cement grout: Mixture of hydraulic cement and water with or without an admixture.

neat cement paste: Mixture of hydraulic cement and water before and after setting.

neat line: Outside boundary of an excavation or structure.

neat plaster: Plaster without sand.

* nebule: Ornamental molding made up of a projecting band with a curved bottom edge.

nebulize: To supply a mist of water or steam into an air stream.

* neck: Portion of a column directly below the capital.

necking: 1. Reduction of inside diameter of a pipe pile as a result of soil pressures. 2. See *neck mold*.

neck mold: Trim member applied horizontally around the top of a column immediately below the capital. NECKING.

needle: See *needle beam*.

needle bath: Shower bath with several perforated pipes used to spray water onto a person from many directions.

needle beam: Horizontal shoring member placed under or through a wall to support the structural load while underpinning or forming a new foundation.

needle nose pliers: Hand tool with long slender jaws for reaching into narrow places and grasping small objects. See *pliers**.

* needle valve: Type of globe valve in which a long tapered needle is moved in and out of a seat to regulate flow of fluid.

negative pressure: Force exerted inside a closed system that is less than atmospheric pressure.

neoclassic architecture: Building design reflecting classic Greek and Roman architectural styles.

neon lamp: Illumination device that produces light by passing an electrical charge through a gas vapor.

neoprene: Synthetic rubber with high resistance to petroleum products. Used for expansion joints, bearing pads, and as a base material for certain types of contact cement.

nest of saws: Collection of various saw blades that fit interchangeably into a common handle. Commonly includes a metal-cutting blade and several blades for cutting irregular shapes and small holes.

net absorption: Amount of preservative absorbed by a piece of wood during treatment.

net cut: Amount of excavation material removed from an area less the compacted fill that is added.

net fill: Amount of fill required in an area less the excavated material.

net floor area: Area that is occupied in a building, not including walls, stairways, bathrooms, and storage areas. Used to determine rental space and fire-code requirements.

net line: See *neat line*.

net size: Actual dimensions of a member or material. Expressed as *thickness × width × length*.

network: Electrical conductors that are interconnected.

neutral axis: Hypothetical plane in a structural member where tension, compression, and deformation do not exist.

neutral conductor: Electrical conductor, in a circuit containing at least three conductors, that carries unbalanced current from other electrical conductors.

neutral flame: Oxyfuel gas flame that contains equal amounts of oxygen and fuel gas. Chemically neutral.

neutralize: To treat concrete or plaster surfaces with an acid solution to minimize effects of lime penetration when painting.

* newel: 1. Vertical post that supports a handrail at the top and bottom of a stairway. See *starting newel*. 2. Central post of a circular stairway. NEWEL POST.

newel cap: Decorative top on a newel.

newel post: See *newel*.

* newel stairway: Stairway in which the treads, risers, and stringers are supported by a vertical post; circular stairway. NEWEL STAIR.

N grade: Highest grade of plywood face veneer; cabinet quality.

nib: Small projection on a piece of material, such as on a varnished surface.

nickel: Metal added to steel to increase strength, hardenability, and corrosion resistance. Also used to electroplate other metals, such as plumbing fixtures.

night setback: Reduction of the temperature setpoint in a conditioned space during evening hours or hours of non-occupancy.

90° elbow: Pipe fitting that forms a 90° angle.

* nipple: 1. Short piece of pipe with male thread on each end. See *conduit fitting**. 2. Steel pin used to connect sections of tubular steel scaffold.

nitramon: Safe explosive that is detonated only by a detonating fuse.

nobble: To shape stone roughly.

node: Point in an electrical system where two or more conductors connect.

no-fines concrete: Concrete containing cement, water, and only course-graded aggregate.

nogging: Fill material such as brick and

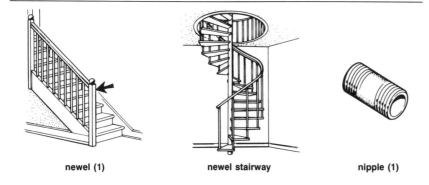

newel (1)   newel stairway   nipple (1)

stone that is mortared in between wood supports such as studs. BACKFILL.

noise reduction coefficient (NRC): Mathematical average of sound absorption quality of a material at 250, 500, 1000, and 2500 hertz.

nominal dimension: Actual masonry dimension plus the thickness of a mortar joint, up to a maximum of $1/2''$.

nominal size: 1. Dimension of a piece of lumber before it is surfaced and dried; e.g., nominal 2 × 4 has actual dimensions of $1^1/2''$ by $3^1/2''$. **2.** Smallest sieve opening through which a specified amount of aggregate can pass.

nominal span: Horizontal distance between the outside edges of supporting members.

nonagitating unit: Truck-mounted container without provisions for agitation. Used to transport concrete a short distance.

non-bearing: To support no load other than its own. NON-LOAD-BEARING.

non-cohesive soil: Material in which the soil particles do not bond together; e.g., sand and gravel.

noncombustible: Unable to support flame and fire; fire-resistant.

noncorrosive flux: Soldering flux composed of rosin or resin-base material that does not chemically attack the base metal.

nondestructive testing: 1. Method used to evaluate the strength of a welded joint without destroying or impairing the weld. Includes visual inspection, radiographic, magnetic-particle, liquid penetrant, and ultrasonic testing. **2.** Method used to evaluate strength of concrete. Includes probe penetration resistance and ultrasonic testing.

nondisplacement pile: Pile formed by drilling or excavating into the earth and placing concrete or a pile.

nonevaporable water: Water that is chemically bonded with cement during hydration.

nonferrous metal: Metal that does not contain iron.

nonflammable: Unable to support flame.

non-load-bearing: See non-bearing.

nonmetallic-sheathed cable: Electrical cable that has two to four conductors with a green insulated or bare conductor used for grounding. Nonmetallic jacket provides protection for the conductors.

nonpotable water: Water not safe for human comsumption.

nonrenewable fuse: Electrical overload protection device that requires replacement after the fusible portion has melted.

nonreturn valve: Combination check

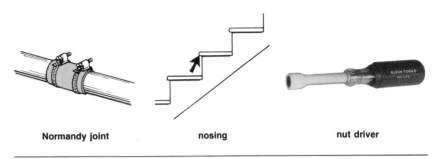

Normandy joint   nosing   nut driver

and globe valve installed in a high-pressure boiler discharge.

nonsiphon trap: Plumbing trap in which the water seal cannot be siphoned.

nontransferred arc: Electrical arc formed between an electrode and constricting nozzle. Base metal or workpiece is not part of the electrical circuit. PLASMA WELDING.

nonvolatile: Portion of paint that remains after evaporation of all carriers and solvents.

noose: Knot that draws tightly around an object as the rope is pulled.

normal consistency: Workability of freshly mixed concrete or plaster, which is affected by the proportions of the various ingredients.

normal recovery capacity: Expression indicating the ability of a water heater to recover from emptying. Expressed as the gallons of water that are heated 100°F per hour when calculated at a thermal efficiency rating of 70%.

normal weight concrete: Concrete with a unit weight of approximately 150 pounds per cubic foot.

Norman brick: Solid masonry unit measuring 4″ × $2^2/3''$ × 12″. See brick*.

* **Normandy joint:** Pipe joint in which two plain end pipes are butted and the joint covered with a sleeve. Ends of the sleeve are sealed with compressed packing rings.

Norwegian brick: Hollow masonry unit that measures 4″ × $3^1/5''$ × 12″ or 6″ × $3^1/5''$ × 12″. See brick*.

nose and miter: Return molding installed on the exposed end of a tread in an open stairway. Molding has same profile as the front of the tread and is cut at a bisecting angle to give the appearance of one unit.

* **nosing:** Portion of a tread that projects beyond the face of the riser; often rounded. For skeleton stairs, the portion of the tread that projects beyond the rear edge of a lower tread. NOSE.

nosing strip: Trim molding with the same profile as the front of a stair tread.

no-slump concrete: Fresh concrete with

a slump of less than $1/4''$.

notch: Recess or indentation in an otherwise continuous surface.

notchboard: See housed stringer.

notch joist: Joist with one end supported by a ledger and the other end notched to fit over a wood girder or beam.

novelty flooring: Flooring that is laid in an irregular, noncontinuous pattern.

nozzle: 1. End of a dispenser from which semiliquid or liquid materials are discharged. **2.** Device in electric arc welding that directs the shielding medium to the proper location.

N truss: See Pratt truss.

nugget: Weld metal joining workpieces in spot, seam, and projection welding.

* **nut:** Piece of metal or other hard material that is female threaded to receive a bolt or other male-threaded device.

* **nut driver:** Screwdriver-shape hand tool with a hexagonal tip. Tip is fitted onto outside of a nut and used to tighten or loosen the fastener.

nylon: Synthetic petroleum by-product.

NUTS

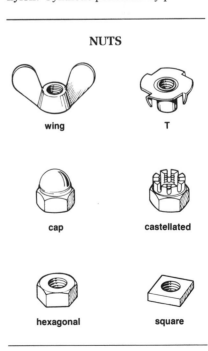

wing   T

cap   castellated

hexagonal   square

* **oakum:** Untwisted coarse rope made from long hemp fibers. Used to retain molten lead when joining bell-and-spigot cast iron pipe.

obelisk: Tapered, four-sided vertical column with a pointed top. Used for a memorial or commemorative marker.

objective lens: Component of a surveying instrument that magnifies the object being viewed through the telescope.

oblique drawing: Type of axonometric drawing in which one face is shown in a flat plane with the receding lines projecting back from the face. Receding lines may be drawn at one-half the scale of the face as in a cabinet drawing, or at full scale as in a cavalier drawing.

oblique joint: Intersection of two members at an angle other than 90°.

obscuration: Covering ability of paint.

obscure glass: Translucent glass, such as frosted or ground glass, that allows passage of light but obstructs vision.

obstruction light: Red light installed on top of tall structures to warn air traffic.

obtuse angle: Angle exceeding 90° and less than 180°.

occupancy: Purpose or intended use of a building.

occupant load: Maximum number of people permitted to occupy a building during a given time.

Occupational Safety and Health Act of 1970: Federal act that protects workers from dangerous job site conditions and monitors unsafe practices.

Occupational Safety and Health Administration (OSHA): Federally funded agency created within the Department of Labor to encourage employers and employees to reduce job site hazards, implement new or revise existing safety programs, monitor job-related injuries and illnesses, and develop job safety and health standards.

octagon: Eight-sided plane figure with all sides and angles equal.

octagon scale: Table on a framing square used to lay out octagonal components. Consists of a series of dots and numbers expressing multiples of 5. Measurements are equal to one-half the length of an octagon side.

octant: One-eighth of a circle; 45° angle.

oculus: Round opening in the top of a dome.

odeum: Small theater.

* **offset: 1.** Change in direction of a flat or straight plane. **2.** Combination of pipe fittings that brings one section of pipe into different but parallel line with other pipes.

offset digging: Excavating with a ladder ditcher when the boom is not centered in the equipment.

* **offset hex wrench:** Hex wrench in which the jaws operate at an angle to the axis of the handle. Used for working in tight locations.

offset pipe: See *offset* (2).

offset pipe wrench: Pipe wrench in which the jaw operates at an angle to the axis of the handle. Used for working in tight locations.

* **offset screwdriver:** Screwdriver with two ends with each end bent at a 90° angle from the axis of the handle. Used for turning screws in tight locations.

offshore lead: Two-piece guide for a pile hammer. Hammer is housed in an upper section and guided when driving by a lower section that fits around the pile.

ogee: Double curve decorative molding resembling the letter S.

ogee bit: Router bit used to form an S-shape along the edge of a member. See *bit* (2)*.

ogive: 1. Diagonal rib in a Gothic vault. **2.** Pointed arch.

Ohm: Measure of electrical resistance. Equal to the resistance encountered when 1 ampere of current is driven by 1 volt. Greater resistance results in less current flow.

ohmmeter: Device used to measure resistance of an electrical conductor.

Ohm's law: Mathematical formula stating that the current in an electrical circuit is directly proportional to the voltage and inversely proportional to the resistance.

oil-base paint: Paint containing resins and other petroleum products. Cleanup is done with a solvent.

oakum   offset (2)   offset hex wrench   offset screwdriver

old English bond

one-point perspective

open cornice

opening leaf (1)

open stairway

oil burner: Furnace that uses oil to generate heat.

oiler: 1. Maintenance assistant for an operating engineer. **2.** Device used to introduce lubrication into a tool such as in a compressed air system for pneumatic tools.

oil separator: Device for removing oil and oil vapor from refrigerant in a compressor discharge line.

oil stain: Wood stain containing pigments combined with oil or oil varnish. Commonly a penetrating stain.

oilstone: Fine-grained whetstone used for sharpening and honing tools. Oil is applied to the surface. OILSLIP, SHARPENING STONE.

oil switch: Switch submerged in oil or other insulating liquid. Controls an electrical circuit. OIL-IMMERSED SWITCH.

oil well cement: Slow-setting cement that is resistant to high temperature and pressure. Used primarily for sealing oil and gas pockets when drilling or repairing oil wells.

* old English bond: Brick pattern with alternating courses of headers and stretchers. Closer (shortened brick) is laid at the corners in each course of headers.

olive knuckle hinge: Hinge with a central ball bearing contained in a hardened steel raceway. See *hinge**.

on center (OC): Distance between center points of framing members and other building components; e.g., studs spaced 16″ OC are placed with their center 16″ apart.

one-hour rating: Measure of fire resistance. Member or surface resists combustion for 60 minutes.

one-part line: Single-stranded cable or rope.

one-pipe system: Hot water or steam heating system in which a single pipe serves as a supply and return piping.

* one-point perspective: Pictorial drawing that utilizes one vanishing point. Receding lines converge to this point.

one-way joist system: Reinforced concrete construction in which a slab is supported by parallel concrete joists. Concrete for the slab and joists is placed monolithically. See *joist**.

on grade: 1. Placed directly on the ground. **2.** Leveled or excavated to the proper elevation.

opacity: Ability of paint to cover a background.

open: 1. Assembly or construction that is exposed. **2.** Noncontinuous construction with spaces or voids between members.

open assembly tie: Time interval between the application of adhesive to joints and the assembly.

open-circuit grouting: Injection of grout into an area without provision for recirculating excess grout to the pumping system.

open-circuit voltage: Voltage between the terminals of an electric arc welding machine while no current is flowing in the circuit.

open coat abrasive: Abrasive paper in which approximately 50% to 70% of the surface area is covered with abrasive material with voids between the abrasive material. Voids allow sanding without excessive filling of the paper. Used for rough sanding, such as paint removal and initial sanding.

* open cornice: Overhang of a roof in which the rafter tails, gutterboard, and roof sheathing are exposed. Bird blocks are installed between the rafters and the members are painted.

open cut: Excavation that is completely uncovered and exposed to the sky.

open defects: Splits, open joints, and knotholes in the surface veneer of plywood.

open-end block: H-shape concrete masonry unit.

open-end wrench: Hand tool with two open ends designed to fit hexagonal nut or bolts. Each end is different in size to allow the wrench to be used on two different size heads. See *wrench**.

open floor: Floor that is exposed on the bottom. Joists are visible from below.

open-graded aggregate: Concrete aggregate with relatively large voids when compacted.

open grain: Wood surface with large pores and widely spaced annual rings.

* opening leaf: 1. Portion of a hinge that moves. **2.** In a door with a pair of leaves, such as a French door, the leaf to which the locking or latching mechanism is attached.

open newel stair: 1. Stairway in which the inside stringers of consecutive flights are not aligned vertically. Short stairways or extended landings create an unobstructed vertical chute between the stringers. **2.** Stairway without newels.

open return bend: U-shape pipe fitting with parallel arms that are not attached to one another.

open shop: Non-organized; non-union; not represented by a collective bargaining agreement.

open socket: Attachment point at the end of a cable consisting of two parallel ears with opposing holes to accept a closing pin.

open soffit: See *open cornice*.

* open stairway: Stair with one or both sides enclosed by a wall. Both sides require a handrail with balusters or intermediate safety rails.

open stringer: Stair stringer in which the ends of the treads are exposed.

open valley: Intersection of two roof surfaces at which the finish materials and each side are positioned so as to leave the valley flashing exposed.

open web steel joist: Framing member

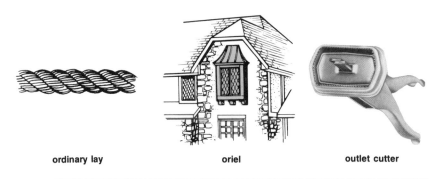

| ordinary lay | oriel | outlet cutter |

constructed with steel angles that are used as chords with steel bars extending between the chords at an angle. Used for light frame construction. See *bar joist*.

open web studs: Light-gauge, non-bearing steel framing member constructed of diagonal wire rods welded to double rod flanges. Used for backing lath.

operable partition: See *accordion door*.

operating efficiency: Ratio of output to input of a device; expressed as a percentage of the input energy.

operating engineer: Tradesperson who operates and performs maintenance on lifting and earth-moving equipment such as road graders and bulldozers.

optical coating: Thin film applied to the glass of a solar collector to reduce reflection of sunlight and maximize amount of light available.

optical fiber: Thin, flexible glass or plastic fiber used to transmit light for control, signal, and communication applications. Consists of a core, cladding, and protective jacket. Three types of optical fiber are nonconductive (no metallic parts), conductive (noncurrent-carrying metallic parts), and hybrid (containing optical fibers and current-carrying metallic conductors).

optical plummet: Device used to center a theodolite or transit over a reference point. Viewer sights through telescope to view the reference point directly below. Used in place of a plumb bob. Preferred in windy applications. VERTICAL COLLIMATOR.

optimum moisture content: Percentage of moisture in soil at which maximum compaction density is possible. Used to determine specifications for embankments.

orange peel finish: 1. Plaster surface texture that creates a slightly rough finish, similar to dried orange peels. **2.** Painting defect resulting from poor flow or application technique. Slightly rough surface similar to dried orange peels.

orbital sander: Hand-held power tool used for finish sanding. Produces a tight circular pattern.

order: Classic architectural styles including Doric, Ionic, Corinthian, Tuscan, and Composite.

order of an arch: Single masonry ring of an arch.

ordinary construction: Structure constructed with masonry or concrete exterior walls with a minimum fire rating of two hours and wood interior non-bearing floor, wall, and ceiling framing.

* **ordinary lay:** Arrangement of strands in a rope in which individual strands are twisted in the opposite direction from the arrangement of strands of the entire piece. Makes rope difficult to untwist.

ordinary portland cement (OPC): Portland cement used for general construction. TYPE I PORTLAND CEMENT.

organic: Derived from living organisms.

* **oriel:** Window that projects outward from the face of a wall. Supported by brackets, cantilevered framing, or corbels.

oriented strand board: Three-ply wood panel product in which wood strands are mechanically oriented and bonded with resin under heat and pressure. Used for subflooring and roof sheathing.

orifice: Small opening.

orifice gas: Gas directed into a torch and that surrounds the electrode when plasma arc welding and cutting.

orifice meter: Device used to determine amount of fluid flowing through a pipe. Plate with a small opening is placed in the pipe and pressure difference on each side of the plate is measured.

O ring: Circular gasket.

orlet: Narrow band forming a frame or border.

ornament: 1. Decorative member. **2.** Decorative design added to a structural member.

orthographic projection: Drawing in which each face of an object is projected onto flat planes generally at 90° to one another. Commonly used for elevation drawings.

orthostyle: Columns in a straight line.

OSHA: See *Occupational Safety and Health Administration (OSHA)*.

outband: Stone cut to accept a frame and laid as a stretcher.

outcrop: Rock that is exposed at the surface of the ground.

outfall sewer: Pipe that receives sewage from collector pipes and carries it to final discharge location.

outfeed rollers: Rotating cylinders on outfeed side of millwork equipment that is used to pull stock through.

outlet: 1. Point in an electrical system where current is taken to supply utilization equipment. **2.** Opening in a piping system for discharge of fluid.

outlet box: Electrical hardware box that houses receptacles and conductors. Electrical connections are commonly made within the box.

* **outlet cutter:** Hand tool used to cut openings in drywall for outlet boxes.

outline lighting: Lighting used to illuminate the perimeter of a structure or object.

output: 1. Capacity; net performance. **2.** Current, voltage, or electrical power delivered by a circuit.

outrigger: 1. See *lookout*. **2.** Equipment accessory that provides additional support and prevents overturning by providing a larger load-bearing area.

outside calipers: Hand tool used to measure outside dimensions and diameters. Legs are turned inward at their tips.

outside corner molding: Trim member installed on the exterior surface of a structure to cover and protect an outside corner. See *molding**.

outside corner trowel: L-shape tool with a smooth steel face used to apply plaster or joint compound to outside corners.

outside stem and yoke valve: Valve in which the position of the valve is determined by the height of the valve stem.

outsulation: Insulating material attached on the exterior side of structural wall components.

oval head: Screw head similar to a flat head with a convex upper surface. See *screw**.

oval point: Rounded point of a set screw.

overall dimension: Total outer dimension of a member.

overburden: Soil or rock covering a deposit of rock, sand, or gravel.

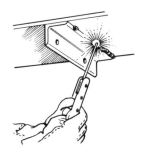

overhead position

overrim bathtub fitting

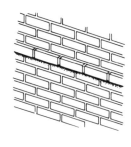

oversailing course

overcloak: Overlapping portion of a metal roofing sheet.

overcurrent: Electrical current in excess of the equipment limit or total amperage load of the circuit.

overflow pipe: 1. Pipe used to channel excess fluid from a plumbing fixture or storage tank when the fluid has reached a predetermined level. **2.** Pipe used to prevent flooding in storage tanks or fittings.

overgrain: To paint a non-wood surface to give it the appearance of woodgrain.

overhand knot: Knot formed by passing the end of a rope over the standing part and passing it through the loop. Simplest of all knots. See knot*.

overhand loop: Crossing one end of a rope over the standing part.

overhand work: Laying masonry units on the exterior of a wall from inside of a building.

overhang: Horizontal projection beyond the vertical face below it.

overhead door: Door which when opened is suspended in a horizontal track above the opening. May be constructed of a single large panel or several narrow stacked panels. Rollers along the ends of the door allow it to travel in tracks. Cables and springs on the back of the door facilitate lifting and lowering the door. See door*.

* overhead position: Position in electric arc welding in which the welding is performed on the underside of the joint.

overhead shovel: Excavating machine that digs at one end, swings the load over the operator's head, and deposits the load on the other end.

overlaid plywood: Plywood panel with metal or resin-treated fiber faces on one or both sides.

overlap: 1. To install material with adjoining edges over one another. **2.** Extension of weld metal beyond the weld area. **3.** Plywood defect in which a double veneer has been applied.

overlay: Layer of concrete or mortar at least 1″ thick placed on and bonded with an existing concrete slab. Used to resurface or improve appearance of an existing surface.

overlay doors: Cabinet doors that cover the edges of a cabinet face frame.

overload: 1. Operation of equipment with greater than the full-load rating or conductors with greater than rated ampacity. May result in overheating. **2.** To place excess weight or stress on a structural member.

* overrim bathtub fitting: Faucet assembly and spout mounted on the drain end of a bathtub. Spout is above the flood level rim of the bathtub.

* oversailing course: Masonry course that projects beyond the course below.

oversanded: Containing excess amount of sand in concrete or plaster.

oversize brick: Masonry unit measuring $3^3/_3″ \times 2^3/_4″ \times 7^1/_2″$. See brick*.

overvibration: Application of excess vibration to fresh concrete, resulting in segregation and bleeding of the mix.

overvoltage: Electrical voltage at a higher level than normal operating range.

overwind: To arrange line on a drum so that it is unwound from the top of the drum.

ovolo: See quarter round.

ovum: Egg-shape ornament.

ox eye: Round or oval window.

oxidize: To chemically combine with oxygen. OXIDATION.

oxidizing flame: Flame produced in oxyfuel gas welding resulting from excessive oxygen. Flame has short, pointed, purple inner cone.

oxyacetylene cutting: Process used to sever metal by combining oxygen with a heated base metal. Temperature to the base metal is maintained by the combination of oxygen and acetylene gases.

oxyacetylene welding: Process of joining metal with heat produced from a gas flame fueled by a mixture of oxygen and acetylene. Pressure and filler material may be used.

oxyfuel gas cutting: General description of processes used to sever metal with heat obtained from a gas flame fueled by a mixture of a fuel gas, such as hydrogen or propane, and oxygen.

oxyfuel gas welding: General description of processes used to join metal with heat obtained from a gas flame fueled by a mixture of a fuel gas, such as hydrogen or propane, and oxygen.

oxygen cutting: General description of processes used to sever or remove metal by means of a chemical reaction at high temperatures between oxygen and the base metal.

pace: Stairway landing.

pack: Shipping bundles for wood shakes and shingles. Pack of wood shakes consists of 9 courses at each end. Pack of shingles consists of 20 courses at each end.

package air conditioner: See *air conditioner.*

* **package boiler:** Boiler unit with all components assembled as one piece of equipment; includes boiler, burner, controls, and auxiliary devices.

packer: Expandable device used to prevent backflow of grout during injection into a hole.

packer head process: Method of casting concrete pipe in which concrete of a low slump is compacted in a vertical manner with a revolving compaction device.

packing: 1. Material used to fill a void, such as in a hollow masonry wall. **2.** Material used to prevent leakage at a joint or shaft.

packing fraction: Cross-sectional core area of fiber optic cable compared to total cross-sectional area of the cable, including cladding and voids. Expressed as a percentage.

* **packing gland:** Sleeve that provides an explosionproof casing around cable or pipe. Used where cable or pipe enters an explosionproof area.

pack set: Cement that has bonded together as a result of prolonged storage. Caused by particle interlocking, compaction, or electrostatic attraction.

pad: 1. Section of a crawler track on earth-moving equipment that contacts the ground. **2.** Flat surface designed to support and distribute a load; a concrete slab.

* **padding:** Depositing layers of weld metal on a worn surface. Used to build worn areas of shafts, wheels, and other machine parts.

* **paddle wheel scraper:** Excavation equipment with a rotating conveyor that scrapes soil into the bowl section.

paddock: Fenced area adjoining or near a stable.

padlock: Security device made of a U-shaped bolt with one pivoting leg and one leg that interlocks with the lock body. The lock may be opened with a combination or keyed device.

pad sander: Portable power tool primarily used for finish sanding.

pad saw: Small compass saw with a narrow, tapered blade. Used for fine woodworking.

padstone: 1. Stone block or concrete placed under a supporting member. Provides additional bearing surface for a concentrated load. **2.** Masonry lintel.

pagoda: 1. Oriental memorial temple or tower. **2.** Decorative furniture ornament with an oriental tower design.

pailing: Concrete form sheathing constructed of individual vertical boards.

paint: Liquid surface covering consisting of a pigment suspended in a vehicle of oil, solvent, or water. Used for weather protection and/or decoration.

painter: Building trades worker who mixes and applies paint, varnish, wallpaper, and other finish coatings.

painter's putty: Pliable material used to fill voids in surfaces to be painted.

paint grade: Lumber classification indicating wood that is unsuitable for a clear finish but suitable for a painted finish.

paint hurling: Process of creating a rough surface by tossing paint-covered stone chips onto a freshly painted surface.

paint remover: Liquid or paste solvent applied to dry paint or varnish to soften the finish and allow for scraping or wiping from the surface.

paint roller: Cylindrical tube covered with porous material. Used to spread paint onto a large flat surface. The tube is used in conjunction with a rotating frame that allows all portions of the roller to contact the area being painted.

paired cable: Electrical cable in which sets of two strands of wire are twisted together.

package boiler

packing gland

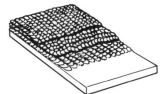

padding

paddle wheel scraper

panel clip

panel form

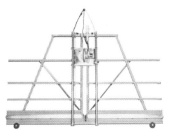
panel saw

panic bar

pairing veneer: Matching full sheets of veneer to reduce amount of handling when constructing plywood panels.

pale: Fence picket.

palisade: Series of posts partially driven into the ground to form a fence or enclosure.

Palladian window: Three-light assembly consisting of a central arched window with tall rectangular windows on each side.

pallet: **1.** Wood strip set in a mortar joint; used as a fastening strip for other wood members. See *ranging bond.* **2.** Platform constructed of wood slats fastened to a frame. Used for a base when transporting material.

pallet brick: Masonry unit with a recess along one edge to receive a wood strip. See *pallet* (1), *ranging bond.*

palmette: Decorative design resembling flower petals radiating from a common base.

pan: **1.** Metal or fiberglass dome used in construction of one- or two-way concrete joist systems. Pans create voids in the bottom of the finished slab. **2.** Masonry or plaster panel in half-timber construction. **3.** Recess for a hinge leaf. **4.** To view a different part of a CADD-generated drawing, other than the one shown on the monitor, without changing magnification.

panache: Segment between two ribs of a groined vault.

pan and roll: Roofing system with flat and rounded tile alternately set in an overlapping pattern.

pane: **1.** Single piece of glass in a door or window. LIGHT. **2.** Hammer peen.

panel: **1.** Section of a wall, ceiling, door, or other surface; e.g., a panel between stiles and rails of a door. **2.** Junction box used to house electrical circuit breakers and main connections. BREAKER PANEL, MAIN PANEL, PANELBOARD, PANEL BOX, SERVICE PANEL. **3.** Length of a truss chord equal to the distance between two struts. **4.** Starter, contactor, or relay assembly that is connected to at least one electrical device

in an air conditioning system.

panelboard: Electrical panel or group of panels designed to form a single panel contained in a cabinet or cutout box. Panelboard is placed in or on a wall and accessible only from the front. Contains buses, automatic overcurrent protection devices, and switches for control of electrical power circuits.

* **panel clip:** H-shaped metal device installed between plywood roof sheathing panels at unsupported joints to add rigidity. Panel clip openings are equal to plywood panel thickness. PLYWOOD CLIP.

panel door: Door with individual panels between stiles and rails. See *door*.

* **panel form:** Prefabricated concrete form consisting of plywood, steel, or aluminum facing attached to a rigid frame. Use of panel forms increases speed and efficiency of form erection. See *form.*

panel heating: Heating an area with pipes or wires concealed in walls and ceilings.

paneling: **1.** Veneer or solid wood interior wall finish. Available in 4' × 8' sheets or 4" to 12" wide solid wood boards. Thickness ranges from $^1/_8$" to $^5/_8$". **2.** Process of constructing a large surface from individual units. **3.** Process of applying molding to a large surface to give the appearance of smaller units.

panel insert: Metal section installed in a half-glass panel door to change it to an all-metal panel door.

panelized construction: See *prefabricate.*

panel length: Center-to-center distance between two joints of a truss chord.

panel load: Force exerted on the panel point of a truss.

panel mold: Trim applied to a large flat surface; used to divide it into smaller sections. See *molding*.

panel pin: See *brad.*

panel point: Intersection of a truss chord and web.

* **panel saw:** Electric saw mounted in a frame that allows for horizontal and vertical cutting of plywood or particle-

board.

panel wall: Non-bearing wall that spans between horizontal and/or vertical supporting members.

pan head: Screw head with a rounded top, vertical sides, and a rounded shoulder. See *screw*.

* **panic bar:** Door hardware with a horizontal bar that releases a latch or bolt when pushed. PANIC EXIT DEVICE, PANIC BOLT, PANIC HARDWARE.

panic bolt: See *panic bar.*

panier: See *corbel.*

panopticon: Building laid out in a radial design with corridors set in a wheel spoke pattern, converging in a central area.

pantile: Curved roofing tile set with an alternating upward and downward curve. The overlap of individual tiles forms a waterproof surface.

pap: **1.** Projection at bottom of a sink or basin. **2.** Vertical outlet from a rainwater gutter.

paper-backed lath: Wire mesh with paper attached to the side opposite plaster application.

paper hanging: See *wallpaper.*

papyriform: Ornamental finish of a column capital using a papyrus flower design.

parabola: Curved geometric shape formed by cutting a cone parallel to one edge.

paraffin oil: Petroleum product used for lubrication.

parallel: **1.** Equidistant, as applied to points or lines. **2.** Precision tool for measuring and laying out parallel lines for close tolerances. Made in individual pieces (fixed distance) or tapered pairs (adjustable distances). PARALLEL TOOL.

parallel application: Installation of gypsum drywall sheets so that the long dimension of the sheet is parallel with the framing members.

parallel-blade damper: Air flow control device in which all fins are in the same position and controlled by a common linkage. Primarily used for an ON-OFF air control.

* **parallel chord truss (PCT):** Prefabricated beam or joist with top and bottom members laid flat and positioned an equal distance apart.

* **parallel circuit:** Electrical circuit with two or more paths for electric current. Electric current passes through all terminals simultaneously.

parallel flow: Hottest portion of one fluid contacting coldest portion of another fluid.

parallelogram: Four-sided geometric shape with opposite sides parallel.

parallel operation: Voltage regulators connected so that electrical outputs are combined to form a common load.

parallel thread: Screw thread with a uniform diameter.

parallel welding: Electric arc welding with electric current divided and routed through workpiece and electrode in similar electrical paths.

parallel-wire unit: Post-tensioning tennon made of a series of parallel wire strands.

* **parapet:** Low wall or a short extension of a wall above a horizontal surface.

parapet skirting: Roofing material that extends a short distance up a parapet to form a waterproof joint where the roof and parapet intersect.

parcel: Single, continuous piece of land without a lot number.

parent metal: See base metal.

paretta: Rough masonry surface consisting of a surface of pebbles.

parge: See parging.

parget: 1. To coat with plaster or mortar. 2. Elaborate decorative plasterwork.

* **parging:** Thin coat of plaster or mortar on a vertical masonry surface. Commonly applied to inside of a wall for waterproofing. PARGE COAT.

paring: Cutting and/or shaving thin pieces of material from a larger member to shape the larger member to a given design of shape.

paring chisel: Cutting hand tool for finish woodworking. Used to slice the wood to create a smooth finished surface.

Parkerized: Iron or steel dipped in boiling manganese dihydrogen phosphate to give it a rustproof coating.

parkway: 1. Roadway with a narrow strip of land between opposing lanes. 2. Strip of land between opposing roadways or between a road and sidewalk.

parliament hinge: H-shaped surface-mounted hinge.

* **parquet:** Wood flooring design in which small strips of wood are arranged in a square pattern with the grain running the same direction. Squares are set with the grain perpendicular to adjoining squares.

parquetry: Flat inlay work with wood or stone.

parquet strip flooring: See strip flooring.

partial discharge: Electric current discharge that does not fully bridge the insulation between electrodes.

partial release: Release of a portion of the prestressing force that was held entirely by the prestressed reinforcing steel in a prestressed concrete member.

particleboard: Wood panel constructed of wood fibers and flakes that are bonded together with a synthetic resin glue; used for underlayment and cabinetmaking.

particle size: Measurement of filtering efficiency based on smallest diameter particle removed by the filter.

particle size distribution: Amount of various size particles in a material. Expressed as a cumulative percentage of particles that are larger or smaller than a specified sieve opening, or percentage of the particles that range between specified sieve openings.

particulate: Solid particle, such as ash, suspended in gas or air.

parting bead: Horizontal member in a double-hung window jamb that separates upper and lower sashes. PARTING SLIP, PARTING STOP, PARTING STRIP.

parting tool: Wood gouge with a V-shaped cutting edge.

partition: 1. Vertical divider that separates one area from another; e.g., an interior wall in a building. 2. To divide land into individual portions.

partition block: Concrete masonry unit, usually 4″ to 6″ thick, used in construction of non-bearing walls. See concrete masonry unit*.

partition plate: See bottom plate, top plate.

partition stud: See stud.

partition tile: Hollow clay masonry unit used to construct interior non-bearing walls.

party wall: Wall between adjoining buildings or structures.

pass: 1. Each layer of weld metal in electric arc welding. 2. Passage of a piece of excavation equipment through desired area. 3. Layer of shotcrete.

passage: Hallway connecting adjoining areas or rooms.

passing: Amount of overlap between sheets of flashing.

passive solar system: Method of heating and cooling a structure in which solar energy is collected and stored directly by the structure. Mechanical methods are not used for transfer or dissipation of the energy.

pass-through: Unobstructed opening in a wall.

paste content: Proportional volume of cement paste in a concrete or mortar mix. Expressed as a percentage of the entire mix.

pat: Sample of cement paste approximately 3″ in diameter and $\frac{1}{2}$″ thick; used to indicate setting time of concrete or mortar. Formed on flat piece of glass with the edges tapered.

patch: 1. Filler material. 2. Temporary connection of electrical circuit with an electrical cord.

patch board: Termination panel for electrical circuits. Temporary electrical connections are made with jacks and plugs.

patch gun: Hand tool for applying texture and patches of joint compound to drywall surfaces. A rubber roller with steel fingers is rotated to distribute joint compound to a drywall surface.

patching: Material applied after initial construction to fill voids in a finished surface.

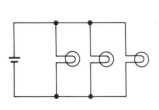

parallel circuit

parapet

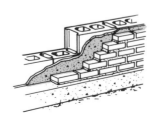

parging

parquet

patented panel system: Prefabricated concrete-forming system used to facilitate construction of concrete forms; commonly includes forms, ties, braces, and other necessary forming hardware.

patent glazing: Installation of glass with any one of a number of puttyless methods.

patera: Round ornament with a floral design used in architectural and furniture work.

patina: Surface film formed by exposure or treatment with an acidic compound.

patio: Outdoor area for dining or entertaining.

patio block: Concrete masonry unit used for walks and ornamental exterior work.

patten: Column base.

pattern: Model or design used as a guide in fabricating or construction of additional duplicate members.

patterned ashlar: Stonework in which a pattern consisting of 6 to 12 pieces of squared stone is repeated. Vertical joints are randomly placed and horizontal joints are semicontinuous. See *bond* (1)*.

patterned lumber: Wood that is shaped or molded to a specific size and design.

patternmaker's saw: Precise wood-cutting hand tool with fine teeth and a thin blade.

paumelle hinge: Hinge with two narrow leafs joined with a cylindrical case containing ball bearings. See *hinge*.

pavement: Layer of hard material, such as concrete; used to cover a horizontal surface for foot or vehicular traffic.

pavement saw: Power tool used to cut expansion grooves in concrete. Uses interchangeable blades and is mounted on wheels for ease in maneuverability and uniformity of cuts.

paver: 1. Machine used to apply and spread hard coatings to a roadbed or horizontal surface. 2. See *paving* brick.

paver tile: Unglazed porcelain or clay tile that is thicker than ceramic tile.

pavilion: Covered area that is open on the sides and may be used for outdoor entertainment.

pavilion roof: Roof style in which all hips are of equal length.

paving: Material used to provide a hard surface for vehicular or pedestrian traffic.

paving brick: Dense masonry unit with high wear-resistance to vehicular and foot traffic. PAVIOR, PAVIOUR.

* **paving train:** Succession of large pieces of paving equipment used to place and finish a concrete roadway.

pawl: Tooth or set of teeth on a wheel in a ratchet designed to interlock with a gear.

Payne's process: Method of fireproofing in which wood is initially injected with iron sulfate, then injected with lime sulfate or soda sulfate.

pea gravel: Gravel that passes through a $^3/_8''$ sieve and is retained on a $^3/_{16}''$ sieve.

peak: 1. Highest point. 2. Largest load.

peak load: 1. Maximum demand for electrical power in a given period of time. 2. Maximum load designed to be supported by any device or structure.

* **pear link:** Tapered ring used in rigging for attaching slings, blocks, and other devices.

peat: Type of soil mainly composed of decayed vegetation.

peavy: Logging tool that has a pointed shaft and a sharpened hook for moving heavy timbers.

pebble dash: Exterior wall finish with exposed gravel embedded in mortar or plaster.

peck: Small section of decayed wood.

pectinate: Having teeth similar to a comb.

pedestal: Vertical compression member; a column.

pedestal floor: Raised floor system supported by short posts and runners. Open area below the pedestal floor is used for routing electrical or computer wiring and ventilation ducts. COMPUTER FLOOR, RAISED FLOOR.

pedestal pile: Caisson with an enlarged base to increase bearing capacity.

* **pedestrian control device:** Device that impedes flow of pedestrian traffic; e.g., gate, turnstile, or railing.

* **pediment:** Decorative unit constructed above a doorway.

peel: 1. To cut a log into sheets of veneer by rotating it against a long knife. 2. To flake off; e.g., paint peeling from a wall.

peel test: Destructive method of testing a weld joint in which a lap joint is separated by pulling it apart.

peen: 1. End of a striking tool opposite the striking face. 2. To shape metal by striking with a hammer.

peen hammer: See *ball peen hammer.*

peg: Short piece of wood used as a fastener.

peg-and-plank: Type of flooring in which wood dowels are left exposed to create a decorative, rustic finish.

pegboard: Tempered hardboard with evenly spaced holes designed to accept storage hooks and hangers. PERFORATED HARDBOARD.

peg point: Tapered rod used to brace an earth-moving machine in position.

pellet molding: Decorative trim consisting of small flat disks.

pelmet: Short member installed at the top inside of a window to conceal curtain rod and fittings.

peltier effect: Absorption of heat when electric current travels through the meeting of two different metals.

penciling: Painting of mortar joints.

pencil rod: Heavy steel wire varying in diameter (normally $^1/_4''$) used as reinforcement in concrete forming. Clamps (catheads) are tightened onto this wire to hold forming in place. Pencil rod remains in the concrete.

pendant: 1. Ornament in shape of a hanging object; a hanging ornament. 2. Device or equipment suspended overhead.

pendant line: Permanently attached rope with an eye on one end and a rigging component on the other end.

pendant post: Timber that projects down onto the wall for holding up part of a truss, beam, or rafter.

pendant switch: Electrical control suspended from an overhead fixture that

paving train

pear link

pedestrian control device

pediment

would otherwise be too high to be easily accessible.

pendative: Design in shape of a spherical triangle.

pendentive: Interior arch or vault that makes the transition between a square building and an octagonal or a domed ceiling.

pendicule: Small supporting pillar.

penetrating stain: Wood coloring, available in many different colors and shades, that enters wood fibers.

* **penetration: 1.** Depth any material travels below the surface of another material. **2.** Depth of fusion below the surface of metal in a welded joint.

penetration rate: Rate at which a pile is driven as compared to the force applied.

penetration resistance: Measure of concrete hardness found by calculating penetration of a needle of a plunger. Expressed in pounds of resistance per square inch.

penny: Measure of nail size designated by the letter d. Based on a 2d nail measuring 1″ long and increasing in $^1/_4$″ increments.

penta: Pentachlorophenol. A chemical liquid wood preservative that protects against termites, rotting, and mildew.

pentagon: Plane geometric figure with five sides. Regular pentagon has equal sides and equal angles.

pentastyle: Architectural design with five columns in front of a structure.

penthouse: Uppermost room or dwelling at top of a building.

percent fines: 1. Percentage of small material in aggregate, usually smaller than a No. 200 sieve. **2.** Percentage of small aggregate in a concrete mix.

perch: Common measure for stonework equal to 16′-6″ × 1′-0″ × 1′-0″.

perched water table: Body of water trapped above otherwise dry soil by an impermeable layer.

perclose: Protective handrail.

percolation: Movement of vapor or liquid through voids in the ground or a filtering material.

percolation test: Measurement of absorption capability of soil. A hole is dug and filled with water, and rate of absorption is measured.

percussion boring: Method for advancing a hole-drilling tool by repeated impact into bottom of the hole.

percussion tool: Power tool that drills or cuts by a repeated reciprocating, striking action.

percussion welding: Joining metal by heating with a rapid flow of electric current. Pressure is applied to this joint during or immediately after heating.

perforated: Having a series of holes.

perforated tape: Thin paper material used with joint compound to help cover joints in drywall construction.

performance-rated panel: Plywood sheets graded for use as sheathing and subflooring. Grade markings indicate grade of panel, span rating, thickness, and other panel information.

pergola: Open-air structure with exposed rafters and/or trusses not covered by sheathing. The rafters and/or trusses are usually supported by posts or columns.

periform: Object with a pear-shaped design.

perimeter: Outside boundary of a space.

perimeter heating: Method of raising temperature in an area with forced air heat from a central plenum fed through ducts with registers at outside walls.

perimeter installation: Method of laying flooring in which adhesive is used only around outer boundaries and seams.

perimeter isolation: Separation around outside edge of structural members from finish and trim components. Limits cracks and distortion of finish material as structural members move with loads.

periodic duty: Intermittent operation under unchanging loading conditions.

peripheral: Around the edge; outside.

periphery: Outside edge of an object.

peripteral: Structure with columns on all sides.

peristalic pump: Squeeze-type device in which a series of rollers are turned in a circular motion and squeeze material through a cylindrical-shaped hose.

peristyle: Complete set of roof-supporting columns encircling or surrounding a central area.

perling: See *purlin*.

perlite: Lightweight material made from crushed and heated volcanic siliceous rock. Used for insulation and as an aggregate in concrete.

perm: Unit of measurement of the ability of a material to allow water vapor to pass through; 1 perm equals 1 grain of

water vapor per square foot hour per inch of mercury vapor pressure difference.

permafrost: Soil that is permanently frozen.

permanent set: Attainment of a physical state by formation or chemical processes that do not allow material to return to its original state.

permeability: 1. Ability of a material to allow water vapor to pass through; expressed as a ratio of the amount of water vapor flow as related to the vapor pressure difference between two surfaces. PERMEANCE. **2.** Relationship between magnetic induction and magnetic force.

permeance: See *permeability*.

* **perpend: 1.** Masonry unit that acts as a header extending to both interior and exterior of a wall. **2.** Joint between shingles. **3.** Vertical joint between masonry units.

perpendicular: 1. At 90°, or a right angle. **2.** Style of architecture characterized by 90° line relationships and right angles. FLORID, RECTILINEAR.

perron: 1. Outside stairway, such as those leading to first floor of a structure. **2.** Stairway in a garden or terrace.

* **perspective:** Representation of an object as it appears when viewed from a given point. Allowances are made for variations in size resulting from appearances caused by distance and viewing location.

pervious soil: Ground that allows relatively free water movement.

petcock: Small valve used for draining.

petrifying liquid: Waterproofing material for masonry.

pH: Measurement of acidity of a material. Measurement of 0 to less than 7 indicates an acid material; 7 indicates a neutral material; greater than 7 indicates an alkaline material.

phantom: Object or portion of an object that cannot be shown or seen on a drawing. Dashed or hidden lines are used to represent these objects.

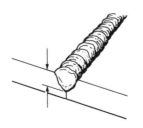

penetration (2)

perpend (3)

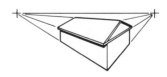

perspective

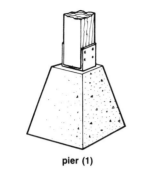

pier (1)

pigtail splice

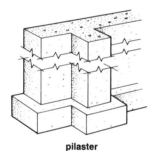

pilaster

pile driver (1)

---

phase: 1. Measurement of number of electric currents alternating at different times in an electrical circuit. **2.** Time span between the beginning of two related occurences.

phenolic resin glue: Powdered and liquid thermosetting adhesive with excellent bonding to wood and paper; has shear strength up to 2800 psi. Commonly used in manufacture of exterior plywood.

Philadelphia pattern: Type of mason's trowel with a square heel.

Philadelphia rod: Ruler divided into hundredths of a foot or eighths of an inch. Used in surveying and establishing elevations. See *surveying**.

Phillips head: Design of top surface of a screw with perpendicular slots shaped like an X. See *screw**.

phon: Measure of sound volume received by the average human ear.

phosphorescence: Light emitted as a result of absorption of eletromagnetic radiation.

phot: Amount of illumination equal to 1 lumen per square centimeter.

photoelectric eye: Electrical device triggered by interruption of a beam of light. PHOTOELECTRIC CELL, PHOTOELECTRIC CONTROL.

photogrammetry: Calculation of measurements from photographs.

photometer: Instrument that measures light intensity and amounts of illumination.

photovoltaic cell: Device for converting light into electric current. Used in solar power systems.

piano hinge: Hardware device that allows for pivoting and moving of adjoining pieces by means of two narrow, continuous leaves joined with a pin.

piano wire: Thin wire with high tensile strength. Used for alignment.

piazza: 1. Enclosed courtyard or colonnade. **2.** Entranceway covered with a porch or roof.

pick: 1. Striking tool with two pointed ends attached to a long handle. Used for digging and loosening hard material. PICKAX. **2.** Number of weft yarns indicating closeness of lengthwise weave in carpeting.

picket: Narrow vertical board, often pointed at the top. Used in wood fencing.

picking: 1. Lifting a load with rigging equipment. **2.** Creating a pitted surface on rubblestone by striking with a steel point. DABBING, STUGGING, WASTING.

pickled: Treatment of a metal surface with an oxidizing agent that cleans and increases corrosion resistance.

pickup load: Amount of heat required to bring pipes and radiators up to proper operating temperature.

pickup point: Predetermined location for lifting a large object with a sling or other rigging hardware.

pictorial diagram: Electrical wiring sketch showing actual positions of wires in a device.

pictorial view: Three-dimensional representation of an object. Axonometric, oblique, and perspective drawings are three-dimensional representations.

picture mold: Narrow trim member fastened horizontally to a wall. Hanging wires are attached to this member and pictures hung below. PICTURE RAIL.

picture plane: Imaginary flat surface on which an object is projected in a perspective drawing.

picture window: Large glassed area in a wall that provides an unobstructed view. Bottom of this area is usually near the floor.

pien: Roof ridge.

pien check: Rabbet in lower front edge of a stone step that interlocks in the adjoining lower step.

* pier: 1. Vertical support that provides bearing in the ground; functions similarly to a column. Bottom of this support may be widened or belled to enlarge the load-bearing area. Upper structural members are set on these supports. **2.** Wharf projecting into a body of water.

pierrotage: Mixture of small stones and mortar; used as filler material between framing members.

pig: Air manifold with several pipes for distributing compressed air supplied by one main pipe.

pigment: Coloring substance mixed with liquid (water or oil) to make paint.

* pigtail splice: Method for joining electrical wires in which ends are stripped of insulation and twisted together. The exposed connection is covered with a wire nut.

pike pole: Tool with a sharp metal point at the end of a handle. Used for holding and working with heavy timbers and wood poles.

* pilaster: Projection from the face of a wall that extends the full height of the wall to provide lateral support.

* pile: 1. Structural member installed in the ground to provide vertical and/or horizontal support. Piles can be wood, steel, concrete, or a combination of these materials. **2.** Upright yarns that are the wearing surface of carpeting.

pile bent: Two or more in-ground structural members close together and joined by a cap or braces.

pile cap: 1. Structural member for distributing loads to in-ground supports. **2.** Temporary protective device on top of an in-ground support that protects it from damage during the driving process.

pile core: See *mandrel*.

* pile driver: 1. Equipment used to force supporting members into the ground by repeated blows or vibration. Consists of a lifting apparatus, hammer, and leads for holding the pile and hammer in place. **2.** Person who works on a crew installing in-ground supports.

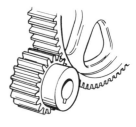
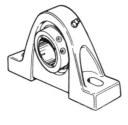
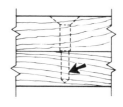

pillow block (1)   **pilot hole**   **pinion gear**

pile extractor: See *extractor*.

pile foot: Tip of a pile.

pile hammer: Device for driving piles into the ground. See *hammer*.

pile head: Upper end of a pile in contact with the pile hammer.

pile height: Distance from top of carpet backing to top of yarn or fabric on the face.

pile load test: See *load test*.

pile shoe: See *shoe* (3).

piling: 1. See *pile* (1). **2.** Thick, uneven paint finish caused by paint that dries too fast.

pillar: Column; a vertical support. May be structurally supporting, decorative, or freestanding.

pillar bolt: Projecting cylindrical fastener that supports another member near the end of the fastener.

* **pillow block: 1.** Piece of rubber encased in metal to allow a certain amount of movement to a supporting or driving member. **2.** Type of bearing.

pilot: See *pilot light*.

* **pilot hole:** Boring made into a material to receive a screw. The hole diameter is slightly smaller than the screw to provide good holding power.

pilot lamp: Small light attached to an electrical circuit to indicate the circuit has power and is in operation.

pilot light: Small flame in a furnace or stove that ignites burners when gas is sent into the system. PILOT.

pilot valve: Automatic device in an air compressor that regulates and keeps air pressure at a fairly constant level.

pin: Slender connector that holds two or more pieces together; e.g., metal piece that holds concrete forms together or cylindrical rod in a hinge.

pincers: Scissors-like tool with two jaws. Used for gripping or cutting. See *end-cutting pliers*.

pinch bar: See *wrecking bar*.

pin hinge: Hardware device that allows movement of two adjoining members by means of a slender rod encased between knuckled plates. See *hinge*, *loose pin hinge*.

pinholing: Blemish in a painted surface caused by air bubbles sealed in the layer of paint; remains after the paint has dried.

* **pinion gear:** Driving gear that is smaller than the driven gear.

pin knot: Small wood defect less than $1/_2''$ in diameter.

pinnacle: 1. Slender vertical member with a cone-shaped top. **2.** Uppermost point of any structure or object.

pinner: Small stone in masonry work that supports a larger stone.

pinning in: Filling opening in masonry

construction with chips of stone.

pin punch: See *driftpin*.

pintle: 1. Metal device used in heavy timber construction to transfer a load from an upper post to a post directly below. A metal shaft spans vertically between a plate at bottom of upper post and top of lower post, keeping the load independent of the intermediate floor structure. **2.** Vertical pin that acts as a rotating, hinging point.

pipe: Seamless cylindrical tube for transporting or containing a material being moved from one location to another. Materials used for pipe include concrete, metals, and plastics.

pipe-and-clamp scaffold: Assembly for creating raised work platforms by means of attaching vertical, horizontal, and angular pipes together with various fixed and pivoting clamps. Clamps are designed for holding adjoining

pipes together in different designs to fit in difficult-to-reach areas.

pipe bend: Pipe fitting that provides a change of direction in the main pipeline.

pipe clamp: Holding device made of a length of pipe with two adjustable fixtures. Both fixtures slide along the bar, and one is fitted with a threaded face to allow for tightening on the material to be held in place. See *clamp**.

pipe cutter: Hand tool with a sharpened disk and an adjustable, clamping, rotating assembly. Clamped onto and spun around soft wall pipe, such as copper, to shear the pipe wall.

pipe die: Tool for cutting external threads on pipe.

pipefitter: Building trades worker who installs piping systems for hot water, steam, gas, and heating and air conditioning.

PILES

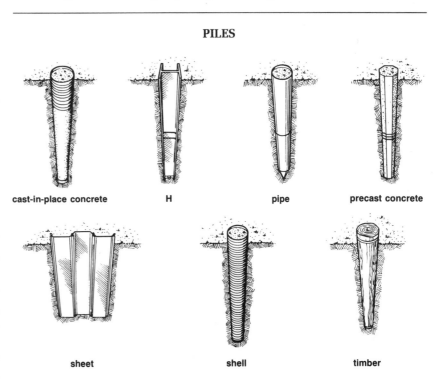

cast-in-place concrete   **H**   **pipe**   **precast concrete**

sheet   **shell**   **timber**

PIPE FITTINGS

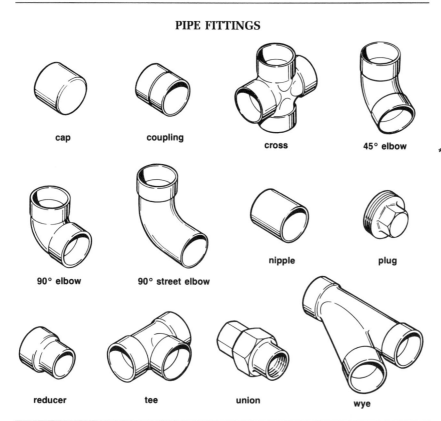

cap   coupling   cross   45° elbow

90° elbow   90° street elbow   nipple   plug

reducer   tee   union   wye

*pipe fitting: Device fastened to ends of pipes to make connections between individual pipes.

pipe hanger: Device fastened to a structure to hold piping in place.

pipe laser: Surveying instrument in which a concentrated beam of light is projected in a given direction at a specified grade or slope. Target is placed into the ends of a series of pipes

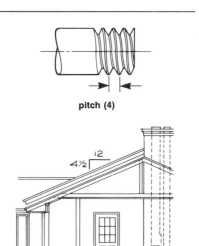

pitch (4)

pitch symbol

and brought into alignment with the light beam to ensure proper alignment of the pipe. See surveying*.

pipe pile: Tubular cylinder of any diameter and wall thickness sufficient to prevent collapse during driving without a mandrel; driven into the ground and filled with concrete to form a foundation support. See pile(1)*. STEEL PIPE PILE, TUBE PILE.

pipe reamer: Conical-shaped tool that is rotated inside the end of a pipe to remove burrs.

pipe schedule: System of classifying pipe sizes by outside diameters and wall thicknesses.

pipe thread: Spiral thread on end of pipe for joining fittings and other pipes. Standard types are regular (used for plumbing work) and dryseal (used for automotive, refrigeration, and hydraulics.)

pipe tongs: Hand tool with a short section of chain for grabbing and turning pipe.

pipe vise: 1. Scissors-type cutting tool with jaws that shear small pipes. 2. Hand tool with a clamping chain assembly that is used to loosen, or break, cast iron pipe. Tightening of chain breaks the threaded connection at the correct spot.

pipe vise stand: Tripod assembly for holding pipe in place during cutting and threading operations.

pipe wrench: Hand tool with serrated, adjustable jaws. The lower jaw has a slight pivoting action that tightens when pressure is applied. See wrench*.

pique: French type of furniture inlay work.

pit: Indentation or a small surface defect in a paint finish.

*pitch: 1. Incline of an object from a horizontal plane, such as a roof rafter or plumbing waste pipe. 2. Spacing between centers of rivets. 3. Center-to-center spacing between turns of a spiral, such as screw threads. 4. Number of threads per inch. 5. Resin substance obtained from tar residues. Used as a waterproofing material. 6. Length of rope or line needed to lay one strand the full distance around the center of a pulley. 7. Slope of a saw tooth or other toothed object in relation to its direction of travel. 8. Distance from center to center for pins in roller or silent chain. 9. Squaring a stone with a chisel.

pitched Howe truss: Prefabricated roof framing member with sloping top chords joined to the bottom chord with a series of vertical and angular struts. Angular struts are inclined outward from top of the vertical struts to bottom chord. Central vertical strut is not used. See truss*.

pitched Pratt truss: Prefabricated roof framing member with sloping top chords joined to the bottom chord with a series of vertical and angular struts. Angular struts are inclined outward from bottom of the vertical struts to top chord. Central vertical strut is used. See truss*.

pitched stone: Rough-faced masonry unit with edges partially exposed as a result of a slight bevel around perimeter of the face.

pitched truss: Prefabricated roof framing member with sloping top chords joined to the bottom chord with a series of vertical or angular struts. See truss*.

pitching chisel: Hand tool with a wide, thick edge for rough-dressing stone.

pitching piece: Horizontal stair member with one end supported by the wall to hold upper end of the rough stringer.

pitch pocket: Opening between annual growth rings, causing a void in lumber filled with resin. See defect*.

pitch streak: Accumulation of resin in one spot on lumber, making a noticeable mark.

*pitch symbol: Triangle that appears on elevation drawings. Located just above roof line and indicates incline of the roof in number of inches of horizontal rise per 12″ of horizontal distance; e.g., a 5 in 12 roof rises 5″ per 12″ of run.

pith: Central portion of a log encased by heartwood.

pitot tube: Measuring device that indicates total pressure, static pressure, and velocity pressure in an air duct.

pit-run gravel: Ungraded stone as it is taken from a pit.

pitting: Surface deformation characterized by many small holes or indentations.

Pittsburgh seam: Sheet metal seam used as a longitudinal corner seam for various shaped pipes and ductwork. Consists of a single lock seam combined with a pocket lock. See *seam* (1)*. HAMMER LOCK, HOBO LOCK.

pivot: 1. Point where a moving object is connected to a solid surface or another moving object; e.g., points where a bi-fold door is connected to the top and bottom of a door opening. **2.** Nonmoving axle.

pivoted bucket conveyor: Mechanical transportation system in which a series of open-topped containers are suspended on a moving chain or other guide. The containers remain freely suspended to carry a load until mechanically tipped at the appropriate drop-off point.

pivot hinge: Hardware device with large leaves and a fixed pin. Normally fastened to the side of cabinet doors.

placage: Thin coating of decorative material applied to a wall.

placing: Deposit of concrete or mortar in forms or the location in which it is to harden. PLACEMENT.

placing drawings: Prints with steel reinforcement information for reinforced concrete construction.

plafond: Ceiling formed by underside of floor above.

plain ashlar: Rectangular stone with a smooth surface.

plain bar: Reinforcing steel rod with a smooth surface.

plain rail: Horizontal member of a double-hung window that separates upper and lower sash and is of equal thickness to rest of the window sash.

* **plain-sawn:** Hardwood lumber cut parallel to the squared side of a log. TANGENT-SAWN.

plan: Drawing of a component or structure; used for and during construction. See *plan view*.

* **plancier:** Soffit; underside of an overhang.

plancon: Hardwood timber that is octagonal in section and approximately

PLANES

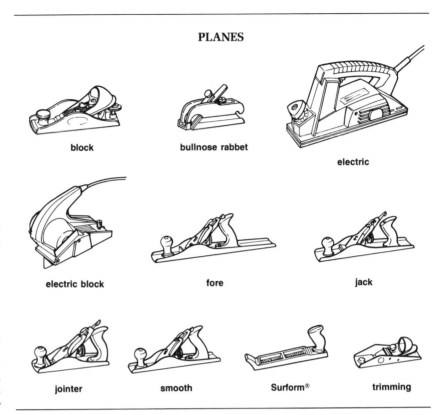

block   bullnose rabbet   electric

electric block   fore   jack

jointer   smooth   Surform®   trimming

10″ across.

* **plane: 1.** Smoothing tool that removes excess wood and smoothes surface of wood. Many types and lengths are available, depending on the application. See *block plane, fore plane, jointer plane.* **2.** Flat surface without curvature.

plane iron: Cutting blade used in a plane.

plane iron cap: Casting that covers a plane iron.

planer: Woodworking machine with rotating knives that smooth wood and cut it to uniform thickness.

plane surveying: Measurement of angles and distances projected onto a horizontal surface.

planimeter: Device that is moved around edges of a map for measuring area on the map.

planing mill: Manufacturing plant for wood products such as structural lumber, doors, trim members, and moldings.

planish finish: Smooth, bright finish on metal.

plank: Piece of lumber wider than 8″ and thicker than 1″ for hardwoods and 2″ to 4″ for softwoods.

plank-and-beam construction: Method of building with large wood members. Members may be solid pieces of lumber or glued and laminated into larger

units. This type of building is mostly used to create large, unobstructed areas.

plank rod: Steel pin inserted in a drilled hole in the ends of scaffold boards or other lumber to prevent splitting.

planned community: Subdivision of land including residential, service, commercial, and possibly some industrial development to create a desired type of environment.

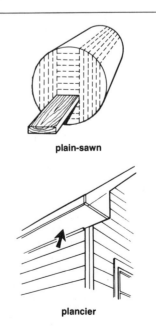

plain-sawn

plancier

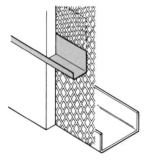

plaster bead

plat

plate (1)

plate tracery

plant: To place a decorative member onto another piece. See *planting*.

planting: 1. Putting an ornament or other decoration on the surface of another member; e.g., applying panel mold to the face of an otherwise flat wall. **2.** Laying of initial masonry courses on foundation.

plant mix: Ingredients that are mixed at a central location and transported to the job site for placement; e.g., mixing of cement, water, and aggregate at a central location to create concrete.

plan view: Drawing of an object as it appears looking down from a horizontal plane; e.g., a floor plan is a plan view looking down at a building from approximately 5'-0" above the floor.

plasma arc cutting: Cutting metal by melting a portion of the metal and blowing molten metal away with a stream of heated, high-velocity ionized gas.

plasma arc welding: Joining metal by melting with heat from an electric arc that is shielded from the atmosphere by heated, ionized gas. Filler metal may or may not be used.

plaster: Mixture of portland cement, water, and sand; used as an interior and exterior wall finish material. Variety of finishes and ornamental designs may be formed with plaster.

plaster base: Backing material that supports a plaster finish.

* **plaster bead:** Metal outside corner trim that acts as a protective member and guide for plaster application. PLASTER HEAD, PLASTER STAFF.

plasterboard: See *drywall*.

plasterer: Person who applies various plaster materials to wall and ceiling surfaces and ornamental applications.

plasterer's putty: Paste composed of sieved pure slaked lime.

plaster grounds: See *grounds*.

plaster guard: Shielding that keeps fresh plaster from entering places where voids are to exist; e.g., plaster guards

are installed on hollow metal door jambs to keep plaster away from hinge screw holes.

plastering machine: Assembly of a pump that forces fresh plaster through a hose, into a nozzle, and onto the surface being plastered.

plaster of paris: Material made of calcined gypsum. Rapid-setting plaster used in ornamental work.

plastic: 1. Mixture in a semiliquid, semisolid state; e.g., concrete that is freshly mixed and not hardened is in a plastic state. **2.** Non-corrosive material made from petroleum products.

plastic anchor: Plastic sleeve used to create a hole for inserting a fastener in a solid or hollow wall. Various types are used, depending on the load to be supported. See *anchor**.

plastic fracture: Break in metal caused by tension.

plasticity: Amount of workability and flexibility of a material.

plasticizer: Chemical admixture that prolongs or enhances workability of concrete or mortar in a plastic condition.

plastic laminate: Surface finish material made in sheets composed of a decorative sheet and kraft paper impregnated with synthetic resins. Protective resin sheet is on the surface of this material. Commonly used for countertops and cabinets. See *high-pressure laminate*, *low-pressure laminate*.

plastic soil: Earth that can be rolled into $1/8$" diameter pieces and not crumble.

plastic wood: Soft, putty-like material for filling holes. Hardens to a wood-like consistency and can be sanded.

* **plat:** Map showing division of a given area of ground into lots, streets, townships, mineral claims, and improvements. Boundaries of several properties are shown on this map. PLAT PLAN.

plat book: Public record of names of owners, locations, and sizes of recorded

pieces of ground within a municipality or county.

* **plate: 1.** Horizontal framing member at top or bottom of a wood-framed wall. Studs bear on bottom plate. Joists and rafters rest on top plate. See *bottom plate, top plate*. **2.** Horizontal support for concrete forms. **3.** Flat piece of steel. **4.** To mechanically, chemically, or electrically cover a material with an adherent layer.

plate cut: See *seat cut*.

plate glass: Polished glass manufactured in large sheets. Thicker and of higher quality than ordinary glass.

platen: Flat piece that transmits electric current in the arc welding process. Fixtures, electrode holders, and other welding backups are attached to this piece.

plate rail: Narrow, horizontal trim member fastened to the interior of walls to support chinaware, or plates, for display.

* **plate tracery:** Type of decorative work formed by cutting a pattern through a flat piece of stone.

platform: 1. Raised portion of floor; commonly used as a floor for speeches or as a stage. **2.** Any floor surface. **3.** Location for operator to work on a piece of large equipment.

platform framing: See *western framing*.

platform stairs: Stairway with landings.

plenishing nail: Large fastener for joining plank flooring to joists.

plenum: 1. Duct or area connected to distribution ducts through which air moves. Air in this duct may be maintained at a higher pressure than normal atmospheric pressure. PLENUM CHAMBER. **2.** Compressed air used to hold walls of an excavation from slumping into the open area.

Plexiglass®: Manufacturer's name for a variety of rigid sheet plastic products.

plied yarn: Several individual yarns entwined to create heavier strands for carpeting.

* **pliers:** Hand tool with opposing jaws for gripping and/or cutting.

plinth: 1. Square portion at base of a column. **2.** Lowest course of stone in a wall.

plinth block: Square or rectangular member at bottom of the casing on a door jamb, between casing and floor. BASE BLOCK.

plot: To record surveying information on a map or plat drawing.

plot plan: Drawing that shows property lines of a building lot, elevation information, compass directions and lengths of property lines, and locations of structures to be built on the lot. Utility services and landscaping may also be shown. See *print**.

plow: 1. Trowel for finishing drywall joint compound or plaster in a corner. ANGLE PLOW. **2.** Groove cut in wood. **3.** Type of wood plane for cutting grooves.

plow bolt: Threaded fastener with a flat, circular head finished in a square design where it joins with the threaded shaft.

plucked finish: Rough stone surface produced by spalling the stone rather than shaving.

plug: 1. Fitting attached at the end of an electrical cable that allows for safe fastening to another cable or receptacle. **2.** Male threaded fitting screwed into the end of a pipe or valve to seal it. See *pipe fitting**. **3.** Piece of wood or some other filler material to fill a hole or cover a blemish. **4.** Stopper for a drain.

plug and feathers: Device for splitting boulders; made of two semicircular pieces of hardened steel and a tapered wedge.

* **plug cutter:** Woodworking tool that cuts cylindrical pieces from a board to be used for filling holes.

* **plug fuse:** Thin metal electrical element contained in a porcelain fitting that is attached to a threaded metal casing.

plugging chisel: Tool with a tapered blade. Used for removing mortar from joints.

* **plug weld:** Joining two pieces of metal by electric arc welding around the inside circumference of a hole in one of the metal pieces.

plum: Large rough stone used as filler in concrete. PLUM STONE.

plumb: 1. True to a vertical plane; perpendicular to a horizontal. **2.** To seal with lead.

plumb bob: Cone-shaped metal weight fastened to a string. The force of gravity on this weight causes the stringline to hang in a true vertical plane. This al-

PLIERS

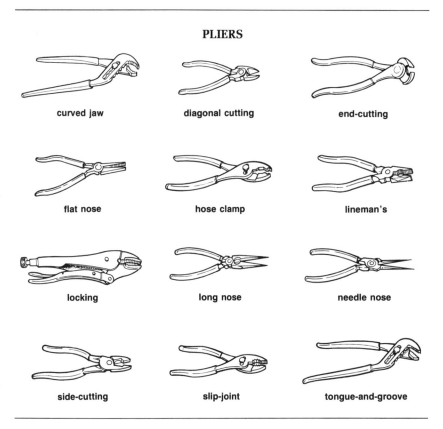

curved jaw   diagonal cutting   end-cutting

flat nose   hose clamp   lineman's

locking   long nose   needle nose

side-cutting   slip-joint   tongue-and-groove

lows for aligning members such as columns and walls in a true vertical plane. See *surveying**.

plumb bond: Laying of masonry units in a way that all head joints are in vertical alignment.

plumb cut: Cut that is in a true vertical plane; e.g., cut made at the top end of a rafter at the point it meets the ridge.

plumber: Person who installs and repairs piping, fixtures, and appliances that are connected to water supply, drainage, and waste systems.

plumber's dope: Soft, pliable material for sealing threaded pipe joints.

plumber's rasp: Hand tool for filing lead.

plumber's solder: Mixture of two parts lead and one part tin that is melted in

making pipe joints.

plumbing: 1. Installing pipes, fixtures, and other devices for supplying fresh water and removing waterborne waste and storm water. **2.** Process of transferring a vertical point or plane to another surface.

plumbing fixture: Receptacle connected to water supply system that requires a supply of water for removing waterborne waste or sewage.

plumbing plan: Print indicating plumbing supply and waste piping and fixtures. Exact placement of piping and fixtures is not specified, but is commonly left to be determined by job site conditions. Commonly denoted with a capital P preceding the drawing sheet number. See *print**.

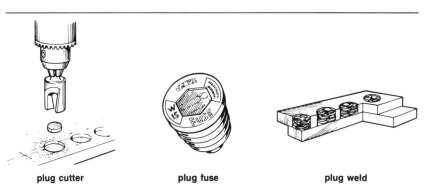

plug cutter   plug fuse   plug weld

plumbing system: Potable water supply pipes and fixtures, and drainage, vent pipes, and waste removal pipes.

plumbing wall: Wall or partition that provides a vertical space for fresh water and waste plumbing pipes. Often thicker than other walls to allow room for piping.

plumb joint: Attaching two pieces of sheet metal by overlapping and soldering flat.

plumb rule: Hand tool with a plumb bob suspended from the top of a narrow board with parallel edges. As the plumb bob is aligned with a given point along the board, the board is plumb.

plummet: See *plumb bob*.

plunger: Flexible rubber cup with a handle. Used to create suction in a plumbing waste system and remove material clogging the pipes. PLUMBER'S HELPER.

* **plunge router:** Power tool with a motor and shaft that holds a cutting bit. Motor and shaft are held in a frame that allows movement of the cutting bit directly downward into the work.

ply: 1. One layer or thickness of a material that is applied in several layers, resulting in a built-up component. 2. Strands of yard entwined into a single strand for carpeting.

Plyform®: Grade of plywood panel used for concrete forms.

plywood: Wood product made in sheets 4'-0" wide and 8'-0", 9'-0", and 10'-0" long. Several layers (an odd number such as 3 or 5) of veneer are glued together, with grain of each layer perpendicular to adjoining layers. Different thicknesses, core materials, and surface thicknesses are available. Face veneer grades are marked with letters, with A being a high grade and D being a low grade.

plywood blade: Circular saw blade with small fine teeth designed to produce a smooth cut in plywood. See *blade**.

plywood joist: Horizontal structural member composed of a top and bottom member made of narrow strips of laminated plywood, and a vertical ply-

wood member connecting them. See *joist**.

pneumatic: Powered or operated by compressed air.

pneumatic roller: Machine for compacting material by rolling over it with a number of smooth, wide-tread, air-filled tires.

pocket: Recess or indentation.

pocket butt: Type of shutter hinge with each leaf in an L shape. L shape allows the shutters to fold and move without jamming.

pocket chisel: Wood cutting and smoothing tool that has a wide blade is sharpened on both edges.

pocket door: Sliding door suspended on a track. Opened by sliding it into a cavity in the wall. See *door**.

pocket level: See *torpedo level*.

pock marks: Small air bubbles that appear as depressions in a finish surface.

podium: Elevated speaker's platform.

point: 1. Apex of an angle. 2. Outlet acting as a source of electric current. 3. Outlet in a gas piping system. 4. Diamond-shaped piece of thin steel for holding glass in a wooden frame. SPRIG. 5. Saw tooth.

point-bearing pile: See *end-bearing pile*.

pointed ashlar: Rough face marking on stonework that is made with a pointed tool, such as a chisel.

pointed style: Architectural design with arches that have a sharp peak.

pointing: Finishing of masonry mortar joints.

pointing trowel: Small hand tool with a steel blade and a handle. Used for finishing mortar joints, cleaning and scraping larger trowels, and other small work.

point of abutment: Point at which concrete meets a structural member or other concrete slab. Isolation joints are formed at this point to limit cracking.

point of beginning (POB): Location on a building lot that acts as a reference point for horizontal dimensions and vertical elevations. POINT OF ORIGIN.

polar coordinates: Means of specifying location of a reference point in a CADD

system by using angles and distances.

polarity: Direction of flow of electric current through a circuit.

* **polarized outlet:** Electrical receptacle designed to allow a plug to be inserted in only one manner.

pole: 1. Cylindrical vertical member or timber. 2. One set of electrical contacts in a switch or other electrical device. 3. Either of the two locations on the earth's surface to which a compass needle points.

pole plate: See *top plate*.

pole sander: Tool with a flat, pivoting surface attached to the end of a rod. Commonly used for sanding drywall joint compound on ceilings and walls.

poling back: Excavating behind shoring.

poling boards: See *lagging*.

polishing varnish: Finish treatment that is hard and dry and can be rubbed with abrasive materials and mineral oils without harming the finish.

* **poll:** Striking face of a hammer.

poly-backed metal lath: Expanded diamond-shaped wire grid with a clear sheet of polyethylene on the back; used as a base for stucco on curtain walls and plaster on interior applications.

polychromatic: Made up of many colors.

polyester: Synthetic resin used as a binder in mortar and concrete, a fiber laminate, or adhesive.

polyethylene: Flexible plastic material formed into thin sheets for use as a vapor barrier, waterproofing, and formed into tubing for conveying natural gas.

polygon: Plane geometric shape with many sides. Regular polygon has equal sides and angles.

polymer concrete: High-strength concrete that is cast and cured by drying with heat and vacuum conditions. Organic chemicals, such as methyl methacrylate, styrene, and urethane are introduced into the concrete during curing process.

polyphase circuit: Electrical circuit with more than one phase.

polypropylene: Synthetic fiber used for carpet backing.

polystrene: Plastic resin beads compressed to form Styrofoam®.

polystyle: Structure supported or surrounded by many columns.

polystyrene: Clear and colorless plastic resin; used in concrete paint and as insulation.

polytropic process: Means by which heat is transferred and exchanged with the surrounding area.

polyurethane: Hard plastic finishing material for coating woodwork. Also used under pressure for foamed-in-place insulation.

plunge router

polarized outlet

poll

polyvinyl acetate (PVA): Colorless thermoplastic used as a binder in latex paint and as a synthetic substitute for hot animal glue.

polyvinyl chloride (PVC): 1. Rigid plastic material used for waterstops and plumbing pipes and fittings. **2.** Rigid nonmetallic conduit, either thin-walled (Schedule 40) or thick-walled (Schedule 80), approved by the National Electrical Code® as a raceway system for conductors.

pommel: Decorative, ball-shaped ornament placed at the top of objects such as finials or pillars. PUMMEL.

ponding: 1. Increasing hardness of a concrete slab by keeping it covered with a layer of water. **2.** Accumulation of water at low points in a flat surface.

pontoon: Floating platform for supporting a structure or equipment.

poor lime: Lime mix containing over 15% impurities.

poppet valve: Mushroom-shaped flow control device that is opened by raising the stem.

poppy heads: Ornamentation in the design of a poppy flower blossom.

pop rivet: Fastener with point that is inserted into a predrilled hole. Head of the fastener is expanded with a special tool to lock the pieces together.

pop valve: Device that opens automatically when pressure exceeds a given amount.

porcelain: Smooth, hard, ceramic waterproof finish for plumbing fixtures and other metal or ceramic pieces.

porch: Covering that projects from the face of a structure to protect an entrance or provide a covered outdoor recreational area.

pore pressure: Water pressure in open areas of soil.

pore water: Free water in soil.

porock: Quick-hardening cement used for anchoring handrails or other hardware into predrilled holes in concrete. POUR-ROCK, POROX.

porosity: Percentage of void area in a material compared to overall volume of the object. Material with high porosity has a large amount of open, unfilled area.

portal: 1. Door or opening. **2.** Nearly horizontal access hole into a tunnel.

port control faucet: Single handle, noncompression device for supplying hot and cold water.

portico: Small roofed porch at entrance to a building.

portland cement: Fine grayish powder made of various minerals (clay, limestone, shale) burned in a kiln. Resulting powder has hydraulic power and is used as base material for mortar and concrete.

positive displacement: Forcing material through the supply hose of shotcrete equipment in a solid mass with either an auger or piston.

positive pressure: Force in a sanitary drainage or vent piping system that is greater than atmospheric pressure.

post: Column or pillar; vertical support.

post-and-beam construction: See *plank-and-beam construction.*

post and petrall: Wall-building style with exposed half timbers and plaster or masonry panels in between. POST AND PAN.

postern: Small inconspicuous entry.

post hole digger: Manual or mechanical device used to bore small holes in the earth. POST HOLE AUGER.

post indicator valve: Extension on a valve stem that projects above grade and shows open or closed position of the valve.

post shore: Vertical temporary support used in concrete formwork.

post-tensioned construction: Building method in which steel cables are placed in a concrete slab or other concrete structure. The cables are stretched after concrete has set, putting overall structure in tension. Since concrete has high compressive strength and steel has high tensile strength, this combination creates a strong structure.

potable water: Pure drinkable water without impurities.

*potato masher: Webbed-shaped wire fastened to the end of a handle. Used for hand-mixing drywall joint compound.

potential difference: Unequal amounts of electrical charge between two points in a circuit.

potentiometer: Device that measures small electromotive forces.

poultice: Paste formed with various ingredients; used for removing stains from masonry.

pound: Measure of weight equal to 16 ounces.

pounds per lineal foot (plf): Measurement of loads on a structural member; equal to pounds per square foot multiplied by spacing of the members, such as trusses.

pounds per square foot (psf): Measurement of loading capacity of structural members, such as plywood.

pounds per square inch (psi): Measure of pressure; e.g., the air pressure required to power a pneumatic nail gun may be 80 psi.

pounds per square inch absolute: Measure of pressure equal to gauge pressure plus atmospheric pressure.

poured fitting: Connection at end of a wire rope made by placing end of the rope in a conical socket, separating the strands, and filling the socket with molten metal (zinc).

pour point: Lowest temperature at which a liquid will flow.

*pour strip: Narrow piece of material placed inside concrete forms at finish elevation of concrete to act as a guide.

*powder-actuated tool (PAT): Device that drives fasteners by means of an explosive charge. A pin or other fastener is placed in the tool along with an explosive shell (commonly .22 caliber). The chamber is closed and charge is detonated, driving the fastener. Used for driving fasteners into hard surfaces such as concrete. Both high-velocity and low-velocity types are available, along with many different strengths of charges. STUD GUN.

powdering: Decorating with a repetition of small figures.

power: 1. Rate of performing work. **2.** Magnification ability.

power and free conveyor: Mechanical moving device in which a load is carried on a trolley. Trolley is propelled mechanically through a portion of the system and moved by the force of gravity or by physical control in other portions.

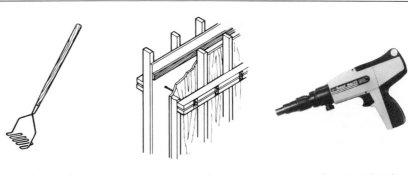

potato masher   pour strip   powder-actuated tool

power buggy

power shovel

power arm: Portion of a lever between the fulcrum and the point where force is applied.

* **power buggy:** Gasoline-powered device with a front-end bucket for moving concrete and other material on a construction site. GEORGIA BUGGY.

power factor: Electrical calculation that determines ratio of watts (true power) to volts multiplied by amperes (apparent power). Expressed as a percentage.

power hammer: Tool driven by compressed air or electricity that moves a shaft rapidly in a reciprocating motion. Wedges and other cutting points are inserted for breaking concrete, asphalt, etc. See *jack hammer*.

power outlet: Enclosed electrical assembly with circuit breakers, fuses, receptacles, and meters as required for distributing power to mobile or temporary equipment.

power plant: 1. Blower or compresser that supplies air to a pneumatic conveyor system. **2.** Structure and all systems that generate electricity. **3.** Prime moving equipment for a crane or other heavy machinery.

power saw: See *circular saw*.

* **power shovel:** Excavation equipment with an open bucket suspended from a boom and driven by a series of cables.

* **power trowel:** Concrete flatwork finishing tool in which a series of blades is rotated by a gasoline motor. Spinning action of the blades smooths and finishes the concrete surface. ROTATING

FLOAT, ROTARY FLOAT.

pozzolan: Siliceous and/or aluminuous material in cement that reacts with calcium hydroxide when finely divided to improve cementitious value. May be mixed with portland cement.

* **Pratt truss:** Prefabricated horizontal support with parallel top and bottom chords. Truss webbing is made of vertical members with angular supports installed in an outward direction from the bottom chord to the top.

pre-apprenticeship: Time a new worker spends gaining basic knowledge about a construction craft immediately before entering an apprenticeship program.

precast concrete: Large sections of walls, floors, beams, or other structures formed and poured in other than their final position. Pouring may occur at a construction yard or job site. Concrete units are set in place with a crane or other lifting machine after hardening.

precast concrete pile: Deep foundation support composed of concrete formed and placed at a location other than its final position. See *pile* (1)*.

precipitate: Solid that is separated from a liquid.

precure: Process of preparing and curing a glue joint before clamping.

precut: 1. Wood stud cut to a specific length at the mill to allow for the thickness of top and bottom plates in western (platform) framing. **2.** Wood product manufactured to a specific length to minimize job-site cutting.

prefabricate: To build stand-alone mod-

ules composed of many different building components. Modules are transported to the job site for final construction. PREFAB.

* **preferred angle: 1.** Slope of a stair providing safest and most comfortable walking incline; between 30° and 35°. **2.** Slope of a ramp providing safest and most comfortable incline; less than 15°.

prefilter: Low-efficiency screening device for purifying air or liquid.

prefinished: Surface that is painted, stained, or otherwise completed before final installation on the job site.

preform: To shape to a fixed position prior to installation.

preheater: Device for raising combustion air temperature in a large furnace system to improve efficiency.

preheat temperature: Specified temperature for a base metal before beginning some types of brazing, cutting, soldering, or welding.

prehung door: Door unit with jamb, door, hinges, and trim members prefabricated.

premises wiring: All interior and exterior electrical wiring and hardware that extend from load end of the service drop or lateral to the electrical outlets. Does not include internal wiring in fixtures and appliances.

premixed: Concrete, mortar, or plaster packaged with proper proportions of ingredients, such as cement, sand, aggregate, and lime, ready to be mixed with water.

prepacked concrete: Placing coarse aggregate in a form and pressure-injecting portland cement and sand grout later to fill voids between aggregate.

prepost-tensioned concrete: Concrete in which reinforcement is placed in tension before and after placing and curing.

prepreg: Reinforcing material in plastic combined with resin before molding.

preservative: Chemical that protects a material from insects, fungi, and other natural forces of decay.

preshrunk concrete: Concrete products mixed one to three hours before placement to reduce shrinkage.

pressed edge: Side of a spread footing that receives the greatest soil pressure when ground overturns.

pressed wood: Panel manufactured by compressing wood fibers under heat.

pressure: Measure of force applied to a given area; e.g., pounds per square inch.

pressure bulb: Area of ground that surrounds and actively supports a friction pile.

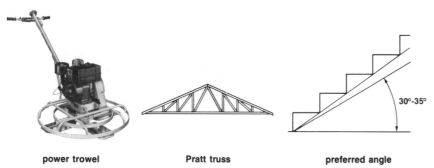

power trowel   Pratt truss   preferred angle

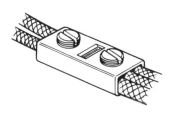

pressure connector

prick punch

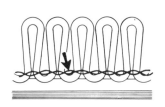

primary backing

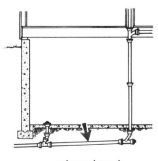

primary branch

pressure cell: Device for measuring forces in a soil mass or soil pressure against another object.

* **pressure connector:** Device that joins two or more electrical conductors or terminals without solder; e.g., a wire nut.

pressure drop: Reduction in force between two points in a closed system.

pressure gas welding: Process of joining metals by heating with a gas flame and pressing them together. Filler metal is not used.

pressure gauge: Instrument for measuring pressure in a closed system.

pressure-injected footing: Building support in which a steel tube is driven into the ground. Enlarged base is formed by compressing concrete at bottom of the hole to a given resistance. Remainder of the hole is filled with compressed concrete as the steel tube is removed.

pressure meter: Device for measuring lateral resistance of soil.

pressure plate: Member that is driven by a flywheel and slid toward the flywheel to engage a disk between the two members.

pressure-reducing valve: Device for lowering force of gas or liquid to a constant lower pressure.

pressure regulator: Device for maintaining constant output pressure.

pressure-relief valve: Device for automatically lowering excessive pressure in a closed system.

pressure switch: Device that opens or closes an electrical circuit at a given pressure.

pressure-treating: Forcing preservative or fire-retardant chemicals into wood cells in a pressurized tank or other closed, pressurized system.

presteaming period: Interval between casting and temperature rise in manufacture of precast concrete items.

prestressed concrete: Building method in which steel cables are placed into a concrete slab or other concrete structure. Cables are stretched and put in tension before concrete is placed. After concrete has set, tension on cables is transferred to the concrete, putting overall structure in tension. As concrete is high in compressive strength and steel is high in tensile strength, this combination creates a strong structure.

pretensioned concrete: Concrete that has steel reinforcing tenons pulled tight and placed in tension in it before the concrete hardens. After concrete hardens, steel remains in tension.

pretest: To measure bearing capacity of piling or footing by maintaining pressure on the piling or footing for a given amount of time.

pricking up: Roughing first coat of plaster for adhesion of subsequent layers.

prick post: Secondary vertical support in framed buildings.

* **prick punch:** Sharp, pointed steel shaft struck with a hammer to mark center points or punch holes in light-gauge metal.

primacord: Detonating fuse with a waterproof casing around the core.

primary air: Air that enters a burner and is mixed with gas before entry into combustion chamber.

* **primary backing:** Material to which carpet threads are tufted.

* **primary branch:** 1. Sloping plumbing drain from a soil stack to the building drain. 2. Main water supply or air supply duct in a structure.

primary excavation: Digging in previously undisturbed earth.

prime: 1. To apply initial coat of paint or liquid material for sealing a bare surface. 2. To supply adequate liquid to activate a machine operation, such as pumping. 3. Lumber grade below superior and above lowest finish grade. 4. Major architect or contractor when more than one is employed on a job.

primer: 1. Initial coat of paint or liquid material applied to a surface to seal it before applying subsequent coats of paint or other finish material. 2. Combination of dynamite and detonating caps in a blasting operation.

princess post: Smaller vertical support installed at the side of a queen post to provide additional support.

principal: Main; central; largest.

principal meridian: Established imaginary line from north pole to south pole; used as reference in surveying and layout of property lines and boundaries.

print: Drawing showing a building structure and all its components. BLUEPRINT.

prism: 1. Reflective device used with electronic surveying equipment to measure distance. See *surveying**. 2. Transparent device consisting of non-parallel faces that disperse and refract a beam of light.

prismatic beam: Horizontal support with flanges parallel to longest axis.

prismatic billet molding: Trim ornamentation with a series of prisms in alternating, staggered rows.

prismatic glass: Glass with many surfaces that refract light rays in different directions.

prismoidal formula: Mathematical formula for figuring volume of an excavation with the length, areas of two ends of the excavation, and area of a plane at the middle.

processed shake: Cedar shingle sawn and shaped into texture of a cedar shake.

Proctor compaction test: Method for measuring optimum moisture content of soil by compacting and weighing.

profile: Outline of an object; a contour drawing.

profile sheet: Drawing on graph paper showing slopes and grades for excavation and roadwork.

profilograph: Instrument which is rolled over a surface to measure a profile.

progressive: Performed in steps or stages.

projected: 1. Extended; carried out to a greater distance; e.g., perspective drawing is made with lines projected from face of the object to create a three-dimensional appearance. 2. Protruding from or jutting out from a flat surface. PROJECTION.

PRINT COMPONENTS

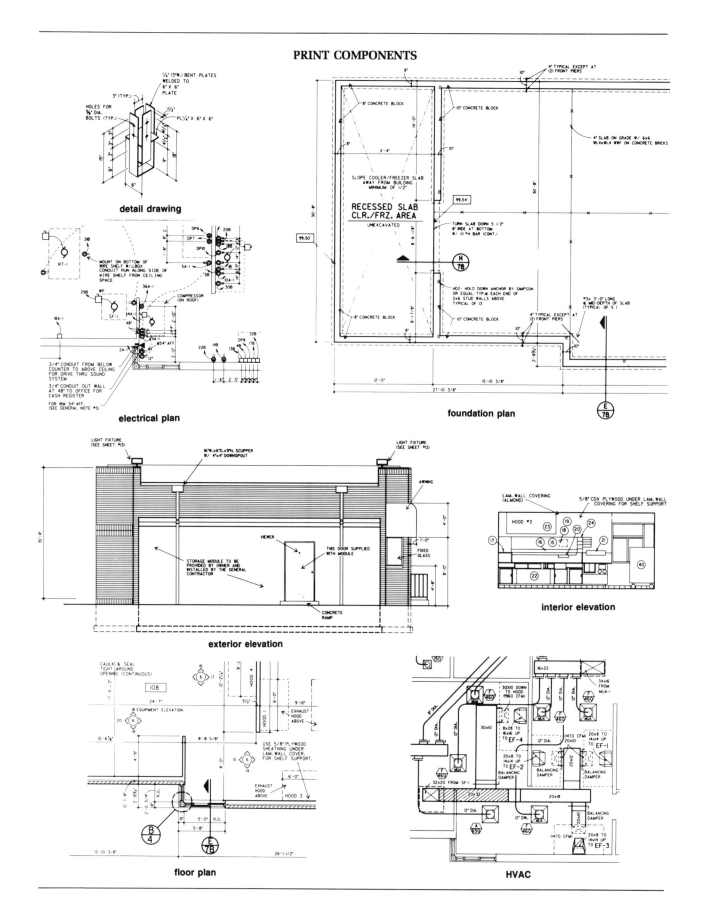

detail drawing

electrical plan

foundation plan

exterior elevation

interior elevation

floor plan

HVAC

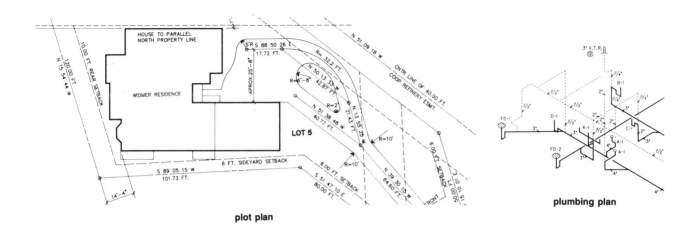

plot plan

plumbing plan

ROOM FINISH SCHEDULE

SPACE DESIG.	ROOM	FLOOR QUARRY TILE	CARPET	CERAMIC TILE	DECOR. TILE	BASE QUARRY TILE	WOOD	CERAMIC TILE	DECOR. TILE	WALLS WOOD PANELING	CERAMIC TILE	LAM. WALL COV'G.	VINYL FABRIC	EXPOSED BRICK	CEILING SUSPENDED	ACOUSTIC	WASHABLE	DRYWALL	HEIGHT (FEET)	TRIM PAINT	STAIN	STEEL DOOR TRIM	DOOR ALUMINUM	PLASTIC LAMINATE
100	VESTIBULE	–	–	–	③	–	–	–	③	–	–	–	–	–	–	●	–	–	9⁰	–	–	–	●	–
101	PUBLIC AREA	–	②⑧	–	③	–	⑤	–	③	–	–	●	●	●	–	●	–	–	9⁸	●	–	–	●	–
102	WAITRESS STATION	–	–	–	③	–	–	–	③	–	–	●	●	–	–	●	–	–	9⁸	●	–	–	–	●
103	HALL	–	–	–	③	–	–	–	③	–	–	–	●	–	–	●	–	–	8⁴	●	–	–	–	●
104	MEN'S ROOM	–	–	–	③	–	–	●	–	–	●	–	–	–	–	–	–	●	⑥	●	–	–	–	●
105	WOMEN'S ROOM	–	–	–	③	–	–	●	–	–	●	–	–	–	–	–	–	●	⑥	●	–	–	–	●
106	SERVING AREA	●	–	–	–	●	–	–	–	–	–	④	–	–	–	–	–	●	9⁰	●	–	–	–	●
107	COOLER-FREEZER	–	–	–	–	–	–	–	–	–	–	–	–	–	–	–	–	–	–	–	–	–	–	–
108	PREPARATION AREA	●	–	–	–	●	–	–	–	–	–	④	–	–	–	–	–	●	9⁰	●	–	–	–	●
109	OFFICE	●	–	–	–	●	–	–	–	●	–	–	–	–	●	●	–	–	8⁰	–	–	–	–	●
110	STORAGE	–	–	–	–	–	–	–	–	–	–	–	–	–	–	–	–	–	–	–	–	–	–	–
111	EMPLOYEE BK. ROOM	●	–	–	–	●	–	–	–	–	–	④	–	–	–	●	–	–	9⁰	●	–	–	–	–
112	WTR HTR CLOSET	●	–	–	–	●	–	–	–	–	–	④	–	–	–	–	–	⑨	–	–	–	–	–	–

1. SEE INTERIOR ELEVATIONS ON SHEET 6 FOR EXTENT OF WALL COVERING
2. SEE FLOOR PLAN FOR EXTENT OF CARPETING
3. SEE FLOOR PLAN FOR EXTENT OF DECORATIVE TILE. INSTALL TILE IN A COMMON BOND PATTERN PARALLEL TO THE FRONT OF BUILDING.
4. FOR LAMINATED WALL COVERING FOLLOW MANUFACTURER'S SPECIFICATIONS. FLAME SPREAD LESS THAN 2000 HRS.
5. ALL WOOD BASES TO BE OAK.
6. DRYWALL CEILING AT 9'-0"-SUSPENDED AT 8'-0"
7. USE LAMINATE WALL COVERING IN TRASH CAN AREA ONLY BELOW TRASH COUNTER. SEAL CONCRETE PAD W/ CONCRETE FLOOR SEALER
8. MOHAWK CARPET FIRE TEST ASTM E-84 RESULTS BETTER THAN OR EQUAL TO: FLAME SPREAD-50 QUALITY-B-7557(SYNTHETIC BACK) FUEL CONTROL-35 SMOKE DENSITY-185
9. CEILING TO BE AT BOTTOM OF ROOF JOIST

schedule

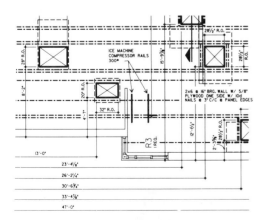

structural plan

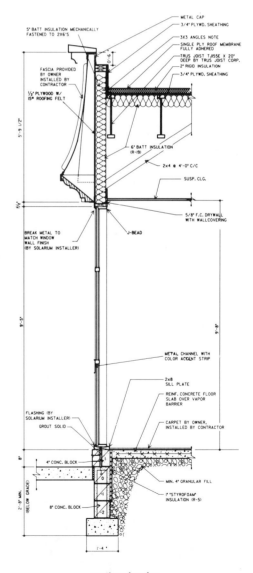

section drawing

projected pipe: Cylindrical tube laid on the ground before excavation is done. Fill is then laid around the pipe to bury it.

projecting belt course: Horizontal band of masonry jutting out from face of a wall.

projection: Horizontal distance from face of a wall to end of the rafter overhang or other cantilevered structure.

promenade tile: See *quarry tile.*

proof stress: Compressive or tensile forces applied to a material to determine amount of force required to permanently deform the material.

property line: Outside perimeter boundary of a piece of ground.

proportioning: Mixing and combining various ingredients in correct amounts to achieve a final blend, such as concrete or plaster.

proprietary compound: Patented chemical mixture for cleaning masonry.

prostyle: Building style in which columns under a front porch roof are freestanding and not attached to the wall.

protected corner: Corner of a concrete slab built so that loads are readily transferable to other parts of the structure.

protected noncombustible construction: Building in which all vertical structural bearing members have a fire-resistance of two hours and horizontal bearing members have a fire-resistance of one hour.

protected ordinary construction: Building in which all structural bearing members have a fire-resistance of one hour.

protected waste pipe: Plumbing sewage removal pipe not connected directly to a drain, soil or vent pipe, or waste pipe.

* **protractor:** Tool for measuring and laying out angles.

* **protractor head:** Attachment for a combination square; used to measure and lay out angles.

proximity switch: Control device triggered by the closeness of another object.

pry bar: Hand tool with both ends curved in different shapes for wedging materials. One end is U-shaped and the other end slightly bent. Both ends are slotted for pulling nails. See *wrecking bar.*

psf: See *pounds per square foot.*

psi: See *pounds per square inch.*

psia: See *pounds per square inch absolute.*

psychometric chart: Table that illustrates thermodynamics of moist air.

* **psychrometer:** Device for measuring relative humidity.

P trap: Plumbing fitting with a lowered area for holding water in the line to form a seal against backup of sewer gases into the waste piping system.

puddle: 1. To vibrate or lightly tamp fresh concrete or mortar to aid in consolidation and bring fines to the surface. 2. Portion of a weld joint that is molten as a result of applying heat. 3. To compact and consolidate loose soil by saturating and drying it.

pugging: Coarse mortar used to provide sound insulation.

pug mill: Excavating machine with several rows of steel arms inside a container that churns soil with a mixture of lime as it is driven across the ground.

pull: Handle or other device that can be grabbed and brought toward the user.

pull box: Metal access box installed in an assembly of electrical conduit in which long runs of wire are pulled. This access point allows for easier pulling through conduit.

pull chain: Hanging light-gauge chain for control of electrical fixture or damper.

pulley: Rotating wheel with cable or line around portion of the circumference; used for lifting or moving loads.

pulley block: Assembly of one or more pulleys.

pulley stile: Vertical member of a window frame in which a sash pulley is installed.

pulling: Resistance to movement in paint caused by high viscosity.

pull shovel: Earth-moving equipment that digs with bucket moving toward the machine.

pull switch (PS): Electrical fixture activated or turned off by pulling a chain or string. Commonly used for basement ceiling lighting fixtures or closets. PULL CHAIN.

pulsed arc transfer: Method of joining metal with an electrical arc. Spray transfer of filler metal occurs at regularly timed intervals created by two power sources. One source is used for preheating; the other source provides peak voltages for filler transfer.

pulverize: To crush into small pieces.

pulvinated: To be swelled or bulged in a convex curve.

pumice: Gritty material of powdered volcanic ash. Used for polishing and as an aggregate in concrete.

pump: Device that moves, raises, or compresses fluid or gas by pressure or mechanical means.

pump jack: Scaffold support made of a frame that travels up and down a pole; controlled by person raising and lifting a foot pedal on the frame.

punch: Hand tool with a pointed or blunt tip for marking or making holes or driving objects when struck by a hammer.

puncheon: 1. Short vertical framing member. 2. Rough split log used for flooring.

purfle: To decorate with an ornamental edging.

purge: 1. To release unburned gases from a combustion chamber. 2. To remove air from a closed liquid piping system.

purlin: Horizontal support that spans across adjacent rafters. May be above or below principal rafters.

purlin roof: Roof in which purlins are supported by walls instead of rafters.

push bar: 1. See *panic bar.* 2. Plate used to open and close door or window.

pushbutton: Electrical control that activates a circuit by depressing a control device.

push drill: Hand tool for boring holes in wood. Interchangeable bits are driven by ratcheting action in the handle when pushed against the work. See *drill* *.

push drill bit: Tool consisting of two flutes at the end of a straight shaft. Used to bore holes in wood and other soft material. See *bit* (1) *.

push penny: Round metal cover for open end of an electrical conduit; prevents debris from entering conduit during installation.

* **push plate:** Flat piece of metal attached to face of a double-acting door to allow for opening without touching the door itself. Provides protection from stain-

protractor

protractor head

psychrometer

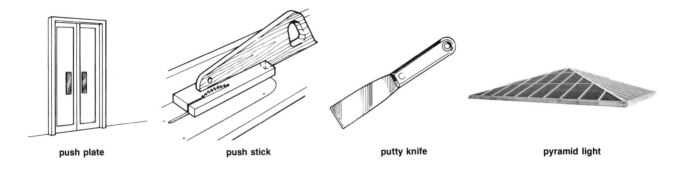

| push plate | push stick | putty knife | pyramid light |

ing or marring the door finish.

push point: Thin metal piece for holding glass in place with one pointed end and a raised shoulder. Pointed end is pushed into a window frame by bearing on the raised shoulder.

push-pull rule: Measuring tool with a graduated metal tape coiled in a case. A spring retracts the tape into the case when not in use. See *steel tape*.

* **push stick:** Piece of material used to help guide and move work through a machine where pushing by hand could be dangerous. Allows hands to be kept farther away from cutting blades.

putlog: Horizontal scaffold member supporting scaffold planks.

putty: 1. Pliable material used to fill nail holes or other depressions. **2.** Plaster cement of lime and water that is allowed to partially harden before it is applied. Sand or plaster of paris are mixed in when applied.

putty coat: Smooth finish coat of plaster.

* **putty knife:** Hand tool with a flat, unpointed blade for smoothing pliable materials. Blade widths vary.

PVA: See *polyvinyl acetate*.

PVC: 1. See *polyvinyl chloride*. **2.** Pigment volume concentration in paint.

PVC cutter: Hand tool for shearing PVC pipe of 2″ diameter or less.

pycnometer: Device for determining specific gravity of a liquid or solid.

pylon: Tower structure for supporting electrical power lines or other equipment overhead.

* **pyramid light:** Skylight formed with a raised point, designed in the shape of a pyramid.

pyramid roof: Roof style in which four equal-pitch roof slopes converge at a pointed peak.

pyrnometer: Device for measuring solar insulation and sunlight intensity. Measurements are made in Btu per square foot per hour.

pyrometer: Device for measuring extremely high temperatures.

quadrangle: 1. Open courtyard partially or fully surrounded by buildings. 2. Buildings surrounding an open courtyard.

* quadrant: 1. Area of a circle equal to a sector with angular measurement of 90°. 2. Instrument with a 90° graduated arc used to measure altitude. 3. Hardware used to secure upper and lower halves of a Dutch door.

quadrel: Square tile.

quadrilateral: Four-sided polygon with sides of different length and angular measurements.

quaggy timber: Timber harvested from marshy land. Produces wood with many shakes.

quaking concrete: Concrete with medium consistency. Used for large walls and abutments.

quality control: System for ensuring specified standards, including accuracy and quality, of manufactured products and materials.

quarrel: Small square- or diamond-shaped piece of glass or tile set in position diagonally. QUARRY.

quarry: 1. Open excavation for mining stone. 2. See quarrel.

quarry-faced masonry: Stone with a rough face and squared at the joints. Rough face is freshly split and not weathered.

quarry run: Stone used in its original condition as it is removed from a quarry.

quarry sap: Moisture in freshly quarried stone. Stone becomes harder to work as the quarry sap evaporates.

quarry-stone bond: Pattern of stones in rubblework.

quarry tile: Unglazed ceramic tile made of clay or shale and formed by the extrusion process. Used for floors or base trim.

quarter: 1. To divide into four equal sections. 2. See cripple.

quarter bend: 1. 90° bend in pipe or conduit. 2. Cast iron soil pipe with a 90° bend.

quarter crown: Area between the center line of a road to the curb or shoulder parallel to it.

quarter hollow: Decorative molding with a 90° concave arc.

quarterpace landing: Stair platform between two flights of stairs that make a 90° turn. Width of landing is equal to the width of a stair tread. QUARTER SPACE LANDING.

quarter round: Trim molding with quarter circle profile. See molding*.

* quarter-sawn: Lumber cut from a log that has been quartered lengthwise. End grain of lumber is at 45° angle or greater. Less shrinkage, twisting, and splitting occur.

quarter section: Plot of land measuring $1/4$ mile square. Equal to 160 acres.

quarter-turn stair: Stairway with top and bottom riser at 90° to each other.

quartz glass: Glass primarily composed of amorphous silica. Highest heat resistance and ultraviolet transmission of any type of glass.

quatrefoil: Decorative design consisting of a four-leaved flower or four-lobed figure.

quay: Wharf constructed parallel to the shoreline. Used as a landing.

Queen Anne arch: Arch design over a Palladian window consisting of a central semicircular arch with a flat horizontal member on each side. VENETIAN ARCH.

queen bolt: Vertical steel rod used in a truss as a tension member. Used instead of a queen post.

queen closure: Full brick cut lengthwise. Used to complete a course or prevent alignment of vertical joints. QUEEN CLOSER.

* queen post: Vertical supporting member in a truss. Extends between the top and bottom chords.

queen truss: Truss with two vertical support posts. QUEENPOST TRUSS.

quetta bond: Brick pattern with vertical spaces for the placement of reinforcement. Spaces are later filled with mortar.

quadrant (1)

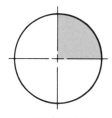

quarter-sawn

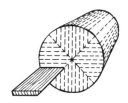

queen post

quick-acting clamp   quick link   quick-opening valve

* **quick-acting clamp:** Short bar clamp designed for fast adjustment. Handle and screw mechanisms are mounted on the movable slide. QUICK CLAMP.

quick condition: Soil weakened resulting from upward movement of ground water. Bearing capacity of soil is reduced.

quicklime: White powder remaining after limestone is heated to high temperature. Used as a plaster component.

* **quick link:** Open chain link with a threaded coupling. Used to fasten sections of chain together.

quick-load test: Compression test for piles in which loads are applied to the piles for short time intervals. Increasingly larger loads are applied to the piles after each interval. QUICK TEST.

* **quick-opening valve:** Lever-operated gate valve used as a sealing device. Requires maximum one-quarter turn to be opened or closed.

quick-release belt: Safety belt with a buckle designed to be rapidly disengaged. Allows the wearer to be immediately freed from the belt.

quicksand: Fine silty soil weakened as a result of upward movement of ground water. Bearing capacity of soil is reduced.

quick set: **1.** Rapid development of rigidity in plaster. **2.** See *false set, flash set*.

quicksilver: Mercury.

quick sweep: Carpentry or joinery work with a small radius.

quilt insulation: Thermal insulation with paper facing on one or both sides.

quirk: **1.** Small groove near the intersection of two surfaces, such as a molding and plaster surface. Reduces possibility of uncontrolled cracking. **2.** Acute angle formed between adjacent trim moldings. **3.** Narrow groove in underside of a drip cap. Prevents water from flowing toward the joint between a water table and drip cap.

quirk molding: **1.** Trim member with a small concave recess. **2.** Trim member with both convex and concave curves and separated by a flat surface. Used as column trim or flat panel mold.

quoin: Large squared stone distinct from adjacent masonry. May be used as a header or installed for ornamental purpose.

quoin bond: Corner interlock in a masonry wall made up of staggered headers and stretchers.

Quonset® hut: Prefabricated building with a semicircular cross section. Covered with thermal insulation and corrugated steel.

R: 1. See *R value*. **2.** Pile capacity expressed in tons.

rab: Pointed rod used to mix and consolidate mortar or concrete.

rabbet: Recess along end or edge of a member to receive another piece. See *joinery**.

rabbeted siding: See *shiplap siding*.

rabbeting bit: Router bit used to cut a recess along the end or edge of a wood member. See *bit* (2)*.

rabbet plane: Wood-cutting tool used to make recess along edge of a material.

race: 1. Inner or outer ring of ball or roller bearing. **2.** Groove in machine part that allows movement of an object.

raceway: Enclosed channel for routing and placing electrical conductors and cables.

rack: 1. To fasten temporary angular brace for securing a member against horizontal forces. **2.** To square a framed wall or floor and fasten with temporary angular brace. **3.** Grate for catching debris before it is conveyed to a waterway. **4.** Flat, slotted bar.

* rack and pinion: Gear arrangement in which flat, slotted bar (rack) and circular toothed gear (pinion) interlock to convert rotary motion to linear motion.

racking: Stepping masonry material back as it approaches a corner. Successive courses are shortened slightly to lock corner together and to prevent vertical joint.

rack saw: Wood-cutting tool with wide teeth.

radial: Extending from the center of a circle toward the circumference.

radial arm saw: Stationary power equipment primarily used to cut wood. Motor and interchangeable blade are suspended from arm (movable horizontal track). Arm is moved vertically or horizontally for various cutting procedures. See *saw**. RADIAL SAW, CUTOFF SAW.

radial gate: Water retention dam component consisting of curved panel mounted on arms that pivot on bearing assembly. Water pressure is imposed on the bearing assembly. TAINTOR GATE, TAINTER GATE.

radial step: Tread in circular or elliptical winding staircase.

* radian: Length of an arc of a circle equal to length of the radius.

radiant glass: Glass with embedded heating elements.

radiant heating: 1. Method of heating an area by distributing heated water or steam through pipes embedded in floor and/or walls of a structure. **2.** Method of heating an area with electrical wires embedded in floor, ceiling, and/or walls of a structure.

radiant panel test: Method for measuring flammability of surface of a material by using radiant heat source.

radiating brick: Masonry unit tapered in at least one direction. Used in curved masonry work.

radiation: Heat transfer in an open area due to electromagnetic waves.

radiator: Device that transfers heat from a source, such as steam or hot water, to the air.

radioactivity: Emission of electromagnetic waves resulting from breakdown of nuclear material.

* radius: Linear distance from the center of a circle to the circumference.

* rafter: Sloped roof structural member that supports roof sheathing and roof loads. See *common rafter, cripple jack rafter, hip rafter, jack rafter, valley rafter.*

rafter anchor: Light-gauge metal framing reinforcement used to join a rafter to a top wall plate. See *anchor**.

rafter plate: Top plate of wall that supports structural roof members.

rafter table: Table etched or printed on a framing square that provides mathematical information for calculation of common, hip, jack, and valley rafter lengths and cuts.

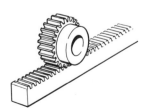

rack and pinion

radian

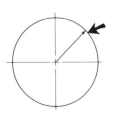

radius

rafter

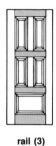

rail (3)

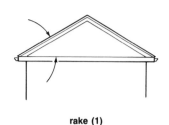

rake (1)

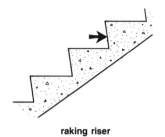

raking riser

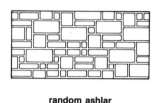

random ashlar

rafter tail: Portion of roof framing member that overhangs the plate and projects from face of the wall.

raft foundation: Foundation system consisting of reinforced monolithic concrete slab and walls that are the main support. See *foundation**.

rag felt: Heavy paper with large percentage of rag fibers. Used in manufacturing asphalt roof coverings.

raggle: 1. Masonry unit manufactured with groove for inserting flashing. **2.** Groove in masonry unit for receiving another material. RAGLET.

rag-rolled finish: Paint texture produced by rolling twisted piece of cloth over freshly painted surface.

rag rubble: Thin pieces of rough stone.

ragstone: Thin blocks or slabs of building stone.

* **rail: 1.** Top member of a stair baluster; e.g., handrail. **2.** Horizontal member of a window frame; e.g., meeting rail. **3.** Horizontal member of a panel door; e.g., bottom rail. **4.** Horizontal member of a face frame.

rail bolt: See *stair bolt.*

rail cup: Ornamental hardware for end of pipe handrail or fence rail.

rail file: Wood-shaping tool with coarse, curved teeth.

railing: See *handrail.*

rail pile: Foundation support consisting of three previously used railroad rails welded along the edges and driven as a single unit.

railroad pick: Striking tool with two pointed ends. Used to break hard materials.

rail sleeve: Connecting pipe section between adjoining pieces of pipe for handrail or fence rail.

Raimann patch: Elliptical filler used in finish grade plywood face veneer.

rainproof: Constructed or protected to prevent rain from interfering with successful operation of equipment under specified conditions.

raintight: Constructed or protected to prevent rain from entering a structure or enclosure.

rainwash: Erosion caused by rainfall.

rainwater leader: See *downspout.*

raise: Vertical shaft dug upward from a tunnel.

raised barrel hinge: Pivoting device in which leaves are bent to position pin away from jamb. Used where doors are set in a deep jamb.

raised face flange: Area around a drilled hole or bolt head that is approximately $^{1}/_{32}$" thicker than surrounding surface.

raised flooring system: See *pedestal floor.*

raised grain: Rough surface on planed lumber caused by hard summerwood extending above soft springwood.

raised mold: Trim member extending above surface of an adjoining area.

raising: Painting defect that creates blistered, wrinkled surface finish.

raising hammer: Hand tool with long head and rounded face for striking and forming sheet metal. See *hammer**.

* **rake: 1.** Angle of slope or incline; e.g., rake of a roof. **2.** To remove mortar from masonry joint. **3.** Tool used to remove mortar from masonry joint.

rake blade: Excavation blade consisting of spaced tines.

raked joint: Mortar between masonry units that has been removed to a specified depth before hardening. See *mortar joint**.

rake molding: Trim member applied at an angle.

raker: Tooth on circular saw blade that removes material left in the kerf by the cutting teeth.

raker pile: See *batter pile.*

raking bond: Angular style of laying brick with diagonal header brick. Similar to herringbone bond. RAKE BOND.

raking course: Course of bricks laid diagonally across the face row of bricks in a masonry wall. Used to strengthen a thick masonry wall.

* **raking riser:** Vertical portion of a step that is sloped away from nose of the tread to allow extra stepping space on lower tread.

raking stretcher bond: Pattern of laying masonry units with vertical joints slightly offset in each successive course.

ram: 1. Movable driving weight in a pile hammer. **2.** Pneumatically or hydraulically powered cylinder. **3.** To consolidate and compress by striking with force.

ramp: 1. Sloped structure that provides access to different elevations without stairs. Ramp incline for pedestrian traffic should be 15° maximum. Ramp incline for handicapped traffic should be 5° maximum. **2.** Short piece of drain pipe installed at a steep angle.

rampant: Arch in which the two lower ends rest on supports of different heights.

rampart: Defensive wall that is part of a fortification.

rance: See *shore.*

ranch-style: Single-story residential construction with low-pitched roof.

* **random ashlar:** Squared stones set in a random pattern without continuous mortar joints.

random coursed ashlar: Stonework pattern in which smooth rectangular stones of various sizes are set with semicontinuous horizontal mortar joints across several stones but without continuous or repeated design. See *bond* (1)**.

random match: Veneer layout in which pieces are joined without specific woodgrain match.

random rough bedded: Stonework pattern in which smooth rectangular stones of various sizes are set without continuous horizontal mortar joints or a repeated design. See *bond* (1)**.

random-tab asphalt shingle: Asphalt shingle manufactured or installed with noncontinuous or nonuniform pattern. See *asphalt shingle**.

* **range: 1.** North-South row of townships. **2.** Operational limits of an electrical control. **3.** Countertop cooking device. **4.** Two points that are aligned with point of observation in surveying. **5.** Row or course of masonry.

range boiler: Reservoir in plumbing supply system for storing water that is heated by auxiliary unit.

ranged rubble: See *rubblestone.*

range hood: Venting device that draws hot air away from stovetop cooking surface.

range masonry: Squared stones laid horizontally in courses of equal height. RANGEWORK.

range number: Manufacturer's code for masonry product that designates blend, color, and texture.

range pile: Foundation support placed as guide for locating other piles in marine work.

ranger: See *waler.*

range rod: Surveying tool used as sighting instrument; 6'-0" to 8'-0" long, graduated in 1'-0" lengths in alternating white and red bands. See *surveying*. LINE ROD, RANGE POLE, SIGHT ROD.

ranging bond: Masonry work with wood projecting from mortar joints. Construction materials are fastened to wood piece.

* **rasp: 1.** Coarse file used to shape and smooth wood. **2.** Piece of expanded metal lath wrapped around wood block to smooth rough edges of drywall.

* **ratchet:** Part of a tool or device that allows repeated turning without removing pivot point from the work. A pawl engages toothed wheel to allow for force to be applied in one direction while allowing free spinning in opposite direction. Ratchets are used with braces, wrenches, and lifting devices.

rated lamp life: Average length of time a lamp fixture will operate.

rated load: 1. Maximum amount of weight a machine is designed to carry. **2.** Maximum amount of weight a crane is designed to lift at specific radius without tipping.

rated speed: Measurement of movement of a device in feet per minute.

rathskeller: Area below ground level.

rating: 1. Maximum capacity of electrical equipment. **2.** See *fire rating.*

rating input: Volume of natural gas that can be safely burned in an appliance. Measured in Btu/hour.

rating output: Measurement of Btu of heat generated by an appliance when operating at its rating input.

ratio: Proportion or relationship between two or more elements. Concrete mixes are based on a ratio of cement, sand, aggregate, and water.

rattail file: Cylindrical tapered file used to shape and smooth holes or curves. See *file*.

raveling: Deterioration of asphalt pavement resulting from segregation of aggregate.

rawl plug: Solid fibrous wall anchor inserted in predrilled hole. Screw is inserted into rawl plug and tightened to secure anchor in wall. See *anchor*.

raze: To tear down; demolish.

reactance: Resistance encountered by alternating current as it changes direction of flow in a circuit.

reaction: Forces acting on structural member that are equal but opposite to total live and dead loads and that stabilize the member.

reaction flux: Oxide removal material that interacts with base metal when heated.

reaction pile: Foundation support driven close to test piles. Used to anchor jacking beam when testing piles. ANCHOR PILE.

reaction wood: Wood formed by abnormal tree growth.

reactive aggregate: Rock or other material with chemical content. When added to concrete mix, it reacts with cement in destructive manner.

reactor: Regulating device in an arc welding circuit to maintain a uniform flow of electrical current.

readily accessible: See *accessible* (2).

ready-mixed concrete: Concrete made of cement, sand, aggregate, water, and other specified ingredients that is mixed at a batch plant. The concrete is delivered to a job site and is placed in forms. READY-MIX CONCRETE.

ream: 1. To enlarge the diameter of a hole. **2.** To smooth the inside diameter of a hole and finish to desired diameter.

* **reamer: 1.** Tool used to remove sharp edges from inside edge of a drilled hole and finish the hole to required diameter. **2.** Device used to straighten or enlarge inside diameter of a hole.

rear arch: Arch that supports inner portion of a wall when outer portion of the wall is supported in some other manner.

rearing: Raising the front of a piece of earth-moving equipment to offset the weight when pulling heavy load.

rebar: See *reinforcing bar.*

rebutted and rejointed: Wood shingles trimmed to produce parallel sides and perpendicular ends.

receiver: 1. Holding tank on air compressor. **2.** Portion of photocell system containing phototube.

receptacle: Electrical device for connecting electrical equipment to power source. Conductors from the power source are attached to back and/or sides of receptacle, and a plug connected to the equipment is inserted in slots on front of receptacle.

receptacle outlet: Electrical outlet where one or more receptacles are installed.

receptor: 1. Plumbing device that receives waste water and channels it into drainage system. **2.** Window adapter used to adapt window frame to the size of the rough opening.

recess: Indentation in a surface.

recessed fixture: Electrical device with the connections enclosed behind the finished ceiling surface. A trim plate is visible at the ceiling line.

reciprocating: Having an alternating forward and backward motion.

reciprocating saw: Electric cutting tool in which a short, stiff, pointed blade is moved forward and backward quickly. Blades are interchangeable for cutting wood, metal, and other material. See *saw*.

reconditioned wood: Hardwood that is steam-dried to correct defects produced in first curing.

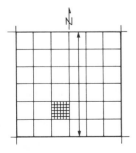

range (1)

rasp (2)

ratchet

reamer (2)

recovery peg: Temporary surveying marker used to relocate permanent peg or marker.

rectangle: Four-sided plane figure with opposing sides parallel and equal in length. Angle formed by the intersecting sides is 90°.

rectangular file: Hand tool with a rectangular cross section that is used to smooth wood or metal. See *file**.

rectangular tie: Bent wire device embedded in horizontal mortar joints. Used to tie two sections of cavity wall together. See *tie* (1)*.

rectifier: 1. Electrical device for converting alternating current (AC) to direct current (DC) by allowing the current to flow in one direction only. **2.** Externally cooled heat exchanger that condenses absorbent and separates it from the refrigerant.

redenture: See *cabling*.

red iron: See *structural steel*.

red top: Marking stake for subgrade.

reduced coarse aggregate concrete: Pumped concrete with the amount of aggregate decreased and the amount of sand and cement increased propor-

reeding

reel fixture

reinforcing bar

tionally.

reducer: 1. Pipe fitting that provides transition from a large pipe to a smaller pipe. **2.** Solvent that lowers the viscosity of paint. See *fitting**.

reducing atmosphere: Chemically active protective air or gas that reduces metal oxides to a metallic state when heated.

reducing valve: Device for controlling and maintaining minimum amount of pressure in a steam heating system. A reducing valve opens when pressure falls too low, thereby increasing pressure to an acceptable level.

reduct: Small piece cut from larger piece of material to make the larger piece the proper size and shape.

reduction gear: Device for reducing rotational speed from one turning member to another.

* **reeding:** Decorative design characterized by series of small, parallel, half-round moldings.

reef knot: See *square knot*.

reel: 1. Circular case for holding and dispensing tie wire for reinforcing steel. See *tie wire reel*. **2.** Flanged spool on which line is wound.

* **reel fixture:** Electrical device attached to spring-loaded case that automatically retracts a cord when not in use.

reentrant angle: Internal angle of less than 90°.

reeve: To thread a line between blocks in a block and tackle so the two blocks operate at right angles to each other.

reference monument: Mark used by surveyors when a corner cannot be permanently marked or may be damaged.

reflected plan: Plan view with the viewing point located beneath the object. REVERSE PLAN.

reflected sound: Sound waves that strike a surface and rebound in another direction.

reflective insulation: Material that limits transmission of heat by means of reflective backing material. Aluminum is commonly used as a backing material.

refractory: Material that can withstand excessively high temperature.

refractory brick: Masonry material that can withstand very high temperature without structural failure.

refractory concrete: Mixture of calcium-aluminate cement and special aggregate. Used in high temperature applications.

refrigerant: Material that produces cooling effect by absorbing heat while expanding or vaporizing.

refrigerant charge: Amount of refrigerant introduced into a closed cooling system.

refusal: Point beyond which a pile cannot be driven deeper.

regenerative heating: Raising temperature by means of heat that has been discarded in part of the heating cycle, recaptured by transfer, and utilized in another part of the cycle.

register: Combination grill and damper in a duct in forced air heating and/or cooling system. The register is release point for air.

regle: Groove or slot that serves as guide for moving or sliding object.

reglet: 1. Flat, narrow strip of wood or metal that separates structural panels; e.g., batten. **2.** Groove or slot in a wall to receive flashing.

reglette: Measuring device used for surveying; graduated in feet and tenths, and hundredths of a foot.

regrate: To clean masonry by removing the surface layer.

regular lay: Winding strands of rope in direction opposite the winding of individual strands.

regulated-set cement: Type of cement having additional active components to give concrete rapid strength development during first few hours after mixing.

regulator: Reducing valve used to control release of pressurized materials at a usable rate or pressure.

reheat: To raise temperature of previously cooled air in air cooling system in order to equalize overall cooling effect.

reignier: Wood inlay work of various colors.

reinforce: To strengthen a material.

reinforced concrete: Concrete with embedded reinforcing bars or fibrous material to provide additional tensile strength.

reinforced masonry: Brick, concrete masonry units, or stone with steel rods or reinforcing bars embedded in the mortar joints to provide additional tensile strength.

reinforcement: 1. Material embedded in another material to provide additional support or strength. **2.** Welding metal built up above surface of base metal.

reinforcement ratio: Measurement of effective area of reinforcement compared to effective area of concrete in a given section.

reinforcing: 1. Steel bars or rods, welded wire fabric, or fibrous material that provide additional tensile strength to concrete or masonry. **2.** Device that strengthens or provides additional support to a member.

* **reinforcing bar:** Steel rod with deformed surface. Used to reinforce masonry or

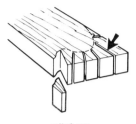

relief cut

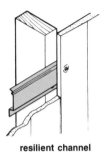

relief valve

resilient channel

resistor

concrete construction. The deformations of the steel rods interlock with the concrete mix to give a structure the required tensile strength.

reinforcing tape: Perforated paper or glass fiber strips used to conceal and strengthen joints between sheets of drywall.

reinforcing unit: Metal box built into hollow metal door to provide support for latch and lockset.

relative compaction: Density of dry soil divided by maximum dry density of soil determined from a standard testing procedure. Expressed as a percentage.

relative humidity: Measurement of amount of moisture in the air. Expressed as the percentage of water vapor present compared to the amount present in a saturated atmosphere at a given temperature.

relaxation: 1. Decrease of final resistance to pile penetration over time due to shifting of soil or other conditions. 2. Decrease in stress of steel due to creep over period of prolonged stress.

relay: 1. Device activated by electrical, pneumatic, or hydraulic means, or temperature that opens, closes, or activates other mechanisms. 2. Valve or switch that restores pneumatic or hydraulic pressure or electrical current to its original level or strength.

release agent: Material used to prevent concrete from bonding to a surface. See *form oil.* BOND BREAKER.

releve: 1. Measured drawing of older work. 2. Restoration.

relief: Variations in heights and elevations of a surface.

* **relief cut:** Cut made into a material to reduce twisting strain when making a curved cut. Relief cuts are made from the edge of the material to points close to the curved design.

relief hole: One of several holes bored into rock to create weakened plane that acts as breaking point during blasting.

* **relief valve:** Device that is activated to open when pressure and/or temperature in a closed system exceeds safe

operating limits. When pressure and/or temperature returns to safe levels, the relief valve returns to closed position. RELIEF PRESSURE VALVE.

relief vent: Plumbing pipe that provides additional air circulation between soil or waste stacks and vent stack.

relieving arch: Additional supporting member over an opening in a wall. Provides further distribution of structural load to other portions of the structure. DISCHARGING ARCH.

relievo: Projection of a design from a flat background. RILIEVO.

remodel: To alter or change existing design.

remote-control circuit: Electrical circuit that controls another circuit through a relay or equivalent device.

Renaissance: Period of time characterized by freedom of expression in architecture, art, and literature. During the Renaissance, classic styles of architecture were imitated.

render: 1. To adjust a sling or choker to proper position for use in rigging. 2. To apply first layer of plaster or mortar. 3. To draw an object with shading and shadows.

rendered masonry: Masonry structure coated with waterproofing material.

renewable fuse: Electrical device used in existing structures in which thin replaceable piece of wire or other material fails when excess current is applied. The wire or other material is replaced to allow the fuse to be reused.

repeat: Distance between identical patterns in a continuous design; patterned wallpaper with a 5″ repeat.

replum: Panel in a framed door.

repousse: Patterned in a relief method.

representative fraction: Drawing scale expressed as the ratio of a distance on the drawing to the same distance on object or area shown.

resaw: To rip lumber on edge to produce two thinner pieces. The thinner pieces of lumber have similar grain patterns.

resealing trap: Plumbing fitting connected to a fixture drain designed so the liquid flow seals the air opening

and prevents self-siphonage.

reservoir: Holding tank or basin for liquid.

reshore: To install temporary supports after the original shores have been removed. Used to avoid deflection of a shored member or damage of partially cured concrete.

residence: Dwelling.

residual deflection: Deviation from the original shape of an object that remains after a load is removed.

residual pressure: Water pressure in a water supply system when the outlets are open.

resilience: Ability of a material to return to its original size and shape after it is bent or pressure is applied.

* **resilient channel:** Light-gauge metal strip installed on ceilings and walls for attaching drywall and to limit sound transmission.

resilient flooring: Floor finish material made of plastic, with the ability to return to its original shape. Available in many patterns and sizes ranging from square tile to large sheets. SHEET GOODS.

resin: Solid or semisolid substance that has high molecular weight and tendency to flow under stress. Available natural or synthetic.

resistance: 1. Opposition of a material to electron flow in an electrical circuit. Resistance, measured in ohms, determines the rate at which electricity is converted to heat. 2. Opposition to the flow of a liquid created by pipe walls in a hydraulic system.

resistance brazing: Joining metal with heat produced from the resistance to electrical current through the workpiece.

resistance welding: Joining metal with heat produced from the resistance to electrical current through the workpieces. Electrodes concentrate the current and apply pressure at required point.

* **resistor:** Electrical device that provides resistance to electrical current in a circuit.

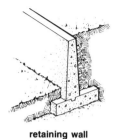

retaining wall

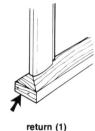

return (1)

return bend

resonance: **1.** Electrical condition in which inductive resistance in an alternating current circuit is equal to capacitive reactance. **2.** Condition reached when applied load is equal to natural frequency of the support of the load.

resonator: **1.** Device or enclosure for increasing volume of a sound. **2.** Hollow metal container used to produce microwaves.

resorcinol resin glue: Waterproof glue used for bonding laminated timbers and exterior applications. Good thermal resistance and bonding properties.

respond: Corbel or pilaster used to support one end of an arch or vault.

ressant: Projecting member or part. RESSAUT.

* **retaining wall:** Structure for sustaining earth, water, and other lateral pressure. Designs are based on amount of lateral pressure, height of material to be supported, and hydrostatic pressure.

retap: To drive additional piles because of relaxation of previously driven piles.

retardation: Reduction in the hardening and setting rate of concrete, mortar, or plaster.

retarder: Chemical admixture that slows the hydration process of concrete, mortar, or plaster.

retemper: To add water and remix concrete or mortar that has lost its workability.

retention tank: Temporary holding vessel for liquid.

reticle: System of graduations, such as crosshairs in a telescope, for sighting and aligning an object or target.

reticulated: Designed in the shape of series of diamonds; series of diagonal lines forming mesh-like design.

retract: Mechanism that activates a dipper shovel bucket on excavating equipment.

retrofit: **1.** Addition of components or building components to a system or building already completed. **2.** Addition of equipment to change power or energy source of a heating plant; changing from oil burning to natural gas burning.

* **return:** **1.** Change in direction of a molding or wall; e.g., a corner mitered back into itself. **2.** Part of the rope or line between a block and tackle. **3.** Pipe in a heating system that returns water to heating source.

return air: Air that has been circulated through a conditioned living area and is drawn back into the heating or cooling source.

* **return bend:** Pipe or fitting with 180° turn.

return main: Pipe or duct that moves heating or cooling medium from point of heat transfer to heating or cooling source.

return rail: Portion of handrail extending from newel post to wall.

return sheave: Pulley set at a distance from a power drum. Line passing around the return sheave allows the power drum to pull in a direction away from itself.

revale: Decorative stone molding that is carved after being set in its final position.

reveal: Portion of window or door jamb not covered by trim molding.

reveal pin: Clamp or threaded device inside a window opening. Used to attach and secure scaffolding to the opening.

revent pipe: Vent pipe connected to waste fixture and main or branch vent. Used to equalize pressure within waste system.

reverse: Plastering template cut to the opposite shape of a decorative pattern. Used to verify accuracy of a pattern.

* **reverse bend:** Threading a line over a drum or sheave in one direction and in opposite direction on adjoining drum or sheave.

reverse board-and-batten: Exterior plywood siding design with grooves cut lengthwise in the face of a plywood panel to give appearance of individual boards. See *siding**.

reverse polarity: Welding cable arrangement where electrode is attached to positive side. The workpiece is negative portion of electrical circuit. DIRECT CURRENT ELECTRODE POSITIVE.

reverse swinging door: Door that swings outward from the inside of a room.

revertible flue: Chimney or flue that momentarily redirects flow of gas downward during its natural upward flow.

revet: To face a sloping surface with rough stone, concrete, or other erosion protection.

revetment: **1.** Facing put on wall or sloping surface to protect against erosion. **2.** Wall with severe slope away from the base.

revolutions per minute (RPM): Orbital movement of a component during one minute; the arbor of a machine operating at 3000 RPM revolves 3000 times per minute.

revolving door: Door consisting of four panels that are attached to central pivot point along the inside stiles. Limits the amount of air passage and drafts. See *door**.

* **revolving shelf:** Spinning trays used in corner cabinets to allow full access of entire cabinet space.

revolving shovel: Excavating equipment in which upper section moves and rotates separately from supporting mechanism.

R factor: See *R value*.

rheology: Study of the ability of plastic substances to flow.

rheostat: Control that regulates electrical current by varying amount of resistance in a circuit.

* **rhomboid:** Four-sided plane figure with oblique angles and opposite sides of equal length.

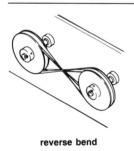

reverse bend

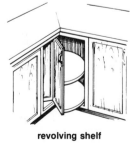

revolving shelf

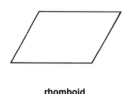

rhomboid

rhombus: Four-sided plane figure with oblique angles and all sides of equal length.

* **rib: 1.** Plain or ornamental supporting member of arch or vault. **2.** Decorative ornament in the shape of a supporting arch member. **3.** Projecting deformation on reinforcing bar. **4.** Raised portion of flat surface to provide additional rigidity.

ribband: Narrow strip of wood fastened to top of ceiling joists with a long span. Used to align and prevent twisting of joists.

ribbed seam: Sheet metal seam in which adjoining pieces are formed over a batten and an interlocking strip of metal is placed over the seam. See *seam* (1)*. BATTEN JOINT, BATTEN SEAM.

ribbing: Raised corrugation.

* **ribbon: 1.** Supporting ledger in framing applied horizontally across studs and is a support for joists. RIBBON STRIP, RIBBON BOARD. **2.** Lead bar for holding edges of stained glass in position. **3.** See *edge form*.

ribbon courses: 1. Alternating rows of tile with long and short exposed depths. **2.** Extra thickness of asphalt shingles applied to alternating courses to add depth to roof's appearance.

ribbon loading: Batching concrete with all solid materials and sometimes water entering the mixer simultaneously.

ribbon saw: See *band saw*.

riblath: Sheet metal backing for plaster. Riblath is more rigid than expanded metal lath and not suitable for contour work.

rice brush: Brush with long, stiff bristles for creating textured surface on acoustical plaster finishes.

rich concrete: Concrete with high cement content.

rich lime: Pure quicklime used for plastering and masonry. FAT LIME.

rich mixture: Mixture of cement, lime, sand, and water with larger percentage of cement than commonly used.

Richter scale: Measurement of magnitude of a seismic disturbance, such as an earthquake. Expressed on a scale of 0 to 8.9, with larger number denoting more severe disturbance.

riddle: Coarse sieve.

ridge: 1. Highest point of a sloping roof; roof peak. **2.** Highest horizontal member of a sloping roof to which rafters are fastened. RIDGE BOARD.

ridge cap: Top course or layer of waterproofing material at roof peak. RIDGE COURSE.

ridge cut: Cut made at upper end of a rafter where it meets a ridge board.

ridge fillet: High point between flutes or other depressions.

ridge rib: Uppermost arched member in vaulted roof.

ridge roll: Metal fitting installed at roof peak for waterproofing.

ridge roof: See *gable roof*.

ridging: Long, narrow blister-like deformation on surface of built-up roof.

riffler: Bent rasp for working on concave surfaces. FILLET RASP.

rifling: Forming a spiral thread on inside of pipe or drilled hole.

rift: Narrow split in rock surface.

rift cut: Cut made by slicing oak veneer to create straight grain pattern.

rig: 1. Mechanical excavation or lifting equipment. **2.** See *rigging*.

rigger: 1. Paintbrush with long bristles and flat end for painting lines of various widths. **2.** Person who prepares heavy equipment or materials for lifting.

rigging: Use of lines and various lifting equipment to transport and move loads.

riggot: Gutter for rainwater.

right angle: 90° angle.

* **right-angle drive unit:** Electric drill attachment for drilling at 90° angle to the motor.

right bank: Edge of stream to the right of the observer when looking downstream.

right-hand door: Door that swings inward with swinging stile on right when facing the outside of the door. See *door**.

right-hand reverse door: Door that swings outward with the swinging stile on the right when facing the outside of the door. See *door**.

right-hand rule: Method of determining direction of electron flow in a motor. A current-carrying conductor (represented by middle finger) is placed in a parallel magnetic field (represented by index finger), and the resultant force or movement is in the direction of the thumb.

right-hand stairway: Stairs with railing on the right while ascending the stairs.

right lay: Twisting of rope strands in clockwise direction around core.

right-of-way: Strip or area of land for construction or maintenance; easement.

right triangle: Three-sided geometric figure with one 90° angle.

rigid conduit: Nonflexible steel tubing that carries electrical wires.

rigidity: Resistance of a material or structure to movement and deformation.

rigidized sheet metal: Sheet metal embossed for additional strength and rigidity.

rib (3)

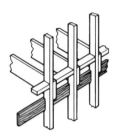

ribbon (1)

right-angle drive unit

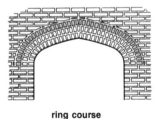

ring course

rigid pavement: Traffic surface that has high bending resistance and good load distribution across relatively large area.

rim: Open edge of plumbing fixture that is unobstructed.

rime: Ladder rung.

rim joist: See *band board, header*.

rim-mounted: Attached to the face of an object.

* **ring course:** Outer row of masonry in an arch.

ring rot: Wood decay following the annual ring pattern.

ring scratch awl: Pointed hand tool used to mark sheet metal.

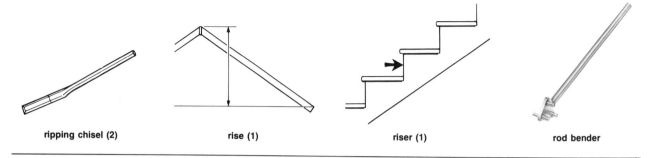

ripping chisel (2)   rise (1)   riser (1)   rod bender

ring shake: Splitting wood by following annual ring pattern.

ring-shank nail: Fastener with series of raised rings around the shank of the nail to improve its holding power and make it less likely to loosen. See *nail**.

rip: To cut wood parallel with the grain.

rip block: Concrete masonry unit cut to less than full height; usually used as starting piece.

rip fence: Adjustable attachment for a saw that guides edge of a material cut at a specified distance from the blade.

ripper: 1. Tooth-shaped extension attached to excavation equipment. Used to break into and dig through extremely hard surfaces. **2.** Device with teeth that score and cut into the ground to a depth of several feet. **3.** Hand tool with hooked end for removing nails from damaged roofing slate.

ripping bar: Hand tool used as a wedge for separating items, pulling nails, and general prying. Both ends have chisel-like tips with one end bent and the other end slightly curved. CLAW BAR, PINCH BAR, WRECKING BAR.

* ripping chisel: 1. Hand tool with split, sharpened end for pulling nails and splitting wood. **2.** Hand tool with curved, sharpened end for cleaning mortise joints.

ripping hammer: See *straight claw hammer*.

ripping size: Width dimension of lumber approximately 1/4″ greater than finished size. See *nominal size*.

ripple: 1. Uneven, wavy surface finish. **2.** Lines formed within deposited weld bead caused by welding process.

ripple amplitude: Variation of voltage output waveform in a direct current (DC) electrical power supply.

riprap: Broken, random size stone used as filling and stabilizing material for roadbed or revetment.

ripsaw: Wood-cutting tool with teeth that cut wood parallel to the grain. See *saw**.

* rise: 1. Vertical distance from top of a wall to top of the roof ridge. See *total rise* (1), *unit rise* (1). **2.** Vertical

distance between adjoining floors or landings in a stairway. See *total rise* (2), *unit rise* (2). **3.** Vertical distance from the springing (lowest point of an arch) to the intrados (highest point). **4.** Vertical distance from the crown to the lowest point of the road. **5.** Overall travel of an elevator.

rise and run: See *pitch symbol*.

* riser: 1. Vertical member of stairway placed between two treads. A trim board set on edge and rabbetted into the treads. **2.** Vertical electrical conduit that routes conductors from floor to floor in a multistory building. **3.** Vertical pipe that conveys hot water or steam for heating from floor to floor in a multistory building. RISER PIPE. **4.** Potable or nonpotable water supply piping installed vertically from one floor to another in a multistory building. **5.** Raised platform for a performer.

rising hinge: Hinge that raises a door when opened. Used where doors are blocked at the bottom by thick floor coverings or thresholds.

rive: To split wood parallel to the grain.

rivet: Permanent fastener made of soft metal. Inserted into a predrilled hole; shank is rounded over to secure rivet.

rivet buster: Pneumatic tool with chisel-like bit. Used to cut off rivet heads before backing rivet out.

rivet collar: Excess metal under head of rivet after driving it.

rivet flushing: Cutting off rivet head with oxyacetylene torch before backing rivet out.

riveting hammer: Striking tool with head that shapes sheet metal or drives rivets in place.

rivet set: Hand tool used to shape rivet head.

roadbed: Portion of highway construction over which pavement is laid.

rock anchor: High tensile steel rod grouted into hole drilled into rock. ROCK BOLT, ROCK HANGER.

rocker shovel: Mechanical excavating device used in tunneling.

rocklath: Gypsum sheets with special

paper coating that is used for applying and bonding plaster.

rock pocket: Area of concrete with large concentration of aggregate and little cement paste. See *honeycomb*.

rock rake: Tractor attachment with series of teeth. Used to clear area that has rocks, trees, and other large debris.

Rockwell hardness test: Measurement of hardness of metal by determining penetration depth of an instrument under specified conditions. High hardness number indicates a hard metal. Soft metal is tested with 1/16″ steel ball penetrator using a 100 kilogram load. The hardness is read on the B scale. Hard metal is tested with diamond cone penetrator using a 150 kilogram load. The hardness is read on the C scale.

rock wool: Lightweight heat and sound insulating material made by blowing steam through molten rock or slag.

rococo: Architectural style characterized by ornate decorations and groups of curves, shells, foliage, and scroll designs.

rod: 1. See *Philadelphia rod, pencil rod, reinforcing bar, story pole*. **2.** Unit of measure equal to 16′-6″. **3.** 5/8″ diameter steel shaft used for tamping and consolidating concrete samples. **4.** See *welding rod, electrode*. **5.** Straightedge used for plastering. **6.** To straighten plaster surface between grounds and screeds. **7.** To clean drains with drain rod.

* rod bender: Tool with adjustable rollers and supports used to shape rebars.

rod cutter: Bench shear for cutting rebars.

rodding: 1. Repeated up-and-down motion of a rod into concrete test cylinder or slump cone to tamp and consolidate sample.

rod man: 1. Worker in surveying crew who positions and holds surveying rod at proper location. **2.** Ironworker who installs reinforcement.

rod saw: Rough abrasive blade that fits in a hacksaw frame. Used for cutting ceramics, glass, and other hard material.

roll: **1.** To compact with mechanical roller or rolling weight. **2.** Semicircular or rounded molding. **3.** Quantity of material packaged in cylindrical form.

roll and fillet: Design in which square fillet projects from middle of convex face.

rolled curb: Edge of pavement that has slight depression and upward curved edge for channeling rainwater.

rolled glass: Glass manufactured with partial obscuring of vision and light transmission.

rolled-shape steel: Structural steel manufactured by passing heated steel through series of rollers that forms it to desired shape.

roller: **1.** Cylindrical weighted device for applying pressure to a surface. **2.** Fabric-covered cylinder used for painting. Attached to handle with rotating head. **3.** Rotating cylindrical device used for moving loads or machinery.

* **Rollerbug® tamper:** Device used to surface-tamp fresh concrete. Steel mesh is attached to rotating frame and rolled across concrete to bring cement paste to the surface.

roller-compacted concrete: Placement of successive layers of low-slump concrete which is consolidated by drawing heavy rollers across the mix. Used for large concrete structures such as dams.

roller conveyor: Series of cylindrical rods that rotate manually, mechanically, or by gravity to move objects and materials.

roller latch: See *bullet catch.*

rolling door: See *roll-up door, sliding door.*

* **rolling scaffold:** Raised platform mounted on casters for easy movement. ROLLING TOWER.

roll joint: Sheet metal joint formed by rolling two edges of sheet metal together and compressing them.

roll marks: Small parallel scratches in glass resulting from difference in speed between the rollers and glass during manufacturing.

roll roofing: Waterproof, asphalt-impregnated material used for roof covering. Common roll size is 3′-0″ wide and 33′-0″ long. Roll roofing is applied to roofs with little or no slope and may be covered with small gravel.

roll-type hand bender: See *hickey.*

roll-up door: Large door with series of horizontal slats tied together and coiled around spring-loaded cylinder for storage. Slats are guided along both sides of door opening by rollers and tracks.

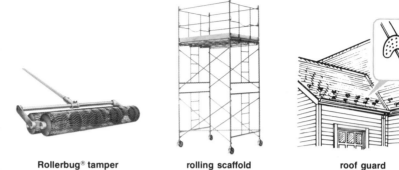

Rollerbug® tamper　　　　rolling scaffold　　　　roof guard

See *door*.*

rollway: See *spillway.*

rolok: See *rowlock.*

Roman arch: Arch with semicircular design.

Roman brick: Solid masonry unit measuring $1^5/_8″ \times 3^5/_8″ \times 11^5/_8″$. Also manufactured in lengths of 16″ or more.

Roman cement: Mixture of slacked lime and volcanic ash.

Romanesque architecture: Type of architecture characterized by large walls and rounded arches.

Roman ogee: Edge design combining convex and concave shapes. See *bit* (2) **.*

Roman tile: Roof tile with tapering, channel shape.

Romex: Trade name for nonmetallic-sheathed cable.

rondelle: Small circular piece of glass installed in leaded window. RONDEL.

* **roof:** Exterior top covering of a structure.

roof cement: Waterproofing material made of various petroleum products. Used to seal leaks in a roof.

roof collar: See *roof jacket.*

roof comb: Projection along peak of roof. ROOF CREST.

roof decking: See *roof sheathing.*

roof drain: Plumbing device installed on flat roof to collect rainwater and discharge it to the storm sewer system.

roofer: Worker who applies final roof coverings such as shingles, built-up roofing, and other waterproof covering.

roof framing: Calculating rafter and ridge lengths, laying out and making cuts, and assembly of roof components.

* **roof guard:** Device installed along lower edge of sloping roof to prevent large masses of snow and ice from sliding off. SNOW GUARD.

roofing nail: Fastener with large flat head and short pointed shaft. Used to fasten asphalt shingles and roll roofing to a roof. See *nail*.*

roofing square: See *framing square.*

roofing tile: Clay or lightweight concrete members installed as weather-protec-

tive finish on a roof.

roof jacket: Sheet metal flashing installed around plumbing vent pipes where they extend through the roof. ROOF COLLAR, ROOF JACK, ROOF FLANGE.

roof leader: Pipe in storm water system between storm drain and roof water collection system or roof terminal.

roof pitch: Roof slope expressed as ratio of total rise to total span (e.g., $^1/_4$, $^1/_2$) or in inches of rise to inches of run (e.g., 6 on 12, 12 on 12). See *pitch.*

ROOF TYPES

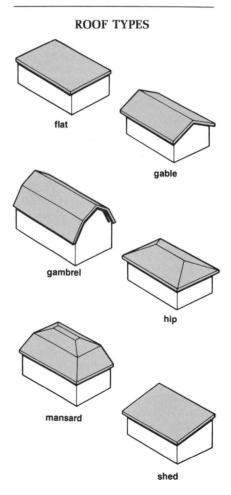

flat

gable

gambrel

hip

mansard

shed

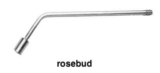

rosebud

rotary cut

rotary hammer

roto-operator

rough service lamp

roof sheathing: Material attached to top of roof rafters for use as base material for finish roofing materials. Plywood is commonly used for roof sheathing. ROOF DECKING.

roof terminal: End of vent pipe in plumbing system extending approximately 12″ above the roof.

roof tie: See *collar tie.*

roof truss: See *truss.*

room: Interior area surrounded and enclosed by walls and/or partitions.

room divider: Temporary partition that divides a room and creates two or more separate areas.

root: 1. See *root of joint, root of weld.* **2.** Widened portion of a tenon. **3.** Portion of a dam built into the ground where the dam meets a sloping surface.

root hook: Large hook pulled across the ground to tear out large tree roots.

root of joint: Part of weld joint where pieces of metal to be welded together are closest to each other.

root of weld: Points at which back of welded joint intersects base metal surfaces.

root opening: Distance between pieces of metal to be welded at root of joint.

root pass: First layer of metal deposited in electric arc welding.

rope: Twisted or braided strands of fiber or wire.

rope molding: See *cabling.*

rose: 1. Decorative metal plate between door knob and door through which door knob shaft is placed. See *escutcheon.* **2.** See *rosette.*

rose bit: See *countersink.*

* **rosebud:** Large tip on an oxyacetylene torch used to heat metal.

rose nail: Wrought nail with a coned, rounded head.

rosette: 1. Rose-shaped ornamentation, commonly circular in design. **2.** Electrical device that allows attachment of flexible cord to ceiling outlet.

rose window: Circular window with glass arranged in circular pattern.

rosin: Amber-colored gum-like by-product of wood. Used for soldering flux and various paint products.

rot: Decay of a material, such as lumber and plaster, resulting from prolonged exposure to moisture.

* **rotary cut:** Continuous sheet of veneer cut from a log by rotating the log against a knife.

rotary float: See *power trowel.*

* **rotary hammer:** Electric tool with cutting bit that is rotated and moved with a percussive action to bore into hard material.

rotary pump: Pump with geared wheels used to move liquid. Moves high volume of liquid at low pressure.

rotating laser: Leveling instrument with spinning head that projects a laser beam in a horizontal plane. The laser beam is detected by a sensor to determine an elevation. See *surveying*.

* **roto-operator:** Mechanical device with geared handle used to operate casement, jalousie, and awning windows.

rotor: 1. Rotating member of electric motor. Consists of slotted sections sandwiched together to reduce eddy currents. **2.** Portion of mechanical device that spins.

rotor-stator pump: Low-pressure pump used to convey drywall finish and plaster.

rottenstone: Soft pulverized limestone used for polishing soft metal and wood.

rotunda: Circular room with domed ceiling and roof.

rough arch: Arch built from rectangular brick using angular mortar joints.

rough ashlar: Undressed stone.

roughback: Unfinished side of a stone.

rough coat: 1. Initial layer of plaster applied to lath. **2.** Initial layer of joint compound applied to drywall surface.

rough floor: Initial layer of flooring material secured to floor joists. Rough grade of plywood or oriented strandboard is commonly used.

rough framing: Structural and backing supports, such as studs, plates, and cripples. Rough framing members are covered by finish materials.

rough grading: Moving and shaping earth around a structure in preparation for finish grading.

rough grinding: Initial cutting of projecting chips from terrazzo surface.

rough hardware: Hardware and fasteners not visible in finished product. Includes joist hangers, hold downs, common nails, and lag bolts.

rough hewn: Lumber with uneven, unfinished surface. ROUGH LUMBER.

rough-in: Initial construction done by tradepeople; e.g., water supply and waste pipes in a plumbing system. Rough-in work is concealed by finish trim members. ROUGHING-IN.

rough opening: Opening in wall designed to receive a finish component such as a finish jamb. Rough opening is larger than the finish component to allow for final adjustment.

rough rubble: Unsquared field stones laid in irregular courses.

rough-sawn: 1. Rustic finish on plywood panels or wood. **2.** Lumber cut to approximate size and not planed.

* **rough service lamp:** Light designed to withstand minor impact without filament failure.

rough string: See *carriage.*

rough work: 1. Masonry surface covered with a finish material. **2.** See *rough framing.*

round: 1. Circular. **2.** One turn of rope around a drum.

round arch: See *Roman arch.*

round dresser: Wood hand tool used to shape plumbing lead, bend lead pipe, and draw lead from one point to another.

round file: Cylindrical file used to shape and smooth holes or curves. See *file**.

round head: Semicircular screw head. See *screw**.

round in: To lift a load with rigging equipment by pulling pulleys of a block and tackle together with a rope.

rounding: Deflector around a bridge pier. Used to prevent damage caused by floating debris.

round joint: Convex finish of a mortar joint.

round turn: To wrap rope or cable around an object twice.

rout: 1. To cut or shape with a router. **2.** To enlarge a crack in preparation for patching and/or sealing.

* **router:** Electric tool used to cut and shape wood. A motor is mounted in a frame that allows depth adjustment of the interchangeable cutter. Handles on the side of the frame are used to guide router across the work. See *plunge router*.

router patch: Oblong wood insert with parallel sides and rounded ends used to repair plywood surface defects.

rover: Member that follows a curved line.

row: Number of tufts lengthwise in 1″ of carpet.

row house: One of a series of similar houses built side-by-side or with adjoining firewalls.

rowlock: Brick position with end exposed in a vertical position. See *brickwork**.

rowlock arch: Arch formed with brick arranged in separate concentric rings.

rowlock course: Course of brick with individual bricks positioned with the ends exposed in a vertical position.

rowlock stretcher: Brick position with largest surface exposed in a horizontal position. See *brickwork**. SHINER.

royal: Wood shingle measuring 24″ long and $1/2$″ thick at the butt end.

rubbed brick: Brick smoothed on its faces. The number of smooth faces is determined by the number of exposed faces.

rubbed effect: Dull luster in a painted surface.

rubbed finish: Semismooth surface on concrete produced by rubbing the surface with abrasive stone or rubbing brick.

rubbed joint: 1. Flat masonry joint formed by rubbing the wet mortar flush with face of the masonry units. **2.** Method of joining boards edge-to-edge. Glue is applied to two edges and rubbed together until tight seal is formed. Clamping is not required.

* **rubber test plug:** Pliable device used to obstruct water or air flow when testing plumbing systems. Rubber test plug is inserted in ends of a sanitary drainage and vent system and other openings, and inflated.

rubbing brick: Fluted silicon carbide material used for smoothing concrete and masonry materials. RUB BRICK, RUBBING STONE.

rubbing stone: See *rubbing brick*.

rubble: Rough, irregular stone. Used as fill or where rustic appearance is desired.

rubble ashlar: Squared stones with rough back.

rubblestone: Unsquared or roughly shaped stone with irregular size and shape. RUBBLE, RUBBLE MASONRY.

rubrication: Background coloring using enamel or other type of paint.

rudenture: See *cabling*.

ruderation: Process of paving an area with small aggregate pressed into the surface.

rule joint: 1. Overlapping joint used between the edge and drop leaf in table construction. One edge is finished with concave quarter-round design and adjoining edge is finished with convex quarter-round design. **2.** Overlapping joint used in a folding rule.

rule of thumb: General calculation that is easily remembered and used to determine approximate amounts.

ruler: Measuring tool divided into even increments. See *architect's scale*, *engineer's scale*, *extension rule*, *folding rule*, *push-pull rule*. RULE.

* **run: 1.** Horizontal distance between outside of a wall and center of a roof ridge. See *total run* (1), *unit run* (1). **2.** Horizontal distance from face of a top riser to face of a bottom riser in stair construction. See *total run* (2), *unit run* (2). **3.** Plumbing pipe or fitting that continues in a straight line with direction of flow. **4.** Continuous series of surveyed and measured elevation differences made in one direction. **5.** To pass plaster or putty through a sieve. **6.** Narrow drip of paint on a surface. **7.** Track for a window sash.

rung: Step of a ladder.

* **runner: 1.** U-shaped channel fastened to floor and ceiling in metal framing system. TRACK. **2.** Device for suspending and guiding a sliding drawer. **3.** Rotating portion of a turbine in a water wheel. **4.** Timber piles driven at the perimeter of area to be excavated.

runner line: Crane line supporting a load.

running block: Movable pulley assembly in a block and tackle.

running bond: Pattern for laying masonry units. Bricks are set as stretchers with vertical joints centered on the brick in the course below. See *bond* (1)*. STRETCHER BOND, STRETCHING BOND.

running dog: Ornamental design with repeated scrolls connected by wave-like band.

running end: Free, working end of line or rope.

running foot: Lineal length equal to 12″.

running inch: Lineal length equal to 1″.

running mold: Template used to form plaster trim molding. Template of desired design is passed across fresh plaster to form trim molding.

running off: Applying a finish coat of plaster to a molding design.

router

rubber test plug

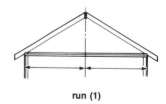

run (1)

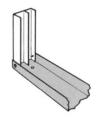

runner (1)

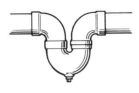

running trap

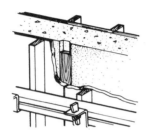

rustication strip

---

* **running trap:** Plumbing trap in which inlet and outlet are aligned horizontally. A lowered area between inlet and outlet retains water, which prevents sewer gases from entering the living area.

run number: Indication that distinguishes one batch of wallpaper from another. Used to assure trueness of color throughout a project.

runoff: Rainwater not absorbed by the ground but that flows into a storm sewer.

run off tab: Piece of metal positioned at end of weld joint to stop the weld on to maintain bead consistency throughout the weld.

run on tab: Piece of metal positioned at beginning of weld joint to start the weld on to allow bead consistency throughout the weld.

runway: Temporary passage for walking and vehicular traffic on a building site.

rupture disk: Valve or other member designed to fail when excessive pressure is placed. Pressure relief device in a pressurized pipe system. RUPTURE MEMBER.

rust: Reddish-brown material formed by oxidation of iron.

rustic: 1. Rough finished; country-style. **2.** Multicolored brick with rough surface.

rusticated: Finish of a stone wall where rough stones project from the face of the wall. Mortar joints are emphasized using deep recesses.

rusticated ashlar: Stonework with smooth-faced stone and mortar joints emphasized with deep recesses.

* **rustication strip:** Piece of wood or other material fastened to the face of a concrete form to produce a groove in the finished concrete.

rustic joint: Deep mortar joint between masonry units.

rustic work: Stonework with rough, uneven surface.

rust joint: Connection between pipes in which an oxidizing chemical mixture is introduced to form a waterproof seal.

rust pocket: Cleanout at lower end of a ventilation pipe to allow for rust removal.

rutile: Mineral consisting of 60% titanium and small amount of iron. Used in paint and coating for welding rods.

R value: Measure of the effectiveness of a material to provide thermal insulation. Higher R values indicate greater insulating capabilities. R FACTOR.

saber saw: Power tool with a short, straight reciprocating blade used primarily to cut curves in thin material. See *jigsaw, reciprocating saw.* SABRE SAW.

sabin: Measure of sound absorption equal to 1 square foot of perfectly absorptive material.

sack: See *bag of cement.*

sack rub: Finish for concrete surfaces used to produce even texture and fill all minor voids. After dampening surface, mortar is rubbed over surface. Before it dries the excess is removed using dry cement and sand rubbed over it with a piece of burlap. SACKING.

sacrarium: Sink connected to a fresh water supply and drained into a pit or directly to the ground.

sacrificial protection: Coating of zinc, cadmium, aluminum, or other protective metal applied to steel. Coating is dissolved by an electrolyte before reaching the steel.

* **saddle: 1.** Short projection above a roof on the upper side of a chimney or other structure; used to divert water around the structure. **2.** Covering for a roof ridge. **3.** Thin metal or rubber strip covering the joint between finish floor and threshold. **4.** Steel channel with flanges installed downward on top of a pile for use as a bearing plate for the pile cap. **5.** Frame that locks a pile hammer into the leads. **6.** Pipe fitting used to hold a corporation stop.

saddle bar: Horizontal steel rod extending across a leaded glass window to reinforce the lights.

saddle bead: Glazing channel used to fasten two panes of glass together.

saddle block: Swiveling device on boom of excavating equipment through which the stick of a dipper shovel slides.

saddle fitting: Fitting used to make a connection to a pipe that has already been installed. Fitting is clamped to a pipe and sealed with a gasket.

saddle flange: Curved flange that conforms to shape of the surface to which it is installed and is threaded to accept a threaded pipe. May be welded or riveted in position.

* **saddle joint: 1.** Stepped masonry joint used to divert water away from the structure. **2.** Sheet metal joint in which one end of a sheet is bent downward to fit over the edge on an adjacent sheet that has been turned up.

saddle scaffold: Temporary work platform constructed over the top of the ridge of a roof.

* **saddle threshold:** Metal hardware attached to the floor immediately under a door to create an airtight seal.

saddle tie: Wire tie used to secure adjacent members together by looping the wire around the members.

safe: 1. Permanent or temporary enclosure used for protection of important material such as jewelry or documents. **2.** Collection pan under a plumbing pipe or fixture that collects leakage or overflow.

safe-carrying capacity: 1. Maximum load that can be supported by a structural member without permanent deformation. **2.** Maximum electric current a conductor can withstand without overheating.

safe edge: Device on leading edge of elevator doors that reopens the doors when contacted by an object when closing.

safe load: Load on a structure that does not produce deformation or exceed maximum allowable stress.

safety belt: Strap worn around the waist and attached to a lifeline to provide protection against injury in the event of a fall.

safety clutch: Equipment clutch that slips when a predetermined capacity is exceeded, thus preventing overload of the motor.

safety factor: Safety stress factor; ratio of yield point of a material to the designed working stress.

safety fuse: Cord that contains a combustible medium used to convey fire or a spark at a uniform rate to detonate a blasting cap.

safety glass: Two or more sheets of flat glass with a layer of transparent plastic laminated between them; used to prevent shattering of the glass.

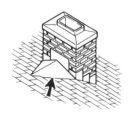

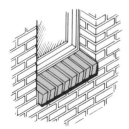

saddle (1)   saddle joint (1)   saddle threshold

safety hook   saltbox construction   sanitary cross

safety goggles: Eye protection that prevents injury to the eyes by a foreign object.

safety hasp: Hardware device with pivoting slotted plate that fits over a U-shaped rod; conceals all fasteners.

* **safety hook:** Metal hook incorporating a swivel stop on the throat to prevent a line or sling from slipping out of the hook.

safety nosing: Rough-textured strip installed along front edge of a stair tread to prevent slipping.

safety shutoff: Device that discontinues flow of natural gas into a burner in the event the initial ignition fails.

safety valve: See *pressure-reducing valve.*

safe working pressure: Maximum allowable internal pressure for a boiler or other closed vessel.

safing: 1. Noncombustible material placed between a floor slab and curtain wall to prevent passage of fire. **2.** Barrier in ductwork that ensures proper air flow through a device.

sag: To deflect downward as a result of gravity or excessive loads.

sailor: Brick set vertically with the largest face exposed. See *brickwork*.*

salamander: Small portable heater used on a job site to heat surroundings and prevent building material from freezing.

salient: Projecting outward; e.g., salient corner projects outward from the corner of the structure.

sally: A projection.

salmon brick: Soft, unburnt masonry unit with a pink color. CHUFF BRICK, PLACE BRICK.

salt-and-pepper brick: Masonry unit with black specks in the finish resulting from iron in the clay.

* **saltbox construction:** Wood-frame residential construction with a short sloping roof in front and a long sloping roof in back. Usually rectangular in plan view and covered with horizontal lap siding.

salt glaze: Clear, glossy finish on clay tile.

* **sand:** Granular material that passes through a $^3/_8''$ sieve, almost entirely passes through a No. 4 sieve, and primarily retained on a No. 200 sieve. Result of natural decomposition and abrasion of rock.

* **sand bending:** Process of bending pipe by partially filling it with sand before bending to prevent collapse.

* **sand blasting:** Process of cutting or abrading a hard surface by forcing sand through a hose at high speed with compressed air.

* **sand catcher:** Plumbing device used to trap and retain material that is heavier than water in a waste system. INTERCEPTOR, SAND TRAP.

* **sand cone test:** Method of determining compaction of soil by filling a pre-drilled hole with sand and measuring the volume and weight.

* **sanded fluxed-pitch felt:** Material composed of felt paper that is saturated with fluxed coal tar and coated with sand on both sides to prevent sticking when rolled.

* **sand equivalent:** Measure of proportions of fine dust or claylike material present in fine aggregate or soil.

* **sander:** Power tool used to remove imperfections and defects from surface of wood and other materials. See *belt sander, orbital sander, pad sander, vibrating sander.*

* **sand filter:** Layer of sand laid over coarse aggregate to remove impurities from a water supply.

* **sand finish:** Rough texture on a surface resulting from addition of sand to the base material.

* **sanding:** Process of smoothing the surface of a material with an abrasive-coated medium.

* **sanding block:** Device used to hold a piece of abrasive material for manual sanding of a surface.

* **sanding sealer:** Liquid finish that fills and seals surface of wood without concealing the grain. Surface is commonly sanded between applications.

* **sand jack:** Container filled with dry

sand on which a timber plunger rests and supports the bottoms of posts used in centering. Sand is released from bottom of the container to allow lowering of the centering. SAND BOX.

sand-lime brick: Masonry unit made from sand and lime; formed under pressure and cured with steam. CALCIUM SILICATE BRICK.

sandpaper: Tough paper or other base material covered with abrasive material such as aluminum oxide or garnet. Coarseness of the abrasive coating is graded by a grit numbering system in which low numbers denote a coarse abrasive and high numbers denote a fine abrasive; e.g., 40 grit abrasive provides a coarse finish, and 400 grit abrasive provides a fine finish.

sand plate: Flat steel plate or strip attached to bottom of reinforcing bar chairs for use on compressed soil.

sandstone: Sedimentary rock composed of many grains of sand joined with silica, iron oxide, or other minerals.

sandstruck brick: See *soft-mud brick.*

sandwich beam: See *flitch girder.*

sandwich panel: Prefabricated composite panel formed by attaching two thin pieces of facing material to a thicker core; e.g., precast concrete panel formed with two layers of concrete fastened to a lightweight insulating core.

sanitary base: Coved finish at intersection of wall and floor. Eliminates a square corner and accumulation of debris around the member. SANITARY COVE, SANITARY SHOE.

* **sanitary cross:** Waste fitting consisting of four intersecting outlets, two of which are perpendicular to the other two. Perpendicular outlets are formed with a small fillet to prevent accumulation of waste within the fitting.

sanitary drainage piping: Piping system used to convey waste water and water-borne waste material from a fixture to the sanitary sewer.

sanitary sewer: Underground pipe or conduit used to convey waste water and waterborne waste material. Does not convey storm water, surface water, and groundwater.

sanitary tee: Waste fitting with three outlets, one of which is perpendicular to the other two. Single outlet is formed with a small fillet to prevent accumulation of waste within the fitting.

sap: 1. Semiliquid material in vegetation, such as trees and shrubs. **2.** Moisture in newly quarried stone.

sapwood: Portion of a tree immediately below the bark. Lighter in color and less decay-resistant than heartwood.

sarking: Thin board used as sheathing.

sash: Fixed or movable framework of a window that contains the glass.

sash balance: Spring-loaded device in a double-hung window used to counterbalance the weight of the sash and facilitate lifting.

sash bar: See *muntin* (1).

sash block: See *jamb block*.

sash brush: Paintbrush with the bristles finished square or at an angle to allow painting in tight locations.

sash center: Pivoting hardware for horizontally pivoting windows, such as hopper windows. Consists of a jamb socket attached to the jamb and a connecting pin on which the sash of the window pivots. SASH PLATE.

sash chain: Metal chain used to connect a single- or double-hung window sash to a counterweight.

sash chisel: Wood-cutting tool with a wide blade that is sharpened on both sides.

sash cord: Rope used to connect a single- or double-hung window sash to a counterweight. SASH LINE.

sash door: Door in which the upper portion is glazed.

sash holder: Hardware used to secure two window sashes together and prevent opening. Usually attached to the meeting rails of a double-hung window. SASH FAST, SASH LOCK.

sash lift: Handle attached to a window sash to provide a means of raising and lowering the window. WINDOW LIFT, WINDOW PULL.

sash pulley: Pulley mortised into the side of a double-hung window frame near the top. Sash cord or chain passes over the pulley.

sash stop: See *window stop*.

sash weight: Cast iron weight used to counterbalance a single- or double-hung window sash.

satin finish: Paint texture with a dull gloss.

saturated: **1.** Soaked to maximum limit; containing maximum amount of moisture. **2.** Paint color that is completely free of white.

sauna: Steam bath in which water is introduced to a heat source to produce steam.

* **saw:** Hand or power tool used to cut material by means of a thin, flexible or rigid metal blade with teeth cut into the edge. Material is cut with a reciprocating or circular motion.

sawbuck: See *sawhorse*.

sawhorse: Four-legged support used to hold material at a convenient height.

saw set: Tool used to bend saw teeth at a predetermined angle or set.

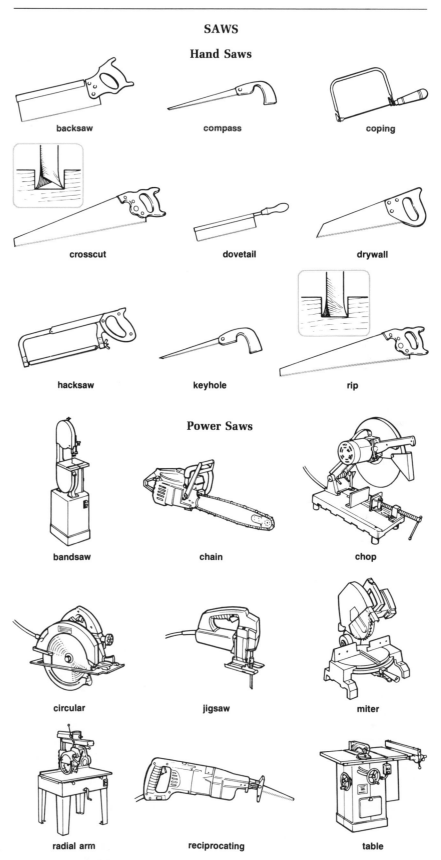

SAWS

Hand Saws

backsaw   compass   coping

crosscut   dovetail   drywall

hacksaw   keyhole   rip

Power Saws

bandsaw   chain   chop

circular   jigsaw   miter

radial arm   reciprocating   table

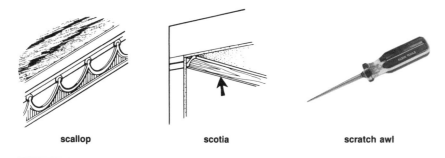

scallop   scotia   scratch awl

sawtooth: Individual cutting point on a saw blade.

sawtoothed: Designed in a pattern of repeated, pointed angles similar to the teeth on a saw blade.

sawtooth truss: Prefabricated roof framing member with a series of parallel sloping top chords alternating with vertical chords. Top and bottom chords are joined with a series of vertical and/or angular struts. See *truss**.

sax: Hammer with a pointed peen used to punch nail holes in slate tile. SLATE AXE.

scab: Short piece of lumber or other material fastened across a butt joint for reinforcement.

scabbard: Sheath on a tool belt used to hold ironworking tools such as a spud wrench.

scabble: To dress stone to produce prominent tool marks. Commonly used prior to finish dressing.

scabbling hammer: Hammer with one pointed end used for rough-dressing stone.

scaffold: Temporary elevated platform used to support workers and materials.

scagliola: Surface finish that resembles marble. Produced by mixing fine gypsum with adhesive and marble chips or graphite dust and applied in a variegated pattern. Mixture is polished when hardened.

scale: 1. System of drawing representation in which the elements are proportional to other elements of the same drawing. Dimensions of the drawing may be larger or smaller than actual size of the object. 2. Layout tool used for determining measurements on reduced and enlarged drawings. See *architect's scale, engineer's scale.* 3. Mineral surface deposits precipitated from water. 4. Device used to determine weight.

scale setting: Temperature to which a thermostat is set.

scaling: 1. Measuring with a scale to determine actual length according to a given reduction or enlargement. 2. Flaking away of a surface coating.

scaling hammer: Hammer with two wedge-shaped heads which are perpendicular to one another. Used for brickwork. See *hammer**.

*** scallop:** Decorative pattern consisting of a series of semicircular segments.

scalping: Removal of oversized particles that are larger than a specified size.

scamillus: Slight bevel at the outer edge of a block of stone.

scantling: 1. Timber member measuring $1\frac{1}{8}$" to 4" thick and 2" to 4" wide. 2. Hardwood that is squared to nonstandard dimensions.

scarcement: Setback in a wall or embankment that provides a ledge or footing.

scarf joint: Wood joint in which the mating ends are chamfered or beveled and joined to form a flush surface.

scarifier: 1. Hand tool used to scratch the surface of fresh plaster to allow for better bonding of subsequent coats. SCRATCHER. 2. Excavating equipment attachment used to scratch the surface of the earth.

scarify: To roughen the finish surface of an area.

scarp: Natural or man-made steep slope.

scavenge: To thoroughly clean.

schedule: Detailed list on a print that provides information about building components such as doors, windows, and mechanical equipment. Numbers and/or letters on the print refer to the schedule. See *print**.

schematic diagram: 1. Drawing that shows electrical system circuitry with symbols that depict electrical devices and lines representing the conductors. 2. Isometric drawing of a plumbing system showing fixtures and piping.

Schmidt hammer: Device used for nondestructive testing of concrete by measuring the rebound of the hammer. Used to determine compressive strength of concrete.

scintled joint: See *skintled joint.*

scissors truss: Roof truss with a steep exterior slope and lesser interior slope.

Commonly used to create a sloping interior ceiling and provide an area for insulation between the roof and ceiling. See *truss**.

S clip seam: Sheet metal seam formed by using an S-shaped piece of metal that forms two pockets for adjoining pieces of metal. Commonly used for joining sections of ductwork. See *seam* (1) *.

sconce: Decorative projecting wall bracket used to support candles or lights.

sconcheon: Reveal of a door or window opening that extends from the frame to the inner face of the wall.

sconcheon arch: See *rear arch.*

score: 1. To scratch or roughen the surface of a member or material to create a rough finish or key for subsequent coats. 2. To cut a narrow channel or groove in the surface of a material to allow for deeper cuts without shattering or splintering the face of the material.

scotch: Masonry tool with a sharp edge on each side for cutting or trimming brick. SCUTCH.

Scotch bracketing: Lath fastened at an angle at the intersection of a wall and ceiling to form a hollow base for a plaster cove.

*** scotia:** Concave trim molding. REVERSE QUARTER ROUND.

scour: Erosion.

scrabbled rubble: See *rubblestone.*

scraffeto: See *sgraffito.*

scraper: 1. Hand tool with a flat, burnished edge; used for final smoothing of woodwork. CABINET SCRAPER. 2. Hardened steel blade attached to a long handle; used to remove excess joint compound and/or plaster from walls and floors. 3. Excavation equipment that performs digging, hauling, and grading operations. Cutting edge mounted under a carrying bowl scrapes soil from surface and deposits into the bowl. 4. Blade used to remove undesirable material from a conveyor belt.

*** scratch awl:** Hand tool with a sharp pointed shaft used for punching holes in soft material and laying out designs on a hard surface such as metal.

scratch cloth: Fine wire brush used to clean and prepare lead pipe for soldering.

scratch coat: Initial coat of plaster with a rough finish that provides for improved bonding of successive plaster coats.

scratch course: Layer of pavement separate from binder course. Laid over a base course to level or smooth the surface.

scratcher: Plastering tool used to roughen a plaster surface to improve bonding of successive plaster coats.

SCR brick: Solid masonry unit with nominal dimensions of $6'' \times 2^2/_3'' \times 12''$. Developed by the Structural Clay Products Institute. See brick*.

*** screed: 1.** Straightedge used to level and grade freshly placed concrete flatwork. STRIKEOFF. **2.** Guide used to maintain predetermined depth of placement for concrete or plaster. **3.** To strikeoff freshly placed concrete that is above the desired plane.

screed chair: Wire support used as a depth gauge guide in concrete flatwork.

screed guide: Grade strips or side forms for unformed concrete used as a guide when screeding a surface to the desired shape.

screeding: Process of forming a concrete surface using guides and a straightedge.

screen: Metal or other durable fabric composed of regularly spaced strands of material, such as wire or nylon, providing uniform sized openings.

screen bead: Narrow, flat molding. See molding*. SCREEN MOLD.

screen block: Concrete masonry unit with a pattern of ornamental voids. Used as a partial solar screen in decorative work. See concrete masonry unit*.

screen door: See storm door.

screenings: Coarse grains in sand that do not filter through a specified size sieve, or fine grains that are smaller than the smallest desired size.

*** screw:** Wood or metal fastener consisting of a tapered male-threaded shank and a head designed to allow turning of the fastener. Shanks are sized by number from 0 to 24, with higher numbers indicating larger diameter.

screw anchor: Expansion device used to fasten screws to masonry walls. Device is inserted in a predrilled hole and expands when a screw is tightened. See anchor*.

screw clamp: Clamp consisting of two wood jaws with steel threaded rods passing through them. May be used to clamp parallel materials or material that is at an angle to each other. HAND SCREW.

screwdriver: Hand tool with a tip designed to fit into a screw head for turning. See spiral ratchet screwdriver.

screw extractor: Tool used to remove threaded fasteners that have broken off below the surface. Consists of a tapered metal rod with a reverse thread that is inserted into a predrilled hole in the top of the fastener.

*** screw eye:** Hardware with a circular end and a threaded shaft designed to be

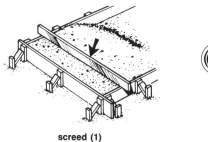

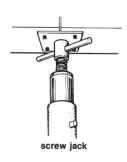

screed (1) screw eye screw jack

secured in wood. EYELET.

screwgun: Power tool with interchangeable tips for turning various types of screws. A slip clutch and a tip guard allow the screws to be inserted to a uniform depth. Commonly used for drywall installation.

*** screw jack:** Lifting mechanism made up of a threaded rod with plates or poles on either side of the threaded portion.

Loads are raised or lowered as the threaded rod is rotated.

Screwmate®: Drill bit used with an electric drill that drills the pilot hole, shank hole, and countersink in one operation. See bit (1)*.

screw nail: Fastener with a round head and helical fluted shaft designed to keep the fastener from backing out. See nail*.

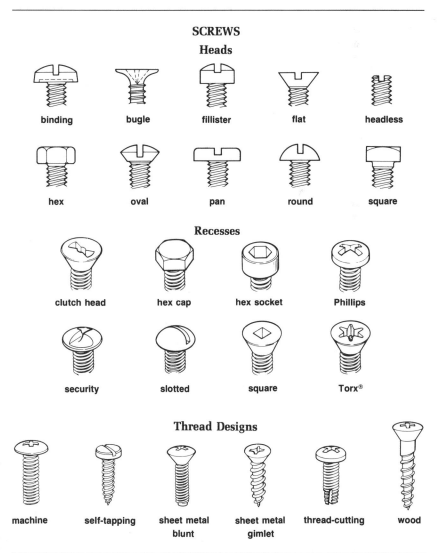

SCREWS

Heads

binding   bugle   fillister   flat   headless

hex   oval   pan   round   square

Recesses

clutch head   hex cap   hex socket   Phillips

security   slotted   square   Torx®

Thread Designs

machine   self-tapping   sheet metal blunt   sheet metal gimlet   thread-cutting   wood

scupper

scuttle

screw plug: Device that obstructs a drain pipe for a water test by compressing a rubber gasket against a steel plate.

screw shell: See *lampholder*.

scribe: 1. Hand tool with two legs attached at a pivot point. One leg has a sharp end and the other holds a pencil. Distance between the two legs is adjustable. Used for fitting finish materials to uneven or irregular shapes. **2.** To fit a finish material to an uneven or irregular surface with a scribe. **3.** To scratch a mark on hard material, such as steel.

scrim: Low-grade tape or fabric used to cover lath joints before plaster application.

scroll: Two-dimensional ornamental design consisting of an expanding and enlarging spiral.

scroll saw: See *coping saw*.

scrubber: Mechanical system used to remove noxious gases and particles from air.

scrub board: See *baseboard*.

scrub plane: Hand tool approximately $9\frac{1}{2}''$ long with a rounded cutting edge. Used to remove large amounts of wood from the surface.

scrub sink: Plumbing fixture used for washing in surgical preparation rooms.

sculpture: Decorative carving or shape.

scum: Deposit that forms on the face of clay bricks; results from formation of soluble salts or deposits during firing.

* scupper: Opening in a wall that allows rainwater to drain. Rectangular fixture channels rainwater into the downspout and storm water removal system.

scutch: Hand tool similar to a small pick used to cut and shape brick. SCOTCH.

scutcheon: See *escutcheon*.

* scuttle: Opening in a ceiling or roof with a removable or movable cover. Provides access to attic or rooftop. SCUTTLE HOLE.

s-dry lumber: Lumber seasoned to a 19% moisture content before it is surfaced.

seal: 1. Air- and/or waterproof joint between two members. **2.** Liquid in lower portion of a plumbing trap that prevents foul gases from entering the living area. **3.** Vertical distance from top of the dip to bottom of the outlet pipe in a plumbing trap.

sealable equipment: Equipment enclosed in a case that can be locked or sealed so that live parts are not accessible without opening the enclosure.

sealant: Liquid or semiliquid material that forms an air- or waterproof joint or coating.

sealed bearing: Bearing in which lubricant is retained to reduce need for repeated lubrication.

sealer: See *sealant*.

sealing ring: Soft circular gasket used to create a watertight seal between a plumbing fixture and pipe.

seal weld: Weld designed to provide a tight seam and prevent leakage.

* seam: 1. Point at which two or more pieces of material are joined; e.g., the edges of two pieces of sheet metal. **2.** Layer of rock, coal, or ore in the earth.

seamless: Continuous; without breaks, joints, or interruptions.

seam weld: Continuous welding process in which heat is produced on the faying surfaces. Continuous weld may consist of a single weld bead or series of overlapping resistance welds.

seasonal performance factor (SPF): Measure of the efficiency of heating equipment during a period of one heating season. Expressed as a ratio between the heating energy output and energy input required.

season crack: Split in metal caused by rolling or other internal stresses.

SHEET METAL SEAMS

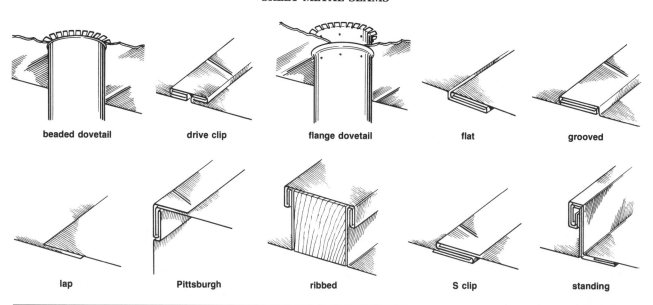

beaded dovetail   drive clip   flange dovetail   flat   grooved

lap   Pittsburgh   ribbed   S clip   standing

seasoning: Process of naturally or mechanically drying wood.

seat: Fixed section of a valve that accepts the moving stem forming a tight seal.

seat angle: Short metal angle fastened to a column to temporarily support a beam during construction.

* **seat cut:** Angular cut in a rafter that allows lower edge to fit tightly to the top plate. FOOT CUT, PLATE CUT.

seawall: Barrier along a shoreline that prevents encroachment of the sea onto the land.

secant: **1.** Circle that intersects true parallel of latitude at the first and fifth mile stations. **2.** Trigonometric function of an acute angle of a triangle equal to ratio of the hypotenuse of which the angle is a part and the leg adjacent to the angle. **3.** Straight line drawn from center of a circle through one end of an arc to a tangent drawn from other end of the arc.

second: **1.** Angular measure equal to $1/_{60}$ of a minute. Denoted as ". **2.** Material or component of less than top quality. **3.** Brick with uneven surface coloring.

secondary air: Air piped into a combustion chamber to promote combustion and increase its efficiency.

secondary beam: Horizontal structural member supported by main beams and that transmits other loads to them.

secondary branch: Plumbing pipe that is not the primary branch.

secondary color: Shade of color produced by mixing two or more primary colors; e.g., orange is produced by mixing red and yellow.

secondary distribution feeder: Feeder that operates at a secondary voltage to supply electrical power to a distribution circuit.

secondary reinforcement: Steel concrete reinforcement other than the primary reinforcement.

secondary winding: Electrical winding connected to the energy output side of a load.

second-cut file: File with medium tooth spacing; coarser than a smooth file and finer than a bastard file. See *file**.

second foot: Measure of liquid flow equal to 1 cubic foot per second.

secret dovetail: See *blind miter*.

secret nailing: See *blind nailing*.

sectilia: Paved surface formed of hexagonal-shaped units.

section: **1.** Area of land with each side measuring 1 mile long. Equal to 1 square mile, or 640 acres; there are 36 sections in a township. **2.** See *section drawing*. **3.** Most desirable piece of wood veneer. Commonly cut to widths of 27″ and 54″.

sectional overhead door: Overhead door composed of several horizontal panels that are joined into one unit; travels upward to open. See *overhead door*.

section drawing: Drawing representation showing a portion of internal features of an object or structure. An imaginary cutting plane is passed through the object or structure to indicate the position of the section. See *print**. SECTION VIEW.

section lines: Lines used to indicate type of material shown in a section drawing. Generally drawn at an angle across the surface.

sector: Horizontal circle with a large radius used in conjunction with a surveying instrument to measure angles.

sectroid: Twisted surface between groins of a vault.

security screw: Threaded fastener with a head design that inhibits removal with a flat-tipped or Phillips screwdriver. See *screw**.

security window: Retractable or removable steel window used for industrial or commercial installations.

sediment: Solid material that settles to the bottom of a liquid.

seepage: Slow movement of water through soil without a definite channel.

seepage bed: See *septic field*.

seepage pit: Covered pit with an open-jointed lining that allows septic tank effluent to leech into surrounding soil.

segment: **1.** Portion or part of an object or area. **2.** Portion of a circle bounded by a chord and an arc.

* **segmented arch:** Curved arch in which the intrados is less than a semicircle. SEGMENTAL ARCH.

segregation: Separation of various ingredients within a mix; e.g., concrete mix.

seismic design: Design that allows a structure to withstand damage from earthquakes.

seismometer: Device used to measure intensity and direction of vibrations such as those produced by an earthquake.

* **seizing:** **1.** Wrapping the end of a rope with string or wire to prevent unraveling. **2.** Damage to a metallic surface as a result of friction.

selective block sequence: Sequence of welding successive portions of a workpiece to create a predetermined stress pattern.

selective coating: Surface covering with high solar radiation absorption qualities.

select lumber: High grade of finish or structural lumber.

selector switch: Manual electrical control with several positions for activating various circuits or devices.

self-centering punch: Hand tool used to mark the center of a small hole, such as for hinge screw installation.

self-desiccation: Removal of free water by chemical action; e.g., hydration in concrete.

self-drilling anchor: Solid wall anchor with a tip designed to bore into a hard surface. After anchor is drilled into position, a threaded fastener is inserted to secure a member in place. See *anchor**.

self-faced stone: Stone that splits along natural lines and does not require additional dressing of the exposed surface.

self-furring metal lath: Expanded metal lath with indentations across the face to position it a short distance from a surface to provide better plaster interlock.

self-scouring flow: Discharge of liquid into a waste system to fill the waste pipe and provide a self-cleaning effect.

self siphonage: Loss of a plumbing trap seal resulting from loss of the liquid in the trap.

self-supporting wall: Vertical partition that supports only its own weight.

self-tapping screw: Screw with a tip that cuts into metal and forms its own hole. Commonly used for light-gauge metal construction. See *screw**. TEC SCREW, TEK® SCREW.

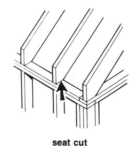

seat cut

segmented arch

seizing (1)

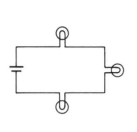

series circuit

service drop

service sink

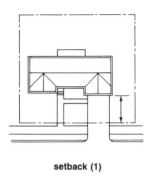

setback (1)

selvage strap: Strap formed by a series of yarns placed parallel to each other. Stronger than a rope strap.

semiautomatic arc welding: Arc welding process in which filler metal is automatically advanced. Advance of the welding is performed manually.

semichord: One-half the length of the chord of an arc.

semicircle: One-half of a circle.

semiconductor: Electrical conductor with electrical resistivity in the range between that of a conductor and an insulator. As temperature of the semiconductor increases, current-carry ability also increases.

semi-direct lighting: Light source that distributes 60% to 90% of the emitted light downward and the remainder upward.

semigloss finish: Surface finish with some luster but not a high shine.

semi-indirect lighting: Illumination from sources that shine 60° to 90° upward and the remainder downward.

semi-lightweight concrete: Concrete with a lightweight aggregate producing a unit weight of 115 pounds to 130 pounds per cubic foot.

semisteel: Cast iron made from steel scrap and molten pig iron.

Sems® screw: Machine screw with a permanently attached lock washer.

sensible heat: Heat that raises the temperature of air without changing the moisture content.

sensor: 1. See *magnetic target.* **2.** Device that detects abnormal conditions, such as smoke or heat, and triggers an alarm.

separation: Moving apart of coarse aggregate from concrete, creating an uneven distribution of ingredients in the mix.

septic field: Area of ground under which perforated pipes are installed for the slow release of waterborne sewage.

septic system: Waste disposal system in which waste is conveyed into a tank for chemical treatment of harmful bacteria

and released into the ground through a series of perforated pipes.

septic tank: Holding tank used to chemically treat waste with bacterial action.

sequence match: Veneer pattern in which panels of veneer are installed adjacent to each other and aligned horizontally.

sequence-stressing loss: Elastic decrease in a stressed tendon resulting from shortening of the member when additional tendons are stressed.

serging: Process of finishing the edges of carpet.

* **series circuit:** Electrical circuit in which the electric current flows through all devices in the circuit.

serrated: Jagged zigzag pattern.

serve: To wrap a rope with fiber cord.

service: Conductors and equipment used to deliver electrical energy from the secondary distribution or street main to the wiring system of the location served.

service cable: Electrical service conductors made into a cable.

service clamp: Saddle-type connection used to secure a water service connection to the water main.

service conductors: Electrical conductors extending from the street main or transformer to the service equipment at the location served.

service door: Door primarily used for delivery or personnel use. May have larger dimensions than a standard door.

* **service drop:** Overhead electrical conductors between the electrical supply and structure served. SERVICE WIRES.

service ell: Malleable iron pipe fitting with 45° or 90° bend. One end is internally threaded and the other end is externally threaded. SERVICE ELBOW.

service entrance: Location for the entry of utilities into a structure.

service-entrance conductors: Electrical conductors between the service equipment and point of connection with the service drop or lateral.

service equipment: Complete electrical

assembly used for main control and as a means for cutoff of the supply. Includes circuit breaker or switch and fuses and accessories. Located near the entrance of service-entrance conductors into a structure.

service lateral: Underground electrical conductors between the street main and point of connection with service-entrance conductors.

service period: Amount of time each day that daylight provides a specified level of illumination.

service pipe: 1. Connecting pipe between a main utility service, such as natural gas or water, and the structure. **2.** Conduit or pipe between the main electrical service and the structure.

service raceway: Conduit or case that encloses electrical service-entrance conductors.

* **service sink:** Deep basin that will accommodate a bucket. Commonly used in commercial maintenance applications. SLOP SINK.

service tee: T-shaped pipe fitting with two openings that are internally threaded and one end of the straight run that is externally threaded.

service valve: 1. Valve used to isolate one apparatus from a main system. **2.** Instrument used to verify pressure and charging of refrigeration units.

set: 1. To bend alternating saw teeth in opposite directions to allow clearance for body of the blade through the work. **2.** Hardening and hydration of cementitious material such as plaster or concrete. **3.** To drive a nail head below the surface. **4.** Depth that a pile penetrates the ground with one blow of a pile hammer. **5.** See *saw set.* **6.** See *nail set.* **7.** Wide chisel for cutting masonry. **8.** Permanent distortion of a member after a load is removed.

* **setback: 1.** Required distance between a structure on a piece of property and the front property line. Established by the local building commissioner. **2.** Recess of a portion of a structure from the

face. **3.** Reduction in heating and cooling level in a structure during the time the structure is vacant.

set of a trowel: Angular relationship of a trowel handle to the blade.

setscrew: Screw used to secure two individual components in a position relative to one another.

setscrew connector: Conduit connector that uses setscrews to secure two pieces of conduit together. See *conduit fitting**.

setting bed: Layer of mortar to which terrazzo topping is applied.

setting block: Small shim at bottom of a piece of glass to support it within a frame.

setting hammer: Hollow-faced hammer used for shaping sheet metal when riveting.

setting shrinkage: Reduction in volume of concrete during hydration.

setting space: Distance from finished surface of a masonry wall to the face of an adjacent wythe.

setting time: Amount of time required for proper hydration and hardening of a cementitious mix such as concrete, plaster, or mortar.

settlement: 1. Sinking or settling of a portion of a structure. **2.** Sinking or settling of solid particles in fresh concrete, mortar, or grout.

settling basin: Basin in a water conduit that allows for settlement of solid particles. Liquid is drained from top of the basin.

set up: 1. See *set* (2). **2.** To drive lead into a pipe joint with a blunt chisel.

sewage: Liquid-borne waste containing animal or vegetable matter in solution or suspension.

sewage ejector: Mechanical device used to raise sewage to a higher elevation.

sewer: Conduit or pipe used to convey sewage or other liquid-borne waste to a treatment or disposal point. See *sanitary sewer, storm sewer.*

sewerage: Entire system used to collect, treat, and dispose of sewage.

sewer brick: Masonry unit with high abrasive resistance and low absorption qualities.

sewer gas: Noxious gases, vapors, and odors found in a sewer. May be poisonous and/or combustible.

sewer laser: See *pipe laser.*

sewer pipe: 1. Pipe used to convey sewage to the disposal point. **2.** See *sewer.*

sewer rod: Flexible metal rod used to dislodge obstructions in a sewer. SNAKE.

sewer tile: Impervious clay pipe used to convey sewage. May be joined with bell-and-spigot joints.

sgraffito: Process of scratching the surface of exposed layers of colored plaster or mortar to produce a decorative effect. SCRAFFETO.

s-green lumber: Lumber that is surfaced with a moisture content greater than 19%.

shackle: U-shaped metal device designed to accept a pin through one end. Used to secure two or more lifting members together. CLEVIS.

shade line: Dividing plane for areas of a structure that are accessible to direct sunlight and those that are shaded by overhanging roofs or other structures.

shadowal block: Concrete masonry unit with decorative recesses in the face.

shadowing: Collection of dust on walls and ceilings in various locations resulting from temperature differential on the surface.

shaft: 1. Vertical portion of a column extending from the base to the capital. **2.** Cylindrical rod used to connect and drive moving parts of mechanical equipment. **3.** Vertical opening for elevators, stairways, or ventilation.

shaft wall: Fire-resistant enclosure around an elevator shaft or stairwell.

shag: Deep-pile carpet with long yarns.

***shake: 1.** Hand-split wood roofing material. See *cedar shake.* **2.** Wood defect characterized by splits between the annual rings. See *defect**.

shampoo bowl: Sink used in beauty salons and barber shops that has a headrest for a person in a reclining position.

shank: 1. Portion of a hand tool between the working end and the handle. **2.** Portion of a drill bit that fits into the drill. **3.** Flat surface between recessed grooves.

***shaper:** Power equipment with a vertical rotating spindle used to cut irregular outlines and moldings. Interchangeable cutters mounted in the vertical shaft allow for various designs and profiles.

sharpening stone: See *oilstone.*

sharp paint: Quick-drying sealing paint.

sharp sand: Coarse sand in which the particles are of angular shape.

shaving: Thin strip of wood or other material sliced from another piece.

shear: 1. Internal force that is tangent to the plane on which it acts. **2.** Scissors-like tool used to sever metal. See *tin snips.* **3.** To cut metal or other material using two opposing blades.

shearhead: Steel member at the top of a column of flat plate or flat slab construction used to transmit loads from the slab to the column.

sheariness: Variation in gloss texture of a surface as a result of variations in film thickness.

shear legs: See *derrick.*

shear pin splice: Heavy timber joint in which overlapped ends are joined with through-bolts and wood or metal rods which are drilled perpendicular to the bolts. Rods are used to prevent slippage of the two members.

***shear plate: 1.** Round plate inserted in the face of timber to minimize shear between adjoining members and provide greater load-bearing capacity than a bolt. **2.** Flat reinforcement plate added to the web of a steel beam to increase its resistance to shear.

shears: See *shear* (2).

shear strength: Maximum shear stress that a material is capable of sustaining.

shear stress: Measure of shear-producing forces; expressed as pounds per square inch of cross-sectional area.

shear stud: See *stud* (3).

sewer rod

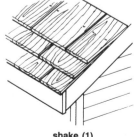

shake (1)

shaper

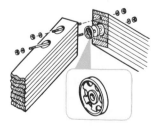

shear plate (1)

sheathed cable

sheathing (1)

sheave

shelf standard

shear wall: Vertical partition that carries shear resulting from earthquakes, wind, or other forces.

sheath: 1. Enclosure in which post-tensioning tendons are placed to prevent adhesion to concrete. **2.** Protective outer covering for insulated cable.

* **sheathed cable:** Electrical conductors that are protected with a nonconductive outer covering.

sheathing: 1. Material used to cover walls and roofs to provide insulation and shear strength to the structure. Includes fiberboard, gypsum board, plywood, polystyrene, and lumber. **2.** See *lagging* (3).

sheathing paper: See *building paper*.

* **sheave:** Grooved wheel used to support or change direction of a line or chain. PULLEY SHEAVE.

sheave block: Pulley with a case.

she bolt: Form tie and spreader device used for concrete forms. Consists of an inner rod that is externally threaded on both ends and joined with two waler rods. Waler rods are secured in position at each end with large hex nuts and washers. Inner rod is available in various lengths for different thicknesses of walls. See *tie* (2) *.

shed: One-story structure commonly used for storage.

shed roof: Roof that slopes in one direction. See *roof*.

sheepsfoot roller: Self-propelled or towed equipment used to obtain deep compaction of fill. Consists of a cylindrical roller with projecting studs that penetrate the ground.

sheepshank: Knot used to temporarily shorten a rope. Made by forming an S-loop in the rope and making a half hitch at each end of the S.

sheet: Thin piece of flat or corrugated material; e.g., sheet of plywood.

sheet bend: Knot for joining ropes of different diameters. Made by forming an overhand loop with one end of the rope, passing the end of the other rope through the loop and behind the standing part, and drawing the loop tight.

See *knot*.

sheet glass: Clear or opaque material manufactured in continuous, long flat pieces and cut to desired sizes and shapes. WINDOW GLASS.

sheet goods: See *resilient flooring*.

sheeting: See *sheathing* (1).

sheet lath: Metal plaster base formed by perforating sheets of metal.

sheet metal: Light-gauge metal ranging from .006″ to .249″ thick; used for various building purposes including flashing, ductwork, and vents.

sheet metal blunt point screw: Self-tapping sheet metal screw used in 18 to 28 gauge sheet metal. Tip of screw is not pointed. See *screw*.

sheet metal gimlet point screw: Self-tapping sheet metal screw used in 6 to 28 gauge sheet metal, plastic, or slate. See *screw*.

sheet metal screw: Coarse-threaded screw used for fastening sheet metal and other thin material. Head is commonly hexagonal and slotted to allow driving with a screwdriver or socket.

sheet metal worker: Tradesperson who forms and installs light-gauge metal devices and components such as ductwork.

sheet pile: Interlocking vertical support commonly constructed of sheet piling driven into the ground with a pile driver and forming a continuous structure. See *pile* (1) *. STEEL SHEET PILE.

Sheetrock®: See *drywall*.

sheet siding: Exterior finish material that is available in sheets 4′-0″ wide and 8′-0″, 9′-0″, 10′-0″ long. Commonly made of plywood or tempered hardboard.

shelf: 1. Horizontal surface used for storage. **2.** Horizontal structural surface.

shelf angle: Metal angle fastened to the face of a wall to support masonry veneer or other wall coverings.

shelf bracket: Projecting structural member fastened to a wall or vertical member to support a shelf.

shelf cleat: Strip of wood fastened to a

wall to support a shelf.

shelf life: Maximum amount of time a material can be stored and still be usable. STORAGE LIFE.

* **shelf standard:** Slotted metal member fastened vertically to a wall or other vertical member to support shelf brackets at various heights.

shell: 1. Structural framework and sheathing of a building. May include exterior finish material. **2.** Thin external layer of material applied to framing members.

shellac: Liquid transparent coating derived from matter secreted by insects. Dries to form a tough, elastic film; used on wood, metal, and other materials. Not suitable for exterior use.

shell pile: Concrete foundation support formed by placing concrete into a thin steel casing that has been driven into the ground with the use of a mandrel. See *pile* (1) *.

sherardize: To coat steel with a thin corrosion-resistant layer of zinc.

shielded carbon arc welding: Carbon arc welding process in which metal is joined with heat produced with an electric arc between a carbon electrode and the base metal. Shielding is produced from combustion of a solid material that is fed into the arc. Pressure and filler metal may or may not be used.

shielded metal arc cutting: Process of severing metal by melting it with heat produced between a covered metal electrode and the base metal.

shielded metal arc welding: Arc welding process in which metal is joined with heat produced with an electric arc between a covered metal electrode and the base metal. Shielding is produced from decomposition of the electrode covering. Pressure is not used and filler metal is obtained from the electrode.

shielded nonmetallic-sheathed cable: Electrical cable consisting of two or more insulated electrical conductors manufactured in an extruded moisture- and flame-resistant core. Assembly is

shim

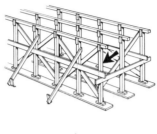

shore

shoulder hook

shoulder nipple

covered with overlapping spiral metal tape and encased in a nonmetallic material such as plastic.

shielding gas: Vapor or gas that protects weld joints from exposure to the atmosphere and possible contamination, which may affect the strength of the weld joint.

* shim: Thin piece of material used to adjust and fill a void between two members or pieces of equipment. Ranges from tapered wood pieces used when installing a door to light-gauge metal strips used for equipment alignment.

shiner: See *rowlock stretcher.*

shingle: Thin piece of wood, asphalt-saturated felt, fiberglass, or other material applied to the surface of a roof or wall to provide waterproof covering. See *cedar shingle.*

shingle mold: Exterior ornamental trim molding used to conceal joints between fascia and roof shingles. See *molding*.*

shingle nail: Nail with a short shank, 3d to 6d long, and a small head. Used with wood shingles.

shingling hatchet: Hand tool used for applying wood shingles. Consists of a small hatchet, hammer, and nail claw.

ship channel: Steel member with a U-shaped cross section. Web is thicker and flanges wider than a standard channel.

shiplap: Wood joint in which edges of adjoining pieces are rabbeted to allow for a fitted overlap between pieces. See *joinery*.*

shiplap siding: Exterior wall covering made of boards with rabbeted edges that allow for a fitted overlap between edges. Applied vertically, horizontally, or at an angle. See *siding*.*

shock arrester: 1. Door hardware applied to the surface to absorb the impact of the door striking a stop or holder. **2.** Device in a piping system used to reduce the shock that occurs when the liquid flow is suddenly obstructed.

shoe: 1. See *base shoe.* SHOE MOLDING.

2. Short angular portion of a downspout that diverts water away from the structure. **3.** Protective device installed on the tip of a pile to prevent damage to the pile tip during driving. **4.** Metal framing component wired to a trussed steel stud and connected to the bottom track. **5.** Ground plate used as a link in the track of excavating equipment.

shoe molding: See *base shoe.*

shoe plate: Flat metal piece at the bearing point of a metal truss.

shoe rail: Trim member installed at the top of a stairway stringer to secure balusters in position.

shoe rasp: Shaping and smoothing hand tool with one flat and one convex side. Each side has a rough and smooth end providing various degrees of coarseness.

shop lumber: Lumber that will be used for further fabrication, such as into moldings or door rails.

* shore: Temporary support for formwork or other structural components. May be adjustable or fixed.

Shore hardness number: Measure of the hardness of material determined by striking the surface with a diamond-pointed hammer and measuring the rebound. Higher values indicate harder materials.

shore head: Wood or metal horizontal member fastened to the top of a vertical shoring member.

shoring: Wood or metal members used to temporarily support formwork or other structural components.

short: See *short circuit.*

short circuit: Abnormal electrical connection, having relatively low resistance, between two points of different potential.

short circuiting transfer: Metal transfer in gas metal arc welding from a consumable electrode that is melted by repeated electrical short circuits.

short cycling: Excessive starting and stopping of an HVAC system.

shortened valley rafter: Roof framing member that extends from the top plate of an inside corner to the supporting valley rafter.

short splice: Method of splicing two pieces of rope in which the ends are intertwined. Produces an enlarged area that will not fit through a properly sized pulley. See *knot*.*

short-time duty: Operation of an electrical device for a relatively short specified length of time.

short ton: Measurement of weight equal to 2000 pounds.

shotcrete: Concrete or mortar applied to a surface with pneumatic pressure at a high velocity. AIR-BLOWN MORTAR.

shot saw: Power tool that uses chilled steel pellets to cut stone.

shoulder: 1. Edge of a roadway. **2.** Portion of a tenon that butts against face of the work surrounding the mortise. **3.** Undesired offset in face of a concrete surface resulting from insufficient bracing and shifting of the formwork. **4.** Enlarged portion of a bolt, rivet, or screw directly below the head.

* shoulder hook: L-shaped hook with one end threaded for fastening in wood. Collar on the shank conceals the point of insertion of the hook.

shouldering: Placing mortar along the lower edge of a slate roof tile to form a watertight seam.

* shoulder nipple: Short pipe threaded on both ends with a short unthreaded section in the middle. SHORT NIPPLE.

shoulder screw: Screw with a hex head and shank that is straight and unthreaded, except for the tip. Tip is reduced in diameter and threaded. Used to fasten wall and roof panels to a structural frame.

shove joint: Masonry joint formed by applying mortar to a brick and forcing it in position against another brick.

shovel: Manual or mechanical digging tool.

shovel dozer: Tractor with a crawler-type track and front-mounted bucket for digging, pushing, or loading.

shower pan

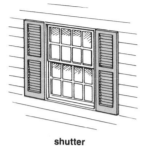

shutter

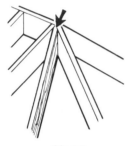

side cut

sight glass

---

shower arm: Bent section of pipe projecting from a wall for attachment of a shower head.

shower bath: Plumbing fixture that discharges water from above a person who is bathing.

shower head: Nozzle that sprays water into a shower bath.

shower mixer: Valve that mixes hot and cold water before being sprayed from a shower head.

* **shower pan:** Noncorrosive shallow tray between the shower finish and the framing under and around a shower. Used to retain leakage and overflow and divert it to the drain.

shower stall: Waterproof enclosure in

which water is sprayed from above the body.

show rafter: Decorative roof member, commonly part of the cornice.

show window: Display window.

shrink: Contraction of a material. SHRINKAGE.

shrinkage loss: Decrease in stress of prestressing steel resulting from contraction of concrete during hydration.

shrink-mixed concrete: Ready-mixed concrete that is initially mixed in a stationary mixer, then in a truck mixer.

shunt: 1. Connection between two wires of a blasting cap. Used to prevent the accumulation of opposite electric potential. 2. Electrical device with a great amount of resistance or impedance when connected in parallel across another device, and diverting some of the current from it.

shunt valve: Valve used to divert a flow of fluid around a component rather than through it.

shute wire: One strand of wire running across the width of wire cloth.

shutoff valve: See *valve*.

* **shutter:** Wood, metal, or plastic panel on either side of a window. May be hinged, but commonly fixed in position. Originally used as a protective covering for windows when swung closed.

shutter bar: Hardware used to secure movable shutters when closed.

shutter butt: Narrow hinge used to fasten movable shutters to a wall.

shutting shoe: Metal or stone hardware that forms a void in the pavement below a gate. Bolt on the gate slides into the void to secure the gate.

Siamese connection: Wye fitting connected to the exterior of a structure providing two inlet connections for fire hoses. Inlet connections are connected to standpipes and fire protection system within the structure.

side arm: Triangular attachment for a scaffold that projects outward from the scaffold frame to create a cantilevered working platform.

side-batter pile: Pile driven at an angle, with tip of the pile initially to the side of the pile driver.

sidecasting: Excavated material piled close to the area from which it was taken.

* **side cut:** Beveled, angular cut on a rafter to fit into another rafter; e.g., the cut at the top of a hip jack rafter to fit into a hip.

side-cutting pliers: Pliers used to cut light-gauge wire. Gripping portion of the jaws allows for crimping and holding work. See *pliers**. SIDE CUTTERS.

side-dump loader: Excavation equip-

ment in which the bucket can be emptied to either side or forward.

sidehill: Slope across the general line of work.

side-hung window: See *casement window*.

side jamb: Vertical member in a door or window frame.

side-knob screw: Setscrew used to secure a doorknob to the spindle.

side lap: Amount that shingles overlap in the same course.

sidelight: 1. Narrow window adjacent to a door. 2. Artificial illumination on an interior wall.

sideline: Line used to hoist light loads on a derrick while the primary lead line supports another load.

side-outlet cross: T-shaped pipe fitting with two outlets at right angles to the run, producing five inlets and outlets.

side-outlet ell: L-shaped pipe fitting with an outlet at right angle to the run, producing three inlets and outlets.

side-outlet tee: T-shaped pipe fitting with an outlet at right angle to the intersection of the T, producing four inlets and outlets.

side set: Difference in thickness between two adjoining edges of metal.

side vent: Vent pipe connected to the drain pipe at an angle of less than 45°.

side view: Elevation drawing showing one side of a component or structure.

sidewalk: Paved surface designed for pedestrian traffic.

sidewall: Continuous vertical support at each end of a culvert.

* **siding:** Exterior wall finish material. Includes wood products, such as shingles, boards, and plywood, interlocking vinyl, aluminum, or steel sheets, and asphalt-saturated shingles.

sieve: Metal plate or sheet, or woven wire cloth with regularly spaced square openings. Mounted in a frame and used to separate and grade loose particles.

sieve analysis: Measure of particle size distribution. Expressed as the weight percentage retained on a series of standard sieves of decreasing size and the weight percentage passing through the finest size sieve.

sieve number: Number used to indicate size of a sieve. Expressed as the approximate number of openings per lineal inch.

* **sight glass:** Transparent section in pipe or tubing that allows for visual inspection of the fluid level.

sight rail: Horizontal board supported by stakes; used to verify the grade of a pipe trench. Grade is measured from the board to bottom of the trench.

sight size: Open, clear area of a window.

signal: See *hand signal*.

signaling circuit: Electrical circuit that activates signaling equipment.

signal light: Electrical device that illuminates when a circuit is energized.

silica: Light-colored or colorless inert material that is very hard and insoluble in water. Used as an ingredient in cement. SILICON DIOXIDE.

silica brick: Refractory brick composed of approximately 96% silica and the remainder alumina and lime.

silica gel: Material that readily absorbs moisture; used as a drying agent.

silicate: Insoluble metal salt.

silicon: Metallic element used in electrical rectifiers.

silicon carbide: Synthetic granules used as a non-skid topping on concrete, an abrasive material in concrete and masonry saws and drills, and as a coating for abrasive paper.

silicone: Water-resistant synthetic resin used in caulking. Remains pliable after setting.

silking: Fine lines in paint film that are parallel to the direction of flow of the paint.

sill: **1.** Lowest horizontal member in a wall opening; e.g., window sill. **2.** Lowest framing member attached to the foundation that provides support to other framing members. SILL PLATE. **3.** Horizontal overflow line of a spillway.

sill anchor: See *anchor bolt*.

sill clamp: Device used to secure a string

silo trowel   single-acting door   single-bevel groove weld

a specified distance from a masonry wall to assist in accurately setting sill brick.

sill cock: Exterior water faucet. HOSE BIBB.

sill connector: Sheet metal framing connector with a lower section that is embedded in a concrete foundation and an upper section that wraps around and is fastened to a wood sill. See *anchor*★.

sill plate: See *sill*.

silo block: Concrete masonry unit with a curved shape.

★ **silo trowel:** Hand trowel used to smooth concrete. Rounded end on the blade prevents digging into curved concrete surfaces.

silt: Granular material produced from the disintegration of rock. Grains generally pass through a No. 200 sieve.

siltation test: Method used to determine the amount of impurities in sand. Sand is immersed in water and the loam that

floats to the top is measured.

silt trap: Basin that allows waterborne soil to settle to the bottom, preventing obstructions in the piping.

silver-lock bond: Brickwork pattern similar to English bond except each stretcher is laid on edge.

silver solder: Solder containing silver and having a high melting point. Used for high-strength solder joints.

simple beam: Horizontal structural member supported only at each end.

simple cornice: Cornice consisting only of a frieze and a molding.

single-acting door: Door that is hinged along one edge and swings in only one direction.

★ **single-acting hammer:** Pile hammer using compressed air or steam to raise the hammer which is exhausted during the fall.

★ **single-bevel groove weld:** Groove weld in which one of the adjoining edges is beveled and joined to a flat edge.

SIDING

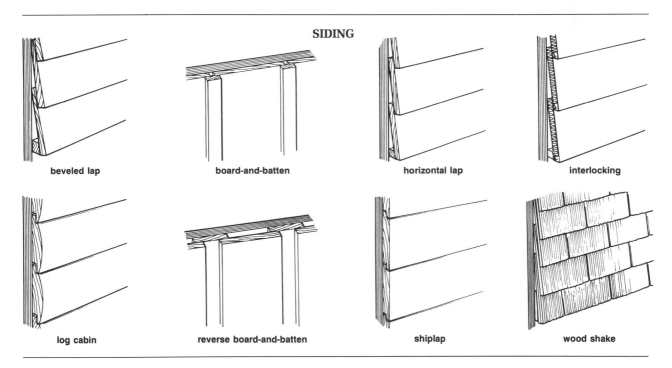

beveled lap   board-and-batten   horizontal lap   interlocking

log cabin   reverse board-and-batten   shiplap   wood shake

single-cleat ladder: Ladder consisting of parallel side rails connected with cleats at regular intervals.

single-cut file: File with serrations cut in one direction only. See *file*.

single Flemish bond: Brickwork pattern consisting of a Flemish bond for the face and an English bond for the body.

single-hub pipe: Pipe with a bell at one end and spigot at the other.

single-hung window: Window in which the lower sash slides vertically and the upper sash is fixed.

single-lever faucet: Fresh water mixing valve in which mixing and flow control is accomplished with a single handle.

single-pitch roof: See *shed roof*.

single-pole switch: Electrical switch with one movable and one fixed contact.

* **single-rabbet frame:** Door jamb that is recessed on one side to receive a door.

single-roller catch: See *bullet catch*.

single sling: Rope with an eye at one end and hook on the other; used for rigging.

single welded joint: Gas or arc weld joint that is welded from only one side.

sink: 1. Plumbing fixture used as a wash basin with a fresh water supply and drainage connection. See *lavatory*. **2.** Ditch or drain pipe used to convey sewage.

sinkage: Depression; lowered area. SINK-ING.

sink bibb: Faucet used to control flow of water into a sink.

sinker: Flat-headed nail with a slight depression in the top. Driven below the surface of the wood. SINKER NAIL.

sinking: 1. Painting defect in which the surface color is absorbed by a porous undercoat, creating an uneven finish. **2.** Removal of material to permit installation of hinges or other hardware. MORTISE.

sinter: To form a material from fusible powder by heating the powder and applying pressure.

siphon: 1. To convey liquid through a pipe or tubing by means of a suction.

2. Pipe or tubing used to convey liquid by means of suction.

siphon trap: Plumbing trap shaped like an S when viewed from the side. Water within the trap is used to create an obstruction and prevent passage of sewer gas.

S-iron: Retaining plate on the ends of a tie rod used to tie masonry walls together.

sisal: Fibrous material used in the manufacture of rope and to reinforce plaster.

site: Building lot or location.

site plan: See *plot plan*.

sitzbath: Bathtub designed for use by a person while sitting. Commonly used in hospitals or in therapeutic treatment.

sizing: 1. Liquid coating applied to an untreated porous surface to fill the pores. Used to prepare the surface for additional adhesive or finishing. **2.** Process of cutting and shaping a material to the desired dimensions. **3.** Estimating the demand of an HVAC system.

skeleton: Framework and supporting members of a structure.

skeleton construction: Construction in which all loads are carried by a framework of metal framing or reinforced concrete. Walls do not have load-bearing capability.

* **skeleton steps:** Stairway composed only of treads and a carriage. Risers are not installed.

skew: 1. Twisted; out of alignment. **2.** Wood-turning tool with an angular, sharpened blade.

skew arch: Curved arch with jambs that are not at right angles to its face.

skewback: Sloping surface that supports the end of an arch.

skew chisel: Woodworking chisel with a sharp cutting edge cut at an oblique angle and beveled sides.

skew corbel: Stone at lowest point of a gable for supporting the coping or wall cornice.

skid: 1. Heavy timber or spacer used under heavy equipment that is moved on rollers. **2.** Blocking placed under

material to hold it slightly off the floor or ground.

skid steel loader: Lifting equipment similar to a forklift; used to set roof trusses on a one-story structure.

skim coat: 1. Finish coat of plaster composed of fine, thin plaster. **2.** Thin layer of liquid or paste material.

skin: 1. Material applied to the exterior of a structural frame to form a curtain wall. **2.** Thin exterior layer, veneer, or coating.

skinning knife: Curved-blade cutting tool used to strip insulation from electrical conductors.

skintled joint: Uneven, irregular masonry pattern in which masonry units are not flush with the face of the wall. Mortar may be squeezed from the joints to enhance the pattern. See *mortar joint*.

skip: 1. Machined lumber defect in which the planer or surfacer does not contact a portion of the surface, leaving a rough area. **2.** Unpainted area on a painted surface.

skirtboard: 1. Finish member installed at the meeting of a staircase and wall. **2.** See *baseboard*. **3.** Vertical member installed along the sides of a conveyor to retain material on the conveyor. SKIRT.

skirting block: See *plinth block*.

skiving: Process of digging in thin layers.

* **Skotch® fastener:** Hardware used to reinforce a wood joint. Consists of a flat metal plate with curved prongs at each end. Prongs are driven into the wood to secure the joint.

skull: Unmelted residue from a liquated welding filler metal.

skylight: Opening in a roof or ceiling that allows entry of natural light. See *window*.

slab: 1. Horizontal or nearly horizontal layer of plain or reinforced concrete. **2.** Large, flat piece of material. **3.** Outermost lengthwise piece cut from a log.

slabbing: Cutting of timber into a square shape.

slab bolster: See *chair*.

slab chair: See *chair*.

slabjacking: Process of raising a concrete slab or filling voids beneath it. Accomplished by pressure-injecting a cementitious, noncementitious, or asphaltic material under the slab.

slab-on-grade foundation: Foundation in which the structure is supported by a concrete slab that is placed directly on the ground. Edges of the slab may be thickened to provide additional foundation support and freeze-thaw resistance. Does not provide basement

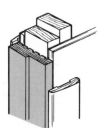

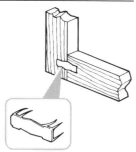

| single-rabbet frame | skeleton steps | Skotch® fastener |

or crawl space beneath the structure.

slag: By-product of smelted iron, copper, or lead ore. Used as a lightweight concrete aggregate.

slag cement: Hydraulic cement consisting of blended blast furnace slag and hydrated lime.

slag hammer: Small hand tool used to chip slag from weld joints.

slag inclusion: Solid nonmetallic material entrapped in weld metal or between base metal and weld metal.

slag strip: See *gravel strip*.

slake: To add water.

slaked lime: See *hydrated lime*.

slaking: Mixing quicklime and water to form a putty.

* **slamming stile:** Upright part of a jamb against which the door swings. SLAMMING STRIP.

slant: Sewer pipe connecting a common sewer to a house sewer.

slat bucket: Digging bucket used to move cohesive, wet soil.

slate: Metamorphic rock easily split into thin sheets and used as roofing tile and flooring material.

slate roll: Cylindrical-shaped slate with a V-notch cut into bottom to fit the ridge on a slate roof.

slave: Device that is driven or activated by another piece of equipment; e.g., valve that is activated by another valve.

sled: Frame that supports and guides a template as it is pushed through fresh plaster to form a plaster molding.

sledgehammer: Heavy striking tool with two striking faces used to drive stakes into the ground and other heavy work. See *hammer**. SLEDGE.

sled runner: Hand tool with a V-shaped or half-round blade attached to a handle. Used to finish masonry joints.

sleeper: 1. Horizontal wood strip placed on a concrete slab to provide a means of attaching other material. 2. Timber on or near the ground that supports loads from posts or framing.

sleeve: 1. Pipe or tubing that fits over another member. 2. Pipe or tubing that forms a void in fresh concrete to allow for future installation of conductors and piping.

sleeve bearing: Anti-friction device made of a round piece of bronze or babbit that is drilled to the diameter of a shaft. Supports the shaft while rotating.

* **sleeve nut:** Elongated hardware made with left-handed female threads on one end and right-handed female threads on the other end. Used between threaded rods to adjust the overall length of an assembly.

sleever bar: Extended connecting bar used in ironwork to provide additional leverage for the connector.

sleeve-type stud-bolt anchor: Fastener used in solid material in which a metal sleeve, one end fitted with a threaded cap, is inserted in a predrilled hole. Sleeve expands as bolt or screw is tightened.

slender beam: Beam that buckles, rather than bends, when loaded without lateral bracing of the compression flange.

slenderness ratio: Comparison of the effective height of a member, such as a column, to the effective thickness.

slewing: Rotating a crane jib to move a load horizontally in an arc.

slicker: 1. Slightly flexible, flat metal or wood device used to smooth plaster. SHINGLE. 2. Thin, double-bladed trowel used to finish masonry joints.

slick hole: Boring that is loaded with explosives for blasting.

slick line: End of pipe from which the concrete is discharged from a concrete pump.

slick sheets: Steel plates placed on the floor of a tunnel prior to blasting to remove debris.

slide pile: Driven pile used to stabilize the soil and prevent erosion.

slidescape: Chute used as a means of quick human egress to street level by sliding down the chute.

sliding door: Horizontal-moving door suspended on rollers that travel in a track that is fixed at the top of the opening or rollers mounted in the bottom of the door. Commonly used for closet and patio doors. See *door**. BYPASS DOOR.

sliding fire door: Fire-rated door that is suspended from a track mounted on a wall. Door automatically closes in the event of a fire by sloping the track or a counterweight.

sliding T-bevel: Hand tool used to lay out angles. Adjustable blade is set to a specified angle and locked into position for repeated layouts.

sliding window: Horizontal-moving window supported by rolling hardware that facilitates movement of the window. See *window**.

slim taper file: Hand tool with a tapered triangular cross section; used for sharpening hand saws.

* **sling:** Cable or rope with an eye or other attachment at each end to hoist material.

sling block: Two sheaves mounted in a block to receive rope or line entering from different directions.

slip: 1. Narrow strip of wood or other material. 2. Thin layer of plaster. 3. Movement occurring between steel reinforcement and concrete in post-tensioned concrete construction.

slipform: Concrete formwork that is moved horizontally or vertically as concrete is placed. Concrete initially sets as the formwork passes over it. Allows for a relatively short form to shape a long area of concrete.

slip-in hinges: Hinges used with metal doors and jambs so that the leaf or leaves slide into a slot in the door and/or frame.

slip joint: 1. Method of joining new and existing masonry. Channel is cut into existing work and new work is interlocked with this channel. 2. Piping connection in which one pipe slides into the other. Joint is secured with gasket or packing.

slip-joint pliers: Pliers with opposing jaws connected by a pin that allows for adjustment to different jaw positions. See *pliers**.

slip knot: Overhand knot that allows the knot to slide on the standing end of the rope or line.

slip match: Veneer pattern in which sheets are joined side-by-side and staggered to give the appearance of a repeated pattern.

slip newel: 1. Newel that is hollowed out and has an open bottom to fit over a short vertical post. 2. Three-sided newel that is installed around the open end of a partition.

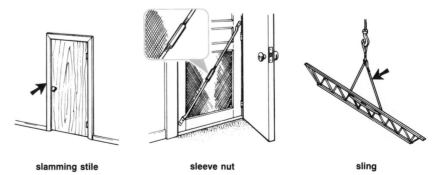

slamming stile   sleeve nut   sling

| slip sill | slump cone | slump test | snap tie |

* **slip sill:** Sill that is equal in length to the distance between the jambs. Allows the sill to be installed after the walls are constructed.

slip stone: Small, wedge-shaped oilstone with rounded edges for sharpening gouges.

slogging chisel: Heavy chisel used to cut off bolt heads.

slope: Incline; at an angle from horizontal. See *pitch* (1).

slop sink: Low, deep basin used to clean mops and empty buckets by custodial or service employees. SERVICE SINK.

slot cutter: Router bit used to cut narrow slots by plunging bit into surface of the workpiece. See *bit* (2)*.

slot mortise: Mortise that does not extend the entire thickness of the stock into which it is cut. SLIP MORTISE.

slot outlet: Air supply grill with the length at least 10 times greater than the width.

slotted-head screw: Externally threaded fastener with a straight recess in the head. See *screw*.

slot weld: Weld made in an elongated hole in one member of a T- or lap joint. One member is joined to the surface of another member that is exposed through the slot.

sloughing: Subsidence of freshly placed shotcrete or plaster as a result of excessive water in the mixture. SAGGING.

sloyd knife: Knife with a single fixed blade used for wood carving or trimming wood members.

sludge: Residue that accumulates at the bottom of a liquid.

sludger: Tool used to remove debris from drilled holes in rock before insertion of explosive charges.

slugging: 1. Process of adding a separate piece of metal to a weld joint before or during welding, resulting in the joint not conforming to the original design. 2. Uneven flow of shotcrete as a result of improper use of delivery equipment. 3. Undesirable reaction of a compressor when liquid refrigerant or oil enters the cylinder.

sluice: 1. Narrow waterway with steep banks. 2. To use rapidly flowing water to excavate or move materials.

slump: Measure of the consistency of freshly mixed concrete. Expressed as inches of fall, to the nearest 1/4", of the fresh mix. See *slump test*.

slump block: Concrete masonry unit formed from a wet mix; designed to sag after removal of the mold.

* **slump cone:** Truncated metal cone with 4" diameter at the top, 8" diameter at the bottom, and overall height of 12". Used to measure consistency of freshly mixed concrete. A 6" high cone is used to measure consistency of freshly mixed mortar or stucco.

* **slump test:** Method used to measure consistency of freshly mixed concrete. Slump cone is filled with fresh concrete from three different intervals and rodded. Cone is removed and inverted, allowing the concrete to sag. Measurement is taken from top of the inverted cone to top of the concrete pile to determine slump.

slurry: Mixture of water and fine insoluble material, such as portland cement or clay. Used as a soil stabilizer or foundation reinforcement.

smalto: Small pieces of colored material used in mosaic work.

smoke chamber: Section of a chimney directly over a fireplace and below the flue. Prevents down drafts from blowing smoke into the living area. SMOKE SHELF.

smoke compartment: Area within a structure enclosed with fire division walls on all sides.

smoke detector: Device that emits a warning signal when amount of smoke in an area reaches a specified density.

smoke-developed rating: Numerical classification indicating the smoke produced by a material when burned.

smoke door: Roof scuttle that opens automatically to confine smoke to a certain area.

smoke shelf: See *smoke chamber*.

smokestack: See *chimney*.

smoke test: Method of testing pipe joints by sealing the system and filling the piping with smoke and identifying leaks by sight and/or smell.

smooth coat: Final coat of plaster.

smoother bar: Metal rod dragged behind excavating equipment to break lumps of soil into smaller pieces.

smooth file: Shaping and smoothing hand tool with fine cutting ridges. Used to produce a smooth finish. See *file*.

smoothing iron: Heated rod used to remove surface irregularities and seal asphalt pavement.

smoothing plane: Wood-smoothing tool approximately 8" long; used to smooth a rough surface after being evened with a larger plane. See *plane* (1)*.

smudge: 1. Mixture of glue and lampblack applied to a lead surface to prevent adhesion of solder. 2. Unwanted surface mark.

snake: 1. Coil spring that is inserted into an obstructed drain pipe and rotated to remove the obstruction. 2. See *fish tape*.

snake hole: Hole drilled under a rock for insertion of an explosive charge.

snap-blade knife: Knife with a thin metal blade that is broken off at the tip to provide a fresh, sharp cutting edge.

snap cutter: Hand tool used to cut cast iron pipe and fittings.

snap header: Half brick set in a wall to resemble a full brick tying two walls together.

snap hook: Quick-connecting hardware used with safety lanyards and other safety lines. Hook quickly snaps shut by spring action to secure it in place after a line or cable has been inserted.

snap switch: See *general-use snap switch*.

* **snap tie:** Concrete formwork tie used to maintain specified distance between form walls. Ends are snapped or broken off after forms are removed.

snatch block: Pulley or block with one side that can be opened to receive the bight of the rope or line. GATE BLOCK.

sneck: Small stone as fill between large stones in rubblework.

snifter valve: Valve with a male thread

on one end and compressed air attachment on the other end. Used to inject air into a piping system for testing.

snipe: Small concave cut at the end of a wood member resulting from improper jointing.

snips: See *tin snips.*

snow guard: See *roof guard.*

snow load: Live load imposed on a roof by the weight of standing snow.

snub: To wrap a line or rope around an object to secure or slowly lower an object. Friction from the wrapping creates small amount of strain on the hand line compared to strain on the load line. SNUB.

snub gable: See *Dutch hip roof.*

soaker: Metal flashing used at a hip or valley of a slate roof, or at the intersection of a masonry wall and a roof.

soap: Masonry unit with a 2″ thick nominal face dimension.

soapstone: 1. Marking medium for steel and masonry products. **2.** Soft rock containing high proportion of talc; used for laboratory sinks and carved ornaments.

socket: 1. See *bell* (2). **2.** See *lamp holder.* **3.** Mortised portion of a dovetail joint. **4.** See *socket wrench.* **5.** See *receptacle.*

socketing: Joining wood timbers by wedging one timber into a mortise in the other.

socket plug: Externally threaded pipe fitting with a hexagonal recess in the head to accept a hex wrench.

socket screw: Fastener with a threaded shank and hexagonal recess in the head to accept a hex wrench.

socket wrench: Short cylindrical tool with a recessed socket on one end designed to fit a bolt head or nut and the other end designed to fit onto a turning handle such as a ratchet. See *wrench*.*

socle: Projecting member at the base of a column, pedestal, pier, or wall.

sod: Uppermost layer of soil containing grass roots and covered with grass.

sodium silicate adhesive: Adhesive with excellent bonding properties with paper and glass but poor weather resistance.

* **soffit: 1.** Exposed underside of an overhead component of a structure; e.g., underside of projecting rafter tails. **2.** Lowered ceiling section; e.g., over the top of wall cabinets eliminating an open area above the cabinets.

soffit block: Concrete masonry unit with flanges that conceal the underside of a beam or other horizontal support member.

soffit spacer: Metal device used to position furring or finish material a speci-fied distance from the bottom of a structural steel beam.

softener: 1. Pliable material placed over the corners or edges of heavy objects when lifting to prevent damage to the lifting slings. **2.** Flat paintbrush with soft bristles.

soft-face hammer: Hand tool with a cushioned striking surface. Used to strike surfaces that must not be marred or dented. See *hammer*.*

soft-mud brick: Brick manufactured by molding clay with a moisture content between 20% and 30%.

soft solder: Alloy of tin and lead that melts below 700°F; used to join metal.

soft water: Water with small amounts of minerals and/or salts in solution.

softwood: 1. Variety of tree classified as conifer or evergreen. **2.** Wood from coniferous or evergreen trees. Classification is not based on actual softness of the wood.

soil: 1. Unconsolidated solid particles generally above bedrock. Consists of decomposed minerals and organic material. **2.** Sewage.

soil anchor: High tensile strength steel tendon secured in the earth with a concrete or pile base. Used to secure a retaining wall or other soil stabilization structure in position.

soil binder: Soil that filters through a No. 40 sieve with little clearance.

soil boring: Hole drilled into the earth to determine composition of the soil and its bearing capacity.

soil branch: Branch soil pipe of a plumbing waste system.

soil cement: Mixture of cement, soil, and water used to stabilize ground or as foundation reinforcement or pavement base.

soil class: Numerical classification of soil: 1—gravel, 2—sand, 3—clay, 4—loam, 5—loam with sand, 6—silt and loam, 7—clay and loam.

soil mechanics: Application of scientific principles of mechanics and hydraulics to engineering problems regarding soil.

soil pipe: Pipe that conveys discharge of water closets or similar fixtures containing sewage, with or without discharge of other fixtures, to the building drain or sewer.

soil pipe cutter: Hand tool used to cut cast iron soil pipe. Consists of two cutting wheels that rotate around the pipe. Cutting wheels are fixed to the end of a shaft and a chain is wrapped around the pipe and rotated; or a chain with cutting wheels that is wrapped around the pipe and tightened.

soil sampler: Hollow tube driven into the ground to remove an undisturbed soil sample.

soil stabilization: Process of treating soil with chemicals or securing it with mechanical devices to maintain and/or increase stability of the soil.

soil stack: Vertical pipe that extends one or more stories and receives discharge of water closets or similar fixtures containing sewage, with or without discharge of other fixtures.

soil vent: Section of vertical waste pipe that extends above the highest waste fixture to which it is connected.

solar cell: See *solar photovoltaic system.*

* **solar collector:** Device used to capture radiation of the sun and convert it to energy. SOLAR COLLECTOR PANEL.

solar constant: Rate at which radiant energy is received from the sun. Approximately equal to 434.9 Btu per hour per square foot. Used in calculating cooling loads for structures.

solar energy: Energy produced by the sun.

solar gain: Portion of a heat load of a structure obtained through solar energy striking and passing through the windows.

solar index: Measure of residential hot water supplied by a solar water heating system. Expressed as a percentage ranging from 0% to 100%.

solar insolation: Total amount of radiant energy available to a given system.

solar irradiance: Rate at which radiant energy is received by the earth's surface.

soffit (1)

solar collector

* **solarium:** Room that is heated directly by the sun. Generally having a large amount of glass and located for maximum exposure to the sun.

solar orientation: Placement of a structure in relation to the sun. Structure may be located to maximize amount of radiation during cold months or minimize amount of radiation during hot months.

solar photovoltaic system: System and components that convert radiant energy into electrical energy.

solar radiation: Direct transfer of energy from the sun to another body.

solar screen: Structure designed to provide shade.

solar transit: Surveying instrument with an attachment that allows establishment and surveying of the astronomic meridian or astronomic parallel by direct observation.

solder: Filler metal alloy, commonly tin and lead, that melts below 800°F to join metal, such as pipes or electrical contacts. Higher percentage of lead results in a higher melting point. Designated as percentage of tin to lead; e.g., 30/70 denotes that the solder is 30% tin and 70% lead.

soldering gun: Electric tool with a pistol-type grip connected to a pointed tip. Heating of the tip to a temperature capable of melting solder is activated by a trigger on the grip.

soldering iron: Tool used to heat metal when soldering. Consists of a wedge-shaped tip that is heated and an insulated handle. Tip may be heated electrically or with a gas source.

solderless connector: See *pressure connector, wire nut.*

solder nipple: Pipe fitting with external threads on one end and unthreaded on the other end for soldering to another pipe.

soldier: 1. Brick installed vertically with the longest, narrowest side exposed. See *brickwork**. **2.** Vertical waler in concrete formwork or in the sides of an excavation.

soldier beam: See *soldier pile.*

soldier course: Horizontal layer of brick set vertically with the longest, narrowest sides exposed.

* **soldier pile:** Vertical steel member driven into the ground at regular intervals to support horizontal lagging. Used to support the side of an excavation.

* **sole:** Horizontal structural member at bottom of a wood-framed wall for attachment of studs and other framing members. BOTTOM PLATE, SOLE PLATE, SOLE PIECE, SOLE PLATE.

solenoid: Electrical control device that opens or closes when current passes through a magnetic coil.

solid block: See *concrete masonry unit.*

solid bridging: Wood members equal in length to the distance between floor joists. Fastened between floor joists to distribute concentrated loads. See *bridging**. SOLID STRUTTING.

* **solid-core door:** Door constructed of solid wood blocks that are glued together and covered with veneer, or made of one or several solid pieces of wood. Voids do not exist in the core.

solid-core plywood: Plywood without voids in the core.

solid loading: Filling a hole completely with explosives for blasting.

solid lumber joist: Intermediate horizontal structural support made of one continuous wood member, commonly a 2 × 10 or 2 × 12. See *joist**.

solid masonry unit: Brick, concrete block, or other masonry unit with a minimum of 75% solid material in a sectional plane parallel with the bearing surface of the unit. See *concrete masonry unit**.

solid modeling: Geometric modeling in a CADD system in which the member is depicted in pictorial form and all of its surfaces are opaque.

solid panel: Portion of a door that is flush with the stiles.

solid punch: Hand tool with a blunt, circular end. Used to strike and loosen bolts and jammed parts.

solids: Components in lacquers, paints, and varnishes that stay on a surface to form a solid coating after the solvent or carrier liquid has evaporated.

solid-state welding: Welding process in which metal is joined at temperatures below the melting point of the base metal without the addition of filler metal. Pressure may or may not be used.

solidus: Highest temperature at which metal or alloy is solid and does not contain molten material.

solute: Dissolved substance.

solvency: Ability of a solvent to hold other material in solution.

solvent: Liquid used to dissolve other material.

solvent adhesive: Adhesive containing a volatile organic liquid as the vehicle.

solvent welding: Process in which plastic members, such as plastic pipe, are fused together by softening adjoining surfaces with a chemical agent. Softened surfaces are forced together to form a solid joint.

sommer: Horizontal wood member that supports other framing members such as joists and studs. SUMMER.

sonic pile driver: Pile driver that uses high-frequency vibrations, less than 6000 times per minute, to drive piles. SONIC DRIVER.

Sonotube®: See *tubular fiber form.*

soot pocket: Flue extension extending approximately 8" to 10" below the air entrance to a smoke pipe. Prevents soot from accumulating in the smoke pipe.

sound: 1. Free from defects. **2.** Oscillating atmospheric pressure that is detected by the ear. Oscillating frequencies heard by the human ear range from 20 to 20,000 cycles per second.

sound-absorbing unit: Measure of soundproofing equal to 1 square foot of surface area that has perfect qualities in deadening sound.

sound absorption: Dissipation of sound energy by converting it into heat or other forms of energy. SOUND ATTENUATION, SOUND INSULATION, SOUND ISOLATION.

sound attenuator: Noise-inhibiting de-

solarium

soldier pile

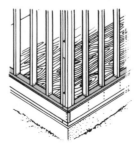

sole

solid-core door

vice installed in ductwork.

sound focus: Small area in a room where sound is concentrated and is louder than in other portions of the same room.

sounding pole: Rod, approximately 15'-0" long, that is graduated with measurements and fitted with a flat metal shoe. Used to measure water depth by probing the bottom of a body of water.

sounding well: Vertical open shaft in a mass of coarse aggregate. Grout is placed in this hole to form preplaced aggregate concrete.

sound knot: Knot in wood that is secure and not likely to loosen.

soundproofing: Material used to limit transmission of sound vibrations from one area to another.

sound-reduction factor: Measure of reduction in intensity of sound as it passes through a wall. Expressed in decibels.

sound transmission class: Numerical value used to indicate the sound insulation quality of a material. Higher value denotes better sound insulation quality.

southern colonial architecture: Architecture characterized by structures with two or more stories and an elaborate raised entrance porch with large columns.

spaced loading: Placing blasting charges with voids between charges.

space frame: Three-dimensional structural framework capable of transmitting loads in three dimensions.

space heater: Mechanical equipment designed to heat a small, confined area.

spacer: Small piece of material or device used to position a larger component a predetermined distance from another.

* **spacer strip:** Metal bar inserted in the root joint of a groove weld to serve as backing material and maintain root opening during the welding operation.

spackling compound: Pliable material used to fill cracks and seams in plaster, drywall, and other paintable wall finishes. SPACKLING, SPACKLE.

spad: Nail with a hook driven into the roof of a tunnel or other excavation for suspending a plumb bob.

spade: Shovel with a flat blade; used for breaking into undisturbed ground.

spade bit: Wood-boring tool with a flat cutting face and central projecting point used to guide the bit. Designed for use in an electric drill. See *bit* (1) *.

spading: Consolidation of freshly placed concrete by repeated insertion and withdrawal of a spade-like tool.

spall: 1. Surface flaking and chipping of concrete caused by a blow, weathering, or expansion of the concrete. SPALLING. **2.** Stone or brick fragment. **3.** To shape a stone or brick to a desired shape by chipping with a spalling hammer. **4.** Hairline cracking of mortar joints.

spalling hammer: Hammer with a chisel edge for rough-dressing stone.

span: 1. Distance between structural supports; e.g., the unsupported distance of a beam between columns. **2.** Distance from the outside of an exterior wall of a structure to the outside of the opposite exterior wall. **3.** Surveying unit equal to 6".

spandrel: 1. Flat area above the curved portion of an arch and below the ceiling. **2.** Area of wall between the head of one window and the sill of the window above.

spandrel beam: Beam in the perimeter of a structure that spans columns and commonly supports a floor or a roof.

spandrel tie: Concrete wire form tie with one end looped to receive a wedge and the other end pointed to be driven into wood decking. Used to secure a vertical perimeter member in position. See *tie* *.

* **Spanish tile:** Clay roofing tile with a semicircular cross-sectional shape and approximately 12" long. See *mission roofing tile*.

spanner: Horizontal brace; e.g., collar tie.

spanner wrench: Hand tool with projecting lugs used to turn nuts and bolts.

spanpiece: See *collar beam*.

span roof: Equal-pitched gable roof.

span web saw: Saw with a flat blade mounted in an H-shaped wood frame. Tension rod extending between the opposite sides of the frame is tightened to maintain blade tension.

spar: 1. See *common rafter*. **2.** Large round timber. **3.** Metal rod used to hold a gate closed.

sparge pipe: Horizontal perforated pipe that releases water to flush a urinal.

spark arrester: See *arrester*.

sparpiece: See *collar beam*.

spar varnish: Varnish composed of durable resins and oils. Has excellent weather resistance and is used on exterior surfaces.

spat: Protective metal member installed at bottom of a door frame to minimize damage.

spatter: Metal particles that are sprayed during the arc welding process and do not form part of the weld.

spatterdash: Mixture of portland cement and sand roughly spread on a surface in a thin, coarse coat. Used as a primary

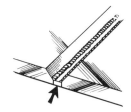

spacer strip

Spanish tile

treatment before rendering.

spattering: Process of finishing wood in which a speckled texture is produced by throwing the finishing material onto the surface from the end of a stiff-bristled brush.

spearing: Process of cleaning out a hole with a drill or reamer.

specific adhesion: Joining of members by a chemical bonding method as opposed to mechanical methods.

specifications: Written information included with a set of prints. Provides additional details that could not be shown on the prints or that require further description. SPECS.

specific gravity: Comparison of the mass of a unit volume of a material at a specified temperature to the mass of the same volume of gas-free distilled water at a specified temperature.

specific heat: Comparison of amount of heat required to raise the temperature of a specified quantity of a substance 1°F to amount of heat required to raise the temperature of an equal quantity of water 1°F.

specific retention: Comparison of amount of water retained by rock or soil to total volume of the rock or soil.

specific volume: Three-dimensional space occupied by a predetermined weight of material. Expressed as cubic feet per pound.

speed reducer: Gear box that reduces the revolutions per minute from one system to another.

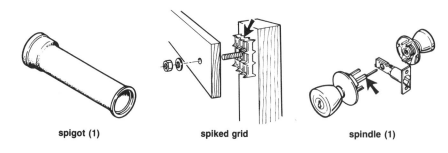

spigot (1)   spiked grid   spindle (1)

speed square: Triangular-shaped tool used to lay out 45° and 90° angles, and other angles required for roof framing pitches.

sphere: Solid geometric figure with all points equidistant from a center point.

spheroid: Solid geometric figure in which points are not exactly equidistant from a center point.

spider: Cap of a guy derrick. Guys are attached to permit free movement of the derrick.

* **spigot: 1.** Straight, untapered end of a pipe that fits into the bell or hub of another pipe. **2.** Faucet used to control flow of liquid.

spigot and socket joint: See *bell and spigot joint.*

spike: 1. Heavy nail 3″ to 12″ long. **2.** To add gypsum to mortar made with cement.

* **spiked grid:** Heavy timber fastener consisting of a steel plate with pins projecting from both sides.

spike knot: Knot in wood that has been sawn parallel to the grain. SPLAY KNOT.

spiking: Uneven joint penetration in an electron beam or laser welding.

spile: 1. Heavy wood stake. **2.** Small wood plug or peg.

spillway: Passage that conveys excess water from a dam.

* **spindle: 1.** Slender rod or shaft upon which a member rotates; e.g., shaft to which a doorknob is attached. **2.** Short cylindrical member that is turned on a lathe, such as a baluster.

spindle sander: Sanding equipment with a small-diameter rotating drum that is covered with abrasive material.

spinning: Shaping sheet metal by rotating it on a lathe and applying pressure.

spira: Decorative moldings at the base of a column or pilaster.

spiral: Geometric pattern consisting of a line continuously wound around a fixed central point.

* **spiral ratchet screwdriver:** Screwdriver in which an interchangeable tip is rotated by pushing the handle inward toward the tip. YANKEE® SCREWDRIVER.

spiral stairway: Winding stairway in which treads radiate from a common central point.

spire: High, tapered tower with a steep roof pitch.

spirit level: Sealed cylindrical transparent tube with a slight curvature nearly filled with liquid, forming a bubble. Used to indicate true vertical and horizontal alignment when the bubble is centered in the length of the tube.

spirit varnish: Liquid wood finish with a highly volatile liquid as a solvent for oil or resin.

splashboard: 1. See *backsplash.* **2.** Protective member installed against the bottom of a wall or other vertical surface to prevent damage from splashed material.

* **splash block:** Flat member, commonly concrete, placed at discharge point of a downspout to prevent erosion of the soil.

splash lap: Portion of an overlapped seam in sheet metal roofing that extends over the adjacent sheet.

splat: Wood strip that covers joints between other boards.

splatter finish: Rough plaster or drywall joint compound finish producing a random raised pattern.

splay: 1. Inclined surface that forms an oblique angle with another member. **2.** To flare outward.

splayed brick: Masonry unit with one side beveled.

splay end: Smaller end of a splayed brick.

splay knot: See *spike knot.*

splice: 1. To join two pieces of material. **2.** Location of the joint between two pieces of material.

splice box: Enclosed connector used to connect short sections of electrical cable to obtain a portable cable of the required length.

splice point: Location at which two truss members are joined.

splice plate: Flat metal plate used to provide additional strength across a butt joint.

* **spline:** Thin strip of wood inserted into grooves in edges of adjoining members to align and reinforce the joint. FALSE TONGUE, FEATHER, SLIP FEATHER, SLIP TONGUE.

split: 1. Separation of wood parallel with the grain. See *defect*. **2.** Masonry unit that is one-half its normal height. **3.** Separation in the surface of a material.

split astragal: Vertical molding attached to the meeting stiles of a pair of doors. Provides a tight joint and allows free movement of both doors.

split batch charging: Process of adding solid concrete ingredients to a mixer individually.

split-bolt connector: Electrical hardware with a conductive threaded shaft with several holes in it. Conductors are placed through the holes and joined by tightening a nut onto the threaded shaft.

split-face block: Concrete masonry unit with one or more rough faces that have the appearance of being fractured. See *concrete masonry unit*. SPLIT BLOCK.

split-conductor cable: Electrical cable in which each conductor is composed of two or more insulated conductors connected in parallel.

split hanger: Pipe support made in two halves that join together around the pipe.

split-image bubble: Configuration of two

spiral ratchet screwdriver

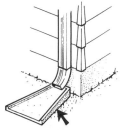

splash block

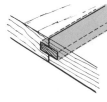

spline

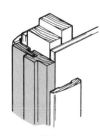

split jamb

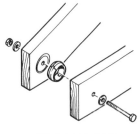

split-ring connector

splitter (1)

prisms over the spirit level in a surveying instrument. Operator views one-half of each bubble adjacent to each other in a mirror. Allows for very precise leveling operations.

* **split jamb:** Jamb for a prehung door in which the inner and outer portions interlock and can be separated when hanging a door. Groove in one-half of the jamb accepts a tongue on the other half to create a solid joint. SPLIT FRAME.

split-level: Residential structure in which the floor levels of one or more rooms are separated by approximately a half story and connected by a stairway.

split loading: See *split-batch charging*.

split-phase motor: Single-phase induction motor with a main and an auxiliary winding; designed to operate with no impedance. Auxiliary winding is only energized when starting the motor.

split pin: See *cotter pin*.

* **split-ring connector:** Heavy timber hardware consisting of a circular metal insert placed in predrilled grooves in the adjoining surfaces and fastened with a bolt. SPLIT RING.

split shake: See *shake* (1), *cedar shake*.

split system: HVAC system in which radiators or convectors are used to heat an area, and cooling and ventilating is accomplished by other means.

* **splitter:** 1. Device consisting of a wedge mounted in a frame; used to cut masonry to desired dimensions. 2. Chisel with a wide cutting edge for splitting stone. 3. Metal plate located directly behind the blade of a table saw; used to prevent wood from binding the blade, causing a kickback.

split: 1. To cut wood with a wedge- or hatchet-type device parallel with the grain. 2. Crack in wood extending completely through a piece of lumber or veneer. See *defect* (1) *. 3. Defect in a painted surface in which the new finish penetrates cracks in the old finish, resulting in a cracked surface.

splocket: See *sprocket* (2).

splush: Carpeting with semi-dense cut pile. Texture is between shag and plush.

spoil: Material excavated from its original location.

spokeshave: Wood-cutting tool used to shape curved surfaces. Consists of a flat blade secured in between two handles.

sponge rubber: Expanded rubber with interconnected cells; used for carpet padding and thermal insulation.

* **sponge rubber float:** Plastering tool used to create a fine sanded finish on plaster.

spongy wood: Timber with a very soft texture, making it unusable for structural purposes.

spontaneous combustion: Self-igniting combustion resulting from the combination of heat and presence of certain chemical conditions.

spool piece: Pipe with flanges on both ends used to replace a valve or other piping device or fixture.

spoon: Hand tool used to finish plaster moldings.

spot: 1. To direct trucks or other equipment to a predetermined area for loading or unloading. 2. To locate a pile tip at the proper location for driving. 3. To set a masonry unit in an approximate location without using a string.

spotter: 1. Pile driver component that connects bottom of the leads to the crane to allow positioning of bottom of the leads. 2. Personnel who direct trucks or other equipment to a predetermined area for loading and unloading.

spotting knife: See *broad knife*.

spotting a brick: See *spot* (3).

spot welding: Process of joining overlapping metal members by a series of short welds rather than a continuous weld joint. Low voltage and high amperage are required.

spout: End of a pipe from which liquid flows.

spray bar: Pipe attached to the back of a truck or other equipment with jets and nozzles along its length. Used to apply road surfacing material.

spray booth: Semi- or fully enclosed area used for applying surface finish. Equipped with a filtered air source and/or waterfall backdrop or filter system to retain overspray, and a fire protection system.

spray gun: Pneumatic tool in which compressed air is used to force paint or other surface coatings through a nozzle in a fine mist.

spray lime: Fine-textured hydrated lime that passes through a No. 325 sieve.

spray painting: Process of applying paint to a surface using a spray gun.

spray pond: System used to cool water by spraying it into the air above a pool of water and collecting it for reuse in a cooling system.

spray transfer: Transfer of metal in electric arc welding in which a consumable electrode is moved axially across the arc in small droplets.

spread: Width of an air pattern at the point of terminal velocity.

* **spreader:** 1. Member placed between two opposing sides of a concrete form, or window or door frame to maintain the predetermined distance during concrete placement or masonry construction. SPREADER BAR. 2. Hand tool used to apply appropriate amount of joint compound or adhesive to a surface. Edges are commonly notched to a given size and spacing. 3. Equipment used to evenly distribute gravel and/or crushed stone on a pavement bed.

spreader beam: Rigging device suspended from a crane hook to support lifting lines at various locations. Used to evenly lift long loads.

sponge rubber float

spreader (1)

spreader box: Equipment used to spread asphalt, concrete, or aggregate. Material is deposited into it and discharged on a flat surface.

* **spread footing:** Lowest foundation support that is wider at the base than at the rest of the foundation.

spreading rate: Measure of capability of 1 gallon of paint to cover a surface. Expressed as the number of square yards per gallon.

sprig: 1. Small brad without a head. 2. See *glazier's point.*

spring: Elastic device that stores mechanical energy as it is compressed and releases the energy as it regains its original shape.

springback: Property of a material to return to its original shape after deformation.

spring bolt: Door hardware with a spring-loaded bolt having a beveled face. Bolt automatically engages when the door is closed.

spring clamp: Clamp with two opposing jaws held closed by spring tension. Pressure applied to the opposite ends of the jaws opens the jaws for securing the clamp in position. See *clamp*.*

springer: Masonry member that rests on top of a column or pier and supports an arch.

spring hinge: Hinge with an integral spring that automatically closes the door to which it is attached. See *hinge*.*

springing: 1. Enlarging the bottom of a drilled hole in the earth by blasting. 2. Point where an arch rises from its supporting members. 3. Angle of an arch rise.

* **springing line:** Imaginary horizontal line extending between the points at which an arch begins to curve. SPRING LINE.

spring latch: Door latch that returns to position automatically when the door is closed.

* **spring snap link:** Hardware similar to a chain link with one side that can be opened and closed. Used for repairing chain.

spring washer: Slightly bent steel ring used as a lock washer.

springwood: Wood formed in the spring and early summer. Characterized by thin, large cells.

* **sprinkler head:** Heat-activated valve that opens to spray water in the event of a fire. AUTOMATIC SPRINKLER.

sprinkler system: Automatic water or chemical distribution system that conveys water or chemical and sprays them at the location of a fire. Consists of piping and heat-activated valves.

sprocket: 1. Gear that interlocks and meshes with a chain or track. 2. Wedge fastened to top of the lower end of a rafter where it overhangs the exterior walls. Provides a change in the slope of the roof. SPLOCKET.

sprung molding: Curved wood trim members.

spud: 1. Steel tube or H-shaped member used to secure dredging equipment or barges in position in the water. 2. Steel rod probed into the sediment at bottom of a body of water to measure thickness of the sediment. 3. Short device driven into the ground to break through hard materials when driving piles. 4. Dowel at bottom of a door post that interlocks with the floor. 5. Short connecting pipe between a water meter and water supply line. 6. Drainage fitting in a plumbing fixture. 7. Hand tool used to strip bark from logs. 8. Hand tool with a sharpened flat point used to remove built-up roofing material.

spud vibrator: See *internal vibrator.*

spud wrench: 1. See *erection wrench.* 2. Smooth-jawed pipe wrench used for tightening chrome-plated nuts and fittings. See *wrench*.*

spur: 1. Sharp, pointed tool that is used to cut wood veneer. 2. Rock ledge that projects from a side wall after blasting. 3. Decorative stone base used to make a transition of shapes in stonework.

spur gear hoist: Chain hoist that provides mechanical advantage.

spur pile: See *batter pile.*

square: 1. Plane geometric figure with four equal sides, opposing sides parallel, and 90° angles. 2. Measure of roofing or siding materials equal to 100 square feet. 3. See *framing square.* 4. To multiply a number by itself; e.g., $5^2 = 5 \times 5 = 25$.

square and flat: Frame containing a flat panel without decorative molding.

square and rabbet: See *annulet.*

square bastard file: Smoothing tool with a square cross section that slightly tapers toward the end opposite the handle.

square billet: Ornamental molding with a series of projecting cubes with space between them.

square bolt: See *door bolt.*

squared splice: Heavy timber joint in which ends are notched and overlapped to interlock.

square file: Shaping and smoothing hand tool with a square cross section. See *file*.*

square foot: Measure of area equal to 144 square inches.

square-head setscrew: Threaded fastener with a square head and soft point to secure two pieces in position.

square-head screw: Fastener with a threaded shaft and square head designed to be turned with a wrench. See *screw*.*

square inch: Area measuring 1″ on each side.

square knot: Knot used to secure two ropes together. See *knot*.* REEF KNOT, SAILOR'S KNOT, FLAT KNOT.

square nut: Internally threaded connecting hardware with a square external shape designed to be turned with a wrench. See *nut*.*

square-recess screw: Fastener with a threaded shaft and flat head with a square recess for turning with a square key or bar. See *screw*.*

square roof: Gable roof with rafters intersecting at the ridge at a 90° angle to each other.

square root: Factor of a number that when multiplied by itself equals that

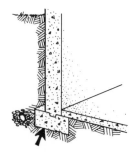

spread footing

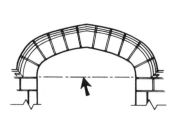

springing line

spring snap link

sprinkler head

number; e.g., square root of 9 is 3 since $3 \times 3 = 9$.

square-shank screwdriver: Screwdriver with a square portion between the handle and tip that allows grasping the screwdriver with a wrench for additional leverage.

square-tip screwdriver: Screwdriver with a tip formed to fit into a screw with a square head recess.

square up: To plane a board or shape material so that all faces are smooth and all corners are perpendicular to one another.

square yard: Measure of area equal to 9 square feet.

squaring: Process of ensuring all corners of a square or rectangular area are 90°. Accomplished by measuring equal diagonals or using the Pythagorean theorem.

squeegee: Tool with a wide rubber blade for applying asphalt sealant to a horizontal surface.

squeeze connector: Flexible metal conduit connector. Connected by placing a flange around the conduit and tightening two adjoining screws. See *conduit fitting* *.

squeezed joint: Wood joint formed by applying adhesive to the surfaces of two members and pressing the members together.

squeeze time: Interval of time between initial application of pressure by an electrode and initial application of current when performing resistance welding.

squib: Explosive detonator that ignites black powder with a firing device and flammable chemical.

squinch: Small arch constructed across an interior corner to support an imposed load. SQUINCH ARCH.

squint brick: Masonry unit manufactured to a special shape.

squint window: Small dormer window.

S-shaped flashing: See *Z-shaped flashing*.

S-shaped member: Structural steel member with a 16.67% slope on its inner flange surfaces. AMERICAN STANDARD BEAM.

S shape structural steel: Steel beam with an I profile. Web depths vary from 3″ to 24″ and flange widths vary from $2^3/_8$″ to 8″. See *structural steel* *.

stab: To roughen brick or masonry surface with a pointed tool to provide a key for plaster finish.

stabilizer: Substance used to increase stability of particles in solution or suspension.

stabilizing core: Central stiffening component of reinforced concrete member or structure.

stab rod: Device that is inserted into fresh asphalt to determine depth of the layer.

stack: **1.** Large chimney containing one or more flues. SMOKE STACK, FLUE PIPE. **2.** Vertical pipe in a waste piping system that extends one or more stories. **3.** Vertical heat supply duct.

stack bond: Masonry pattern in which all joints are aligned vertically. See *bond* (1) *. VERTICAL BOND.

stack group: Several plumbing fixtures located adjacent to the stack so that with the use of proper fittings, venting may be kept to a minimum.

stackhead fitting: Sheet metal duct connection between the riser and register. One end fits the duct and the other end fits the register.

stack height: Overall distance from bottom of a gravity convector heating unit and top of the outlet opening.

stack partition: See *plumbing wall*.

stack vent: Extension of a plumbing waste or soil pipe above the highest horizontal drain connected to the stack.

stadia: Method of measuring distance in which a graduated rod is used to calculate distance from the surveying instrument to the rod.

stadia rod: Graduated rod used in conjunction with a surveying instrument to measure distance.

staff: **1.** Decorative strip applied to an outside corner to protect finished surfaces from damage. **2.** Graduated rod used for surveying. **3.** Ornamental precast plaster moldings. **4.** Filler piece between a wood jamb and masonry wall.

staff bead: Molding applied between a masonry wall and window frame to provide weather protection.

stage grouting: Placing grout in a hole in separate steps rather than at one time.

stage screw: Coarse, externally threaded hardware used to secure scenery to a stage.

stagger: To arrange in an alternating or uneven pattern.

staging: See *scaffold*.

staging nail: See *double-headed nail*.

stain: **1.** Pigmented wood finish used to change color of wood but allow grain to be exposed. **2.** Discoloration of a finished material.

stained glass: Glass that is colored in a molten state and used in decorative windows and other ornamental work.

stainless steel: Hard, corrosion-resistant metal that contains a high percentage of chromium and may include small amounts of nickel or copper.

stair: See *stairway*.

stair gauge clamp

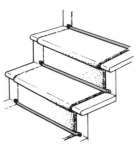

stair rod

stair bolt: Hardware with a wood thread on one end and a machine screw on the other end. Used to fasten a handrail to a volute. See *bolt* *.

stair bracket: Decorative supporting member fastened to the face of an open stringer to reinforce a tread.

stair carriage: See *stringer*.

staircase: See *stairway*.

stair flight: See *flight*.

* **stair gauge clamp:** Small framing square accessory fastened in position to assist in laying out stairway stringers for treads and risers. Used in pairs.

stairhead: Highest tread in a stairway.

stair headroom: See *headroom* (1).

stair horse: See *stringer*.

stair landing: See *landing*.

stair ratio: Relationship between unit rise and unit run in a stairway. Unit rise plus unit run should be greater than 17″ and less than 18″ for the proper incline.

stair riser: See *riser* (1).

* **stair rod:** Metal bar attached at the back of a tread and the lower edge of a riser to hold carpeting in place. STAIR WIRE.

stair stringer: See *stringer*.

stair tread: See *tread* (1).

stairway: Complete flight of steps or series of flights extending from one story of a structure to another. Includes treads, risers, landings, handrail components, and structural supports.

stairwell: Opening in a floor or through a series of floors in a multistory structure through which stairs are installed.

standard matched

star drill

starter strip

starting newel

stake: **1.** Wood or metal member pointed on one end so it can be driven into the ground. Used as a layout marker or brace. **2.** Small anvil used in sheet metal work.

stall: **1.** Bathroom compartment or partition. **2.** Stationary seat enclosed at the back and sides.

stallboard: Sill and framing members below a storefront window.

stamba: Single pillar built as a memorial.

standard air: Air that weights 0.075 pound per cubic foot. Similar to air at 68°F, 50% relative humidity, and barometric pressure of 29.9″ of mercury.

standard and better: Mixture of lumber grades suitable for general construction.

standard brick: Masonry unit measuring $3^3/_4″ \times 2^1/_4″ \times 8″$. See brick*.

standard deviation: Statistic used as a measure of distribution. Equal to the square root of the average of the squares of deviations from the mean.

standard knot: Lumber defect $1^1/_2″$ or less in diameter.

standard lumber: Lumber graded as No. 2.

* **standard matched:** Lumber with a tongue and groove offset from the center.

standard modular brick: Masonry unit that measures $4″ \times 2^2/_3″ \times 8″$. See brick*.

standard penetration test: Method used to determine compaction of soil. Sampling spoon is driven to a depth of 1′-0″ with a 140 pound weight falling 30″. Expressed as the number of blows required.

standard sand: Ottawa sand that passes through a No. 20 sieve and is retained on a No. 30 sieve.

standee: U-shaped rebar with additional legs extending outward at 90° from the top. Used to support upper layer of rebars.

standing block: Fixed pulley assembly.

standing finish: Permanent vertical interior trim members.

standing leaf: Fixed, immovable portion of a hinge.

standing panel: Door with a greater height than width.

standing part: Inactive length of rope or line in a knot. STANDING END.

standing seam: Sheet metal roofing seam in which the edges are turned up, then folded over. See seam (1)*.

standing waste: Vertical pipe connected at bottom of a tank or fixture; used to control water overflow by allowing overflow to rise up the pipe.

standoff distance: Distance between a nozzle and base metal in a welding or cutting process.

standpipe: Tank or pipe used for storing water that may be required for future or emergency use. **2.** Pipe or tank connected to a closed piping system to absorb pressure surges.

staple: U-shaped metal or wire fastener with pointed ends.

staple gun: Tool used to drive staples into place by manual, electrical, or pneumatic operation. STAPLER.

staple hammer: Hand tool used to drive staples. Head pivots into position and drives the staple. Used for fastening light material such as insulation.

* **star drill:** Hand tool used to drill $1/_4″$ to 1″ holes in concrete and masonry. Composed of a high-carbon steel tip that is fin-shaped and a butt designed for striking with a hammer.

star expansion bolt: Concrete fastener consisting of two semicylindrical members that are forced apart as a bolt is inserted. As members spread they grip the wall of the predrilled hole.

star shake: Wood defect in which splits radiate from the center of a timber. Results from drying too quickly.

starter: **1.** Steel rod used to initially make an indentation in rock in preparation for drilling a hole. **2.** Electrical control device used to activate and halt a motor. **3.** Electrical device used with a ballast to light an electric discharge lamp.

* **starter strip:** Initial course of asphalt shingles attached to a roof. Fastened in a reverse direction from the remainder

of the roof and overlapped to provide a waterproof seal.

starting capacitor: Electrical control device that remains in the circuit instantaneously and is used to start windings in a single-phase motor.

* **starting newel:** Post that supports the handrail at the bottom of a stairway.

starting punch: Long tapered pin used for light structural ironwork.

starting step: Lowest tread in a stairway.

star transformer: Electrical device in which the ends of each winding are connected to a common point.

starved joint: **1.** Wood joint between two members in which all the adhesive has been forced out by excessive clamp pressure. **2.** Defect in drywall joint finishing caused by insufficient drying time between applying coats of joint compound.

static cone penetration test: Measure of bearing capacity of soil determined by thrusting a cone into the soil to a depth of 3″ and calculating approximate force required.

static head: Pressure created by the difference in water levels in two separate but connected sections. PRESSURE HEAD.

static load: Weight of a stationary mass exerting only downward pressure.

static pressure: Outward force of air or water in a closed piping system.

station: **1.** Reference point located by surveying. **2.** Stake or point that indicates distance from a point of beginning or other reference point.

stationary mixer: Fixed equipment used to consolidate material such as concrete. Not moved from job site to job site. STATIONARY HOPPER.

stator: **1.** Stationary portion of an electrical motor. Includes station components of an electrical circuit and associated windings and leads. **2.** Set of fixed vanes in a torque converter used to change direction of fluid flow.

staunching piece: Vertical opening left between adjacent sections of concrete. After full shrinkage has occurred, the opening is filled with concrete.

stave: **1.** Support for foliage and leafage

on Corinthian and Composite column capitals. **2.** Ladder rung. **3.** One of a series of narrow vertical members that compose a curved surface.

stay: Brace.

stay lath: Permanent or temporary brace for trusses, floor joist, or other wood structural members.

steam boiler: Closed pressurized vessel in which water is heated and converted from liquid to vapor.

steambound: Condition in a steam heating system in which the vacuum or feedwater pump cannot deliver water to the boiler because temperature in the vacuum tank or open feedwater heater is too high.

steam curing: Curing concrete or mortar in water vapor at a minimum of atmospheric pressure and a temperature between 100°F and 420°F.

steamfitter: Construction worker who installs pipes that carry steam or hot water used for heating a structure. PIPEFITTER.

steam heating system: Heating system that transfers heat to the living area by condensed steam supplied by a boiler.

steam quality: Percentage of vapor in a mixture of liquid and vapor.

steam shovel: Excavating equipment with a digging bucket powered by an integral boiler.

steam trap: Device that automatically removes air and condensate from steam lines and heat exchangers.

steel: Alloy of iron, carbon, and small amounts of other material that are added to obtain desired properties. Carbon content ranges from 0.02% to 1.7%.

steel erector: Contractor or individual who builds with structural steel.

steel-frame construction: Building in which the structural components are made of steel.

steel sheet: Cold-formed metal panel used as a structural member in a lightweight concrete roof.

steel sheet pile: See *sheet pile*.

steel square: See *framing square*.

steel stud: See *metal stud*.

steel tape: Measuring tool consisting of a long, continuous strip of steel that is graduated in regular increments of length measurements. Steel strip is coiled inside a case and rewound manually or by a spring.

steel wool: Polishing and cleaning material made of fine strands of steel that are matted together. Various grades are available.

steelworker: See *ironworker*.

steening: Brick installed without mortar; used for cisterns or wells.

steeple: Tower that is topped with a tapering structure such as a spire.

steeple tip: Decorative end of a hinge pin resembling a steeple.

stemming: Inert material placed in a blast hole.

stem wall: **1.** Vertical wall supported on a footing or other load-bearing surface. **2.** Short wall.

stench trap: See *trap* (1).

step: **1.** Unit consisting of one tread and one riser. **2.** One unit of stair or rafter length when laid out with a framing square. See *stepping off*.

step bolt: Threaded fastener with a large pan head and squared section directly below the head. See *bolt**.

step brazing: Brazing process in which successive joints on a part are made by using successively lower brazing temperatures so as not to disturb the previous joints.

step-down transformer: Transformer in which the power is transferred from a higher voltage to a lower voltage circuit.

step flashing: Short L-shaped piece of sheet metal commonly installed at the intersection of an asphalt shingle roof and a vertical surface.

step joint: Heavy timber joint in which one member is notched to receive another member to join the members at an angle.

* **stepladder:** Portable device with flat rungs built into a supporting frame.

* **stepped footing:** Foundation support consisting of a series of horizontal steps

at succesive elevations to accommodate a sloped grade.

stepping: Lumber used as stair treads.

stepping off: Method of determining rafter length by marking each increment of angular length as it relates to 1'-0″ of horizontal run. Framing square is moved along length of the rafter to lay out individual units.

* **step pulley:** Pulley with various diameter wheels to allow for positioning a belt to regulate rotational speed.

step soldering: Soldering process in which successive joints are made by using successively lower soldering temperatures so as not to disturb the previous joints.

* **step-taper pile:** Pile constructed of corrugated steel shells of increasing diameter that are fastened together. Mandrel is fitted inside the shell to maintain the shape as it is driven into the ground. Mandrel is removed and the shell filled with concrete.

step-up transformer: Transformer in which the power is transferred from a lower voltage to a higher voltage circuit.

stereobate: **1.** Masonry platform that does not support columns. **2.** Foundation or platform on which a structure is built.

stereotomy: Art of properly cutting and placing stone.

stick: Rigid bar in an excavating dipper shovel that is hinged to the boom and also fastened to the digging bucket.

sticker: **1.** See *molder*. **2.** Spacing board between layers of lumber used when drying.

sticking: **1.** Molded edge formed on the inside portion of door rails around the panels. **2.** Setting of ornamental work into a fresh bed of plaster. **3.** Using a molder for manufacturing wood moldings.

stiffback: Vertical brace on concrete formwork.

stiffener: Structural member that provides rigidity to a component or structure.

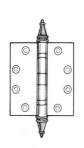
steeple tip

stepladder

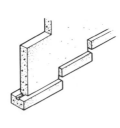
stepped footing

step pulley

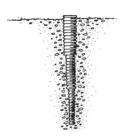
step-taper pile

stile (1)

stilt

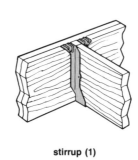

stirrup (1)

stiff leg derrick: Lifting device in which a mast rests on sills and is braced by a structural member. Mast is shorter than the boom and can swing only 270°.

* **stile: 1.** Vertical member of a door, window sash, or cabinet face frame. **2.** Set of steps over a fence or wall.

stile and rail door: See *panel door*.

stilling pool: Enlargment of a river channel to reduce the flow and reduce scour.

stilted member: Member that is higher than its normal elevation or location; e.g., an arch that is above the normal height of its support.

* **stilt:** Leg extension that allows a worker access to higher levels.

stinger: Electrode holder used in arc welding.

stipple: To finish with a rough texture as a decorative effect.

* **stirrup: 1.** Supporting strap for a beam or other horizontal structural member where it butts into another horizontal or vertical member; e.g., a joist hanger. **2.** Support for rebars. **3.** Device used to straddle sheet pile during driving operations. **4.** S-, U-, or W-shaped rebar.

stitch: Number of lengthwise tufts of yarn in 1 lineal inch of carpet.

stitched veneer: Random width wood veneer that is sewn together with heavy thread and used for inner plies of plywood.

stock: 1. Tool handle. **2.** Readily available, standard-sized building mate-

rials.

stock bill: Itemized listing of all cabinet components and their dimensions.

stock brush: Brush used to apply water to a plaster base coat that has dried.

stockpile: To store material in a location for future use.

stock plans: Basic residential prints that are available for purchase and not designed for any specific site.

stoker: Mechanical feed device that feeds coal and other materials into a combustion chamber.

stone: 1. Natural rock deposit. **2.** Individual building unit formed of rock.

stone bolt: Threaded metal rod that is mortared into a predrilled hole in a stone to allow for the fastening of another member.

stonecutter's chisel: Hand tool used to dress soft stone. TOOTH CHISEL.

stonemason: Tradesperson who builds and works with stone or other masonry materials.

* **stool:** Interior horizontal shelf member at the bottom of a window frame.

stoop: Raised entrance platform outside an entrance door.

* **stop: 1.** Molding that is fastened to the inside of the jamb of a double-hung window to secure the bottom sash. **2.** See *doorstop*(1). **3.** See *stop-and-waste valve*.

* **stop-and-waste valve:** Plumbing valve with a side opening that allows for drainage of the connected pipe.

stop block: Piece of material temporarily attached to a saw table to act as a guide

when making a cut that does not extend from one edge to the other.

stop box: Adjustable length cast iron sleeve with a removable cover. Sleeve forms a protective covering for a curb cock.

stopcock: Plumbing valve with a tapered plug and opening that controls flow of liquid by turning the plug.

stope: Below-grade excavation in the form of a series of levels or steps.

stoper: Pneumatic drill mounted on a column or other support; used for drilling rock and other excavating work.

stop mortise-and-tenon: Wood joint in which a recess is cut partially through a member (stop mortise) to receive the projecting portion of an adjoining member (tenon). End of the tenon is concealed. See *joinery**.

stopoff: Material applied on either side of a brazed or soldered joint to prevent filler metal from spreading.

stopped miter: Joint used to connect two wood members with different thicknesses.

stopper knot: Overhand or figure eight knot tied in the end of a rope to prevent it from slipping out of a hole or pulley.

stopping knife: Hand tool with a blade that has one rounded edge and one angular edge. Used to apply spackling and glazing compound.

stopwork: Mechanism in a lock that secures the bolt in a position so that the lock cannot be opened from the outside.

storm cellar: See *basement*.

storm clip: Clip on the outside of a glazing bar that prevents a glass pane from moving outward in the frame.

storm door: Wood or metal exterior door commonly consisting of an interchangeable screen and window unit. Door swings outward to avoid hitting the main entrance door.

storm drain: See *storm sewer*.

storm sash: See *storm window*.

storm sewer: Sewer used to convey groundwater, rainwater, and/or surface water.

storm window: Additional window installed on the outside of a window opening to provide a double layer of glass as insulation and storm protection.

story: Level of a building that extends from the upper surface of a floor to the upper surface of the floor or roof above.

* **story-and-a-half:** Residential construction in which a portion of the second story ceiling is sloped at approximately the same angle as the roof rafters. Exterior walls under the rafters are rela-

stool

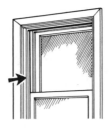

stop (1)

stop-and-waste valve

tively short.

story pole: Rod or board that is graduated along its length and used for repetitious measurements. STORY ROD.

stove bolt: Fastener with a threaded shaft for receiving a nut and a slightly rounded head with a recess. See *bolt**.

straight arch: Arch with a flat upper surface but with masonry members placed at an angle. FLAT ARCH.

straight claw hammer: Hammer with a straight claw and a tapered slot for removing nails. Claw design facilitates splitting wood. RIPPING HAMMER.

straight-cut bit: Router bit that is straight and perpendicular to the face of the work. See *bit* (2) *.

straightedge: **1.** True, straight member used as a guide for marking or measuring straight lines. **2.** True, straight device for screeding plaster or concrete to a smooth finish.

straight-flight stair: Stairway with no turns. STRAIGHT RUN STAIR.

straight-grain lumber: Lumber in which the cell structure is parallel with the length.

straight jacket: Timbers fastened to a wall to reinforce the wall and secure it in a true, straight position.

straight joint: **1.** Joint formed between two members with square ends that do not overlap. **2.** Continuous vertical masonry joint.

straight-line wiring diagram: Drawing of an electric wiring system in which the circuits are shown as they are to be wired. Actual physical location of the conductors are not indicated.

straight peen hammer: Striking tool consisting of a wedge-shaped striking surface with a rounded edge. Wedge is parallel with the hammer handle.

straight pipe thread: Continuous external thread on a pipe in which the minor diameter is equal throughout its length.

straight polarity: See *direct current electrode negative.*

straight tee: T-shaped pipe fitting in which all openings equal in diameter.

strain: Change in shape of an object as result of external forces. Expressed as ratio of the linear unit deformation to the distance in which the deformation occurs.

strainer: Screen, mesh, or perforated obstruction used to separate a solid from a liquid.

straining beam: Short member in a truss that secures the rafters and struts in place. STRAINING PIECE.

strake: **1.** One row of clapboard. **2.** Row of steel plates in a steel chimney.

strand: **1.** Line or rope made from twisting several fibers or wires together.

2. Prestressing tendon consisting of several wires twisted together around a central wire or core.

stranded conductor: Electrical conductor consisting of several wires twisted or braided together.

S trap: S-shaped waste pipe fitting that obstructs the pipe and prevents sewer gases from entering the living area.

strap: Strip of metal or other material used to provide additional tensile reinforcement to a joint. STRAP ANCHOR.

strap footing: Continuous footing in which all loads are applied directly from above without lateral pressure.

strap hinge: Hinge with elongated plates for use in holding heavy doors. See *hinge**.

strapping: Battens attached to the inside of a wall to support lath and plaster work.

strap pipe hanger: Metal strap with a series of uniform perforations used to suspend a pipe from a ceiling or joist. Strap is wrapped around the pipe and secured in position.

strap wrench: Hand tool used to turn irregularly shaped objects and pipe. Consists of flexible strap that is wrapped around the object and secured to turn the object. See *wrench**.

strata: Layer of soil or rock between other similar layers. STRATUM.

stratification: Horizontal layering of different types of fluids or materials with increasingly lighter material toward the top.

streaking: Varying light and dark strips of finish caused by improper application.

street elbow: Pipe fitting elbow with female thread on one end and male thread on the other end. See *pipe fitting**. SERVICE ELL, STREET ELL.

street tee: T-shaped pipe fitting with two openings with female threads and the other opening with a male thread.

stress: Intensity of internal force exerted by one or more adjoining members. May result in strain, deformation, or separation of one or both members.

stress grading: Laboratory rating of the ability of a material to support a load without failure or permanent distortion.

stressed-skin construction: Construction in which a thin surfacing material is used to support loads and provide lateral support.

stress-relief heat treatment: Uniform heating and cooling of a material to relieve residual stresses.

stretcher: Position of a masonry unit in which the edge perpendicular to the end and face is exposed. See *brick-*

story-and-a-half

strike (1)

work*.

stretcher block: Concrete masonry unit with two or three cores and a recessed area on each end; used in the body of a wall. See *concrete masonry unit**.

stretcher bond: Masonry pattern in which all masonry units are set as stretchers with vertical joints offset.

stretcher course: Row of masonry units set as stretchers.

stretcher leveling: Straightening sheet metal by stretching.

stretcher sash block: Concrete masonry unit with two or three cores, a recessed area on one end, and a vertical channel on the other end to accept a sash. See *concrete masonry unit**.

stria: Narrow groove or flute.

striated: Surface texture composed of closely spaced shallow vertical grooves or flutes.

*strike: **1.** Hardware fastened to a door jamb to receive the bolt from a lock. STRIKER, STRIKE PLATE, STRIKING PLATE, KEEPER. **2.** To finish a mortar joint by passing edge of the trowel over the joint to remove and smooth the mortar.

strike block: Wood-smoothing plane that is shorter than a jointer plane.

strike jamb: Vertical portion of a door jamb against which the door closes; lock side of the jamb.

strikeoff: **1.** To remove excess concrete, plaster, or mortar to maintain an even surface or add material to a void to make the surface even. **2.** Screed or straightedge used to maintain or make an even surface.

strike plate: See *strike.*

striking: **1.** Cutting a molding with a wood plane. **2.** Removing temporary braces and shores.

string: See *stringer.*

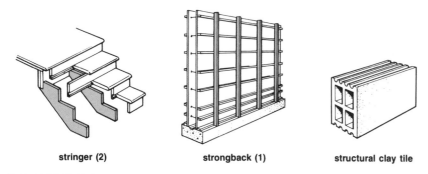

stringer (2)   strongback (1)   structural clay tile

string board: Built-up facing member in a stairway that conceals ends of treads and risers and the stringer.

string course: See *oversailing course.*

* **stringer: 1.** Beam that rests on other supports such as beams and joists. **2.** Inclined structural support for stair treads and risers. CARRIAGE.

stringer bead: Weld bead made without weaving motion.

stringing: Process of spreading mortar with a trowel on a masonry wall.

string level: Spirit level mounted in a frame with prongs at either end for hanging on a string.

strip: 1. Narrow piece of material. **2.** To damage the threads on a bolt or nut by overtightening or other abuse. Damaged area does not allow nut to engage or bolt to be secured in a threaded hole. **3.** To remove concrete forms after the concrete has reached sufficient hardness. **4.** To remove thin layers of soil or overburden. **5.** To remove paint or other surface coatings.

strip flooring: Narrow boards laid edge-to-edge to form a finish floor surface.

strip foundation: Continuous foundation in which the length is greater than the width.

stripper: Hand tool with two opposing serrated wheels for cutting narrow slots in gypsum board.

strip shingle: Asphalt roof covering which is 36″ long and 12″ wide without notches, tabs, or decorative shapes. See *asphalt shingle*.*

stroke: 1. Vertical travel distance of the ram of a pile hammer. **2.** Travel distance of a piston in an engine.

stroked masonry: Masonry with a finely fluted surface.

* **strongback: 1.** Vertical support attached to concrete formwork behind the horizontal walers to provide additional strength against deflection during concrete placement. STIFFBACK. **2.** Additional reinforcement placed behind joints. **3.** Lateral wood brace in wood truss installation. Installed vertically, perpendicular to the trusses, and fastened to the vertical truss web mem-

bers.

strop: Soft, smooth-surfaced hand tool used to apply the final razor-sharp edge on a wood-cutting tool.

struck joint: Mortar joint that is angled to form a recess at the bottom. See *mortar joint*.*

structural bond: Bond produced by setting masonry units in a pattern to create a single supporting unit.

* **structural clay tile:** Hollow or solid building member molded from clay. Set in mortar and used in masonry construction.

structural concrete: Concrete designed to support loads or form an integral portion of a structure.

structural failure: Collapse of a structure or inability to support the design loads.

structural glass: 1. Vitreous finish material used to cover masonry walls. **2.** Glass cubes or rectangles used as wall surfacing.

structural insulation board: Sheet material that provides both strength and insulation qualities.

structural lightweight concrete: Structural concrete made with lightweight aggregate. Unit weight ranges from 90 to 115 pounds per cubic foot.

structural lumber: 1. Timber that is greater than 5″ in both thickness and width. STRUCTURAL TIMBER. **2.** Lumber that is graded for strength; greater than 2″ in thickness and 4″ in width and used to support framing members.

structural plan: Portion of a print that indicates sizes, types, and placement of structural steel members. See *print*.*

* **structural steel:** Load-bearing steel members in a structure.

structural T: Structural steel member with a T-shaped profile. Commonly cut from a W shape or S shape structural steel member. See *structural steel*.*

structural Z: Structural steel member with two offset parallel flanges joined by a perpendicular web. See *structural steel*.*

structural wrench: See *erection wrench.*

structure: 1. Any building or construction. **2.** Several members fastened together to form a continuous unit.

strut: 1. Member fixed between two other members to secure them a specified distance apart; e.g., members between the chords of a truss. **2.** Short column.

stub: Short projecting member.

stub mortise: Recess that does not pass completely through a piece.

stub tenon: Tenon that is shorter than the width of the member into which it is inserted and does not completely pass through the member. JOGGLE TENON.

STRUCTURAL STEEL MEMBERS

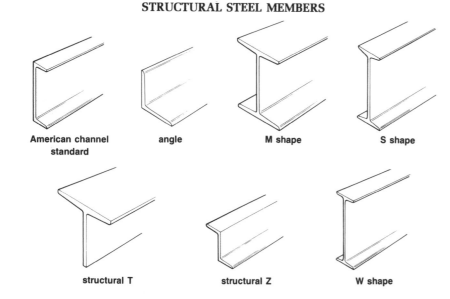

American channel   angle   M shape   S shape
standard

structural T   structural Z   W shape

stub wall: Short vertical structure usually less than 2'-0" high.

stucco: Exterior finish material composed of portland cement, lime, sand, and water.

stucco brush: Tool used to apply stucco to an exterior surface. Stucco is applied by dipping the bristles into the stucco and throwing it onto the surface.

* **stud: 1.** Wood or metal vertical framing member in a wall. **2.** Threaded fastener that is threaded on both ends and has no head. See *mill stud*. **3.** Bolt that is anchored to another member at one end.

stud arc welding: Arc welding process in which metal is joined by heat produced between a metal stud and the base metal. Pressure and/or shielding gas and flux may or may not be used.

stud bolt: See *mill stud*.

stud bolt anchor: Expansive anchor used in concrete or masonry surfaces. Fastener is inserted through the predrilled hole and a sleeve expands as the bolt is tightened. See *anchor*.

stud crimper: Hand tool used to pierce metal studs and runners to join them together.

stud driver: See *powder-actuated tool (PAT)*.

stud gun: See *powder-actuated tool (PAT)*.

stud welder: Electric arc welding equipment used to join bolts to beams and other metal members for use in concrete bridge work.

stud welding: Process of fastening shear studs to a steel member using resistance welding.

stuffing box: Packing gland that provides a watertight connection between a shaft or rod and the member to which it is connected.

stylobate: Masonry platform that supports one or more columns.

Styrofoam®: Lightweight material made of expanded chemical polystyrene plastics; used for insulation.

subbase: 1. Lowest portion of a structural support composed of two or more horizontal members; e.g., base of a column. **2.** See *baseboard*. **3.** Layer in pavement between the subgrade and the base or pavement. **4.** Heating/cooling device to which a thermostat is attached. May contain switches, terminals, and other components.

subbasement: Level of a structure immediately below the basement.

subcasing: Concealed trim member applied over the rough frame and covered with the finish casing. BLIND CASING.

subdivision: 1. Tract of land that is surveyed and divided into parcels

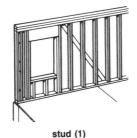

stud (1)

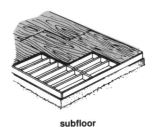

subfloor

subrail

for individual sale and development. **2.** Dividing of a township into sections and portions thereof.

subfeeder: Feeder that originates at a point in the circuit other than the main service panel; supplies electricity to one or more branch circuits.

* **subfloor:** Rough surface applied to the top of floor joists and used as a base for the finish floor. Plywood, oriented strandboard, or particleboard are commonly used.

subframe: 1. Structural framework of wood or metal channels that support a finish door or window frame. **2.** Light structural frame used to support a curtain wall.

subgrade: 1. Compacted soil used to support a concrete slab or other structure. **2.** Elevation at the bottom of a pipe trench.

submerged arc welding: Arc welding process in which metal is joined using heat obtained from an arc between a bare metal electrode and base metal. Arc and molten base metal are shielded by a granular material. Pressure is not used and filler metal may or may not be used.

submerged inlet: 1. Opening of a water supply pipe below the flood level rim of a fixture. **2.** Terminal end of a discharge pipe from a plumbing fixture to the drainage system.

submersible pump: Device used to convey liquid from one level to a higher level. Device operates when completely submerged in the liquid.

subpurlin: Secondary structural support used to support formwork in lightweight concrete roof construction on which concrete is placed.

* **subrail:** Trim member applied to the top edge of a stringer in a closed stringer stairway to receive the bottom end of the balusters. SHOE.

subsidence: Settling or sinking.

subsill: 1. Window member fastened to the sill and used as a stop for screen windows. **2.** Secondary door sill below the threshold.

subsoil: Material immediately below the surface of the ground.

subsoil drain: Pipe or piping system that collects subsurface water and conveys it into the storm sewer.

substation: Assembly of electrical equipment, including switches, circuit breakers, buses, and transformers used to switch power circuits and convert power from one type of voltage to another.

substrate: Underlying material, surface, or support to which a finish surface is applied.

substructure: Foundation that supports the upper portion (superstructure) of a building.

subsurface course: Wearing surface of pavement.

subsurface sewage disposal system: See *septic system*.

suction: 1. Adhesion of plasters or mortar to the base coat or masonry surface when applied. **2.** Atmospheric pressure quickly entering and filling a partial vacuum.

suction head: Height to which a pump can convey fluid.

suction valve: Device that allows refrigerant to flow into a cylinder and prevents backflow.

sullage: 1. See *silt*. **2.** Sewage.

summer: 1. Horizontal supporting member; e.g., a beam. **2.** Support over an opening; e.g., a lintel. **3.** Pier or column cap.

summerwood: Portion of the annual growth of a tree that occurs during the summer; thick-walled cells and denser than springwood.

sump: Pit where water or liquid accumulates.

sump pump: Device used to remove liquid from a sump and convey it into a pipe for dispersion in another location. Commonly used in basements to prevent accumulation of water. Commonly equipped with an automatic switch to turn it on when the liquid reaches a given level.

sun deck: Platform that is open and exposed to the sun; used as a recreational area.

sunk draft: Face of a building stone with a portion removed to give the remainder of the stone a raised appearance.

sunk fillet: Groove in a flat surface.

sunk molding: Decorative features cut or formed into a flat surface. SUNKEN MOLDING.

sunk panel: Area recessed below the surrounding members. SUNKEN PANEL.

super: Continuous slope in a roadway.

supercapital: Stone member above a column capital.

supercilium: Molded lintel above a doorway.

supercolumniation: Second level of columns placed on top of a lower level of columns.

supercooling: Changing the temperature of a material to a point below its normal freezing point. SUPERCHILLING.

superelevation: Raised portion of the outer edge of a roadway that offsets and balances centrifugal force of a moving vehicle traveling around a curve.

superheating: Temperature of a vapor that is above the boiling point of liquid at a specified pressure.

superimposed: Set or laid on top of another surface or member.

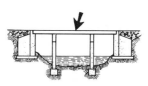

superstructure (2)

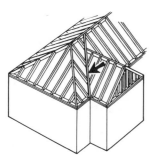

supporting valley rafter

suspended ceiling

superimposed load: Load on a structural member other than its own weight.

supersaturation: Condition in which vapor is at a pressure above the saturation pressure for a specified temperature.

* **superstructure:** **1.** Building that is above the foundation (substructure). **2.** Portion of a bridge above the bridge seats on the abutments and columns.

supply air system: Assembly of ductwork, fittings, grills, plenums, and registers through which heated or cooled air is supplied to designated areas.

supply fixture unit: Measure of the fresh water supply demand on a system. Based on the flow rate, time duration of each use, and time between uses per fixture.

supply line: Pipe extending from the water heater to the heating units in a hot water heating system.

supply main: **1.** Primary pipe for the supply of water or natural gas. **2.** Primary pipe used to supply heating and cooling medium between the source and registers or heating/cooling units.

* **supporting valley rafter:** Roof framing member that extends at an angle from the top wall plate at an inside corner to the ridge; acts as a structural support for valley jack rafters and/or cripple rafters.

surbase: **1.** Trim molding directly above a baseboard. **2.** Trim molding at the top edge of a pedestal base.

surbased arch: Curved arch design with a rise less than one-half its span.

surcharge: Live or dead load at the upper level of a retaining wall.

surface: **1.** To finish the exterior of a material such as wood or stone. **2.** Exterior of any material, component, or structure.

surface-active agent: Additive in plastic concrete that modifies the surface tension and allows water to penetrate the mix for further working of the concrete.

surface bonding: Process of joining masonry units by parging with a layer of fiber-reinforced mortar.

surface gauge: Metalworking tool used to lay out and scribe lines on vertical and horizontal surfaces.

surface hinge: See *strap hinge*.

surface moisture: Free water on the surface of aggregate that becomes part of the mixing water in concrete. SURFACE WATER, FREE WATER.

surface planer: See *planer*.

surfacer: Equipment used to polish stone.

surface tension: **1.** Condition created by impurities at the top of the water supply in a boiler drum containing steam and water. **2.** Attraction between liquid

molecules that maintains the continuous state of a liquid.

surface vibrator: Machine that consolidates freshly placed concrete by vibration applied at the surface. Includes pan, plate, or grid vibrators and rolling and vibrating screeds.

surface water: Rainfall or precipitation that remains close to the surface of the ground.

surfacing: **1.** Application of brazing, welding, or other metal deposits on the surface of a material for purposes other than to form a joint. **2.** Smoothing and truing the faces of a board.

Surform® plane: Cutting and shaping tool that removes the surface of a material by means of a series of small, sharp cutting teeth. See *plane* (1)*.

surge: Sudden increase in liquid pressure or electrical current.

surge arrester: Electrical device used to limit excessive voltage by discharging surge current and allowing the normal amount of current to continue to flow through the circuit.

surge tank: Container in a water system that retains excess water in case of a sudden rise in pressure and supplies water in case of a sudden drop in pressure.

survey: **1.** Orderly determination of the physical properties and characteristics of the earth. **2.** Plan that shows dimensions and location of a piece of ground. **3.** List of load factors and job conditions that are used to estimate and calculate head loads.

* **surveying:** Process of measuring land by recording elevations and noting directions and lengths of property lines and sizes of structures.

surveyor: Tradesperson who accurately measures land according to standard established procedures and landmarks.

suspended: Hung from a support.

* **suspended ceiling:** Upper portion of a room formed by hanging a flat surface with wires or other means below the structural ceiling members.

suspension belt: Device worn around the waist and torso for allowing a person to work while suspended at the end of a line or rope.

suspension bridge: Roadway that extends across an open area by means of cables that are supported and anchored above the roadway. Vertical rods hang down from these cables and support the roadway.

suspension saddle tie: Wire tie used to suspend light-gauge framing members from other members. Wire tie is wrapped around the framing member twice. See *tie* (1)*.

swage: 1. Hand tool used to bend saw teeth in alternate directions; similar to a saw set. 2. Hand tool used to shape sheet metal. 3. Ironworking tool used to shape hot or cold metal. 4. To shape material with force. 5. See *swaging*.

swage pile: Vertical foundation support made of thin-walled pipe with a tapered point. Driven into the ground with a mandrel and filled with concrete.

swaging: Slight offset in the hinge leaf at the barrel of the hinge. Allows leaves to close tighter.

swale: Depression or channel in the ground, often designed for channeling surface water.

sward: Turf-covered ground.

sway brace: Diagonal members used to provide additional lateral support to a structure.

sweat finish: See *swirl finish*.

sweating: 1. Process of joining metal by covering the adjoining surfaces with solder, heating the solder into a liquid state, and holding the material in position until cool. 2. Condensation of liquid on a cool surface.

sweatout: Plaster that retains moisture and will not harden.

swedge bolt: Fastener that is threaded at one end and irregularly shaped at the other end to form a good bond when embedded in concrete or mortar. See *bolt**.

sweep: 1. Curved member. 2. See *sweep strip*.

sweep lock: Window security device consisting of a sash latch with a pivoting, circular moving lever that engages a keeper on the upper sash.

sweep strip: Flexible weatherstripping at the bottom of a door.

swell: To bulge outward; to increase in size.

swift: Reel for handling and placing prestressing tendons in concrete forms.

swing: To move the superstructure of excavating or lifting equipment.

swing angle: Distance an excavating shovel moves between digging and dumping; expressed in degrees.

swing check valve: Fluid control device that opens and closes in relation to fluid pressure. Allows for fluid flow in one direction only.

swinging door: Door that is hinged along one side and pivots. See *door**.

swinging scaffold: Work platform that is suspended with cables or lines that are fastened to beams above. SWING STAGE SCAFFOLD, SWINGING STAGE.

swing jib crane: Crane with a horizontal counterweighted boom that pivots in a full circle.

SURVEYING INSTRUMENTS

builder's level

electronic distance measuring system

pipe laser

rotating laser

theodolite

transit-level

Accessories

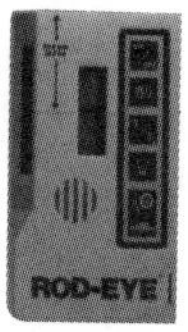

laser target

magnifying glass

Philadelphia rod

plumb bob

prism

range rod

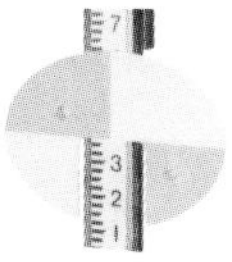

target

tripod

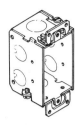

switchbox

switch plate

swing joint: Pipe connection in which a 90° elbow is attached to a street elbow. Allows for movement of the adjoining pipes. SWIVEL JOINT.

swing leaf: Active, moving portion of a hinge.

swing line: Rope or cable that controls the swing of a derrick.

swing saw: Electric power saw that is suspended from a frame and cuts the material by swinging down into it. PENDULUM SAW.

swing stage scaffold: See *swinging scaffold*.

swirl finish: Non-skid, rough texture applied to concrete, plaster, or sand-finish paint. Trowel or brush is moved in repeated circular, overlapping motions to produce the texture.

switch: Electrical device used to control flow of electricity through a circuit or change the connection of a circuit.

switchboard: Panel or assembly of panels containing electrical switches, meters, buses, and other overcurrent protection devices.

* **switchbox:** Container attached to framing or structural members to provide an enclosure for the connection of electrical conductors to a switch.

switchgear: Electrical device that switches or interrupts another device or circuit.

* **switch plate:** Trim piece that conceals a switchbox and has a hole to allow the control portion of an electrical switch to project.

swivel spindle: Device in a door latch that allows one knob to spin and operate the bolt while the other knob is locked in position.

swole: see *swale*.

symbol: Pictorial representation of a structural or material component used on prints; commonly standardized. Used to conserve space and provide as much information as possible.

symmetrical: Evenly proportioned; balanced.

synchronous motor: Electrical motor that converts electrical power to mechanical power by direct current. Average speed of the motor is directly proportional to the frequency of the electrical system.

system lag: Difference in temperature between the closing of a thermostat and the point at which the thermostat begins to open.

tab: Portion of a composition shingle between slots.

tabby: Building material composed of lime and water mixed with shells, gravel, or stone.

tabia: Mixture of compacted earth, lime, and small stones.

table: **1.** Flat horizontal surface projecting from the face of a wall. **2.** See *water table*.

tabled joint: Stonework bed joint in which a projection in one stone fits into a corresponding recess in the adjoining stone.

table saw: Power saw in which a circular blade is mounted on an arbor below the table. Blade projects above the table through a slot. Blade height and angle are adjustable. See *saw**.

tab tie: Masonry tie with two heavy-gauge wire members that are placed parallel to the length of a wall and U-shaped members that project perpendicularly and tie an adjoining wythe to the facing wythe. See *tie* (1) *.

tachometer: Device used to measure rotational speed of an object. Expressed as revolutions per minute (RPM).

tachymeter: Surveying instrument used to determine distance, direction, and difference of elevation with a single observation. TACHEOMETER.

tack: **1.** Short fastener with an enlarged head. **2.** To fasten temporarily. **3.** To intermittently fasten using glue or fasteners. **4.** See *tack weld*.

tack cloth: Cloth impregnated with slow-drying varnish or resin. Used to remove foreign material from the surface prior to applying a finish. TACK RAG.

tack coat: Asphaltic emulsion applied to an existing roadway to create a bond between it and the new layer of pavement.

tack hammer: Lightweight hammer used to drive small fasteners in finish work or upholstery. See *hammer**.

tackle: See *block and tackle*.

tackless strip: Narrow piece of wood or metal with small projections; nailed around the perimeter of a room for securing carpet.

tack track: See *tackless strip*.

tack weld: Weld made to secure parts of a weldment in position until final welds are made.

tacky: Slightly sticky.

taenia: Raised band or fillet between the architrave and frieze in the Doric order of architecture.

tag: Strip of copper or sheet metal folded over to create a wedge. Inserted in a masonry joint to secure metal flashing.

tag line: **1.** Line that extends from a crane boom to a clamshell bucket and secures the bucket firmly in position during operation. **2.** Rope attached to a truss, structural steel member, or other large member being lifted. Used to control the member by personnel on the ground. **3.** Safety line attached to a safety belt or other safety equipment to prevent injury in case of a fall.

tail: **1.** Portion of a roof rafter that extends beyond the wall line. RAFTER TAIL. **2.** Portion of a stone step built in-to a wall. **3.** Rear deck of a mechanical shovel. **4.** Exposed portion of a slate shingle.

tail bay: Area beneath a floor between an outside wall and the closest beam or girder.

tail beam: See *tail joist*.

tail cut: See *seat cut*.

tailing: Portion of a projecting member, such as a cornice, that is embedded in a wall.

tailing in: Fastening one end of a member.

tailing iron: Steel member embedded in a wall to support the upward thrust of a cantilevered member.

tailings: **1.** Stone that does not pass through a sieve. **2.** Waste material or residue of a product.

tailing the lead: Aligning the ends of brick courses at a corner.

tail joist: Horizontal supporting member that fits against a header joist. TAILPIECE, TAIL BEAM.

tailpiece: **1.** Short horizontal support member, such as a beam, joist, or rafter, supported by a header joist at one end and a wall at the other end. **2.** Short pipe used to connect a fixture to a trap.

tail rope: Rigging line that passes around a return sheave.

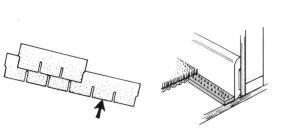

tab   tackless strip   tail (1)

tambour door

tamper (2)

tang

tap (1)

Tainter gate: Control gate with a curved face that pivots vertically to control flow of water through a dam.

take-up: Device that removes slack from a belt or line.

talus: **1.** Loose gravel and soil at the foot of a steep rock slope formed by disintegration of rock. **2.** Slope of a wall.

tambour: Cylindrical member that forms the shaft of a stone column.

* **tambour door:** Sliding door constructed of thin strips of material, such as wood, fastened to a flexible backing. Ends of strips slide within a recessed track. Commonly used on rolltop desks.

tamp: To consolidate and compact a material or surface with repeated blows.

* **tamper:** **1.** Implement with an expanded steel mesh base used to consolidate fresh concrete and bring the paste to the surface for finishing. See *Jitterbug® tamper.* **2.** Manual or motorized device used to compact soil, asphalt, or other material. Flat plate attached to bottom of the device is used to repeatedly strike the surface. See *compactor.*

tamping rod: Steel bar, approximately 24″ long and ⅝″ in diameter, used to consolidate concrete for a slump cone test.

tandem: Pair of devices or units in which one member follows the other; e.g., tandem truck.

* **tang:** Projecting shaft of a tool that secures it to the handle, such as a screwdriver.

tangent: Straight line or curve that contacts an arc or curve at one point only.

tangent arch: Arch used to span a large area without center supports. Sides are slightly inclined inward and joined with sloped members forming the top of the arch. Common design for structural metal buildings.

tangential shrinkage: Lumber shrinkage across the grain.

tangent sawed: See *plain-sawn.*

tank ball: See *float* (4).

tankless heater: Device that heats water by passing it through a series of heated coils. Heater does not store water other than than the water contained in the coils.

tank sprayer: Portable spray equipment that stores liquid, such as form oil, under pressure for application through a hose and nozzle.

* **tap:** **1.** Tool used to cut internal threads in a predrilled hole. **2.** To cut internal threads in a predrilled hole. **3.** Connection to a transformer winding used to vary the transformer turn ratio and control current and voltage. **4.** See *faucet.* **5.** Hole drilled in a water main for the connection of a corporation cock.

tap bolt: See *hex cap screw.*

tape: **1.** See *duct tape, electrical tape, reinforcing tape, steel tape.*

tape creaser: Tool that stores a roll of reinforcing tape and folds it in half lengthwise while dispensing it. Creased tape is fitted to inside corners.

tape measure: See *steel tape.*

taper: **1.** Person who applies reinforcing tape and joint compound to joints between sheets of gypsum drywall. **2.** Gradual reduction in thickness, width, or diameter of an object.

taper file: Triangular file with fine teeth; used to sharpen saw teeth.

taper pile: See *step-taper pile.*

tapersplit shake: Wood shingles with split faces on each side. Produced by cutting with a mallet and a froe (sharpened steel blade).

taper shank: Tapered end of a tool, such as lathe center or end mill, that fits into the spindle or socket of corresponding taper. Provides accurate alignment of tool and less resistance for driving the tool. Common tapers include Morse Taper, American National Standard Machine Taper, and Brown and Sharpe Taper.

taper thread: Screw thread that tapers approximately ¾″ per foot. Commonly used on pipes and fittings to form an airtight joint.

taper tie: Concrete form hardware made of a steel rod with a larger diameter at one end than at the other. Used to secure wall forms in place. After concrete sets, device is removed by striking small end and driving tie through the void created by larger end. See *tie* (2)*.

taping compound: Drywall joint compound applied as an initial filler and a base for reinforcing tape. Not designed for finish coat. See *joint compound, topping compound.*

taping strip: Narrow piece of roofing felt applied over joints of precast concrete roof members. Used to prevent roofing material, such as asphalt, from seeping into area below the roof.

tapped tee: Cast iron pipe fitting with a tapped outlet to receive an externally threaded pipe.

tapping screw: See *self-tapping screw.*

tap wrench: Hand tool used to hold a tap. See *tap* (1).

tar: Dark, heavy oil used in roofing, pavement, and other water-resistant applications. Derivative of coal, peat, shale, and resinous wood.

tar-and-gravel roofing: See *built-up roofing.*

tare: Weight of a rail car, truck, or other means of transportation when empty. Deducted from the weight of the vehicle and its load when determining freight rate.

target: Red and white disk that slides along a surveying rod to facilitate reading the rod graduations. See *surveying*.

tar paper: See *building paper.*

tarpaulin: Waterproof canvas or plastic sheet commonly used to protect building material exposed to weather.

taut: Without sag or slack; tight.

T bar: **1.** T-shaped metal device used to support a line over a trench. **2.** T-shaped, light-gauge metal suspension member in a suspended ceiling. See *main tee.*

T-bar tie: Heavy-gauge, T-shaped steel strap used to connect perpendicular masonry walls. Set in the mortar joints during construction.

T beam: T-shaped reinforced concrete or rolled metal member.

T bevel: See *sliding T-bevel.*

teagle post: Post used to support one end

of a tie beam.

teazle: Angular support member in timber construction.

tee: **1.** T-shaped pipe fitting. See *pipe fitting**. **2.** See *main tee*.

tee handle: T-shaped handle used to operate the bolt on a door lock. Replaces a doorknob.

tee iron: T-shaped, heavy-gauge sheet metal with predrilled, countersunk screw holes. Used to reinforce joints in wood construction. T PLATE.

tee nut: Fastener consisting of an internally threaded cylinder and disk with prongs projecting from the perimeter of the disk. Prongs are embedded in wood to secure the nut during insertion of a bolt. See *nut**.

tegula: Irregularly shaped tile.

Tek® screw: See *self-tapping screw*. TEC SCREW.

telescope: **1.** Component of a surveying instrument that magnifies the field of vision. Mounted in a frame that secures it in a stationary position. **2.** To increase or decrease in length by members sliding inside one another.

telltale: Device used to indicate movement of formwork.

tellurometer: Electronic distance measuring device that measures the time between transmission and reception of microwaves aimed at an object. Time measurement is converted to lineal feet and inches.

temper: **1.** To add water to a concrete or mortar mix to retain its workability. **2.** To relieve stress in metal by heating it to a certain temperature and cooling it at a specific rate. **3.** To treat a wood surface with oil or other resins to improve water resistance.

tempera: Quick-drying, water-soluble paint composed of egg, gum, pigment, and water.

temperature and pressure relief valve: Safety valve used to protect against development of excessively high temperature and/or high pressure in a water heater.

temperature reinforcement: Reinforcement used in a concrete slab to resist stresses resulting from temperature changes.

temperature indicating crayon: Marking stick used to test a steam trap or other thermal equipment. Crayon mark melts if specified temperature is surpassed. TEMPERATURE STICK.

tempered air: Outside air that is heated prior to distribution.

tempered glass: Glass that is prestressed by heating and rapidly cooling. Used to increase strength of the glass.

tempered hardboard: Pressed wood panel in which the surface is impregnated with a compound of oil and resin and baked to form a water-resistant finish.

tempered water: Water with a temperature between 85°F and 110°F.

template: **1.** Guide or pattern commonly used when forming multiple identical components or designs. TEMPLET. **2.** Stone, metal, or wood member embedded in a wall to receive a beam or other horizontal structural member and distribute the load. TEMPLET.

template hinge: Hinge with holes in the leaves designed to fit the recess and screw hole arrangement of hollow metal doors and frames.

templet: See *template* (1).

tender: Person who mixes mortar and transports mortar and masonry material to the desired location for a bricklayer or mason. MASON TENDER.

tendon: Steel rod, cable, or wire used to apply compressive stress to a precast concrete member.

* tenon: Projecting end of a member, commonly wood, inserted into a corresponding recess of another member.

tenoner: **1.** Accessory for a table saw used to shape tenons. **2.** Woodworking equipment used to shape tenons.

tensile strain: Elongation of a member subjected to tension.

tensile strength: Maximum unit stress a member is capable of withstanding under axial tensile loading. Based on cross-sectional area of the member before the load is applied.

tension: Condition of being pulled or stretched.

terminal: **1.** Electricity-conducting connection of a circuit or equipment used for the connection of an external conductor. **2.** Device attached to the end of a conductor to facilitate connecting to another conductor. **3.** End of a structural or ornamental member.

terminal box: Box on electrical equipment in which connections between equipment windings and incoming supply leads are made. Provided with a removable cover for access.

terminal element: Means by which water, electricity, and heat are ultimately delivered to the desired area; e.g., faucets, light fixtures, or registers.

terminal reheat system: Air conditioning system in which a heating element is installed in individually controlled zones to provide additional control of air temperature.

terminal velocity: **1.** Average velocity of an air stream as it is discharged from a grill. Indicates comfort level and degree of draftiness. **2.** Maximum speed

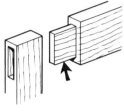

tenon

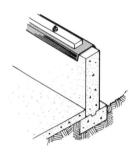

termite shield

an object attains when allowed to fall through a liquid.

* termite shield: Sheet metal or inorganic material installed between the sill plate and top of a masonry or concrete wall or around pipes entering the structure. Used to prevent passage of termites from the ground to the structure.

terne metal: Lead alloy with 20% lead; used as a surface coating for roofing material.

terrace: **1.** Embankment with a level top. **2.** Flat roof or elevated platform adjoining a structure.

terra-cotta: Hard-baked clay product with an unglazed surface. Used for ornamental work and floor and roof tile.

terras: Defect in marble.

terrazzo: Mixture of cement and water with colored stone, marble chips, or other decorative aggregate embedded in the surface. Surface is ground and polished when hardened to produce a hard, durable surface.

tertiary beam: Beam that transfers its load to a secondary beam at one or both ends.

tesselated: Formed with small pieces of stone, marble, or glass, as applied to a surface.

tessera: Small squared piece of stone, marble, or glass used to make a mosaic pattern.

test cylinder: Concrete sample that measures 6″ diameter and 12″ high; used to determine compressive strength. Sample is allowed to cure for a predetermined amount of time and is subjected to compressive force until it fractures.

test pile: Driven pile used to determine driving conditions and approximate bearing capacity of the soil.

test pit: Shallow excavation used to examine existing soil conditions.

* **test plug:** Device used to obstruct openings in a piping system during a pressure test.

test pressure: Amount of hydrostatic or pneumatic pressure exerted on a piping system during a pressure test.

test tee: Pipe fitting used to provide access for a test plug when conducting a pressure test on a piping system.

tetrastyle: Architectural design characterized by four columns.

tewel: Ventilation opening.

texture: Visual or tactile quality of a surface.

textured finish: Rough or irregular finish; surface finish that is not smooth.

texture float: Plastering tool used to impart various textures to a surface.

* **T head: 1.** T-shaped shore formed with a short horizontal member positioned and braced on top of a vertical post. **2.** Portion of a precast girder across the top of an interior column.

theodolite: Precision surveying instrument consisting of an alidade, a leveling device, and a graduated horizontal circle. May incorporate a graduated vertical circle. Used to establish and verify horizontal and vertical angles. See *surveying**.

therm: Quantity of heat equal to 100,000 Btu.

thermal break: Material or member with low thermal conductance placed between two pieces of material or members with higher thermal conductance. Used to limit or prevent thermal transmission.

thermal coefficient of expansion: Measure of the tendency of material to expand when heated. Expressed as fractional inch of expansion per inch of length per degree Fahrenheit.

thermal conductance: Amount of time required for heat to flow through a given area of a material with a specified size and shape. Expressed as Btu-inch

per hour-square foot-Fahrenheit degree.

thermal conductivity: Rate of heat transfer by conduction. Represented by the letter k. K FACTOR.

thermal conductor: Material that conducts thermal energy.

thermal cutout: Overcurrent protection device containing a heating element that affects a renewable fuse that opens the circuit.

thermal cutting: Cutting process in which the metal to be cut is melted.

thermal delay: Time between energization of a forced air or radiant heater and when measurable heat is produced.

thermal expansion: Elongation of a material or object when subjected to heat.

thermal insulation: Material with a high thermal resistance and that limits transfer of heat. Common materials include mineral wool, foamed plastic, and fiberglass.

thermal load: Load on a structure resulting from temperature variations.

thermally protected motor: Electrical motor equipped with a thermal protector that prevents damage to the motor that would be caused by overheating as a result of overload or failure to start.

thermal overload protection: Method used to detect excessive heating of capacitors caused by current, ambient temperature, and solar radiation by initiating an alarm signal or closing the capacitor bypass switch.

thermal protector: Device that is an integral part of a motor or motor compressor that protects the motor against overheating as a result of overload or failure to start.

thermal radiation: Transmission of heat energy by means of electromagnetic waves.

thermal regenerator: Device installed in a heating duct to recover excess heat and distribute it to other areas of the system.

thermal resistance: Ability of a material to resist the transfer of heat. Opposite of thermal conductance. See *R value*.

thermal transmittance: Rate at which heat flows through a material. Based on

the number of Btu per hour that pass through 1 square foot of material when air temperature is 1°F different from one side to another. See *U value*.

thermal unit: See *British thermal unit (Btu)*.

thermistor: Thermally sensitive resistor in which electrical resistance changes with a change in temperature. TEMPERATURE-SENSITIVE RESISTOR, THERMAL RESISTOR.

thermit welding: Welding process in which metal is joined by heating it with superheated liquid metal produced by a chemical reaction between aluminum and a metal oxide, with or without applying pressure. Filler metal is obtained from the liquid metal.

thermocouple: Device in which two dissimilar conductors are joined at two points so that voltage is produced across the conductors when at two different temperatures. Used to measure temperature.

thermodynamics: Physics dealing with mechanical action and heat energy.

thermography: Process used to detect and measure heat emitted from a structure. Emitted heat is converted to a color picture of the structure in which various rates of thermal transmission are color coded.

thermometer: Device used to measure temperature. Graduated in degrees Celcius or degrees Fahrenheit.

Thermopane®: Two panes of glass sealed around the edges with an air space between them, used to create an energy efficient window.

thermoplastic: Synthetic material that becomes soft and pliable when heated and hardens when cooled. Includes nylon, polystyrene, urethane, and vinyl.

thermosetting plastics: Synthetic material that is permanently formed with application of heat and pressure. Additional heat does not soften the plastic. Includes epoxy, polyester, and silicone.

* **thermostat:** Automatic temperature-sensitive device that opens and closes a circuit to regulate the temperature of an area. Used in an HVAC system.

thermostat bulb: Temperature-sensing apparatus in a thermostat.

thermostatic steam trap: Device used to purge air and condensate from steam lines and heat exchangers without losing steam. When steam contacts the element, it expands and closes the valve. When air or condensate contacts the element, it contracts and opens the valve, allowing air or condensate to pass.

thermosyphon: Natural circulation of a

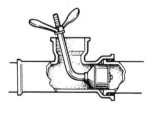

test plug

T head (1)

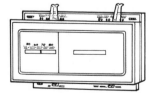

thermostat

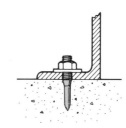

| thimble (2) | threaded stud | threshold | thumb screw |

fluid in which fluid rises as it is heated and falls as it cools.

thickness gauge: See *feeler gauge.*

*__**thimble:**__ **1.** Protective metal sleeve that passes through the wall of a chimney to hold the end of a stovepipe. **2.** Tear-drop-shaped rigging hardware inserted in the eye of an eye splice to prevent wear on the rope or line. **3.** Short sleeve through which a bolt passes, such as a bushing or coupling.

T-hinge: Hinge with a wide inactive leaf and a long, narrow active leaf. See *hinge*.*

thinner: Volatile liquid used to dilute and lower viscosity of paint and adhesive.

thin-wall conduit: See *conduit (1).*

tholobate: Circular substructure that supports a dome or cupola.

thread: Continuous, uniform helical groove on the internal or external surface of a cylinder. Used to hold parts together, adjust parts with reference to one another, and transmit power.

threaded and coupled pipe: Pipe with external threads on both ends with an internally threaded coupling on one end.

thread-cutting screw: Fastener with a slotted tip; used to cut threads in and join metal up to $1/4$" thick. See *screw*.* SELF-TAPPING SCREW.

threaded insert: Device that is turned into a predrilled hole to form screw threads. Screw or bolt is fastened directly into the hole.

*__**threaded stud:**__ Fastener that is pointed on one end and threaded on the other end. Pointed end is driven into a hard material, such as concrete, with the threaded end projecting above the surface. Structural members are fastened to the stud.

thread escutcheon: Small metal plate placed around a small opening such as a keyhole.

#3 clear: Highest grade of shop lumber. THIRD CLEAR, FACTORY SELECT.

three-coat finish: Plaster coating applied in three separate layers. Required on expanded metal lath and clip-supported gypsum lath.

three-hinged arch: Glued-and-laminated wood arch constructed in two parts that are connected with hinges or pins at the job site.

three-inch brick: Masonry unit measuring 3" × 2⅝" or 2¾" × 9⅝" or 9¾". See *brick*.*

three-part line: Single length of rope that is doubled back on two sheaves.

three-phase motor: Induction motor in which the voltage differs in each phase by 120°.

three-point lock: Device that secures the active leaf of a pair of doors at three locations. Used on doors requiring a three-hour fire rating.

three-prong plug: Electrical plug in which two flat projecting metal prongs carry the current and a cylindrical prong is used to provide a ground. GROUNDED PLUG.

three-quarter turn stairway: Stairway which turns 270° in its total rise.

three-tab shingle: Asphalt composition roof covering that is 36" long and 12" wide. Two slots, approximately 5" long, are cut into the edge along the 36" length at 12" intervals. See *asphalt shingle*.*

three-way switch: Electrical switch used to control a fixture from two different locations, such as two ends of a hallway.

three-way valve: Valve that has two inlets and one outlet or two outlets and one inlet.

three-wire system: Electric supply system consisting of three conductors: a neutral wire and two outer conductors. Portion of the load is connected between the outer conductors and the remainder may be split as evenly as possible between the neutral and one outer conductor. Two distinct supply voltages are available, one which is twice the other.

*__**threshold:**__ Horizontal member at the bottom of a door opening.

thriebeam: Galvanized steel plate with three raised sections extending lengthwise. Commonly used as a bridge railing.

throat: **1.** Section of a chimney between the fireplace and smoke chamber. **2.** Groove along the underside of an exterior member used to divert water away from the structure. **3.** Thinnest area of support of a stair stringer. **4.** Thinnest area of a concrete stairway; between the intersection of the back of the tread and bottom of the riser and underside of the concrete soffit. **5.** Thinnest portion of a weld.

6. Opening in a wood plane through which shavings pass.

thrombe wall: Masonry wall used to store thermal energy in a passive solar system. 8" to 16" thick wall is constructed and painted a dark color to absorb the sun's rays. Glass is placed approximately 4" from the wall on the side that receives most of the sunlight.

through bond: Transverse bond formed by masonry units extending through a wall.

through cut: Excavation between parallel embankments that starts and ends at the same elevation. Commonly performed in roadwork.

through shake: Shake that extends between two faces of timber.

through stone: Transverse bond formed by stone extending through a wall. BONDSTONE.

through tenon: Tenon that extends completely through the member in which the mortise is cut.

throw: **1.** Distance that a stream of air travels from an outlet to the point of terminal velocity. **2.** Number of movable contact positions in a multiposition switch. **3.** Longest distance that a reciprocating or rotating part travels; e.g., the throw of a door bolt. **4.** Distance between a light fixture and the area being illuminated.

thrust: Force exerted on or by a structure.

thrust bearing: Shaft support designed to resist its end thrust.

thrust washer: Washer that secures a rotating part in position and prevents lateral movement in relation to its bearings.

thumb latch: Door hardware consisting of a handle fitted with a pivoting bar that extends through a door and lifts a straight bar attached to the door face.

thumb molding: Narrow convex molding.

thumb plane: Small, narrow wood plane.

*__**thumb screw:**__ Threaded fastener with a flattened, knurled head that facilitates turning without a tool.

TIES

Wire

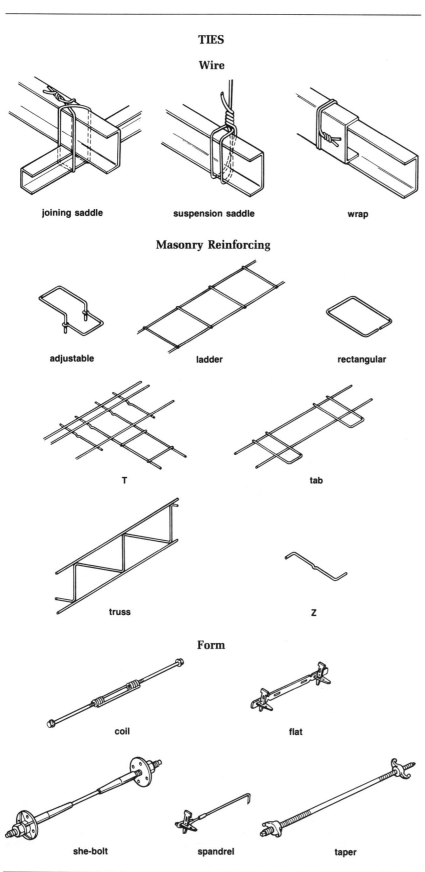

joining saddle   suspension saddle   wrap

Masonry Reinforcing

adjustable   ladder   rectangular

T   tab

truss   Z

Form

coil   flat

she-bolt   spandrel   taper

thurm: To fabricate moldings by cutting across the grain of the wood with a saw and chisel.

* **tie:** **1.** Member or device used to secure two or more members together; e.g., wall tie. **2.** Tensile unit used to secure concrete forms against lateral pressure of unhardened concrete. SNAP TIE. **3.** Loop of rebars encircling longitudinal reinforcement in columns. **4.** Surveying connection from an established reference point to a point whose position is desired.

tieback: Rod attached to a deadman, rigid foundation, or rock or soil anchor used to prevent lateral movement of formwork, retaining walls, or bulkheads.

tie bar: Short rebar used to join adjacent slabs or concrete members.

tie beam: **1.** See *collar beam*. **2.** Concrete beam that connects and distributes loads to pile caps or spread footings.

tier: **1.** Row or series of rows placed one above the other. **2.** See *wythe*. **3.** Series of contiguous townships in an east/west direction.

tie rod: See *tie* (2).

tie wire: 16-gauge annealed steel wire used to secure reinforcing steel, metal lath, or other members in position.

* **tie wire reel:** Cylindrical metal container with a hinged cover; used to store and dispense tie wire.

tige: Shaft of a column.

tight coat: Thin layer of plaster.

tight knot: Knot that is securely embedded in the grain of the wood.

tight side: Side of sliced wood veneer that is opposite the blade during the slicing operation.

tigna: See *collar beam*.

TIG welding: Tungsten inert gas welding. See *gas tungsten arc welding*.

tile: Thin building material made of cement, fired clay, glass, plastic or other resilient material, or stone and used as a finish for walls, floors, or roofs. See *acoustical tile, ceramic tile, clay tile, drain tile, quarry tile*.

tile creasing: Weatherproof covering installed on top of a masonry wall in which two courses of tile project beyond both faces of the wall.

* **tile cutter:** Hand tool consisting of a blade used to score the face of ceramic tile and a breaking ridge used to snap the tile. CERAMIC TILE CUTTER.

tile hammer: Striking tool similar to a brick hammer but smaller. Used to cut glazed brick and tile.

tile handler: Device used to move and lift smooth tile by means of suction.

tile nipper: Hand tool used to cut small

amounts from the edge of a ceramic tile.

tile pick: Striking tool with a pointed end used to form holes in tile.

tilting mixer: Small concrete or mortar mixer consisting of a rotating drum mounted in a frame that allows the drum to pivot and discharge its contents. Material is introduced into the drum when it is in an upright position.

* **tilt-up construction:** Method of concrete construction in which members are cast horizontally at a location close their final position, and tilted in place after removal of the forms.

tilt-up door: See *overhead door*.

timber: **1.** Square-sawn lumber with minimum overall dimensions of 5″. **2.** Generic description of heavy wood members.

timber construction: See *half-timbered construction*.

timber carrier: Tool with a long handle attached to hooks or tongs designed to grip into a log. Used by two people to carry logs or timber.

timber connector: Metal device used with a bolt to join and reinforce timber joints. Common connectors include a spiked grid and shear plate.

timber hitch: Knot used to tow or hoist cylindrical objects. Made by passing a rope around an object, wrapping the end around the standing part, and wrapping the end around the rope several times. See *knot*.

timber pile: Large, straight log driven into the ground for foundation support. Logs are debarked and pressure-treated before being fitted with steel bands and/or a drive shoe. See *pile* (1)*.

time delay fuse: Fuse that holds five times its rating for 10 seconds. Allows motor to start without opening the electrical circuit.

time lag heating: Passive solar heating method in which solar energy is absorbed by a dark material during the day and released as thermal energy at night.

time of haul: Period of time from initial contact between water and cement until termination of discharge of fresh concrete.

tin: Soft white metal that is malleable and has a low melting point. Used for alloys, solder, and a sheet metal coating.

tine: Tooth or projection of a scraping or digging tool.

tingle: See *trig*.

tinker's dam: Small built-up area of sand or other noncombustible material that retains molten solder in the desired location.

tie wire reel

tile cutter

tilt-up construction

tinner: See *sheet metal worker*.

tin snips: Shears used to cut light-gauge metal.

tint: Light color obtained by mixing a pure color with white.

tip: **1.** Point of a pile. **2.** Thinnest end of a wood shingle or shake. **3.** See *cutting tip*.

* **tip cleaner:** Wire device used to remove carbon or other debris from the tip of an oxyacetylene torch.

title block: Area on a working drawing or print used to provide written information about the drawing or print. Commonly includes architect's name and seal, name of the structure, date and location of construction, revision dates, and sheet numbers.

title sheet: First page of a set of prints that provides general written information about a structure. Commonly includes a pictorial rendering, list of sheets and contents in the set, architect's name and seal, and name of structure.

T-joint: **1.** T-shaped pipe fitting. **2.** T-shaped weld joint in which two members are at approximately a 90° angle to each other. See *weldment*.

T-lap: See *middle lap*.

toe: **1.** Tip of a bored hole in which an explosive is placed. **2.** Base of a retaining wall on the opposite side of the retained material. **3.** Widened base that gives an object a wider bearing base. **4.** Lower portion of the lock stile of a door.

* **toeboard:** **1.** Strip of wood attached to

perimeter of a scaffold deck. Prevents workers and objects from falling. TOE PLATE. **2.** Recessed member along bottom of the front of a base cabinet or vanity. TOEKICK.

toehold: Device or batten temporarily fastened to a sloping surface used as a footing for workers.

* **toenail:** To drive a nail at an angle. Should be driven so that one-half the nail is in each member.

toe of weld: Joint between the face of a weld and base metal.

toeplate: **1.** See *toeboard* (1). **2.** See *kick plate*.

toe wall: Low wall constructed at the bottom of an embankment.

toggle bolt: Anchoring device consisting of a bolt and nut with pivoting wings that close against a spring when inserted through a hole and open when emerging from other side; used to fasten objects to a hollow wall. See *anchor*.

toggle switch: See *general-use snap switch*.

toilet: See *water closet* (2).

tolerance: Permissible deviation from a given value or dimension.

ton: **1.** Unit of weight equal to 2000 pounds. **2.** Measure of the effectiveness of refrigerant. Equal to removal of heat at the rate of 12,000 Btu per hour.

tondino: Circular molding.

tongs: **1.** See *brick tongs*. **2.** Device with pivoting arms that are used to grasp and lift objects. **3.** Long-handled tool used to hold and work with hot metal.

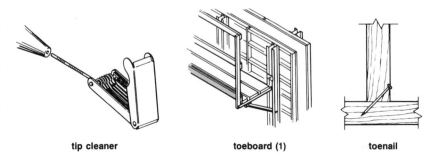

tip cleaner   **toeboard (1)**   **toenail**

tongue: 1. Continuous projection along edge of a wood member that fits a corresponding groove in an adjoining piece. **2.** 16″ long narrow portion of a framing square. **3.** Elongated rod or drawbar on a towed vehicle.

tongue-and-groove: Joint between two wood members in which one member has a continuous projection along the edge that fits a corresponding groove in an adjoining member. See *joinery**.

tongue-and-groove pliers: Pliers with serrated jaws that are adjustable to various sizes. Jaws are self-locking for additional control of the workpiece. CHANNELOCK®. See *pliers**.

tongue joint: Joint between two wood members in which one member has a continuous tapered projection along the edge that fits a corresponding tapered groove in an adjoining member.

tool brush: Small brush used to apply water to small areas and finish ornamental plasterwork.

tooled finish: Stone surface with 2 to 12 concave grooves per inch. TOOLED SURFACE, TOOLING.

tooled joint: Mortar joint that has been compressed and formed before the mortar rigidly sets.

tooth: 1. Single cutting point on a saw or other cutting tool. **2.** Fine texture in paint film produced by pigments or abrasives used in sanding. Provides a good surface for adhesion of subsequent coats of paint.

tooth chisel: See *stonecutter's chisel.*

toothed ring: Steel ring with serrated edges used as a timber connector.

toothing: Removing alternate courses of old brickwork to form a bond with new masonry.

toothing plane: Plane with a serrated blade used to roughen a surface prior to application of veneer.

top beam: See *collar beam.*

* top chord: Upper member of a truss.

top-hung window: See *awning window.*

toplap: Amount that an upper member overlaps a lower member.

top maul: Lightweight sledgehammer used to drive large spikes.

topographical map: Configuration of a surface in which natural and fabricated features are shown in relief. Contour lines are used to indicate elevations.

top out: To install the highest structural member or complete the highest masonry course of a structure.

topping: 1. Layer of concrete or mortar applied to a concrete base to form a floor. **2.** Mixture of marble chips and mortar, which produces a terrazzo surface. **3.** Structural cast-in-place surface used with precast floor and roof systems. **4.** Thin finish layer of material.

topping compound: Drywall tape applied as second and third coats over taping and all-purpose compounds. Also used as a skim coat. Not suitable for embedding tape or for use as a first coat. See *joint compound, taping compound.*

topping lift: Crane lines used to raise and lower the boom.

* top plate: Uppermost horizontal member of a framed wall.

top rail: Highest horizontal member of a panel door or window sash.

topsoil: Uppermost layer of earth that supports plant growth.

torch: 1. Device used to direct preheating flame that is produced by combustion of fuel gases, and also direct and regulate the control of cutting oxygen. **2.** Device used in oxyfuel gas welding, torch brazing, and soldering to direct heating flame produced by combustion of fuel gases. **3.** Device used in gas tungsten and plasma arc welding processes to control electrode position, transfer current to the arc, and direct flow of shielding gases.

torpedo level: Small hand level used to level and plumb short members.

torque: 1. Force that produces rotational motion. **2.** Measure of the force applied to a member to produce rotational motion. Determined by multiplying amount of applied force by distance from the pivot point to the point where force is applied.

torque wrench: Wrench used to determine rotational force applied to a fastener, such as a nut or bolt. Gauge indicates the foot-pounds of force applied to the fastener. See *wrench**.

torsel: Structural member that supports the end of a beam or joist and distributes the load.

torsion: Twisting of a structural member about its longitudinal axis by two equal but opposite torques.

Torx® recess: Fastener recess in the shape of a six-pointed star. See *screw**.

total force: Amount of steam pressure pushing against the disc of a safety valve.

total heat: Sum of sensible and latent heat.

total pressure: Sum of velocity pressure and static pressure. Expressed as inches of water.

total pressure loss: Amount of air pressure lost in a duct system as a result of air friction with internal duct surfaces.

* total rise: 1. Vertical distance from surface of a lower floor to the surface of the floor above. Used to determine unit rise in stair construction. **2.** Vertical distance from the top plate or top of the wall to the top of the roof ridge.

* total run: 1. Horizontal distance from the face of the bottom riser to the face of the top riser. Used to determine unit run in stair construction. **2.** Horizontal distance from the outside of a wall to the intersection of roof rafters at the ridge in an unequal pitch roof. **3.** Horizontal distance from the outside of a wall to the center of a structure in an equal pitch roof. Equal to one-half the span.

touch latch: Door hardware in which the latching mechanism is concealed behind the door. Latch is released by lightly pushing on the door. Used on cabinet doors. TOUCH CATCH.

touch-sanded panel: Structural wood panels that are sized to uniform thickness by lightly sanding the surface. Commonly used for C-plugged faces.

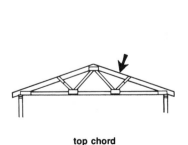

top chord

top plate

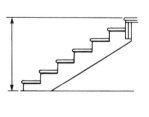

total rise (1)

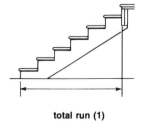

total run (1)

tow conveyor: Endless movable chain suspended by an overhead track or recessed in a floor track. Fittings along the chain allow for attachment of carts or dollies.

tower bolt: See *barrel bolt*.

tower crane: Crane consisting of a fixed vertical mast with a pivoting boom at the top. Winch moves along the length of the boom to provide access to any point within the diameter of the boom. Crane lines are attached to the winch for hoisting loads. LINDEN CRANE.

tower harness: Safety harness that distributes impact forces of a fall to the thighs, buttocks, chest, and shoulders. Harness fits around the shoulders, thighs, and legs of a worker. Used for applications such as transmission towers. Lifeline is attached to a ring on the back of the harness.

townhouse: Two- or three-story one-family dwelling connected to a similar structure with a common sidewall.

township: Square area of land measuring 6 miles on each side, with boundries formed by meridians and baselines.

toxic: Poisonous material or substance.

* **T-plate:** T-shaped steel reinforcement with predrilled holes to accept fasteners. Used to reinforce perpendicular butted wood members.

* **TPT pile:** Pile consisting of an enlarged precast concrete tip with a thin-shelled corrugated shaft. Driven with a pile driver equipped with a mandrel.

trabeate: Architectural style characterized by horizontal beams and lintels supported by columns and posts.

track: 1. Metal channel used to guide and support sliding or bi-fold doors. 2. Light-gauge, U-shaped metal member fastened to the floor and ceiling for a light-gauge metal-framed wall. Metal studs are fastened between the tracks. RUNNER. 3. Heavy equipment component used to support and convey the equipment.

tracking collector: Solar energy collector that automatically repositions itself to obtain maximum amount of sun's rays.

track roller: Wheel that supports and guides the track on heavy equipment.

traction: Friction between two contact surfaces, such as the track of heavy equipment and soil.

traction steel: Steel used to manufacture wire rope. Strength ranges from 180,000 to 190,000 pounds per square inch.

tractor: Mechanical equipment that tows heavy excavating equipment.

trammel: Cross-shaped device used to lay out circles and ellipses for plaster moldings. Consists of two arms that move in perpendicular channels.

* **trammel point:** Device used to lay out circular designs. Consists of a pointed tip that is attached to a clamping frame. Frame is attached to a rod or beam. Used in pairs. TRAMMEL HEAD.

transducer: Device that converts one type of energy into another; e.g., solar energy system that converts solar energy to thermal energy.

transfer bond: Stress of a pretensioned concrete member resulting from stress transferred from the tendon to the concrete.

transfer grill: Grill that allows air to flow from one area to another.

transfer medium: Material that absorbs and transports thermal energy from a solar collector to storage or living areas.

transfer strength: Compressive strength required for a concrete member before stress is transferred from the stressing mechanism to the concrete.

transformer: 1. Electrical device designed to increase or decrease voltage of an alternating current circuit. 2. See *air transformer*.

transistor: Semiconductor with three terminals; used to amplify electrical signals. Terminals are the emitter, base, and collector.

transit: Surveying instrument with a telescope that rotates vertically and horizontally. Used to establish grades and elevations, and lay out vertical and horizontal angles. See *surveying**.

transition belt: Short belt conveyor that moves material from a loading point to the main conveyor.

* **transition piece:** Sheet metal fitting used between two ducts with different shaped mating surfaces.

transit-mixed concrete: Concrete that is at least partially mixed in a truck mixer.

translucent: Opaque finish that allows passage of light but diffuses it sufficiently so as not to allow clear sight through it.

transmission: Passage of energy, such as thermal or electrical, through a material or structure.

transmissivity: Ability of a material to transmit thermal energy.

* **transom:** Small window or panel over a door or window.

transom bar: Horizontal member that separates a transom from a window or door.

transverse: Placed perpendicular to the length or primary pattern of a member or structure.

transverse section: View of an object made by a cutting plane extending across it on its shorter dimension.

transverse wire: Welded wire fabric member that extends across the width of the fabric. Commonly spaced 4″, 6″, 8″, 10″, or 12″ apart.

trap: 1. Downward bend in a waste pipe used to retain liquid to prevent sewer gas from entering the living area. 2. See *trapdoor*. 3. See *traprock*.

trap arm: Portion of a fixture drain between the vent pipe and trap weir.

trapdoor: Door that is set flush with a ceiling, wall, or floor.

traprock: Fine-grained, dark-colored igneous rock. Used as base under or between layers of asphalt paving.

trap seal: Water between the crown weir and upper dip of a trap. Used to prevent sewer gas from entering the living area.

trass: Natural pozzolan with cementitious properties; derived from volcanic rock.

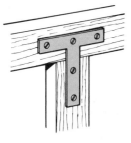

T-plate

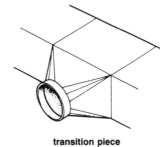

trammel point

transom

transition piece

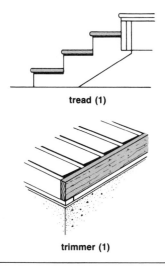

tread (1)

trimmer (1)

T-rated switch: Switch used for a tungsten filament lamp.

trave: 1. Horizontal structural member extending across a structure. **2.** Recessed area in a ceiling formed by intersecting beams.

travel angle: Angle formed between an electrode and a reference line perpendicular to the axis of the weld and in the plane of the weld axis. Also applicable to welding guns, torches, high-energy beams, and welding rods.

traverse: 1. To plane wood across the grain. **2.** Barrier across an opening to allow entry by officials but not unauthorized personnel.

traverse rod: Curtain rod.

travertine: Sedimentary stone used for interior flooring and trim.

*tread: 1. Horizontal portion of a stairway including the nosing; a step. **2.** High friction coating on a sheave. Used to minimize slippage of the belt.

tread return: See return.

tread shoe: Hinged steel pad joined together with other shoes in a continuous ring to support a crane or other heavy equipment.

tread width: Distance from front of the nosing to rear edge of the tread in a stairway.

treated: Coated or saturated with a stain or chemical to retard fire, insect damage, or decay resulting from exposure to weather.

trefoil: Ornamental three-lobed design.

trellis: Wood or metal latticework used to support vines.

tremie: Pipe or tube through which concrete is placed into vertical formwork or under water. Hopper at the upper end facilitates filling the tremie. Lower end is immersed in plastic concrete when placing concrete under water.

trench: Narrow excavation in the ground.

trench box: Reinforced wood or metal assembly used to shore the sides of a trench. Moved along the trench as work progresses.

trench duct: Metal duct embedded in a concrete floor to carry electrical conductors. Access cover is flush with floor surface.

trench excavator: Equipment used to cut narrow trenches in the earth. A continuous chain mounted on an arm rotates to make the cut. TRENCHER.

trench jack: Hydraulic or screw jack used as cross bracing to shore the sides of a trench.

trestle: 1. Support for a work platform. **2.** Bridge constructed of heavy timber posts supported by footings and covered with heavy timber caps. Girders and decking are laid or constructed over the caps.

trial batch: Small batch of concrete mixed to establish or verify proportions of ingredients.

triangle: Three-sided plane figure in which sum of the angles equals 180°.

triangular file: Smoothing hand tool with a triangular profile. Used to reach into recessed areas for smoothing or sharpening. See file*.

triangulation: Surveying method in which reference points form a triangular pattern. Angles and length of sides are based on mathematical computation. Initial data for computation is taken from direct measurements of selected angles or sides.

tribrach: Adapter used to attach a surveying instrument to a tripod that has a different attachment system.

trig: 1. Brick laid in the center of a wall to align and secure a long line. **2.** Metal device used to secure a line and prevent sag in the center of a long wall. TINGLE, TWIG.

trilateration: Surveying method in which the lengths of the sides of geometric shapes are electronically measured and the angles are computed based on the measurements. See triangulation.

trim: 1. Wood, metal, or plastic finish members and moldings used to conceal or protect joints and edges of another material. Installed on the interior and exterior of a structure. Includes door and window trim, baseboards, casings, and cornices. **2.** Water supply and drainage fittings installed on a fixture to regulate flow of water into the fixture and flow of waste material from the fixture. **3.** To cut or otherwise remove material from a member to achieve a tight fit.

trim hole: Hole not loaded with an explosive charge when blasting rock.

Series of holes is used to limit breakage.

*trimmer: 1. Single or double floor joist into which a header is framed. Used to reinforce areas with a heavy load. TRIMMING JOIST. **2.** Vertical support between a bottom plate and bottom of a header. JACK. **3.** Ceramic tile used along the end or top of a tiled area.

trimmer arch: Relatively flat arch used to support a fireplace hearth.

trimming plane: Small, light-duty hand tool with a blade set at a slight angle. Used for smoothing wood or other soft material. See plane (1)*.

trip: Pile driver component that releases a drop hammer at a specified height.

trip coil: Solenoid-operated control device used to open a circuit breaker.

triple brick: Masonry unit measuring $4'' \times 5\frac{1}{3}'' \times 12''$. See brick*.

triple corner brace: Reinforcement consisting of two L-shaped metal brackets joined at 90° to each other. Used to reinforce three adjacent surfaces of a cabinet.

triple-sealed gypsum sheathing: Panel material consisting of a noncombustible gypsum core that is sealed between brown water-repellent paper on all surfaces. Used as an initial exterior covering for wood-framed and steel stud curtain walls.

triplex cable: Electrical cable composed of three single-conductor cables twisted together, with or without a protective coating.

tripod: 1. Three-legged adjustable support for a surveying instrument. Threaded or cup assemblies are used to fasten the instrument to the tripod head. See surveying*. **2.** Rigging support made of three legs lashed together. Lifting hardware is suspended from the upper joint.

triptich: Three carved panels joined side by side to form one decorative unit. TRIPTYCH.

trisect: To divide into three equal portions.

trivet: Low support for a surveying instrument.

troffer: 1. Long recessed light fixture with the opening flush with the ceiling. **2.** Light fixture connected to an HVAC system in which heat generated by the lamp is used to heat an area or discharged into the system.

trolley: Assembly of bearings, brackets, and wheels used to support and transport suspended loads. Travels in a channel or suspended from a track. Cable or chain is fastened between the load and trolley.

trombe wall: See thrombe wall.

TRUSSES

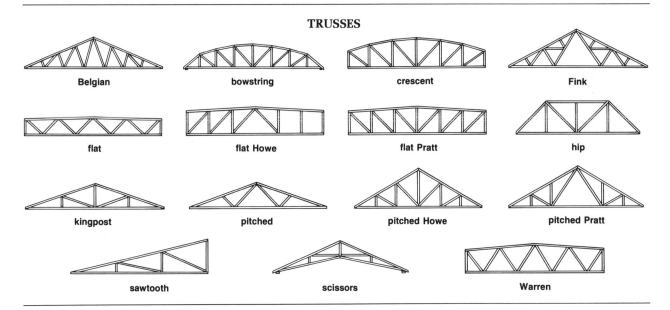

Belgian

bowstring

crescent

Fink

flat

flat Howe

flat Pratt

hip

kingpost

pitched

pitched Howe

pitched Pratt

sawtooth

scissors

Warren

trough: Narrow channel or gutter, commonly used to enclose electrical conductors.

troughing: Process in which a dozer makes several passes through an area in the same track. Side embankments prevent the excavated material from being pushed to the sides.

* **trowel: 1.** Hand tool with a broad, flat blade used to smooth and finish concrete or plaster. **2.** Hand tool with a flat, triangular bade used to apply mortar to masonry.

truck agitator: Vehicle used to transport fresh concrete from a batch plant to the job site. Maintains plasticity of the concrete without mixing.

truck mixer: Vehicle used to mix and transport fresh concrete from a batch plant to the job site. READY-MIXED TRUCK.

true wood: See *heartwood.*

trunk line: Main supply duct in a forced air system.

trunk sewer: See *main sewer.*

trunnion: Pivot consisting of two cylinders projecting from the main section of the pivoted object.

* **truss:** Structural member constructed of components commonly placed in a triangular arrangement. Commonly used to support a roof.

truss clip: See *gang nail.*

truss head: Fastener head with an oval upper surface and a flat bottom surface.

truss joist: Prefabricated structural member with horizontal top and bottom chords. Commonly used to support a floor. See *joist*.*

truss tie: 1. Light-gauge metal framing connector used to secure a wood roof truss to the top wall plate. **2.** Masonry joint reinforcement made of two parallel heavy-gauge wires joined by a series of angular cross wires. See *tie* (1)*.

try cock: Valve used as a secondary means of determining water level in a boiler. With a normal operating water level, water is discharged from the lower try cock, steam and water from the middle try cock, and steam from the upper try cock.

* **try square:** Hand tool with a graduated blade attached at 90° to the handle. Used to measure, lay out, and verify squareness of members.

T-shore: See *T head* (1).

T-square: 1. Tool used to lay out and guide a 90° cut on gypsum drywall. Body of the square is 4'-0" long to facilitate cutting out a 4'-0" wide sheet of drywall. **2.** Drafting tool used to lay out horizontal lines.

T-stud: Aluminum or steel stud used in a movable drywall partition. T-shaped section is inserted in a slot along the edge of the drywall. Allows mechanical work, such as plumbing and electrical, to be performed prior to erection of the opposite side.

T-tie: Masonry reinforcement used to tie a double-wythe wall to a perpendicular masonry wall. See *tie* (1)*.

tube-and-clamp scaffold: See *pipe-and-clamp scaffold.*

tube-and-coupler scaffold: See *pipe-and-clamp scaffold.*

tubercle: Corrosion deposit on inside of iron pipe.

tubing: Extruded flexible tubular material used to convey fluids. Wall thickness is less than that required for a standard pipe thread. Copper tubing is classified as Type K, L, M, or DWV. Type K tubing has the thickest wall.

tubing bender: Section of spring steel placed around tubing to prevent kinking while bending.

tubing cutter: Hand tool consisting of a cutting wheel mounted in a frame with at least one adjustable pressure wheel. Tubing is pinched between cutting wheel and pressure wheel and cutter is rotated to obtain cutting action. TUBE CUTTER.

tubular fiber form: Cylindrical concrete form made of compressed and resin-impregnated paper.

tubular lock: Lock consisting of a cylindrical tube in which lock tumblers and bolt are enclosed. Hole must be drilled in door for installation.

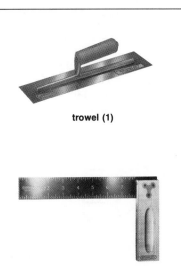

trowel (1)

try square

*tubular scaffold: Scaffold constructed of galvanized steel or aluminum tubing welded together to form prefabricated panels. Panels are fastened together with coupling pins. SECTIONAL SCAFFOLD.

tubular shore: Shore constructed of tubular steel that is open at both ends. Top section fits inside lower section and is secured in position with a locking pin. Final adjustment is made with a threaded adjustment collar.

tuck: Recess in a bed joint formed by raking the mortar out. Used to prepare joint for tuckpointing.

tuckpoint grinder: Electric power tool with a thin abrasive wheel mounted on a rotating shaft. Used to remove hardened mortar from mortar joints.

tuckpointing: Process of filling in cutout or defective mortar joints with fresh mortar.

Tudor architecture: Architectural style characterized by steep gable roofs, small windows, shallow moldings, and tall chimneys.

*tuft: Cut or uncut loop comprising the face of a carpet.

tuft bind: Force required to pull a tuft of carpet loose.

tumbler: Portion of a lock that secures the bolt in position until operated by a key.

tumbling course: Sloping row of masonry that intersects a horizontal course.

tung oil: Drying oil used as a clear wood finish. Resists water, acids, and alkalis.

tungsten electrode: Nonconsumable metal electrode used in arc welding or cutting. Primarily composed of tungsten.

tungsten-halogen lamp: Incandescent lamp in which a tungsten filament is surrounded by halogens.

tungsten inert gas (TIG) welding: See *gas tungsten arc welding*.

tunnel vault: See *barrel vault*.

turbine: Equipment that converts kinetic energy of a moving fluid to mechanical energy. Vanes are used to drive a central shaft of the equipment.

*turnbuckle: Coupling between two threaded rods in which one end has female left-hand threads and the other end has female right-hand threads. Used to adjust length of a brace, rod, or cable.

turned bolt: Machine bolt in which the threads are manufactured to close tolerances.

turning: Process of cutting and shaping members by applying a sharpened cutting tool against material rotating on a lathe.

turning gouge: Hand tool with rounded corners used for rough-cutting wood on a lathe.

turning piece: Template used to guide or form a masonry arch that does not require centering.

turning point: Temporary surveying bench mark.

turning saw: Hand saw with a thin blade mounted in an H-shaped frame. Used to cut intricate patterns.

turning vane: Short, curved fin at a change in direction in ductwork. Used to direct air around a bend and reduce pressure drop in the system.

turn tread: See *winder*.

turnstile: Barrier that rotates on a vertical axis and allows passage in one direction.

turn wheel: Conveyor component around which the conveyor belt changes direction.

turpentine: Volatile fluid obtained from distillation of wood or pine resin. Used as a thinner for oil-base paint.

turret: Small tower.

Tuscan: Order of architecture characterized by plain columns without decorative ornamentation.

tusk: Masonry unit projecting from the face of a wall.

tusk tenon: Tenon reinforced by stepped shoulders along its bottom edge.

T-wall: See *inverted-T foundation*.

twelfth scale: Graduations on a measuring instrument that divide an inch into 12 equal sections. Facilitates scaling objects at 1″ = 1′-0″. Commonly located on the back of a framing square.

twig: See *trig* (2).

twin cable: Cable composed of two insulated conductors laid parallel and attached to each other with insulation or encased in a common covering.

twin carbon arc welding: Process of joining metal with heat produced with an electric arc between two carbon electrodes. Pressure and filler material may or may not be used.

twist: **1.** Wood defect in which the four edges of the member are not in the same plane. See *defect* (1)*. **2.** Direction and shape of yarn design in carpet.

twist bit: Drill with one or more spiral flutes used to drill holes in wood, metal, or plastic. TWIST DRILL. See *bit* (1)*.

two-coat finish: Plaster finish applied in two separate layers. Used on nonperforated gypsum lath and rough concrete block, clay tile, or brick surfaces.

two-height split stone bond: Stonework pattern in which two thicknesses of squared stone are laid in a wall. Larger stone is twice as thick as smaller stone, creating noncontinuous mortar joints. See *bond* (1)*.

two-leg sling: See *bridle hitch*.

two-part line: Single length of rope that is doubled back on two sheaves.

two-stage thermostat: Thermostat that regulates two separate circuits in a given sequence or at two different set points.

two-tab shingle: Asphalt composition roof covering that is 36″ long and 12″ wide. One slot, approximately 5″ long, is cut into the edge along the 36″ length. See *asphalt shingle**.

two-way joist system: Reinforced concrete construction in which a slab is supported by perpendicular concrete joists. Allows for relatively thin slabs to span large areas. Concrete for the joists and slab is placed monolithically.

two-way reinforced footing: Foundation footing with reinforcement extending in two directions, commonly perpendicular to one another.

two-way system: System of reinforcement in which rebars, rods, and wire are placed perpendicular to one another in a slab. Used to resist stress resulting from deflection of the slab in two directions.

T-wrench: See *tee handle*.

tympanum: Recessed triangular area in a pediment.

type X: Classification of lath or gypsum drywall that is fire resistant.

typical: Similar throughout a structure.

tyrolea: Porous plaster used as a waterproofing material.

tubular scaffold

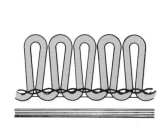

tuft

turnbuckle

U bolt: Bolt shaped like letter U with ends of the legs threaded to accept nuts. See *bolt**.

U factor: See *U value*.

ultimate load: **1.** Maximum stress, pressure, or tension a member or structure can withstand before it fails. **2.** Load at which a member or structure fails.

ultimate strength: Maximum resistance to stress, pressure, or tension a member or structure develops before it fails. MAXIMUM STRESS.

ultrasonic soldering: Process of joining metal by high frequency vibrations transmitted through molten solder. Undesirable surface film is removed and base metal is coated.

ultrasonic testing: **1.** Method of testing structural integrity of a pile. High-frequency waves are transmitted to the pile and the time required for a reflected wave to return to the top of the pile is measured. **2.** Method of nondestructive welding inspection in which high-frequency waves are used to locate and measure defects.

ultrasonic welding: Welding process in which metal is held together under pressure and joined by local application of high-frequency waves to produce heat and energy.

umber: Natural brown or reddish-brown material used as a pigment in paint.

unbalanced backflow valve: Backflow valve equipped with air supply inlet to equalize pressure. Valve remains open until fluid contacts and closes the gate.

unbalanced heating system: Heat distribution system that delivers varying amounts of heat to different areas.

* **unbalanced partition:** Wall with single layer of drywall on one side and two layers on opposite side; e.g., wall requiring fireproofing on only one side.

unbonded: Unattached.

unbonded member: Post-tensioned, prestressed concrete member in which the tendons move freely and tension is applied to end anchorages only.

unbraced length: Distance between lateral structural supports.

* **uncased pile:** Pile formed by drilling a hole into the ground and filling the hole with concrete without using a liner or tube. May be reinforced. BORED PILE, DRILLED PIER, CAISSON PILE.

uncoursed fieldstone: Stonework pattern in which rough, unfinished stones of random sizes are set in a random pattern without continuous vertical or horizontal mortar joints. See *bond* (1)*.

uncoursed roughly squared: Stonework pattern in which semidressed, semi-shaped stones are set in a random pattern without continuous vertical or horizontal mortar joints. See *bond* (1)*.

uncoursed rubble: Random pattern of laying stone.

uncoursed web wall: Stonework pattern in which rough, unfinished stones are set in a random pattern with serpentine, web-like mortar joints. See *bond* (1)*.

underbead crack: Crack in a welded joint that does not extend to the surface of the base metal.

underbed: Layer of base mortar that supports dividing strips and on which the finish layer of terazzo material is applied.

undercloak: **1.** Initial row of wood shingles, which is covered by the lowest exposed row. Used to provide proper slope to the lowest exposed row. UNDER COURSE. **2.** Portion of a lower sheet in a sheet metal roof that is used to make a seam.

undercoat: Initial layer of paint or other material used to seal the surface and provide a base for the top coat.

undercut: **1.** Cut or molded to provide overhanging portion. **2.** Groove formed in the base metal adjacent to root or toe of the weld and left unfilled. Results from excessive welding amperage or insufficient deposit of weld metal.

underdrain: Perforated pipe installed in porous fill under a slab. Designed to divert surface and ground water.

underfill: Depression in the face of a welded joint that extends below the surface of the base metal.

* **underlayment:** Material applied over subfloor and directly below the finish flooring material. Covers imperfections in subfloor and creates smooth surface for carpeting or resilient flooring.

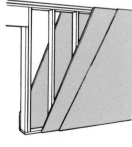

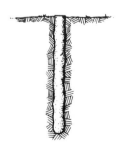

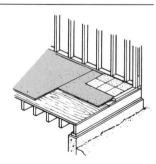

unbalanced partition   **uncased pile**   **underlayment**

underpinning: 1. Material added to and/or that replaces a foundation substructure without removing superstructure. Used to increase load capacity. **2.** Process of constructing additional substructure without removing superstructure.

underream: To widen the base of a hole drilled in the earth. Used to increase bearing area of a pile or pier when filled with concrete or grout and prevent uplift caused by wind or heaving.

underthroating: Cornice cove used as drip molding.

Underwriters' Laboratories, Inc.: Not-for-profit organization that examines and evaluates devices, systems, and materials to determine their degree of safety.

undisturbed sample: Soil sample in which the soil structure is minimally deformed.

undressed lumber: Sawn lumber before it is planed and surfaced. ROUGH LUMBER.

unfaced insulation: Batt or blanket insulation without a covering or coating.

ungauged plaster: Plaster consisting of lime, sand, and water.

unified thread series: Standard method of forming threaded portion of a fastener. Three divisions are unified national coarse (UNC), unified national fine (UNF), and unified national extra fine (UNEF). Number of threads per inch varies with outside diameter of the fastener shaft or inside diameter of the threaded hole.

uniform grading: Distribution of aggregate based on particle size. Sizes of various particles are relatively equal with proportional amounts of each size present.

uniformity coefficient: Ratio related to size distribution of a granular material. Determined by dividing sieve size that 60% of the weight of the material passes through by sieve size that 10% of the weight passes through. COEFFICIENT OF UNIFORMITY.

uniform load: Load such as stress or pressure that is distributed equally over an area.

uninterruptible power supply: Electrical power system that provides continuous current in the event of a power outage.

union: Pipe fitting used to connect or disconnect two pipes that cannot be turned. Consists of three parts: female-threaded section with left-hand thread, female-threaded section with right-hand thread, and central section that connects the two end sections. See *fitting**.

union elbow: Pipe elbow equipped with a union at one end. Used to connect or disconnect the elbow to a pipe without turning or disturbing the pipe.

union ferrule: Pipe fitting threaded on one side and joined to a pipe on the other side by compression.

unit: Construction or mechanical component manufactured as a single prefabricated piece or constructed of several pieces that arrive at the job site as one piece.

united inches: Sum of length and width of a pane of glass. Expressed in inches.

unit heater: Direct heating, factory-built, encased heating assembly. Primary components include a heating element, fan and motor, and directional outlet.

unit of bond: Shortest distance before a brick course repeats itself.

***unit rise: 1.** Vertical distance that a rafter will rise in 12″ of run. **2.** Distance from the top of one tread to the top of an adjoining tread in a stairway.

***unit run: 1.** Horizontal roof framing measurement of 12″. **2.** Width of a tread. Determined by measuring from the face of a riser to the face of an adjoining riser in a stairway.

unit vent: See *common vent*.

unit ventilator: Operable air inlet damper for providing outside air to interior areas. May be equipped with a filter and heating and/or cooling coils.

unit water content: Quantity of water per unit volume of concrete. Expressed as gallons or pounds of water per cubic yard of concrete.

unit weight test: Weight measurement of concrete or aggregate in which a container of a given size is filled and weighed. The weight of the filled container is compared with other samples.

universal: Hardware, such as a closer or lockset, that is used on either a left- or right-hand door.

universal drill: Power tool similar to a drill press, except has added feature allowing drill head to be rotated, tilted, and moved to various positions.

***universal joint:** Connecting device between two shafts that allows them to turn and swivel at various angles.

universal motor: Electric motor that can operate on either direct or alternating current.

universal plane: Wood-cutting hand tool with interchangeable blades for cutting intricate surface designs, such as moldings. Guide and depth fences allow for accurate cutting.

unjointed: Continuous concrete slab without expansion or contraction joints.

unloader: Control device for electric motor-driven compressor that regulates the pressure head of the compressor. Allows the motor to be started at a lower torque by removing the load from one or more cylinders during initial start-up.

unrestrained: Permitted to rotate or move freely.

unsound: Structurally or ornamentally defective.

unstable: Material or structure that is apt to fail when loads are applied.

uplift: Upward force on a structure resulting from forces such as water pressure or frost heave.

uplift test: Method of measuring support capabilities of a pile. After placing a pile, tension is applied to the pile to remove it from the supporting material.

***upperbeam bolster:** Chair used to support top layer of reinforcing steel for a

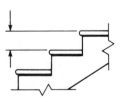

unit rise (2)

unit run (2)

universal joint

upperbeam bolster

horizontal support member.

upper chord: See *top chord*.

upright: Any structural or ornamental member installed in a vertical direction.

upset: 1. Buckling of wood as a result of compressive force. **2.** To thicken cold or heated metal by deforming the end of the member. **3.** Deformation of metal when welding, resulting from too much pressure applied.

upset welding: Resistant welding process in which metal is joined simultaneously over abutting surfaces or along a joint by heat generated from resistance to electric current in the contact area.

* **upstand:** Portion of fabric or metal roof covering that is turned up alongside a wall without being tucked in. Usually covered by flashing. UPTURN.

urea resin glue: Thermosetting adhesive that is heat- and moisture-resistant. Available as a powdered adhesive that is mixed with a catalyst.

urethane: Plastic foam of rigid polyurethane made of a series of joined, closed cells.

urinal: Water-flushed sanitary fixture that collects and drains human urine.

utility grade: Softwood lumber classification equal to a Number 3. Can be used structurally with minimum waste.

* **utility knife:** Hand tool with short blade protruding from a handle. Handle may be used to store additional blades or house the blade when retracted.

utility pole: Outdoor vertical member used to support overhead electrical, telephone, and cable television conductors.

utility sheet: Sheet metal cut into variety of sizes. Suitable for general building construction.

utility vent: Vent pipe that rises above the highest water level in a fixture and then turns downward before its connection to a stack or main vent.

utilization equipment: Equipment that uses electricity to produce motion, heat, light, or perform similar functions.

U value: Measure of heat transmission determined by the rate at which heat flows through a material of known thickness. Based on the number of Btu/hr. that passes through 1 square foot of a structure when the air temperature is 1 °F different from one side to the other. Opposite of R value.

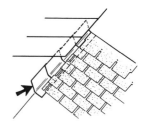

upstand

utility knife

vacuum: Air pressure that is less than atmospheric pressure.

vacuum and blower fish line system: Method of feeding a fish line through conduit by means of air pressure.

* **vacuum breaker:** Backflow preventer installed in a water supply system to prevent a vacuum and causing backflow.

vacuum concrete: Concrete from which excess water is removed by vacuum action before hardening occurs. Produces high-strength concrete in less time.

vacuum lifting: Raising an object by means of a vacuum.

vacuum pump: 1. Centrifugal pump that removes air and condensation from the return main of a steam heating system. **2.** Centrifugal pump that removes air from a given area to create a vacuum.

vacuum tank: Container that collects condensate from heating equipment by drawing it into the boiler by using a vacuum pump.

valance: 1. Ornamental frame at the top of a window to conceal drapery hardware and lighting. **2.** Window drapery.

* **valley: 1.** Sloped intersection of two roof surfaces. **2.** Concave angle formed by the meeting of two inclined surfaces. **3.** See *valley rafter*.

valley jack rafter: Roof framing member extending from the ridge to the valley rafter.

valley rafter: Roof framing member extending from the ridge to the intersecting corners of a building.

valley tile: Trough-shaped roofing tile used to channel water down the valley of an intersecting roof.

valve: Control device for regulating the flow of fluid.

valve area: Surface area contacted by a material flowing through a valve. Surface area varies depending upon the amount of flow.

valve bag: Paper container made up of four or five layers of kraft paper and a self-sealing paper valve. Used for cement or other powdered materials.

valve box: Enclosure allowing access to a valve below grade.

valve flutter: Uncontrolled movement of a compressor valve during opening and/or closing.

valve lift: Distance that a valve is moved to allow passage of material.

valve plug: Movable portion of a valve that fits against a seat to control the flow of fluid.

valve seat: Stationary portion of a valve which, when in contact with the valve plug, controls the flow of fluid.

* **valve stem:** Threaded spindle that guides a valve plug during operation.

vanadium: Hardest known metal.

vane: 1. Blade or diverter for directing the flow of a fluid. **2.** See *weathervane*.

vaneaxial fan: Belt- or direct-driven blower consisting of a disk-type wheel with vanes on one side and installed in a cylinder.

vane lines: Tackle for holding and turning a guy or stiffleg derrick. VANG LINES.

vane test: Method of measuring the consistency of soil by inserting a four-bladed shaft into a bored hole and rotating it. Soil consistency is determined by the resistance against the blades.

vanity: Bathroom fixture consisting of a lavatory and a base cabinet.

vapor: Gaseous substance that is normally in a solid or liquid state; e.g., water vapor.

vapor barrier: Impervious chemical or material that retards movement of vapor or moisture through a structure or other surface. Used in conjunction with insulation in walls and under concrete slabs. MOISTURE BARRIER.

vapor lock: Formation of vapor in a fluid-carrying system that interrupts the flow of liquid through pipes.

vara: Unit of lineal measure approximately equal to 33″. Varies with geographic location.

variegated: Surface with irregular finish of various colors.

variety saw: See *table saw*.

varnish: Clear or colored solution consisting of resin dissolved in alcohol or oil. Applied as a liquid surface coating which dries to a clear, hard, protective surface.

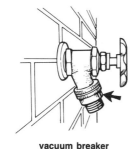

vacuum breaker

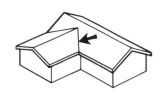

valley (1)

valve stem

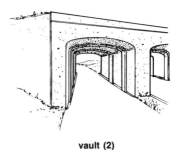

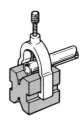

vault (2)   V belt   V block

varnish stain: Colored solution consisting of a pigment and resin dissolved in alcohol or oil. Applied as a liquid surface coating to add color to the surface of wood.

varying duty: Operation of an electrical device with varying loads and time intervals.

* **vault: 1.** Room or enclosure with limited access and high security. **2.** Arched structure forming a ceiling or roof.

vaulting course: Masonry course formed by the springers of an arch. Usually projected out from the surface of the wall.

vaulting shaft: Short, upright support for a rib in a vault or arched structure, such as a column or pillar.

* **V belt:** Continuous reinforced rubber belt used to connect pulleys and motors and transfer rotation. Cross-sectional shape is a letter V or truncated letter V.

* **V block:** Device for holding circular material during drilling, grinding, and milling operations.

V brick: Masonry unit with vertical perforations.

vehicle: Liquid medium in which other materials are suspended for application.

vein: Layer of soil or rock that is different from the surrounding materials.

veining bit: Cutting tool used in an electric router that forms a concave radius. Used for ornamental work. See *bit* (2) *.

velocity: 1. Speed; quickness of motion.

2. Force that includes magnitude, speed, and relative direction.

velocity pressure: 1. Force of air within ductwork. **2.** Fluid pressure capable of causing an equal amount of speed or movement. VELOCITY HEAD.

velocity riser: Vertical pipe used to increase the speed of flow of a material. Vertical pipe has a smaller internal diameter than adjoining pipe.

velvet carpet: Floor covering material in which the yarns are woven in a manner similar to cloth. Usually finished with a smooth surface.

* **veneer: 1.** Thin surface layer of material; e.g., thin layer of wood on the exterior of a hollow core door. **2.** See *brick veneer.*

veneer core: Most common method of plywood construction in which the inner plies are constructed of wood veneer.

veneer saw: Power tool with a fine-toothed blade. Used for cutting wood veneer.

Venetian: Type of terrazzo containing large stone chips.

Venetian arch: Arch design with a pointed top.

* **Venetian window:** Three-light window in which two vertical dividers form a large central window and two smaller flanking windows.

vent: 1. See *vent pipe.* **2.** Opening that allows passage of air.

vent cap: Covering on the open end of a vent, soil, or waste stack to prevent

objects from entering the stack.

vent connector: Pipe that connects the exhaust of a fuel-fired appliance to a chimney or other source of outside air.

ventilated: Provided with sufficient air circulation to remove heat, fumes, and vapors, as applied to an area.

* **ventilating jack:** Sheet metal cover installed over an inlet to a vent stack to increase the air flow. VENT JACK.

ventilating rate: Number of times air is recycled in an area within a given time.

ventilation: Natural or mechanical process of supplying and removing air in an area. Air may be heated, cooled, and/or purified in the process.

ventilator: Device used to circulate air.

vent pipe: 1. See *vent stack.* **2.** Pipe installed to provide free air flow from one area to another.

vent sash: See *awning window.*

vent stack: Vertical vent pipe installed to provide air circulation to the drainage system and prevent loss of water seals from traps. Projects through the roof of a structure. VENT FLUE.

vent system: Piping system that provides circulation of air in a drainage system and expels vent gases. Equalizes pressure in the system and protects against siphonage and back pressure.

venturi: Reduction in the diameter of a pipe or duct to increase velocity of the material passing through and lower the static pressure. May be used for measuring flow of fluids through a system.

verandah: Covered porch along the outside of a structure. VERANDA.

verdigris: Green protective patina formed on copper due to exposure to air.

* **verge: 1.** Edge of roofing material projecting over the gable of a roof. **2.** Unpaved edge of a roadway. SHOULDER. **3.** Column shaft.

verge board: Trim member installed directly under the verge of a gable roof. VERGE RAFTER, BARGEBOARD.

vermiculated work: Finish design consisting of irregular, wavy recesses and giving the surface a rustic appearance.

vermiculite: Expanded natural mica

veneer (1)

Venetian window

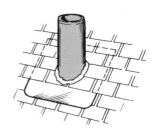

ventilating jack

verge (1)

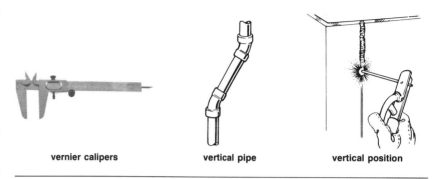

vernier calipers   vertical pipe   vertical position

used for thermal insulation and as a lightweight aggregate in concrete.

vernier: **1.** Graduated auxiliary scale that slides along and is used in conjunction with a primary scale for accurate reading of the primary scale. Commonly used on surveying instruments and precision measuring instruments to lay out angles and measure lengths. **2.** Brake adjustment on an earth spudding drill that automatically controls the length of the line as the chisel tip is repeatedly dropped and the hole deepened.

* **vernier calipers:** Precision tool for measuring inside and outside lineal measurements. A graduated scale incorporated into a telescoping handle is used to take accurate readings.

vertex: **1.** Highest point; pinnacle. **2.** Point where two lines intersect to form an angle.

vertical: In an upright position; plumb.

vertical arc: Portion of a graduated circle on a surveying instrument used to measure vertical angles.

vertical bond: See *stack bond.*

vertical circle: Round, graduated surface on a surveying instrument used to measure vertical angles.

vertical firing: Mechanical arrangement of a furnace fuel supply system in which fuel is discharged up from the lower burners or down from the upper burners.

vertical grade plastic laminate: Surface covering material, approximately $1/32''$ thick, made of layers of paper and plastic overlays. Used for cabinet fronts and other vertical applications.

* **vertical pipe:** Plumbing pipe at an angle of less than 45° from vertical.

* **vertical position:** Welding position in which axis of the weld is approximately vertical.

vestibule: Small enclosed foyer that adjoins a larger room.

V grooving bit: Cutting tool used in an electric router that forms a groove with a pointed bottom and approximately 45° sides. Used for ornamental work. See *bit* (2) *.

viaduct: Bridge supported on narrow piers and arches between abutments.

vibrating conveyor: Trough or tube that is oscillated at a high frequency. Used for moving bulk material and objects.

vibrating sander: Electric power tool in which a flat plate covered with abrasive paper is oscillated at high speed. Used to improve efficiency of the sanding process.

vibrating screed: Power tool used to consolidate and level concrete slabs and other flatwork to the desired height

by means of vibrating action.

vibration: Agitation of fresh concrete mix during placement to help consolidation.

vibration isolator: Device used to support machinery or building components in such a manner that vibration created by the machine does not transfer to the surrounding structure.

* **vibrator:** Power tool used to agitate and consolidate freshly placed concrete and produce close contact with the form or mold. See *external vibrator, internal vibrator.*

vibratory driver: Pile driver that applies a dynamic force to a pile to drive it into the ground. A pair of eccentric weights are rotated to create necessary force to drive the pile.

vibratory extractor: Device incorporating eccentric, rotating weights that apply dynamic force to a pile or casing to remove it from the ground.

vibratory plate: Gasoline- or diesel-powered compaction equipment used to compact the surface of pavement and/or pavement underlayment materials.

vibratory rammer: Hand-held power tool in which a flat plate moves in reciprocating motion to compact soil, pavement underlayment materials, or pavement.

vicat apparatus: Device that penetrates hydraulic cement or similar materials to determine setting characteristics.

* **vice:** **1.** Staircase with steps winding around a central newel post. VIS. **2.** Central post of a spiral staircase.

victaulic pipe: Pipe with movable joints that remains watertight during and after movement.

viga: Solid log beam.

vignette: Decorative ornament in the form of a vine with grape clusters and leaves.

vinyl: Strong thermoplastic resin compound made from vinyl acetate, vinylide chloride, or polymerized vinyl chloride. Used for a variety of building materials including drywall covering, siding, and piping.

vinyl knife: See *hook bill knife.*

vinyl tile: Floor covering material composed of polyvinyl chloride, mineral fillers, pigments, and plasticizers. Manufactured in thin flat squares or continuous sheets.

viscose: Fiber manufactured from cellulose.

viscosity: Measurement of ability of a solid or liquid to resist flow.

viscous filter: Device that cleans the air by trapping airborne particles in a surface coated with a fluid or oil.

* **vise:** Portable or stationary clamping device used to firmly hold work in place. Two movable jaws are adjusted using threaded rod or lever.

vise clamps: Soft metal covers used on the jaws of a vise to prevent damaging the work.

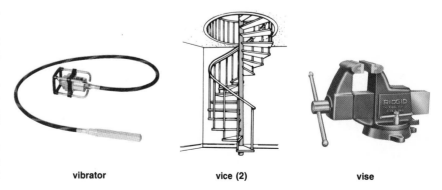

vibrator   vice (2)   vise

V joint   voissoir   volute (1)

Vise grips: See *locking pliers.*

visual grade lumber: Wood member rated by appearance for structural capabilities.

vitreous: 1. Characterized by impermeability; often a material having low water absorption. Less than 3.0% absorption in floor and wall tile, and low voltage electrical porcelain, and less than 0.3% in all other materials. **2.** Glasslike; changed to a glassy surface with heat.

vitrified brick: Glazed masonry unit that is waterproof and resistant to chemical damage.

vitrified tile: Clay pipe that has been glazed to produce a hard, waterproof surface. VITRIFIED-CLAY PIPE.

vitruvian scroll: Decorative design made of a series of scrolls joined by irregular, curved waves. VITRUVIAN WAVE.

vivo: Vertical shaft of a column.

vix bit: Wood drilling tool used with an electric drill that has a tapered casing that automatically centers the drill point.

* **V joint:** Joint formed by adjoining surfaces of two members with both members chamfered on the edges.

void: Open area.

void ratio: Ratio of volume of voids to volume of solid particles in a granular material.

* **voissoir:** Wedge-shaped masonry unit used in forming an arch. Converging sides of the masonry unit are cut as radii of one of the centers of the arch. VOUSSOIR.

volatile: Evaporates readily.

volatile flammable liquid: 1. Liquid with a flash point below 100 °F. **2.** Flammable liquid with a temperature above its flash point. **3.** Class II combustible liquid with a maximum vapor pressure of 40 psia at 100 °F.

volatile solvent: Liquid solvent that vaporizes easily.

volt: Measure of potential electrical difference that causes current to flow through a circuit. Equal to the electrical force applied to move the current of 1 ampere over a resistance of 1 ohm.

voltage: 1. Force that causes electrons to flow through a conductive material. **2.** Potential electrical difference between two individual conductors. Calculated as the greatest root-mean-square difference of potential between two conductors.

voltage drop: Voltage loss between two points in a conductor.

voltage regulator: Electrical control device that maintains constant voltage supply to equipment, such as a welding transformer.

voltage relay: Electrical control device that senses varying levels of electrical current and operates contacts based on the electrical current level.

voltage to ground: 1. In a grounded electrical circuit, the voltage between an electrical conductor and the point or conductor that is grounded. **2.** In an ungrounded electrical circuit, the voltage between any two conductors in a circuit.

volt ampere: 1. Electrical equivalent of 1 volt multiplied by 1 ampere. Equal to apparent power in alternating current circuits and 1 watt in direct current circuits. VOLT AMP.

volume: Three-dimensional capacity of an object. Expressed in cubic units, such as cubic feet or cubic yards.

volume batching: Measuring the ingredients for a concrete mix by volume.

volumetric test: Field testing method of determining air content of fresh concrete. Used especially for lightweight concrete mixes.

* **volute: 1.** Spiral fitting at the bottom of a handrail. Used to join the straight section of a handrail to the newel post. **2.** Spiral architectural ornament.

V-shaped joint: Shallow, recessed mortar joint formed with a steel jointing tool. Effective for resisting rain penetration. See *mortar joint*.

V tool: Gouge with blade angle of less than 180°.

vulcanization: Chemical process in the manufacture of rubber that makes it less plastic, more elastic, and more resistant to swelling by organic liquids.

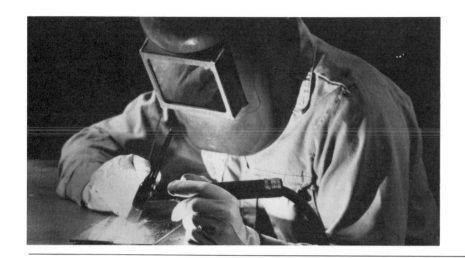

wadding: Fabric or other material that covers explosives in a blasting hole.

waferboard: Wood panel material made of wood wafers and bonded with a phenolic resin under heat and pressure. Used for sheathing, paneling, and crating.

waffle flat plate construction: See *two-way joist construction.*

Wagner fineness: Measure of the fineness of a material. Expressed as the total surface area in square centimeters per gram.

wagon drill: Movable mast and carrier used to hold and position a pneumatic drill.

wagon-headed dormer: Dormer with semicircular roof. Used to provide light and/or ventilation to attic or upper levels of the structure.

wagon-headed vault: Semicircular arched vault.

wagtail: See *parting bead.*

* **wainscot:** Wall finish in which lower portion of a wall, approximately 32" from the floor, is finished with a material different from the upper portion.

waist: Narrowest portion of an object; e.g., distance across the thinnest portion of a column.

waist belt: Safety equipment consisting of a strap fixed around a person's waist and attached to a lifeline. See *safety belt.*

wakefield pile: Timber pile having three planks bolted together with the middle one offset. Offset forms a tongue-and-groove joint with the adjoining pile.

* **waler:** **1.** Horizontal member used to align and brace concrete forms or piles. Placed behind studs or framework of forms. RANGER, WALE, WHALER. **2.** Protective fender placed along the edge of a waterside dock.

walking edger: Tool used to round edges of a concrete slab while in an upright position.

walking line: See *line of travel.*

walkway: Permanent wood or metal platform on a roof; fitted with handrails.

wall: Vertical structure used as a partition to enclose or divide areas. May be load-bearing or non-bearing.

wall anchor: **1.** Metal device with a T- or L-shaped end to interlock with a masonry wall and support wood floor joists. WALL BEAM. **2.** See *hollow wall anchor, molly, toggle bolt.*

wall bearer: Wood member attached to a wall below a stringer to reinforce the stringer and support treads and risers.

wall bed: Bed that is hinged and that stands vertically in a closet or recess in a wall.

wallboard: See *drywall.*

wallboard lifter: See *lifter.*

wall box: Metal or iron frame embedded in a masonry wall to support a beam, joist, or other horizontal supporting member. WALL HANGER.

wall cabinet: Cabinet installed approximately 18" above a base cabinet or 54" above the floor.

wall form: See *form.*

wall hook: Large nail used to anchor a beam or support a beam plate.

wall-hung: Attached to a wall, as applied to plumbing or other fixture.

walling: Material used to construct a vertical partition or wall.

wall iron: Metal bracket fastened to an exterior wall to support downspouts or other exterior members.

wallpaper: Interior patterned-paper or vinyl-sheet wallcovering backed with water-soluble adhesive.

wall plate: **1.** See *bottom plate, plate, top plate.* **2.** Covering of an electrical outlet box attached to the surface of the box or directly to the wall. Used to protect and cover conductor connections.

wall stringer: Stairway stringer attached to a wall. WALL STRING.

wall stud: See *stud.*

wall washing: Lighting a wall by placing illumination fixtures close to the intersection of the ceiling and wall.

wane: Lumber defect along the edge or corner as a result of presence of bark or lack of wood. See *defect*.*

ward: Baffle in a lockset that prevents full action of a key.

warehouse pack: Hardening of cement that has been in storage for an extended time. Eliminated by rolling the bag of cement.

warm air heating system: System used to heat an area by circulating heated air through a series of ducts and dispensing it.

wainscot

waler (1)

WASHERS

cut

external tooth

internal/external tooth

internal tooth

split lock

---

warning pipe: Overflow pipe with a visible outlet that allows for observation of discharge.

warp: 1. Lumber defect in which the ends of the member are in the same plane and the area between deviates from the plane. Caused by improper seasoning. See *defect**. **2.** Lengthwise yarn in a carpet fabric.

warping joint: See *expansion joint*.

Warren truss: Prefabricated wood truss in which top and bottom chords are connected and supported by webs placed in an alternating inclined pattern, similar to a W. See *truss**. WARREN GIRDER.

wash: 1. Sloping upper surface of an exterior member that diverts rainwater away from the building. **2.** See *wash coat*. **3.** To separate weld metal from base metal by melting the weld metal with a torch without melting the base metal.

washbasin: See *lavatory*.

washboard: See *baseboard*.

wash boring: Drilling into the earth by using a rotary drill that emits water or slurry at the drill point to stabilize the sides of the hole.

wash coat: Thin layer of paint; applied as a sealer or primer.

washdown closet: Water closet that uses flushing action of water jets along the rim of the bowl to create head pressure.

* **washer:** Thin metal disk installed between a nut and fastened member to distribute the pressure of the nut onto the member.

washer head: Design of screw head in which a flat washer is permanently joined at the intersection of the head and the threaded shaft. Eliminates the need for a separate washer.

wash water: Water carried on a concrete truck mixer. Used to remove fresh concrete from the discharge assembly and chute after use. FLUSH WATER.

waste: 1. Liquid discharge from a plumbing fixture that does not contain fecal matter. **2.** Undesirable material.

* **waste-and-overflow fitting:** Fitting that provides an outlet for the bathtub drain and has provision for draining water rising above flood level rim.

waste bank: Pile of excess excavation material.

waste disposal unit: See *garbage disposal*.

waste pipe: Pipe used to remove liquid discharge not containing fecal matter.

waster: 1. Face brick with imperfections. Used as a backing brick. **2.** Mason's chisel with a cutting edge used to remove unwanted stone.

* **waste stack:** Vertical pipe that connects a waste pipe to a building drain. Does not connect to toilets or urinals. SOIL STACK.

water base paint: See *latex paint*.

water-borne preservative: Surface treatment used to prevent wood decay, consisting of ammoniacal copper arsenate or chromated copper arsenate.

water-cement ratio: Comparison of the weight of water to cement in a concrete or mortar mix. Expressed as decimal.

water closet: 1. Room that contains a toilet. **2.** Plumbing fixture used to receive human waste and convey it to a waste pipe.

water column: 1. Boiler device that reduces fluctuation of water to facilitate reading the water level. **2.** Water contained in a vertical pipe or tube.

water cooling tower: See *cooling tower*.

water crack: Surface crack in plasterwork caused by excess water in the mix.

water distribution pipe: Pipe that conveys water from a water main to the point of usage.

water gel: Explosive used for blasting; contains water and ammonium nitrate. May be sensitized by explosives such as smokeless powder, or by metal such as aluminum.

water hammer: Water supply system defect in which a loud noise is created when faucet is shut off. Results from sudden water stoppage and improper cushioning of the supply pipes.

* **water heater:** Appliance used to heat water for purposes other than heating a structure.

water joint: Stonework joint that is sloped to allow rainwater runoff.

water level: Hand tool used to transfer elevations across short distances. Clear vials are attached to each end of a hose and the assembly is filled with liquid, commonly colored water. Vials are positioned to allow the action of water to seek its own level.

water lime: Cement that hardens when it comes in contact with water; hydraulic cement.

water main: Main supply pipe used to convey water in a community water supply system.

* **water meter:** Device used to measure the volume of water passing through a pipe. Installed at entrance of water supply pipe in structure.

water outlet: Opening in a water distribution system that discharges water.

waterproof: 1. Impervious to water. **2.** To apply material that resists the flow of moisture.

waste-and-overflow fitting

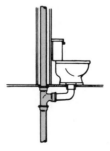

waste stack

water heater

water meter

waterproofing: Material which resists the flow of moisture.

water putty: Powdered material that is mixed with water to produce a wood filler.

water-reducing admixture: Concrete admixture that increases the slump of the mix without extra water but still maintains good workability.

water repellent: Material that prevents penetration of water.

water retentivity: Property of concrete or mortar in which moisture is retained by the concrete or mortar and not discharged into the forms or adjacent masonry units.

water ring: Perforated fitting at the nozzle of shotcrete equipment through which water is introduced to the cement and aggregate.

water seal: Water retained in a trap to prevent sewer gas from entering habitable areas.

water service: Pipe that connects a water main to the water distribution system in a structure.

watershed: See *water table*.

water softener: Device installed in water supply system that removes minerals such as calcium and magnesium from water.

waterstone: Synthetic material used to sharpen cutting edges of tools. Soaked in water for 30 minutes before use and flooded regularly with water during use to prevent accumulation of metal particles.

*waterstop: Flexible rubber or plastic strip installed across a construction joint in concrete or other masonry material to form a moisture-impervious joint.

waterstruck brick: Masonry unit with a smooth finish created by lubrication of the mold with water.

water supply fixture unit: Measure of estimated water demand of a plumbing fixture.

water supply system: Entire system of fittings, piping, and valves that distributes water in a structure.

water table: 1. Exterior trim member with sloping upper surface designed to facilitate rainwater runoff. Bottom edge is grooved to divert water before it reaches the structure. See *molding**. DRIP CAP, DRIP STRIP. **2.** Upper surface of groundwater.

water test: See *hydrostatic water test*.

watertight: Of tight construction so as to prevent entry of water into an enclosed area.

water tube boiler: Boiler in which water is circulated through pipes that are surrounded by combustion gases used to heat the water.

water-white glass: Glass used in a solar collector.

watt: Measure of electrical power equal to amps multiplied by volts. 1 watt equals 1 ampere of current flowing with an electromotive force of 1 volt.

wattle: Screen or wall formed of woven material.

wavelength: Distance between two identical points in a wave or cycle.

wax: Organic material obtained from animal, vegetable, and mineral matter. Used as a protective finish or lubricant.

waxing: Filling cavities in a marble surface with a material colored and patterned to blend into the surface's finish.

W beam: See *W shape beam*.

wearing course: Upper layer of pavement.

weather bar: Material along the intersection of wood and stone members to seal the joint. WATER BAR.

weatherboard: Wood siding.

weather cap: Cover over an exterior electrical connection to prevent moisture damage.

weathered: 1. Colored or textured to resemble a surface that has been exposed to the elements over a period of time. **2.** Shaped with a sloping upper surface to shed rainwater.

weathered joint: Mortar joint finished with a downward slope to shed rainwater. See *mortar joint**. WEATHER-STRUCK JOINT.

weatherproof: Constructed or protected so that exposure to the weather does not affect operation or planned use.

weatherstrip: Material such as metal or neoprene placed between adjoining members to seal the joint and prevent precipitation or draft from entering a structure. WEATHERSTRIPPING.

weathervane: Rotating device that indicates direction of the wind; often installed as an ornament on a cupola.

weave: To interlace surface and backing carpet yarns in a single operation.

weave bead: Weld metal deposited in a zigzag pattern.

*web: 1. Portion of a steel beam between the flanges. **2.** Diagonal support member in a truss. **3.** Thin panel or plate between ribs. **4.** Joining member between two faces of a concrete masonry unit.

web clamp: Clamp that exerts pressure on an object using a nylon or canvas strap attached to a cam device. Strap is manually wrapped around the object and tightened using a ratchet mechanism.

wedge: V-shaped member.

wedge gate valve: Compression valve used to regulate fluid flow by means of a tapered valve plug that fits into a tapered opening.

wedge socket: Attachment at the end of a wire rope in which a wedge secures the rope in the socket.

*weephole: Small opening in a masonry wall that allows for the passage of moisture. Located at regular intervals near bottom of the wall to allow condensation behind face of the wall to drain.

weep wick: Short piece of absorptive material embedded in a masonry joint that allows condensation behind face of the wall to drain. No opening in the wall is required.

weft: Lengthwise yarn in a carpet.

weight batching: Measuring the ingredients of a concrete mix by weight.

weir: Structure across a stream or ditch used to divert or measure flow of water.

weld: To join metal or nonmetallic material through fusion by applying heat and/or pressure with or without the addition of filler material.

weldability: Capacity of metal or nonmetallic material to be welded and to perform satisfactorily in the intended use.

weld axis: Imaginary line through the length of a weld perpendicular to and at the geometric center of the cross section of the weld.

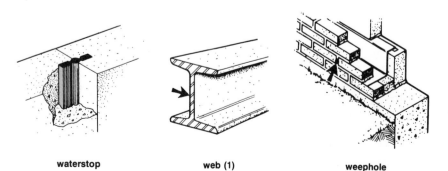

waterstop   web (1)   weephole

welded butt splice: Method of joining reinforcing steel by arc welding the butted ends.

welded wire fabric: Heavy gauge wires joined in a grid used to reinforce and increase tensile strength of concrete. WELDED WIRE MESH, WIRE MESH.

welder: Person who performs welding operations.

welder certification: Written document certifying that a welder can produce specific welds that meet established strength and quality standards under specific conditions.

weld face: Exposed surface of a weld on the side that the welding operation is performed.

weld gauge: Device used to measure size and shape of a weld.

welding electrode: See *electrode*.

welding head: Part of a welding machine into which a welding gun or torch is incorporated.

welding hood: Protective covering worn during welding operations to prevent injury to the face of a welder. Filler plate protects the eyes against injury from ultraviolet, infrared, and invisible radiation.

welding machine: Machine used to produce welds.

welding position: Position in which a weld is to be made. Four basic welding positions are flat, horizontal, vertical, and overhead.

welding rectifier: Electrical device in a welding machine that converts alternating current to direct current.

welding rod: Nonconductive filler metal used for welding and brazing.

welding screw: Threaded fastener with projections under the head that allows attachment to another member by welding.

welding symbol: Symbol developed by the American Welding Society used to give information about a weld.

* **weldment:** Assembly of parts joined by welding.

weld metal: Portion of a weld melted during welding.

weld pass: Single progression of welding or surfacing along a joint, weld deposit, or metal or nonmetallic material, resulting in a weld bead, layer, or spray deposit.

weld pool: Molten metal in the weld prior to cooling and solidification. WELD PUDDLE.

weld root: See *root of weld*.

weld symbol: See *welding symbol*.

well: 1. Enclosed vertical chute used to provide light, air, or passage between floors or levels of a structure. See *stairwell*. WELL HOLE, WELL OPENING. 2. Wall around a tree trunk that protects it during excavation. WELL CURBING. 3. Hole bored into the ground to bring water to the surface or into a water supply system.

well casing: Lining around a well.

well-graded aggregate: Aggregate with a particle size distribution that produces minimal amount of voids.

well point: Pipe with a perforated point that is driven into the ground to allow underground water to drain. Perforated end is covered with a screen to prevent solid particles from accumulating within the pipe.

Welsh arch: Flat arch with a keystone and single horizontal member on each side. Side members are shaped to fit the keystone.

welt: 1. Flexible metal roofing seam formed by folding the edges of two adjoining sheets, interlocking the edges with each other, and hammering the joint flush. 2. Wood strip that covers and strengthens a flush joint.

westa: Area of land equal to 60 acres.

western framing: See *platform framing*.

wet-bulb depression: Difference between dry bulb and wet bulb temperatures.

wet-bulb temperature: Air temperature measured by a thermometer in which the bulb is covered with a dampened gauze-like material.

wet galvanizing: Method of coating steel with zinc by passing it through a reservoir of molten zinc.

wet location: Electrical installations underground, in concrete slabs, or in other areas that may be saturated with water or other liquids.

wet sand: To apply abrasive action while the abrasive material or surface is wet. Reduces dust and lubricates the abrasive surface.

wet standpipe: Vertical pipe in a fire extinguishing system that contains the chemicals or other extinguishing material.

wettest stable consistency: Condition of concrete or mortar containing the maximum water content without sloughing.

wetting: Spreading and bonding of flux or liquid filler metal on base metal in a continuous layer.

wetting agent: Material that lowers the surface tension of liquids, allowing penetration of liquids into capillaries. Used to improve concrete adhesion to an adjoining surface.

wet vent: Ventilating pipe that also acts as a drain pipe.

whaler: See *waler*.

wharf: Structure parallel with a shoreline that provides a berth for ships and large water vessels.

wheelabrating: Method of surface abrasion in which steel grit is discharged at high velocity against the surface.

wheelbarrow: Large open-top container mounted on a single-wheel frame. Used for mixing and/or transporting bulk material.

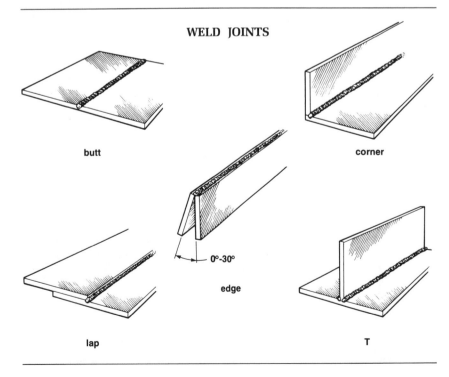

WELD JOINTS

butt

corner

0°-30°

edge

lap

T

wheel dresser: Hand tool with a series of teeth formed on the perimeter of a rotating wheel and attached to a handle. Used to remove metal particles and renew the surface of a grinding wheel.

wheel trencher: Excavating equipment with series of earth-cutting buckets attached to a circular wheel. As the wheel rotates, soil is removed to form a trench, lifted by the buckets, and deposited at the sides of the trench.

wheel window: Circular window with mullions placed as spokes in a wheel. ROSE WINDOW, CATHERINE WHEEL WINDOW.

whetstone: Natural or synthetic material used to sharpen the edges of cutting tools. See *oilstone*.

whetting: Putting the final edge of a cutting tool by drawing the edge repeatedly across an oilstone or whetstone.

whip: 1. To bind the end of a rope with twine or yarn to prevent the end of the rope from unraveling. See *knot**. **2.** Light cable or rope.

whirlpool bathtub: Plumbing fixture equipped with water circulation equipment.

white coat: Top, final layer of plaster. FINISH COAT.

white glue: Fast-drying liquid adhesive used to bond wood, paper, and other porous materials. Water soluble, softens when heated, and dries clear. POLYVINYL RESIN GLUE.

white shellac: Wood finish material with poor moisture resistance properties that gives a clear finish.

whitewash: Mixture of quicklime or slaked lime and water that is used as a thin paint.

Whitney punch: Hand tool used to punch holes in metal.

whole brick wall: Brick wall with its thickness equal the length of one brick.

whorl: Spiral scrollwork design in furniture.

wicking: Process of absorbing water by capillary action into the cells of a material.

wide flange beam: See *W shape beam*.

wide throw hinge: Hinge with elongated leaves. Swings door farther away from jamb than conventional hinge.

Wilton carpet: Velvet cut-pile carpet woven with loops. Good wearing characteristics.

winch: Drum or roller around which rope or cable is wrapped. Rotation is controlled by a system of gears to improve the mechanical advantage and provide hoisting power.

wind: Bend or turn.

wind: Movement of air.

wind-brace: Member used to reinforce a structure against wind load.

winder: Wedge-shaped tread in a circular stairway, or angular transition of a stairway.

winding stairway: Staircase constructed primarily of winders. See *circular stair*.

windlass: Simple winch that uses a crank or lever to rotate the drum.

wind load: Force exerted on a structure by air movement. WIND FORCE.

*window: Glazed opening in an exterior wall designed to provide light and/or ventilation into a structure.

WINDOWS

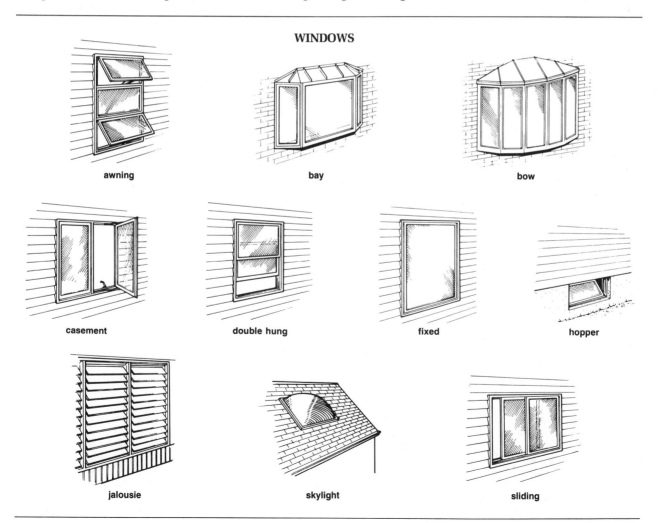

awning

bay

bow

casement

double hung

fixed

hopper

jalousie

skylight

sliding

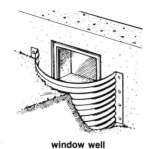

window well

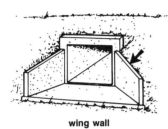

wing wall

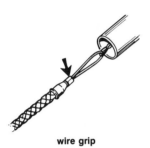

wire grip

wire nut

wireway

window bar: See *muntin*.

window catch: See *window lock*.

window frame: Fixed part of a window assembly consisting of the top and side jambs and the sill. Designed to accept and hold the sash and all required hardware.

window jack scaffold: Scaffold supported by a bracket that projects through a window opening.

window jamb: See *window frame*.

window lift: See *sash lift*.

window lock: Hardware that secures a window sash closed. WINDOW CATCH.

window pane: See *light*.

window pull: See *sash lift*.

window sash: See *sash*.

window schedule: See *schedule*.

window seat: Seat approximately 18″ high constructed along the recessed area inside a bay or bow window. May be hinged to provide storage below the seat.

window sill: Lowest horizontal member of a window frame.

window stool: See *stool*.

window stop: Narrow horizontal or vertical strip in a window frame that holds the sash in the proper position within the frame. SASH STOP.

window wall: Wall primarily composed of fixed glass or other transparent material to allow for the passage of light and/or air.

* window well: Open area directly outside of a basement or below-grade window. Corrugated metal frame or concrete wall retains the soil to allow operation of the window.

windrow: Loose ridge of material that falls away from the outer edges of a bulldozer or grader blade.

windshake: Lumber defect in which wood fibers separate along circumference of the annual growth rings. Produced by wind stress on a tree during growth. CUPSHAKE.

wind stop: Permanent insulating strip or member that prevents passage of air into a structure.

wing: 1. Section of a building projecting outward from the main structure. 2. One panel of a revolving door.

wing divider: Hand tool with two pivoting legs and an arc-shaped piece fastened to one leg for setting and securing the legs. Used to scribe and lay out circular work.

wing nut: Internally threaded fastener with two projecting ears to facilitate turning by hand. See *nut**.

* wing wall: Short section of wall along the edge of an abutment. Used to retain soil, stabilize the abutment, and divert water into an opening, such as a culvert.

wiped joint: Joint formed by pouring molten solder on a joint and shaping the joint with a cloth pad or paddle.

wiping solder: Alloy consisting of 60% lead and 40% tin. Used to form pipe joints.

wire: Flexible strand of metal.

wire brad: See *brad*.

wire brush: Hand tool with a series of wire bristles attached to a handle. Used to clean a surface or apply a rough texture to plaster or concrete.

wire cloth: Fabric formed of thin woven wires. Used as reinforcement in plaster.

wire connector: See *wire nut*.

wired glass: Glass in which wire mesh is embedded between the two faces. Used to prevent the glass from shattering if it breaks. WIRE GLASS.

wiredrawing: Thin concave deformation in a valve seat caused by high-velocity fluid flowing through a nearly closed valve over a long period of time.

wireframe modeling: CADD representation of a three-dimensional member in which only the edges are represented.

wire gauge: 1. Diameter of a wire. The higher the number, the smaller the diameter of the wire. 2. Device used to measure the diameter of wire. Made up of a flat piece of metal with standard-size notches along the perimeter.

* wire grip: Device used to pull conductors through conduit and raceways and temporarily secure them until permanent connections have been made.

wire hardware: Device made from bent wire, such as screw eyes and hooks. WIRE GOODS.

wire lath: See *expanded metal lath*.

wire mesh: See *welded wire fabric*.

wire nail: See *wire brad*.

* wire nut: Electrical connector used to join two or more conductors. Threaded metal fitting is enclosed in a nonconductive casing, commonly plastic. Conductors are twisted together and nut is turned onto the connection.

wire rope: Rope made of high tensile strength wires laid around a flexible core material. Used for rigging and hoisting.

wire saw: Equipment used to cut stone in which a continuous wire carrying abrasive material is rapidly moved to produce the cutting action.

wire stripper: Hand tool with opposing jaws notched to cut and pull insulation from electrical conductors.

wire tie: Short wire wrapped around structural members to secure them. See *anchor**, *tie**.

wire track: Metal reinforcement placed in a masonry joint for tying two wythes of masonry together and forming one structural unit.

* wireway: Sheet metal enclosure with an openable cover to provide access to conductors inside. Used to house and protect electrical conductors.

wire winding: Wrapping high tensile strength wire around circular concrete walls, domes, or other tension-resisting structural components.

withe: See *wythe*.

witness corner: Surveying marker set at a recorded distance from an inaccessible corner. WITNESS MARK.

wobble coefficient: Measure of the loss of friction when post-tensioning concrete.

wood: Hard fibrous product of trees. Used for structural and ornamental work. Classified as softwood and hardwood.

wood chisel: Hand tool with a cutting blade attached to a handle. Used to cut wood by striking the end of the handle with a hammer or mallet.

wood dough: Synthetic filler material containing wood fibers.

wood fiber plaster: Mixture of gypsum products with wood fiber. Used as a lightweight plaster scratch coat, which provides greater fire resistance than gypsum plaster.

wood frame construction: Structure in which all loads are supported by wood members such as wall plates and studs, posts, joists, and rafters.

wood oil: See *tung oil*.

wood rasp: See *rasp*.

wood sash jamb block: Concrete masonry unit with one end shaped to receive a wood jamb.

wood screw: Fastener with a helical threaded shaft designed to grasp and hold in wood. Threads extend from the tip for 70% of the length toward the head. Length varies from $1/4''$ to $6''$. Gauge varies from 0 to 30. See *screw**.

work angle: Position of a welding electrode in relation to the surface of the base metal and weld axis.

work arm: Portion of a lever between the fulcrum and the working end.

work edge: First edge of a board to be straightened and squared. Other edges and faces are measured and trued from this surface. WORKING EDGE.

worked lumber: Wood product that has been dressed and shaped to interlock with another member, such as shiplapped siding.

work end: First end of a board to be straightened and squared.

work face: First face of a board to be straightened and squared.

working drawing: Drawing that shows information needed for construction of a component or structure. Intended for use by architect and contractors.

working load: Load that a structure is intended to support. See *live load*, *dead load*.

working weight: Total weight of a piece of equipment, such as a crane or excavating equipment.

work light: Lamp in a protective cage attached to an extension cord. Used to provide temporary illumination in work areas. TROUBLE LIGHT.

work plane: Comfortable height for a person to work when standing; commonly 30″ high.

worm gear saw: Electric saw in which rotational direction of the motor is transferred 90° by means of a worm gear and applied to the saw arbor. WORM DRIVE SAW.

wormhole: See *insect hole*.

worsted yarn: Carpet material made of long wool fibers that are all combed in the same direction to remove short fibers.

woven carpet: Carpet formed by interlacing the yarns.

woven valley: 1. Valley formed by overlapping roof shingles in an alternating pattern on each successive course.

woven-wire fabric: Steel reinforcement made up of cold-drawn steel wire twisted into a design with hexagonal openings.

wrapping: Process of installing steel reinforcement around a structural steel column to provide additional tensile strength for concrete or plaster installed around the column.

wrap tie: Wire used to join overlapping light-gauge metal members by wrapping around the members several times. See *tie**.

wreath: See *volute*.

wrecking ball: Heavy metal or concrete weight suspended on a crane line and swung to demolish structures.

wrecking bar: Hand tool used to pry members and pull nails. Commonly $3/4''$ diameter and 24″ to 30″ long. Ends are hooked, split, or U-shaped to provide mechanical advantage. See *pinch bar*, *pry bar*.

wrecking strip: Narrow strip or panel placed in concrete formwork to be removed ahead of the main forms. Used to facilitate stripping of the main forms.

***wrench:** Hand tool with jaws at one end that are designed to turn bolts, nuts, or pipes. Both jaws may be fixed or one may be movable.

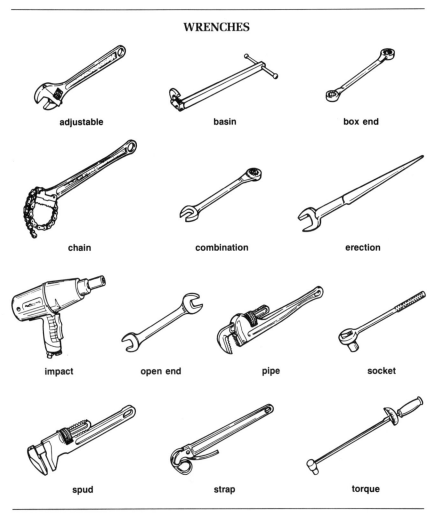

WRENCHES

adjustable   basin   box end

chain   combination   erection

impact   open end   pipe   socket

spud   strap   torque

wythe

wrought iron: Soft metal with a low carbon content. Commonly used for decorative work.

wrought nail: Fastener made from wrought iron. Used where a bendable, cinchable fastener is desired, such as crating.

wrought steel nipple: Nipple made of carbon steel pipe cut to less than 1'-0" in length and threaded on both ends.

wrought timber: Wood that has been planed on at least one surface.

W shape beam: Structural steel member with parallel flanges that are joined with a perpendicular web. Intersection of web and flanges is filleted. See *structural steel**. W-BEAM, WIDE FLANGE BEAM.

W truss: See *Fink truss*.

wye: Y-shaped pipe fitting. One arm is at a 45° angle to the main run and may be reduced in diameter. See *fitting**.

wye level: Surveying instrument with the telescope resting in a Y-shaped frame. Telescope may be removed for turning it end to end.

* **wythe:** Single, continuous vertical masonry wall one unit thick.

X brace: Pair of diagonal sway braces.

X-ray examination: Method of determining the quality of a welded joint by subjecting the joint to radiographic inspection.

* **xyst:** Shaded or covered walk.

yacht rope: Highest quality manila rope.

yankee screwdriver: See *spiral ratchet screwdriver*.

yard: 1. Lineal measurement equal to 3′. See *cubic measure, square yard*. **2.** Open, unoccupied area on a building lot.

yardage: 1. Volume of excavated material expressed in cubic yards. **2.** Surface area expressed in square yards.

* **yard catch basin:** Submerged tank with a grate that prevents large objects from entering the sewer system.

yard lumber: Lumber less than 5″ thick. Used for general construction.

yarn: Natural or synthetic material used for carpeting.

yarning iron: Plumbing hand tool with a long offset blade. Used to insert and compact packing yarn and jute into cast iron soil pipe joints.

Y branch: Y-shaped plumbing fitting. Used where a change in direction is required. Y FITTING, WYE FITTING.

yellowing: Change in color of a material or finish from clear or white to off-white or yellow due to age or exposure.

yield: 1. Volume of concrete produced from a known quantity of ingredients. **2.** Volume of concrete produced per sack of cement. **3.** Permanent deformation of a structural steel member produced by bending or stretching.

yield point: Highest tensile limit a structural member can be subjected to and return to its original shape. Tensile stress beyond the limit will result in permanent deformation.

Y level: See *wye level*.

yoke: 1. Upper horizontal member of door or window frame. **2.** Y-shape soil pipe fitting. **3.** Collar or clamp installed around outside of a concrete column form to prevent spreading during concrete placement. **4.** Mechanical assembly for a slipform, which prevents the forms from spreading and tranfers the load of the forms to jacks. **5.** Curved excavating drawbar on a scraper. **6.** Collar for supporting lead pipe.

* **yoke vent:** Pipe connected upward from soil or waste stack to vent stack. Used to prevent pressure differences in stacks.

Yorkshire light: Window with horizontal sliding sash and at least one fixed sash.

zax: Hand tool for cutting and punching nail holes in slate tile.

Z-bar: 1. Z-shaped wire reinforcing member embedded in mortar joints to tie two sections of masonry wall together. See *tie*(1)*. Z TIE. **2.** Z-shaped metal channel, such as that used for runners in an acoustical ceiling. Z-FLASHING, Z-MOLD.

zee: Drawn, extruded, or rolled metal member with modified Z shape. Internal angles are approximately 90°.

zero slump concrete: Concrete with stiff or dry consistency. Concrete sample does not measurably drop after removal from slump cone.

* **Z-furring channel:** Z-shaped, 26 gauge steel member used to attach blanket insulation and gypsum drywall to concrete or masonry walls.

zigzag bond: See *herringbone bond*.

zigzag molding: Molding design consisting of continuous diagonal lines. See *chevron*.

zigzag rule: See *folding rule*.

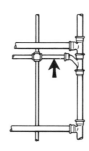

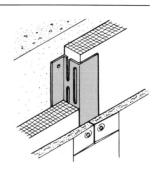

| xyst | yard catch basin | yoke vent | Z-furring channel |

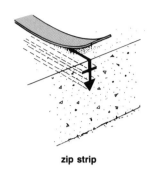

zip strip

zinc: Non-corrosive metal used as a galvanizing material.

* zip strip: T-shaped vinyl member used to form control joints in concrete slabs. Inserted into the slab immediately after finishing. Top of strip is peeled away to expose concealed member.

zone: 1. Area within a structure with separate heating/cooling systems and controls. **2.** See *comfort zone.*

zone control: Heating/cooling control system that has more than one thermostat. Used to regulate air temperature in different areas of a structure.

zonolite: Insulating material used for loose-fill insulation and as an aggregate in insulating concrete.

Z-shaped flashing: Z-shaped sheet metal flashing used at the intersection of a roof and wall. Used to prevent water from leaking into a structure. See *flashing*.*

Contractors' Terms

abstract: 1. Legal summary of important points in a document. **2.** Short, itemized materials list used for bidding purposes.

abstract of title: Summary of conveyances such as deeds, wills, and legal proceedings, including names of parties, property description, and ownership agreements.

accrued depreciation: Decreasing value of an asset during a certain time period or by a given date.

activity: Single task or work item as defined in critical path method scheduling.

activity duration: Estimated amount of time needed to perform a single task or work item as defined in critical path method scheduling.

acquired lands: Federal lands to which ownership was gained by condemnation, gift, or purchase.

acreage: Measure of land. Expressed in acres.

addendum: Written document or drawing issued before the execution of a contract that modifies original drawings and/or specifications.

adjustable-rate mortage (ARM): Mortgage loan in which the interest rate changes within certain limits, depending on the prevailing loan interest rates.

advertisement for bids: Public notification of the request for cost estimates for a construction job.

affidavit of noncollusion: Sworn and notarized statement stating that bids and estimated prices for the same work performed by various bidders were obtained independently. NONCOLLUSION AFFIDAVIT.

agreement: 1. Written or verbal contract between two or more parties. **2.** Understanding regarding mutual interests.

air rights: Legal right to utilize air space above a piece of property to a given elevation.

allowance: 1. Cost figure built into a construction bid that covers contingency costs. **2.** Cost figure written into a contract to specify the amount that may be spent for a particular portion of construction; e.g., lighting fixture allowance.

alternate bid: Offer stating that costs will be added or deducted from an original bid if the corresponding changes in the work are agreed upon.

amortization: Scheduled payment of a debt by a series of payments, each payment being made up of a sliding scale of principal payment and interest payment. In a typical amortization schedule, principal payments are low at the beginning of the schedule and high at the end of the loan period.

annexation: Act of attaching or joining one area or territory to another.

application for payment: Contractor's written request for payment on a completed portion of work and/or materials.

appraisal: Determination of the value of an item or property through estimation and calculation of the value of similar property.

appreciation: Increase in value or worth.

appropriated public land: Original public domain land.

approved equal: Material, equipment, or methods agreed to by the architect or engineer of a building project as being acceptable for an equivalent or substitute for the material, equipment, or methods specified in the construction contract documents.

arbitration: Method of settling disputes in which the opposing parties agree to abide by the decision of a noninvolved third party. Avoids need for litigation.

assessed value: Estimated price of property as determined by an authorized agency. ASSESSED VALUATION.

assessment: 1. Tax or fee levied against a piece of property. **2.** Determination of the value of a piece of property or item.

assignment: 1. Contract document stating that payment for completed work and/or material is made to a party other than the party specified in the original contract. **2.** Appointing a certain craft or trade to perform a particular type of construction work.

balloon payment loan: Loan arrangement in which regularly scheduled payments are made to pay a portion of the loan prior to the maturity date. Balance of the loan (balloon) is due on the date of maturity.

bar chart: Drawing used to schedule construction work. Each part of the job is represented by a bar or line, with one end representing the starting date and the other end the completion date. Progress of the job is visually represented. BAR GRAPH.

base bid: Original cost amount stated in a construction bid prior to adjustments for alternate bids.

base bid specifications: Construction specifications that describe the basic job, excluding alternates.

bid: Offer by a contractor to a potential client to construct a structure or provide equipment that meets the conditions specified in the plans and specifications for a given price.

bid abstract: Summary of bids for a given construction project.

bid bond: Form of security executed by the bidder. Designed to guarantee that the bidder will actually sign a contract for the project and provide other required bonding. BID SECURITY.

bid call: Public announcement of the bid acceptance time and location.

bid date: Final date that bids will be accepted.

bidder: Person or contractor who develops a cost estimate and offer for a construction project and submits the offer to an architect for consideration in awarding a construction contract.

bidding period: Length of time from the issuance of prints and specifications for estimating purposes to the bid date.

bid guarantee: Security required to be submitted with the bid to ensure proper execution of the contract and furnishing of the required bonds.

bid opening: Occasion on which all bids are opened and compared to determine if they meet given criteria. Bidder may be awarded the contract at this time.

bid shopping: Comparison of one bid against another with the intent of reducing the price.

bill of materials: List of construction supplies needed to build a project, giving individual quantities, descriptions, and sizes.

binder: Check of deposit to validate a contract.

blight: Deterioration of the value and/or condition of a section of real estate or property.

bona fide bid: Complete and acceptable bid including all necessary documents.

bonding agent: Independent entity that represents surety companies and is a liaison between contractors and the surety company.

bonding capacity: Maximum contract amount a bonding company extends to a particular contractor.

bonus and penalty clause: Contract arrangement in which a contractor is monetarily rewarded if work is completed prior to a specified date or charged an additional amount for each day the work extends beyond a specified date.

builder's risk insurance: Specific type of insurance coverage designed for a construction project.

building code: Regulations adopted by a federal, state, county, or city government for the construction of buildings in a safe and structurally sound manner to protect the health, safety, and general welfare of those within or near the buildings.

building inspector: Building department official who determines if construction is in accordance with the approved plans and applicable building codes.

building official: Person designated to administer the building code.

building permit: Release form required from state or local governments that allows the construction of permanent structure. Fee is usually required to obtain this release, and regular inspections are required and noted on the permit in many cases.

business agent: Trade union representative elected by the membership to negotiate contracts, check jobs in progress for compliance to contracts, and act as a liaison between tradesworkers and management. BUSINESS REPRESENTATIVE.

business district: Area of property zoned for commercial use such as retail and manufacturing.

buyer: Person or agent who purchases material or equipment.

callback: Return to a construction site or portion of work to correct work that was improperly completed.

call loan: Loan paid in full at the demand of either the lending agency or borrower.

certificate for insurance: Statement from an insurance company indicating the dates, types, and amounts of coverage included in a builder's risk insurance policy.

certificate for payment: Confirmation from the architect to the owner concerning contractor's fee for work and/or materials.

certificate of occupancy: Document that verifies that the local building authority approves the finished construction of a building and its zoning. After issuance, the building is ready for use.

certificate of title: Legal proof that the title to property is clearly in the name of the current owner or purchaser.

change order: Variation from original plans and specifications directed by the owner or architect in written form.

city planning: Design of urban growth with business, industry, manufacturing, residential, transportation, aesthetic, and utility considerations. TOWN PLANNING, URBAN PLANNING.

clerk of the works: Architectural representative who oversees a construction project.

closed-bid list: Restricted list of contractors who may bid on a given construction project.

closed specifications: Written portion of a construction document that details particular products and processes that must be used.

closing costs: Fees charged by lenders for securing a loan.

code: Set of regulations governing building construction practices. Includes building, electrical, plumbing, and mechanical installation.

collective bargaining: Meeting of union labor and employer representatives to establish work rules and pay rates by mutual agreement.

completion: Point at which a building is ready for its intended occupancy or use.

completion bond: Bonding company guarantee that a construction project will be fully completed and free of encumbrances and liens.

completion date: Date established by an architect upon which a construction project is to be ready for occupancy or use.

completion list: See *punch list.*

comprehensive general liability insurance: Insurance that covers claims for bodily injury and property damage from known or unknown hazards.

condemnation: 1. Legal declaration that a structure is unfit for occupancy or use. 2. Legal declaration that property is being confiscated by a governmental agency for public use.

contingency loan: Loan to a property buyer that depends upon the sale of some other property or real estate currently owned by the buyer.

contingent agreement: Agreement between an architect and owner stating that a portion of the architect's payment depends upon the owner being successful in obtaining necessary money to fund the project, or some other condition.

contract documents: All written and graphic materials used for design and construction of a project, including the agreement, general conditions, prints, specifications, addenda, and other specific written stipulations.

contract for deed: Contract to sell real estate on an installment payment basis. Ownership is transferred upon receipt of final payment.

contractor's bond: Monetary commitment set aside by a contractor before construction begins to ensure that the contractor fulfills all obligations set forth in the contract documents.

construction contract documents: All legal and approved information concerning a building project, including request for bids, specifications, general conditions, prints, and addenda.

construction loan: Monetary arrangement between a lending agency and building contractor. Money is loaned for construction of a building and payments are made based on certain completion points of the project.

construction site: Property on which a structure is erected. JOB SITE.

convey: To transfer title of property from one owner to another.

cost-plus-fee agreement: Agreement in which a contractor and/or architect receives payment for the direct costs of performing construction services or work and receives an additional fee, stated as either a percentage of the overall cost or a fixed amount.

creditor: Person or organization that extends a loan or credit to another party.

critical path method (CPM): Scheduling method that graphically shows the relationship between the parts of a construction project. A charting system identifies the jobs that must be completed before other parts of construction can start and which are independent of others.

CSI format: Division of various construction operations, materials, and services into a numerical system. Developed by the Construction Specifications Institute (CSI) to assist architects in categorizing various building services for the purpose of specifications writing.

date of substantial completion: Date stated by the architect on which a construction project is ready for occupancy or use by owner. TIME OF COMPLETION.

deed: 1. Legal agreement that transfers the title of a property from one person to another. **2.** Contract document stating that a contractor will perform construction work according to prints, general conditions, addenda, and specifications, and the owner will pay for this construction work.

deed restriction: Specific-use limitations on property stated in the deed.

demand loan: Loan that is payable in full upon demand by the lender. DEMAND MORTGAGE LOAN.

deposit: Payment made to an architect or public agency to obtain a set of prints, general conditions, specifications, and addenda. May be refundable or nonrefundable.

depreciation: Reduction in value or worth.

design-build: Construction arrangement in which the contractor provides the design work and building construction services.

dummy: Restraint with no activity and no time in critical path method scheduling.

earliest expected time (EET): Scheduling term indicating the earliest time a portion of a project can be expected to be completed. EARLIEST EVENT OCCURRENCE TIME.

early finish: Scheduling term indicating the first day additional work is not required to be performed on a particular activity.

early start: Scheduling term indicating the first day work can begin on an activity if all other preceding activities are completed as soon as possible.

earnest money: Initial payment from a buyer to a seller to bind a contract sale.

eminent domain: Right of a governmental body to condemn private property for public use.

employee: Person who works for another person or organization for wages and/or salary.

employer: Person or organization that hires individuals to perform a specific job for wages and/or salary.

employers' liability insurance: Insurance that covers the employer against employee claims for damages resulting from sickness or injury. Protects against claims other than those covered by workmen's compensation insurance.

encroachment: Portion of a building or structure that intrudes on another property.

encumbrance: 1. Right or interest in land, such as restrictions, reservations, and easements, which reduces the value of a fee but does not prevent conveyance of the fee by the owner. **2.** Restriction on the use of a piece of property. **3.** Lien or other claim against a piece of property.

environmental impact statement: Analysis of the probable significant environmental effects of a construction project.

equity: Ownership or interest an individual has in a piece of property after allowances are made for any liens.

escrow: Payment or documents held by a third party.

estimate: Projected cost of materials and labor for a construction project or portion of a project.

event: Critical path method scheduling term that indicates the starting time of an activity that must occur after all appropriate prior activities are completed.

eviction: Act of removing tenants from property.

extended coverage insurance: Construction site insurance that protects against loss from a variety of natural or uncommon catastrophies such as storm damage, riots, vehicular damage, or explosion.

extra: Additional work or service added to a construction project after a contract

agreement has been reached; usually requires additional costs.

fair market value: Reasonable price for property based on current market conditions.

fee: 1. Payment received by professionals such as architects or engineers. **2.** Estate of inheritance in real property.

field order: Written statement from an architect to a contractor for minor work change that does require adjustment to the contract documents.

final acceptance: Complete acceptance of the project by the owner.

final payment: Completion of the contract payment from the owner to the contractor.

first mortgage: Primary mortgage that takes precedence over other mortgages.

fixed-rate mortgage: Mortgage loan in which the interest rate is established at a given percentage for the life of the loan.

float time: Open spots in a schedule using critical path method scheduling.

foreclosure: Transfer of a deed or title to a lending agency or creditor because the borrower has failed to make payments.

free float: Critical path method scheduling term indicating the time that the completion of an activity can be delayed without affecting other activities.

GANTT: See *bar chart*.

general conditions: Written agreements included as part of a set of prints that detail and describe building components and procedures.

general contract: Agreement between an owner and general contractor for completion of an entire construction project.

general contractor: Person or company that agrees to fulfill an entire building agreement with various items or types of work to be completed such as carpentry and electrical and plumbing work. Subcontractors may be hired to do some of the actual building.

general requirements: Division 1 of the CSI Format; covers equipment costs and overhead.

guaranty bond: Arrangement in which a person or organization makes a financial commitment that certain work or requirements will be completed. Includes bid bonds, labor and material payment bonds, performance bonds, and surety bonds.

hazardous waste site: Work area in which special precautions must be taken to protect workers and surrounding environment from contamination.

hold harmless: Clause in which an insurance carrier agrees to take on a

254

improvement • notice to proceed

client's contractual obligations under certain conditions that might otherwise be another party's obligations.

improvement: Change in property that increases its value.

indenture: 1. Formal agreement between a bond issuer and the bondholder. **2.** Formal agreement between an apprentice and employer or an employer organization that describes the terms of apprenticeship. APPRENTICE AGREEMENT.

ingress: Enter; to go into.

initiation: Critical path method scheduling term indicating beginning of an activity.

inspection: Examination of work to ensure it is done in accordance with plans and/or codes.

instructions to bidders: Portion of a bidding document that indicates the procedures for preparing and submitting a bid on a particular project.

invitation to bid: 1. Portion of a bidding document that requests bids. **2.** Notification by a contractor to selected subcontractors that a particular portion of a project is open for bidding.

invited bidders: Selected individuals or companies that are asked to bid on a particular project or portion of a project.

job conditions: Portion of a contract that describes rights and responsibilities of contractors on a project.

job site: See *construction site*.

Joint Apprenticeship and Training Committee (JATC): Organization formed of an equal number of members from labor and management. Responsible for administration and development of a craft training program for beginning workers (apprentices) and/or experienced workers (journeymen).

joint tenancy: Holding of property by several persons. When one of the owners dies, the ownership of the entire property transfers to the remaining owner(s).

joint venture: Project in which two or more people or organizations work together as partners.

jurisdictional dispute: Disagreement between building trades as to the proper craft to perform a given task.

jurisdictional offer: Written notice from a public agency to a private owner that informs the owner of the intent of the agency to purchase certain property for a given amount. Prerequisite to condemnation.

labor and material payment bond: Agreement in which a person or company makes a financial commitment to guarantee that a contractor will pay for

labor costs and materials. PAYMENT BOND.

last expected time (LET): Critical path method scheduling term indicating the last time at which an activity can be completed without delaying the starting time of subsequent activities. LATEST EVENT OCCURRENCE TIME, LATEST FINISH DATE.

latest start date: Critical path method scheduling term indicating the last date an activity can be started without delaying the starting time of subsequent activities.

lease: Agreement in which a person conveys property to another person for a given period of time.

leasee: Person receiving temporary occupancy from an owner.

leasor: Person owning property and conveying the occupancy to another person.

lending agency: Organization or group that provides loans to various individuals or groups based on their ability to repay the loan amount in addition to applicable interest charges and loan fees.

letter of intent: Written statement indicating an intention to enter into a formal agreement.

liability: 1. State of being obligated or responsible for an activity or item according to law or equity. **2.** Debt; monetary obligation.

license: Privilege granted by a state to an individual to practice a particular profession; e.g., engineering, architecture, or contracting.

lien: Legal charge against property; made for securing a payment of debt or performance of an obligation.

life-cycle costing: Calculation of the cost of a system as figured over its entire projected time of use or installation (including maintenance) as opposed to the initial cost only.

limit of liability: Maximum amount an insurance company will pay in a settlement.

liquidated damages: Amount stated in a construction contract that will be paid by the contractor to the owner in the event the project is not completed by a given date.

litigate: To prosecute or make the subject of a lawsuit.

litigation: Lawsuit; engagement in a legal contest using the judicial process.

loan: 1. Money lent for payment of interest. **2.** Item lent for temporary use.

loss-of-use insurance: Insurance coverage used to prevent losses during time required for the repair of damaged or destroyed property.

low bid: Cost estimate which has the lowest price for performance of work described in the construction contract documents.

maintenance bond: Arrangement in which a person or company guarantees a contractor's work for a specified period of time after completion; normally one year.

maintenance period: Time after the completion of a construction job during which a contractor will repair defects without additional charges.

market value: Highest price that property will bring when sold on the open market.

material safety data sheet (MSDS): Written statement of the hazardous nature of a material that describes the hazards, proper protective work clothing, and emergency procedures.

mechanic's lien: Claim for payment of labor, materials, or professional services, which restricts clear title to the property pending full payment.

modification: Written work order change.

mortgage: Conditional conveyance of property contingent on the failure of the purchaser to make timely debt payments.

mortgagee: Person or company lending money to another party.

mortgage lien: Claim for payment of labor, materials, or professional services using the property as security.

mortgagor: Person or company borrowing money from another party.

negligence: Failure to use the required care or proper procedures for a particular operation.

negotiated work: Construction work assigned by an owner to a particular contractor or subcontractor without open bidding. Price is negotiated between owner and contractor/subcontractor.

network: Critical path method scheduling term indicating a graphic representation of activity relationships.

node: Critical path method scheduling term indicating a junction between arrows showing the early start and late finish dates.

noncollusion affidavit: See *affidavit of noncollusion*.

note: Written commitment to repay a loan.

notice to bidders: Written portion of the bidding documents that gives bidders the proper procedures for bid submission.

notice to proceed: Written statement from an owner to a contractor authorizing the beginning of work on a construction project.

occupancy: 1. Acquiring title to property. **2.** Inhabiting a structure.

occupancy permit: See *certificate of occupancy*.

occupant: Person or organization that acquires rights of habitation of a structure by legal title.

office expense: Overhead cost of operating a construction office that is not charged to any particular job.

open bidding: Bidding procedure in which bids are accepted from all interested and qualified bidders.

open-end mortgage: Loan arrangement in which additional money can be borrowed for property repair and upkeep.

open shop: Construction work arrangement in which collective bargaining is not utilized to establish wages or work rules. MERIT SHOP.

option: Financial arrangement between an owner and a user that grants the user the right to buy or rent a piece of property within a specified time limit.

overhead: Business expense, such as heat or electricity, not charged to a specific project.

overtime: Work performed beyond normal work hours; normally paid at a higher rate of pay.

owner's liability insurance: Insurance that protects an owner against claims arising from work performed by a contractor on behalf of the owner.

penal sum: Contractual or bonded amount of payment to be made in the event of failure to meet conditions of the contract or bond.

penalty clause: Contract provision that establishes a monetary charge for failure by the contractor to complete a project by a given date.

performance bond: Arrangement in which a person or company guarantees that a contractor will execute a job in the manner described in the contract documents.

permit: Written license issued by an authorized agency or person.

personal injury insurance: Insurance coverage for personal injury or damage to other persons caused by the insured.

picket: To demonstrate at a construction site or construction site entrance to inform the public and/or other construction workers of complaints against a contractor.

planning commission: Citizens' committee formed to review proposed development of an area, making suggestions and changes, and approving final plans for development.

post-construction services: Additional work performed by a contractor after completion of the original contract and issuance of final payment due according to the contract.

premise: 1. Tract of land with improvements. **2.** Structure or portion of a structure including the lot and peripheral structures.

prequalification of bidders: Investigation of bidders' availability, capabilities, capacities, and/or experience as they pertain to a given construction project. Performed prior to acceptance or solicitation of bids.

prime contract: Written agreement between the owner and contractor for a construction project.

prime contractor: Contractor with a written agreement directly from the owner.

professional liability insurance: Insurance coverage that protects against damages arising from alleged negligence or errors in the performance of a professional service; e.g., structural engineering.

progress chart: Graphic representation of the schedule and/or completion of various phases of a construction project. PROGRESS SCHEDULE.

progress payment: Partial payment made to a contractor during a construction project based on the amount of work accomplished or materials supplied.

project: Planned and designed undertaking; e.g., construction of a building.

project application for payment: Certified request for payment from a contractor; compiled by the architect and presented to the owner.

project certificate of payment: Written statement from the architect to the owner indicating payment amounts due to contractors.

project engineer: Person responsible for design and management of the engineering aspects of a construction project.

project evaluation and review technique (PERT): Construction scheduling method in which individual events are charted and various portions of the project anticipated as the job progresses.

project manager: Person or company managing administrative and technical responsibilities of a construction project.

prosecute: To pursue for payment or punishment of a crime or violation of the law by instituting legal proceedings against the alleged offender.

protective covenant: Contract between a subdivider and lot purchaser stating that the value of the lot and surrounding property will be maintained and not diminished by other purchasers.

property damage insurance: Insurance coverage against claims resulting from damage to tangible property by fire, theft, vandalism, or natural causes.

property insurance: Insurance coverage against claims resulting from loss or damage to the construction. May be based on the complete property value or a graduated scale of value based on the present value of the construction project.

property tax: Governmental assessment for payment based on a tangible object or possession such as land, structures, or equipment.

protective covenant: Written agreement restricting the use of property.

proximate cause: Reason for an injury or damage based on injury or damage resulting from some party and stating that this party should be held accountable.

public domain: Land or property owned by a governmental agency.

public liability insurance: Insurance coverage against claims resulting from bodily injury, disease, or death of non-employees.

punch list: List compiled by the owner and/or architect on a walk-through of the completed construction project. Indicates items that must be completed before finalization of the project.

purchase order: Written document of the details of a purchase, often coded with a specific number.

quality control: System that ensures a project is completed according to the prints.

quantity survey: Estimation of amounts of materials and labor required for the construction of a given structure.

quotation: Established price for materials and/or work in the form of a proposal.

real estate: Land and all buildings, fences, and improvements on the land. Water and mineral rights are also considered part of real estate.

real property: See *real estate*.

record drawings: Prints that are marked to indicate changes to a project and provide a historical record of these changes.

reserve for depreciation: Sum of money charged against income and set aside into a reserve fund. Used to offset the depreciation of fixed assets.

restrictive covenant: Agreement between two parties that states the particular use for a piece of land.

retainage: Amount of money withheld from payments as agreed to by the owner and contractor. Paid in full when project is complete. RETENTION.

riparian rights: Legal right of an owner to use water from a source that borders the owner's property.

satisfaction: Relief of an encumbrance on property, commonly because of receipt of payment.

sealed bid: Bid submitted in a closed form. Opened at the time and place of the bid opening.

search of title: Examination of public documents concerning ownership and claims against a piece of property.

selected bidder: Bidder chosen by the owner to be considered for final award of the construction contract.

seller: Person or organization that offers property or materials for purchase.

set aside: Process in which a percentage of a construction project is designated for bid by a particular category of contractors; e.g., small businesses or disadvantaged businesses.

special conditions: Section of the contract conditions describing work unique to a particular construction project.

special hazards insurance: Insurance coverage against claims resulting from additional risks not stated in standard insurance coverage; e.g., water damage.

speculation: Construction of a project on the premise that a buyer will be found during or after completion of construction.

standard net assignable area: Portion of a building that is available for use by an occupant.

starting date: Time at which a construction project is approved for the beginning of construction.

statute of frauds: Statement that certain contracts are unenforeable without signatures and specific written terms.

statute of limitations: Statement that a given length of time exists during which legal action can be brought against any party.

stipulated sum agreement: Construction contract containing a given amount as the total payment.

stop work order: Statement from an owner that work on a project shall not continue until certain conditions are satisfied.

strike: Work stoppage.

subcontractor: Person or party that performs part of the work on a construction project under an agreement with the general contractor.

sublease: Legal agreement that transfers occupancy of property from a lessee to an additional, subsequent lessee.

subordinate lien: Mortgage lien subject to the first lien.

subrogation: Substitution of one party for another in respect to legal rights.

successful bidder: Bidder awarded a construction contract by an owner.

superintendent: Contractor's field representative responsible for supervision of an entire project.

supplementary conditions: Portion of the contract documents that modifies a section or sections of the general conditions. SUPPLEMENTAL CONDITIONS.

surety: Person or parties that make a written promise to pay the debt or default of another person or party.

surety bonds: Legal agreement in which one party agrees to be responsible to another party for a performance failure by a third party.

take-off: 1. Written list of types and quantities of materials from a print. **2.** Development of a written list of types and quantities of materials from a print.

tax abatement: Reduction of property taxes by reducing assessed value. Commonly used to attract business or industry to a particular municipality.

timely completion: Performance of construction work according to the dates and schedule set forth in the contract agreement.

time of completion: See *date of substantial completion.*

title: Legal means whereby an owner has possession of a piece of property.

title insurance: Insurance against any defective ownership records (titles) concerning a piece of property.

title search: Investigation conducted to determine the proper ownership of a piece of property and freedom from liens.

total float: Critical path method scheduling term indicating the difference between the time available to perform an activity and the estimated required time.

travel time: Paid time spent by construction workers moving between jobs or from their residences to a construction site.

turnkey contract: Building arrangement in which a contractor is totally responsible for all design, financing, and construction until the building is delivered to the owner as a completed project.

umbrella liability insurance: Insurance coverage against claims resulting from items not covered by existing liability insurance. Provides excess liability coverage for otherwise uninsured losses.

unbalanced bid: Bid package in which higher prices are charged early in the construction project and lower prices later in the project. This arrangement is designed to obtain payment at the beginning of a construction project to finance future work.

union: Trade organization representing workers of a similar craft to establish standard wage rates, working conditions, and training.

unit price: Cost per item of material or work.

value: Marketable price of property.

value engineering: Study of prints and specifications with the intention of making slight modifications that will improve production and cost without changing the final product.

variance: Written authorization by the appropriate authority to deviate from standard building code or ordinance.

volume estimating: Calculation of construction costs based on multiplying total volume of a structure by a fixed cost per unit of volume; e.g., a fixed amount per cubic foot.

waiver of lien: Document in which a person with the rights of a mechanic's lien gives up these rights for a particular job.

walk-through: On-site inspection of a construction project by the owner, architect, and/or a representative of the contractor.

warranty: Assurance of quality of a product or project.

warranty deed: Legal guarantee that the title being transferred is free of encumbrances.

workers compensation insurance: Insurance for employees covering injury, illness, or death that occurs while working on the job site. WORKMEN'S COMPENSATION INSURANCE.

zone: Portion of a city or governmental subdivision set aside for a specific use.

zoning: Division of land or subdivisions of land into areas that are regulated for the different types of usage: one-family residential; multifamily residential, commercial, industrial, etc.

Appendices

APPENDIX A—MATH SYMBOLS

MATH SYMBOLS	
+	Plus; add
−	Minus; subtract
×	Times; multiply
÷	Divided by; divide
Σ	Sum of
=	Equals; is equal to
≅	Approximately equals; approximately equal to
>	Is greater than
<	Is less than
√	Square root
$\sqrt[3]{}$	Cube root
E^x	Power of
( )	Parentheses } operations indicated
[ ]	Brackets } within are to be
{ }	Braces } performed first

APPENDIX B—SYMBOL AND UNIT ABBREVIATIONS

SYMBOL AND UNIT ABBREVIATIONS			
β	Arc length	l_b	Bisector length
A	Area	′	Minute; minutes
A_b	Area of base	n	Number of sides
A_s	Total surface area	p	Bisector length
A_t	Area of top	P	Perimeter
b	Base	P_b	Perimeter of base
C	Circumference	P_t	Perimeter of top
°	Degree	π	Pi; 3.1416
d	Diameter (circular figures)	r	Radius
d	Depth (linear figures)	″	Second; seconds
d_b	Depth of base	s	Side length
d_t	Depth of top	t	Top length
h	Height	V	Volume
h_s	Slant height	w	Width
l	Length	w_b	Width of base
		w_t	Width of top

APPENDIX C—GEOMETRIC FIGURES

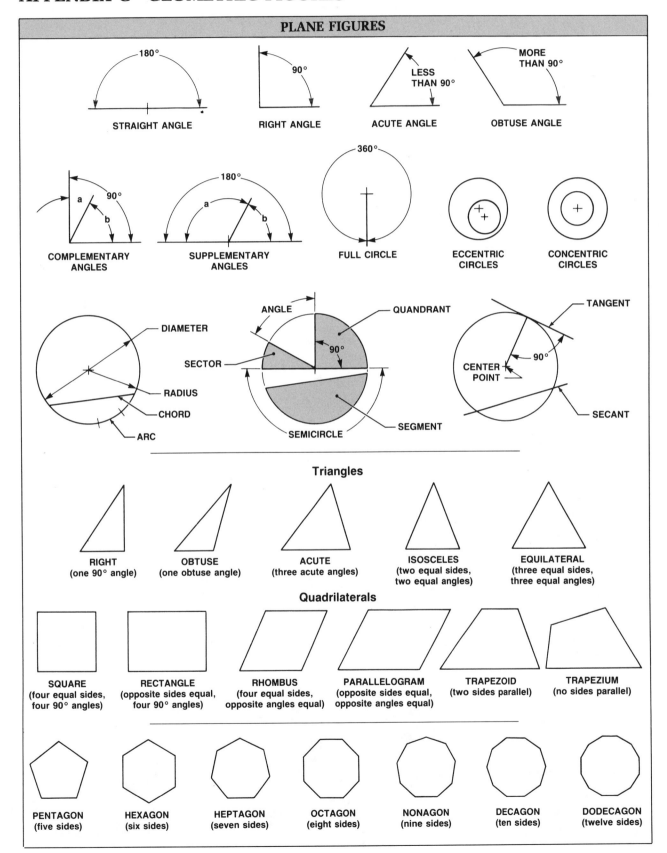

PLANE FIGURES

STRAIGHT ANGLE — 180°

RIGHT ANGLE — 90°

ACUTE ANGLE — LESS THAN 90°

OBTUSE ANGLE — MORE THAN 90°

COMPLEMENTARY ANGLES — 90°

SUPPLEMENTARY ANGLES — 180°

FULL CIRCLE — 360°

ECCENTRIC CIRCLES

CONCENTRIC CIRCLES

DIAMETER, SECTOR, RADIUS, CHORD, ARC

ANGLE, QUANDRANT, 90°, SEMICIRCLE, SEGMENT

TANGENT, CENTER POINT, 90°, SECANT

Triangles

RIGHT
(one 90° angle)

OBTUSE
(one obtuse angle)

ACUTE
(three acute angles)

ISOSCELES
(two equal sides,
two equal angles)

EQUILATERAL
(three equal sides,
three equal angles)

Quadrilaterals

SQUARE
(four equal sides,
four 90° angles)

RECTANGLE
(opposite sides equal,
four 90° angles)

RHOMBUS
(four equal sides,
opposite angles equal)

PARALLELOGRAM
(opposite sides equal,
opposite angles equal)

TRAPEZOID
(two sides parallel)

TRAPEZIUM
(no sides parallel)

PENTAGON
(five sides)

HEXAGON
(six sides)

HEPTAGON
(seven sides)

OCTAGON
(eight sides)

NONAGON
(nine sides)

DECAGON
(ten sides)

DODECAGON
(twelve sides)

SOLID FIGURES

Regular Solids

TETRAHEDRON	HEXAHEDRON	OCTAHEDRON	DODECAHEDRON	ICOSAHEDRON
(4 Triangles)	(6 Squares)	(8 Triangles)	(12 Pentagons)	(20 Triangles)

Parallelepipeds   Prisms

RIGHT SQUARE   RIGHT RECTANGULAR   OBLIQUE RECTANGULAR   RIGHT TRIANGULAR   RIGHT PENTAGONAL   OBLIQUE HEXAGONAL

Pyramids

  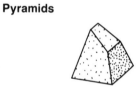

RIGHT TRIANGULAR   RIGHT SQUARE (Frustum)   RIGHT SQUARE (Truncated)   OBLIQUE PENTAGONAL

Cylinders   Cones

RIGHT CIRCULAR   OBLIQUE CIRCULAR   OBLIQUE CIRCULAR (Truncated)   RIGHT CIRCULAR   OBLIQUE CIRCULAR (Frustum)   OBLIQUE CIRCULAR (Truncated)

 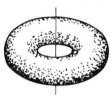 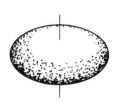 

SPHERE   TORUS   OBLATE ELLIPSOID   PROLATE ELLIPSOID

APPENDIX D—PLANE GEOMETRIC FIGURES—MATH

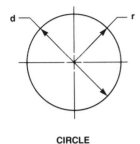

CIRCLE

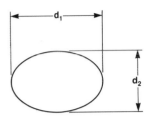

ELLIPSE

CIRCLE

$Circumference = \pi \times diameter$
$$C = \pi d$$
$Circumference = 2\pi \times radius$
$$C = 2\pi r$$

$Area = .7854 \times diameter^2$
$$A = .7854d^2$$
$Area = \pi \times radius^2$
$$A = \pi r^2$$

Example
Determine the circumference (C) and area (A) of the circle using the diameter (d) and radius (r).

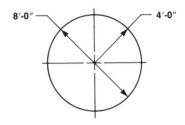

ELLIPSE

$Circumference = 1.5708 \times (diameter_1 + diameter_2)$
$$C = 1.5708 (d_1 + d_2)$$
$Area = .7854 \times diameter_1 \times diameter_2$
$$A = .7854 d_1 d_2$$

Example
Determine the circumference (C) and area (A) of the ellipse.

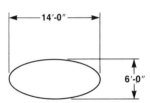

Solution
$$C = 1.5708(d_1 + d_2)$$
$$= 1.5708(14'\text{-}0'' + 6'\text{-}0'')$$
$$= 1.5708(20)$$
$$C = \mathbf{31.416'\text{-}0''}$$

$$A = .7854 d_1 d_2$$
$$= .7854 \times 14'\text{-}0'' \times 6'\text{-}0''$$
$$A = \mathbf{65.974 \ sq \ ft}$$

Solution using Diameter
$$C = \pi d$$
$$= 3.1416 \times 8'\text{-}0''$$
$$C = \mathbf{25.133'}$$

$$A = .7854d^2$$
$$= .7854 (8'\text{-}0'')^2$$
$$= .7854 \times 64$$
$$A = \mathbf{50.266 \ sq \ ft}$$

Solution using Radius
$$C = 2\pi r$$
$$= 2 \times 3.1416 \times 4'\text{-}0''$$
$$C = \mathbf{25.133'}$$
$$A = \pi r^2$$
$$= 3.1416 (4'\text{-}0'')^2$$
$$A = \mathbf{50.266 \ sq \ ft}$$

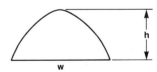

PARABOLA

PARABOLA

$Arc \ length = 2 \sqrt{\dfrac{(width)^2}{4} + \dfrac{4(height)^2}{3}}$

$$\beta = 2 \sqrt{\dfrac{w^2}{4} + \dfrac{4h^2}{3}}$$

$Area = .667 \times width \times height$
$$A = .667wh$$

Example
Determine the arc length (β) and area (A) of the parabola.

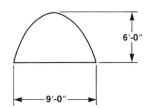

Solution

$$\beta = 2\sqrt{\frac{w^2}{4} + \frac{4h^2}{3}}$$

$$= 2\sqrt{\frac{(9'\text{-}0'')^2}{4} + \frac{4(6'\text{-}0'')^2}{3}}$$

$$= 2\sqrt{\frac{81}{4} + \frac{4(36)}{3}}$$

$$= 2\sqrt{\frac{81}{4} + \frac{144}{3}}$$

$$= 2\sqrt{20.25 + 48}$$

$$= 2\sqrt{68.25}$$

$$= 2 \times 8.26$$

$$\beta = \mathbf{16.52'\text{-}0''}$$

$$A = .667\,wh$$

$$= .667 \times 9'\text{-}0'' \times 6'\text{-}0''$$

$$A = \mathbf{36.02 \text{ sq ft}}$$

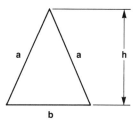

TRIANGLE

TRIANGLES

Perimeter = sum of side lengths
$$P = \Sigma s$$

Area $= \frac{1}{2}$ *base* $\times$ *height*

$$A = \frac{1}{2}bh$$

Example
Determine the perimeter (P) and area (A) of the triangle.

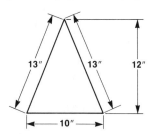

Solution

$$P = \Sigma s$$
$$= 13'' + 13'' + 10''$$
$$P = \mathbf{36''}$$

$$A = \frac{1}{2}bh$$

$$= \frac{1}{2}(10'' \times 12'')$$

$$= \frac{1}{2}(120)$$

$$A = \mathbf{60 \text{ sq in}}$$

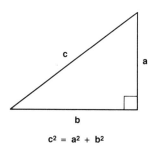

$$c^2 = a^2 + b^2$$

Pythagorean Theorem

The square of the hypotenuse of a right triangle is equal to the sum of the squares of the other two sides. Common combinations of side lengths derived from the Pythagorean theorem include the following.

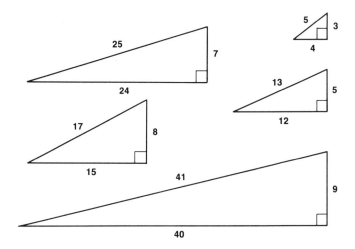

Example
Determine the length of the hypotenuse of the right triangle.

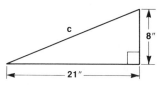

Solution

$$c^2 = a^2 + b^2$$

$$= (8'')^2 + (21'')^2$$

$$= 64 + 441$$

$$\sqrt{c^2} = \sqrt{505}$$

$$c = \mathbf{22.47''}$$

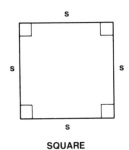

SQUARE

SQUARE

Perimeter = 4 × side length
$$P = 4s$$

Area = (side length)^2
$$A = s^2$$

Example

Determine the perimeter (P) and area (A) of the square.

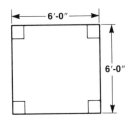

Solution

$P = 4s$
$\quad = 4 × 6'\text{-}0''$
$P = \textbf{24'-0''}$

$A = s^2$
$\quad = (6'\text{-}0'')^2$
$A = \textbf{36 sq ft}$

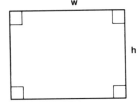

RECTANGLE

RECTANGLE

Perimeter = (2 × width) + (2 × height)
$$P = 2w + 2h$$

Area = width × height
$$A = wh$$

Example

Determine the perimeter (P) and area (A) of the rectangle.

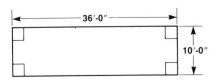

Solution

$P = 2w + 2h$
$\quad = (2 × 36'\text{-}0'') + (2 × 10'\text{-}0'')$
$\quad = 72 + 20$
$P = \textbf{92'-0''}$

$A = wh$
$\quad = 36'\text{-}0'' × 10'\text{-}0''$
$A = \textbf{360 sq ft}$

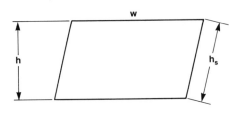

PARALLELOGRAM

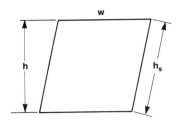

RHOMBUS

PARALLELOGRAM/RHOMBUS

Perimeter = (2 × width) + (2 × slant height)
$$P = 2w + 2h_s$$

Area = width × height
$$A = wh$$

Example

Determine the perimeter (P) and area (A) of the parallelogram.

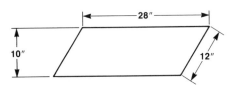

Solution

$P = 2w + 2h_s$
$\quad = (2 × 28'') + (2 × 12'')$
$\quad = 56 + 24$
$P = \textbf{80''}$

$A = wh$
$\quad = 28'' × 10''$
$A = \textbf{280 sq in}$

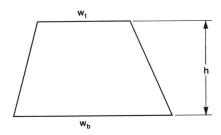

TRAPEZOID

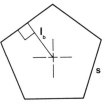

PENTAGON **HEXAGON**

TRAPEZOID

Perimeter = *sum of side lengths*
$$P = \Sigma s$$

Area = $\frac{1}{2}$ *height (width of top + width of base)*

$$A = \frac{1}{2} h(w_t + w_b)$$

Example

Determine the perimeter (P) and area (A) of the trapezoid.

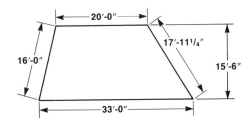

Solution

$$P = \Sigma s$$
$$= 20'\text{-}0'' + 17'\text{-}11\frac{1}{4}'' + 33'\text{-}0'' + 16'\text{-}0''$$
$$P = \mathbf{86'\text{-}11\frac{1}{4}''}$$

$$A = \frac{1}{2} h(w_t + w_b)$$
$$= \frac{1}{2} \times 15'\text{-}6''(20'\text{-}0'' + 33'\text{-}0'')$$
$$= 7.75(53)$$
$$A = \mathbf{410.75 \ sq \ ft}$$

PENTAGON/HEXAGON

Perimeter = *sum of side lengths*
$$P = \Sigma s$$

Area = *number of sides* $\left(\dfrac{\text{side length} \times \text{bisector length}}{2}\right)$

$$A = n\left(\frac{sl_b}{2}\right)$$

Example

Determine the perimeter (P) and area (A) of the pentagon.

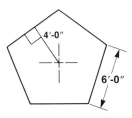

Solution

$$P = \Sigma s$$
$$= 6'\text{-}0'' + 6'\text{-}0'' + 6'\text{-}0'' + 6'\text{-}0'' + 6'\text{-}0''$$
$$P = \mathbf{30'\text{-}0''}$$
$$A = n\left(\frac{sl_b}{2}\right)$$
$$= 5\left(\frac{6'\text{-}0'' \times 4'\text{-}0''}{2}\right)$$
$$= 5\left(\frac{24}{2}\right)$$
$$= 5(12)$$
$$A = \mathbf{60 \ sq \ ft}$$

APPENDIX E—SOLID GEOMETRIC FIGURES—MATH

SPHERE

SPHERE

$Total\ surface\ area\ =\ \pi\ \times\ diameter^2$

$$A_s\ =\ \pi d^2$$

$Total\ surface\ area\ =\ 4\ \times\ \pi\ \times\ radius^2$

$$A_s\ =\ 4\pi r^2$$

$Volume\ =\ \dfrac{\pi\ \times\ diameter^3}{6}$

$$V\ =\ \dfrac{\pi d^3}{6}$$

$Volume\ =\ \dfrac{4\ \times\ \pi\ \times\ radius^3}{3}$

$$V\ =\ \dfrac{4\pi r^3}{3}$$

Example

Determine the total surface area (A_s) and volume (V) of a sphere using the diameter (d) and radius (r).

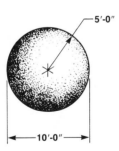

Solution using Diameter

$A_s\ =\ \pi d^2$
$\quad =\ 3.1416\ \times\ (10'\text{-}0'')^2$
$\quad =\ 3.1416\ \times\ 100$
$A_s\ =\ \mathbf{314.16\ sq\ ft}$

$V\ =\ \dfrac{\pi d^3}{6}$

$\quad =\ \dfrac{3.1416\ \times\ (10'\text{-}0'')^3}{6}$

$\quad =\ \dfrac{3.1416\ \times\ 1000}{6}$

$\quad =\ \dfrac{3141.6}{6}$

$V\ =\ \mathbf{523.6\ cu\ ft}$

Solution using Radius

$A_s\ =\ 4\pi r^2$
$\quad =\ 4\ \times\ 3.1416\ \times\ (5'\text{-}0'')^2$
$\quad =\ 4\ \times\ 3.1416\ \times\ 25$
$A_s\ =\ \mathbf{314.16\ sq\ ft}$

$V\ =\ \dfrac{4\pi r^3}{3}$

$\quad =\ \dfrac{4\ \times\ 3.1416\ \times\ (5'\text{-}0'')^3}{3}$

$\quad =\ \dfrac{4\ \times\ 3.1416\ \times\ 125}{3}$

$V\ =\ \mathbf{523.6\ cu\ ft}$

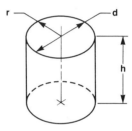

CYLINDER

CYLINDER

$Total\ surface\ area\ =\ 2(area\ of\ base)\ +$
$\qquad\qquad\qquad\qquad (\pi\ \times\ diameter\ \times\ height)$

$$A_s\ =\ 2(A_b)\ +\ \pi dh$$

$Total\ surface\ area\ =\ 2(area\ of\ base)\ +$
$\qquad\qquad\qquad\qquad 2(\pi\ \times\ radius\ \times\ height)$

$$A_s\ =\ 2(A_b)\ +\ 2\pi rh$$

$Volume\ =\ .7854\ \times\ diameter^2\ \times\ height$

$$V\ =\ .7854d^2h$$

$Volume\ =\ \pi\ \times\ radius^2\ \times\ height$

$$V\ =\ \pi r^2h$$

Example

Determine the total surface area (A_s) and volume (V) of the cylinder using the diameter (d) and radius (r).

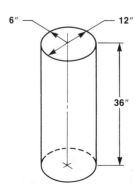

Solution using Diameter

$A_s = 2(.7854d^2) + \pi dh$

$\quad = 2[.7854 \times (12'')^2] + 3.1416 \times 12'' \times 36''$

$\quad = 2(.7854 \times 144) \times 1357.171$

$\quad = 2(.113.098) + 1357.171$

$\quad = 226.196 + 1357.171$

$A_s = $ **1583.367 sq in**

$V = .7854d^2h$

$\quad = .7854 \times (12'')^2 \times 36''$

$\quad = .7854 \times 144 \times 36$

$V = $ **4071.514 cu in**

Solution using Radius

$A_s = 2\pi r^2 + 2\pi rh$

$\quad = 2 \times 3.1416 \times (6'')^2 + 2(3.1416 \times 6'' \times 36'')$

$\quad = (2 \times 3.1416 \times 36) + 2(3.1416 \times 6 \times 36)$

$\quad = 226.196 + 1357.171$

$A_s = $ **1583.367 sq in**

$V = \pi r^2 h$

$\quad = 3.1416 \times (6'')^2 \times 36''$

$\quad = 3.1416 \times 36 \times 36$

$V = $ **4071.514 cu in**

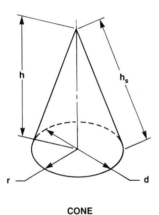

CONE

CONE

$Total\ surface\ area = \dfrac{\pi \times diameter \times slant\ height}{2} + area\ of\ base$

$A_s = \dfrac{\pi dh_s}{2} + A_b$

$Total\ surface\ area = (\pi \times radius \times slant\ height) + area\ of\ base$

$A_s = \pi rh_s + A_b$

$Volume = .2618 \times diameter^2 \times height$

$V = .2618d^2h$

$Volume = \dfrac{\pi \times radius^2 \times height}{3}$

$V = \dfrac{\pi r^2 h}{3}$

Example

Determine the total surface area (A_s) and volume (V) of the cone using the diameter (d) and radius (r).

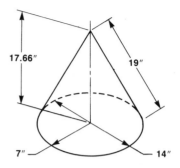

Solution using Diameter

$A_s = \dfrac{\pi dh_s}{2} + A_b$

$\quad = \dfrac{3.1416 \times 14'' \times 19''}{2} + [.7854 \times (14'')^2]$

$\quad = \dfrac{835.67}{2} + (.7854 \times 196)$

$\quad = 417.835 + 153.938$

$A_s = $ **571.773 sq in**

$V = .2618d^2h$

$\quad = .2618 \times (14'')^2 \times 17.66''$

$\quad = .2618 \times 196 \times 17.66$

$V = $ **906.184 cu in**

Solution using Radius

$A_s = (\pi rh_s) + A_b$

$\quad = (3.1416 \times 7'' \times 19'') + [3.1416 \times (7'')^2]$

$\quad = 417.835 + (3.1416 \times 49)$

$\quad = 417.835 + 153.938$

$A_s = $ **571.773 sq in**

$V = \dfrac{\pi r^2 h}{3}$

$\quad = \dfrac{3.1416 \times (7'')^2 \times 17.66''}{3}$

$\quad = \dfrac{3.1416 \times 49 \times 17.66}{3}$

$\quad = \dfrac{2718.552}{3}$

$V = $ **906.184 cu in**

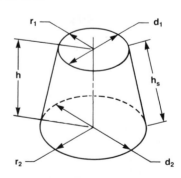

FRUSTUM OF A CONE

FRUSTUM OF A CONE

$Total\ surface\ area =$

$\dfrac{slant\ height \times \pi \times (diameter_1 + diameter_2)}{2} + area\ of\ top + area\ of\ base$

$$A_s = \frac{h_s \pi (d_1 + d_2)}{2} + A_t + A_b$$

Total surface area = slant height × π × (radius₁ + radius₂) + area of top + area of base

$$A_s = h_s \pi (r_1 + r_2) + A_t + A_b$$

Volume = height × π × [(diameter₁)² + (diameter₂)² + (diameter₁ × diameter₂)]/12

$$V = \frac{h\pi \, [(d_1)^2 + (d_2)^2 + (d_1 d_2)]}{12}$$

Volume = height × π × [(radius₁)² + (radius₂)² + (radius₁ × radius₂)]/3

$$V = \frac{h\pi \, [(r_1)^2 + (r_2)^2 + (r_1 r_2)]}{3}$$

Example

Determine the total surface area (A_s) and volume (V) of the frustum of a cone using the diameter (d) and radius (r).

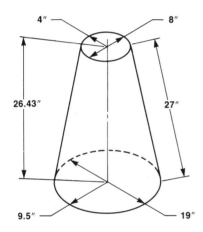

Solution using Diameter

$$A_s = \frac{h_s \pi (d_1 + d_2)}{2} + A_t + A_b$$

$$= \frac{27 \times 3.1416 \, (8'' + 19'')}{2} + [.7854 \times (8'')^2] + [.7854 \times (19'')^2]$$

$$= \frac{27 \times 3.1416 \times 27}{2} + 50.266 + 283.529$$

$$= \frac{2290.226}{2} + 333.795$$

$$= 1145.113 + 333.795$$

A_s = **1478.908 sq in**

$$V = \frac{h\pi \, [(d_1)^2 + (d_2)^2 + (d_1 d_2)]}{12}$$

$$= \frac{26.43'' \times 3.1416 [(8'')^2 + (19'')^2 + (8'' \times 19'')]}{12}$$

$$= \frac{26.43 \times 3.1416 (64 + 361 + 152)}{12}$$

$$= \frac{26.43 \times 3.1416 \times 577}{12}$$

$$= \frac{47,909.746}{12}$$

V = **3992.479 cu in**

Solution using Radius

$$A_s = h_s \pi (r_1 + r_2) + A_t + A_b$$

$$= 27'' \times 3.1416 \times (4'' + 9.5'') + [3.1416 \times (4'')^2 + 3.1416 \times (9.5'')^2]$$

$$= 27'' \times 3.1416 \times 13.5 + [(3.1416 \times 16) + (3.1416 \times 90.25)]$$

$$= 27 \times 3.1416 \times 13.5 + (50.266 + 283.529)$$

$$= 27 \times 3.1416 \times 13.5 + 333.795$$

$$= 1145.113 + 333.795$$

A_s = **1478.908 sq in**

$$V = \frac{h\pi \, [(r_1)^2 + (r_2)^2 + (r_1 r_2)]}{3}$$

$$= \frac{26.43 \times 3.1416 [(4'')^2 + (9.5'')^2 + (4'' \times 9.5'')]}{3}$$

$$= \frac{26.43 \times 3.1416 (16 + 90.25 + 38)}{3}$$

$$= \frac{26.43 \times 3.1416 (144.25)}{3}$$

$$= \frac{11,977.436}{3}$$

V = **3992.479 cu in**

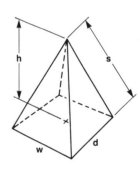

PYRAMID

PYRAMID

Total surface area = $\dfrac{perimeter\ of\ base \times side\ length}{2}$ + area of base

$$A_s = \frac{P_b \times s}{2} + A_b$$

Volume = $\dfrac{area\ of\ base \times height}{3}$

$$V = \frac{A_b \times h}{3}$$

Example

Determine the total surface area (A_s) and volume (V) of the pyramid.

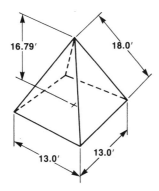

$$Volume = .33 \times height \, [(area \, of \, top + area \, of \, base) +$$

$$\sqrt{area \, of \, top \times area \, of \, base}]$$

$$V = .33h \, [(A_t + A_b) + \sqrt{A_t A_b}]$$

Example

Determine the total surface area (A_s) and volume (V) of the frustum of a pyramid.

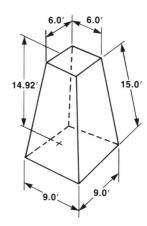

Solution

$$A_s = \frac{P_b \times s}{2} + A_b$$

$$= \frac{[(4 \times 13.0') \times 18.0']}{2} + (13.0')^2$$

$$= \frac{52 \times 18}{2} + 169$$

$$= \frac{936}{2} + 169$$

$$= 468 + 169$$

$$A_s = \textbf{637 sq ft}$$

$$V = \frac{A_b \times h}{3}$$

$$= \frac{(13.0')^2 \times 16.79'}{3}$$

$$= \frac{169 \times 16.79}{3}$$

$$= \frac{2837.51}{3}$$

$$V = \textbf{945.84 cu ft}$$

Solution

$$A_s = \frac{s \, [(P_t + P_b)]}{2} + A_t + A_b$$

$$= \frac{15.0' [(4 \times 6.0') + (4 \times 9.0')]}{2} + (6.0')^2 + (9.0')^2$$

$$= \frac{15.0(24 + 36)}{2} + 36 + 81$$

$$= \frac{15(60)}{2} + 117$$

$$= \frac{900}{2} + 117$$

$$= 450 + 117$$

$$A_s = \textbf{567 sq ft}$$

$$V = .33h \, [(A_t + A_b) + \sqrt{A_t A_b}]$$

$$= .33 \times 14.92' [(6.0')^2 + (9.0')^2 +$$
$$\sqrt{(6.0')^2 \times (9.0')^2}]$$

$$= 4.924 [(36 + 81) + \sqrt{36 \times 81}]$$

$$= 4.924 [117 + \sqrt{2916}]$$

$$= 4.924 [117 + 54]$$

$$= 4.924 \times 171$$

$$V = \textbf{842 cu ft}$$

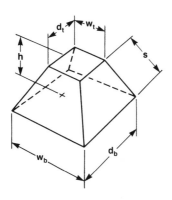

FRUSTUM OF A PYRAMID

FRUSTUM OF A PYRAMID

$Total \, surface \, area =$

$$\frac{side \, length \, (perimeter \, of \, top + perimeter \, of \, base)}{2} +$$

$$area \, of \, top + area \, of \, base$$

$$A_s = \frac{s \, (P_t + P_b)}{2} + A_t + A_b$$

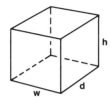

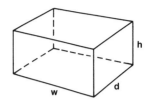

CUBE

RECTANGULAR PRISM

CUBE

Total surface area = 6 (side length)²

$$A = 6s^2$$

Volume = width × depth × height

$$V = wdh$$

Example

Determine the total surface area (A_s) and volume (V) of the cube.

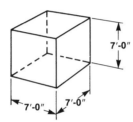

Solution

$$A_s = 6s^2$$
$$= 6(7\text{'-}0\text{''})^2$$
$$= 6 \times 49$$
$$A_s = \textbf{294 sq ft}$$

$$V = wdh$$
$$= 7\text{'-}0\text{''} \times 7\text{'-}0\text{''} \times 7\text{'-}0\text{''}$$
$$V = \textbf{343 cu ft}$$

RECTANGULAR PRISM

Total surface area = 2(width × height) +
2(height × depth) + 2(width × depth)

$$A_s = 2wh + 2hd + 2wd$$

Volume = width × depth × height

$$V = wdh$$

Example

Determine the total surface area (A_s) and volume (V) of the rectangular prism.

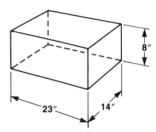

Solution

$$A_s = 2wh + 2hd + 2wd$$
$$= 2(23\text{''} \times 8\text{''}) + 2(8\text{''} \times 14\text{''}) + 2(23\text{''} \times 14\text{''})$$
$$= 2(184) + 2(112) + 2(322)$$
$$= 368 + 224 + 644$$
$$A_s = \textbf{1236 sq in}$$

$$V = wdh$$
$$= 23\text{''} \times 14\text{''} \times 8\text{''}$$
$$V = \textbf{2576 cu in}$$

APPENDIX F—TRADE-RELATED MATH

CARPENTRY

Roofs

Theoretical length of common rafter =
bridge measure of common rafter × run

Theoretical length of hip or valley rafter =
bridge measure of hip or valley rafter × run

Example

Determine the theoretical length of a common rafter for a roof with a 7 on 12 pitch and a run of 15′-6″.

Solution

Theoretical length of common rafter =
$$\textit{bridge measure of common rafter} \times \textit{run}$$
$$= 13.89 \times 15'\text{-}6''$$
$$= 13.89 \times 15.5'$$
$$= 215.295''$$
Theoretical length of common rafter = **17′–11¹/₄″**

Example

Determine the theoretical length of a hip rafter for a roof with a 10 on 12 pitch and a run of 16′-0″.

Solution

Theoretical length of hip rafter =
$$\textit{bridge measure of hip or valley rafter} \times \textit{run}$$
$$= 19.70 \times 16'\text{-}0''$$
$$= 19.70 \times 16.0'$$
$$= 315.2''$$
Theoretical length of hip rafter = **26′–3¹/₈″**

Stairs

1 unit rise + 1 unit run > 17″ and < 18″
2 unit rises + 1 unit run > 24″ and < 25″
Number of treads = Number of risers – 1

Stair Calculations

1. Determine the number of risers by dividing the total rise by 7.5″. If a decimal or fraction is obtained, round to the nearest whole number.
2. Determine the unit rise by dividing the total rise by the number of risers.
3. Determine unit run by subtracting unit rise from 17.5″.
4. Determine the number of treads.

Example

Determine the number of risers and treads, and the unit rise and unit run for a stairway with a total rise of 8′-2″ (98″).

Solution

Step 1. Determine the number of risers.
$$\textit{No. of risers} = \textit{total rise} \div 7.5''$$
$$= 98'' \div 7.5''$$
$$= 13.07$$
No. of risers = **13**

Step 2. Determine unit rise.
$$\textit{Unit rise} = \textit{total rise} \div \textit{no. of risers}$$
$$= 98'' \div 13$$
$$= 7.54''$$
Unit rise = **7⁹/₁₆″**

Step 3. Determine unit run.
$$\textit{Unit run} = 17.5'' - \textit{unit rise}$$
$$= 17.5'' - 7.54''$$
$$= 9.96''$$
Unit run = **9¹⁵/₁₆″**

Step 4. Determine number of treads.
$$\textit{No. of treads} = \textit{no. of risers} - 1$$
$$= 13 - 1$$
No. of treads = **12**

BRIDGE MEASURES			
Pitch	Angle of Incline (°)	Bridge Measure of Common Rafter	Bridge Measure of Hip or Valley Rafter
2 on 12	9¹/₂	12.16	17.09
3 on 12	14	12.37	17.23
4 on 12	18¹/₂	12.65	17.44
5 on 12	22¹/₂	13.00	17.69
6 on 12	26¹/₂	13.42	18.00
7 on 12	30¹/₄	13.89	18.36
8 on 12	33³/₄	14.42	18.76
9 on 12	37	15.00	19.21
10 on 12	40	15.62	19.70
11 on 12	42¹/₂	16.28	20.22
12 on 12	45	16.97	20.78
13 on 12	47¹/₄	17.69	21.38
14 on 12	49¹/₂	18.44	22.00
15 on 12	51¹/₂	19.21	22.65
16 on 12	53¹/₂	20.00	23.32
17 on 12	54³/₄	20.81	24.02
18 on 12	56¹/₄	21.63	24.72

ELECTRICITY

Ohm's Law and Power Formulas

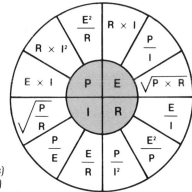

I = current (amps)
E = voltage (volts)
R = resistance (ohms)
P = power (watts or volt-amps)

Example—Voltage

Determine the voltage of a circuit pulling 15 A with a resistance of 8 ohms.

Solution

$E = R \times I$
$\quad = 8 \text{ ohms} \times 15 \text{ A}$
$E = \textbf{120 volts}$

Example—Resistance

Determine the resistance of a 240 V circuit pulling 20 A.

Solution

$R = \dfrac{E}{I}$

$\quad = \dfrac{240 \text{ V}}{20 \text{ A}}$

$R = \textbf{120 ohms}$

When calculating three-phase (3ϕ) problems, the voltage is multiplied by $\sqrt{3}$. To eliminate one step in the calculations, use the following values.

for 208 volts × 1.732, use 360
for 230 volts × 1.732, use 398
for 240 volts × 1.732, use 416
for 440 volts × 1.732, use 762
for 460 volts × 1.732, use 797
for 480 volts × 1.732, use 831

PHASE	UNKNOWN VALUE	USE FORMULA
1ϕ	I	$I = \dfrac{P}{E}$
	P	$P = I \times E$
	E	$E = \dfrac{P}{I}$
3ϕ	I	$I = \dfrac{P}{(E \times \sqrt{3})}$
	P	$P = I \times E \times \sqrt{3}$
	E	$E = \dfrac{P}{(I \times \sqrt{3})}$

I = current (amps)
P = power (watts or volt-amps)
E = voltage (volts)

Example—1ϕ

Determine the current that a 32,000 VA feeder circuit will pull on a 240 V, 1ϕ system.

Solution

$I = \dfrac{P}{E}$

$\quad = \dfrac{32,000 \text{ VA}}{240 \text{ V}}$

$I = \textbf{133.3 A}$

Example—3ϕ

Determine the current that a 46,800 VA feeder circuit will pull on a 208 V, 3ϕ system.

Solution

$I = \dfrac{P}{(E \times \sqrt{3})}$

$\quad = \dfrac{46,800 \text{ VA}}{360 \text{ V}}$

$I = \textbf{130 A}$

Horsepower Formulas

$$\text{Horsepower} = \frac{\text{voltage} \times \text{current} \times \text{efficiency}}{746}$$

$$HP = \frac{E \times I \times E_{FF}}{746}$$

Example—HP

Determine the horsepower of a 240 V motor pulling 15 A, and having an 85% efficiency.

Solution

$HP = \dfrac{E \times I \times E_{FF}}{746}$

$\quad = \dfrac{240 \text{ V} \times 15 \text{ A} \times .85}{746}$

$\quad = \dfrac{3060}{746}$

$HP = \textbf{4.1}$

$$\text{Current} = \frac{\text{horsepower} \times 746}{\text{voltage} \times \text{efficiency} \times \text{power factor}}$$

$$I = \frac{HP \times 746}{E \times E_{FF} \times PF}$$

Example—Current

Determine the current of a 10 HP, 240 V motor with a 90% efficiency rating and a power factor of 88%.

Solution

$I = \dfrac{HP \times 746}{E \times E_{FF} \times PF}$

$\quad = \dfrac{10 \text{ HP} \times 746}{240 \text{ V} \times .90 \times .88}$

$\quad = \dfrac{7460}{190.08}$

$I = \textbf{39.25 A}$

AC/DC Formulas

UNKNOWN	DIRECT CURRENT	ALTERNATING CURRENT		
		1φ, 115 or 120 V	1φ, 208, 230, or 240 V	3φ—All Voltages
Current When horsepower is known	$\dfrac{HP \times 746}{E \times E_{FF}}$	$\dfrac{HP \times 746}{E \times E_{FF} \times PF}$	$\dfrac{HP \times 746}{E \times E_{FF} \times PF}$	$\dfrac{HP \times 746}{1.73 \times E \times E_{FF} \times PF}$
Current When kilowatts is known	$\dfrac{kW \times 1000}{E}$	$\dfrac{kW \times 1000}{E \times PF}$	$\dfrac{kW \times 1000}{E \times PF}$	$\dfrac{kW \times 1000}{1.73 \times E \times PF}$
Current When kilovolt-amps is known		$\dfrac{kVA \times 1000}{E}$	$\dfrac{kVA \times 1000}{E}$	$\dfrac{kVA \times 1000}{1.73 \times E}$
Kilowatts	$\dfrac{I \times E}{1000}$	$\dfrac{I \times E \times PF}{1000}$	$\dfrac{I \times E \times PF}{1000}$	$\dfrac{I \times E \times 1.73 \times PF}{1000}$
Kilovolt-Amps		$\dfrac{I \times E}{1000}$	$\dfrac{I \times E}{1000}$	$\dfrac{I \times E \times 1.73}{1000}$
Horsepower	$\dfrac{E \times I \times E_{FF}}{746}$	$\dfrac{I \times E \times E_{FF} \times PF}{746}$	$\dfrac{I \times E \times E_{FF} \times PF}{746}$	$\dfrac{I \times E \times 1.73 \times E_{FF} \times PF}{746}$

I = current (amps) HP = horsepower kVA = kilovolt-amps
E = voltage (volts) PF = power factor
E_{FF} = efficiency KW = kilowatts

Example—Current, 1φ
Determine the current of a 10 HP, 1φ motor with an efficiency rating of 78% and a power factor of 90%. The motor is connected to a 120 V system.

Solution

$$I = \frac{HP \times 746}{E \times E_{FF} \times PF}$$

$$= \frac{10\ HP \times 746}{120\ V \times .78 \times .90}$$

$$= \frac{7460}{84.24}$$

$$I = \textbf{88.56 A}$$

Example—kVA, 1φ
Determine the kVA rating of a 120 V, 1φ motor pulling 2.5 A.

Solution

$$kVA = \frac{I \times E}{100}$$

$$= \frac{2.5\ A \times 120\ V}{1000}$$

$$= \frac{300}{1000}$$

$$kVA = \textbf{0.3}$$

HEATING, VENTILATING, AND AIR CONDITIONING

Absolute Pressure
Absolute pressure = atmospheric pressure + gauge pressure

Example
Determine the absolute pressure when the gauge pressure is 27 psi.

Solution
Absolute pressure = atmospheric pressure + gauge pressure
= 14.7 psi + 27 psi
Absolute pressure = **41.7 psi**

Work
Work is the amount of force required to move an object a specified distance.
Work = force × distance through which force travels

Example
Determine the work performed when hoisting 1500 pounds of water a vertical distance of 8′-0″

Solution
Work = *force × distance through which force travels*
= 1500 lb × 8′-0″
Work = **12,000 ft•lb**

Unit of Heat
A unit of heat is the amount of heat required to raise the temperature of one pound of water 1°F.
British thermal units = weight of water (lb) × temperature difference (F°).
$Btu = wt \times \Delta T$

Example

Determine the amount of heat required to raise the temperature of 80 pounds of water from 70°F to 120°F.

Solution

$$Btu = wt \times \Delta T$$
$$= 80 \text{ lb} \times (120°F - 70°F)$$
$$= 80 \times 50$$
$$Btu = \textbf{4000}$$

Specific Heat Capacity

Specific heat capacity is the amount of heat added or removed from a material to change the temperature of one pound of the material 1°F.

Amount of heat added or removed (Btu) = weight of substance (lb) × specific heat capacity (Btu/lb/ °F) × temperature difference (F °)

$$Btu = wt \times sp \ ht \ capacity \times \Delta T$$

Example

Determine the amount of heat that must be removed from 36 pounds of R-502 to cool it from 70°F to 25°F.

Solution

$$Btu = wt \times sp \ ht \ capacity \times \Delta T$$
$$= 36 \text{ lb} \times 0.255 \text{ Btu/lb/°F} \times (70°F - 25°F)$$
$$= 36 \times 0.255 \times 45$$
$$Btu = \textbf{413}$$

SUBSTANCE	SPECIFIC HEAT CAPACITY
	Btu/lb/ °F
Wood	0.327
Water	1.000
Ice	0.504
Iron	0.129
Mercury	0.0333
Alcohol	0.615
Copper	0.095
Sulphur	0.177
Glass	0.187
Graphite	0.200
Brick	0.200
Glycerine	0.576
Liquid ammonia at 40°F	1.100
Carbon dioxide at 40°F	0.600
R-502	0.255
Salt Brine 20%	0.85
R-12	0.213
R-22	0.26

PLUMBING

Pipe Bends

Length of bend = .01745 × radius × bend angle

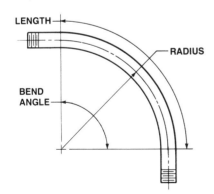

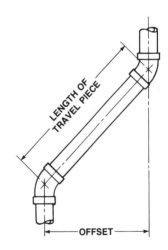

Example

Determine the length of the bend when a 15″ radius is used with a bend angle of 90°.

Solution

$$Length \ of \ bend = .01745 \times radius \times bend \ angle$$
$$= .01745 \times 15″ \times 90°$$
$$Length \ of \ bend = \textbf{23.56″}$$

Single Pipe Offsets

Length of travel piece = offset × angular constant

CONSTANTS FOR SINGLE PIPE OFFSETS		
Fitting	Fitting Angle (°)	Angular Constant
$^1/_{32}$	11¼	5.126
$^1/_{16}$	22½	2.613
$^1/_8$	45	1.414
$^1/_6$	60	1.155
$^1/_5$	72	1.051

Example

Determine the length of the travel piece with an offset of 10″ and ¹/₆ bend fittings.

Solution

> Length of travel piece = offset × angular constant
> = 10″ × 1.155
> Length of travel piece = **11.55″**

Multiple Pipe Offsets (Equal Spreads)

> Difference in length = spread × length constant
> Length of travel piece = offset × angular constant

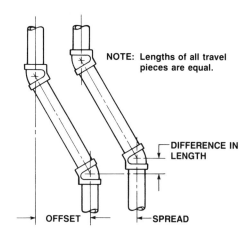

NOTE: Lengths of all travel pieces are equal.

DIFFERENCE IN LENGTH

OFFSET   SPREAD

Example

Determine the difference in length of the two runners and length of travel piece. The runners have a 10″ spread and a 24″ offset. The offsets are joined with 45° elbows.

Solution

> Difference in length = spread × length constant
> = 10″ × 0.414
> Difference in length = **4.14″**
>
> Length of travel piece = offset × angular constant
> = 24″ × 1.414
> Length of travel piece = **33.94″**

CONSTANTS FOR MULTIPLE PIPE OFFSETS			
Fitting	Fitting Angle (°)	Length Constant	Angular Constant
¹/₃₂	11¹/₄	0.099	5.126
¹/₁₆	22¹/₂	0.199	2.613
¹/₈	45	0.414	1.414
¹/₆	60	0.577	1.155
¹/₅	72	0.727	1.051

APPENDIX G—CONVERSIONS

DECIMAL EQUIVALENTS OF AN INCH			
Inches	**Decimal of an Inch**	**Inches**	**Decimal of an Inch**
$1/64$	0.015625	$33/64$	0.515625
$1/32$	0.03125	$17/32$	0.53125
$3/64$	0.046875	$35/64$	0.546875
$1/16$	0.0625	$9/16$	0.5625
$5/64$	0.078125	$37/64$	0.578125
$3/32$	0.09375	$19/32$	0.59375
$7/64$	0.109375	$39/64$	0.609375
$1/8$	0.125	$5/8$	0.625
$9/64$	0.140625	$41/64$	0.640625
$5/32$	0.15625	$21/32$	0.65625
$11/64$	0.171875	$43/64$	0.671875
$3/16$	0.1875	$11/16$	0.6875
$13/64$	0.203125	$45/64$	0.703125
$7/32$	0.21875	$23/32$	0.71875
$15/64$	0.234375	$47/64$	0.734375
$1/4$	0.250	$3/4$	0.750
$17/64$	0.265625	$49/64$	0.765625
$9/32$	0.28125	$25/32$	0.78125
$19/64$	0.296875	$51/64$	0.796875
$5/16$	0.3125	$13/16$	0.8125
$21/64$	0.328125	$53/64$	0.828125
$11/32$	0.34375	$27/32$	0.84375
$23/64$	0.359375	$55/64$	0.859375
$3/8$	0.375	$7/8$	0.875
$25/64$	0.390625	$57/64$	0.890625
$13/32$	0.40625	$29/32$	0.90625
$27/64$	0.421875	$59/64$	0.921875
$7/16$	0.4375	$15/16$	0.9375
$29/64$	0.453125	$61/64$	0.953125
$15/32$	0.46875	$31/32$	0.96875
$31/64$	0.484375	$63/64$	0.984375
$1/2$	0.500	1″	1.000

DECIMAL EQUIVALENTS OF A FOOT					
Inches	**Decimal of a Foot**	**Inches**	**Decimal of a Foot**	**Inches**	**Decimal of a Foot**
$1/16$	0.0052	$4 1/16$	0.3385	$8 1/16$	0.6719
$1/8$	0.0104	$4 1/8$	0.3438	$8 1/8$	0.6771
$3/16$	0.0156	$4 3/16$	0.3490	$8 3/16$	0.6823
$1/4$	0.0208	$4 1/4$	0.3542	$8 1/4$	0.6875
$5/16$	0.0260	$4 5/16$	0.3594	$8 5/16$	0.6927
$3/8$	0.0313	$4 3/8$	0.3646	$8 3/8$	0.6979
$7/16$	0.0365	$4 7/16$	0.3698	$8 7/16$	0.7031
$1/2$	0.0417	$4 1/2$	0.3750	$8 1/2$	0.7083
$9/16$	0.0469	$4 9/16$	0.3802	$8 9/16$	0.7135
$5/8$	0.0521	$4 5/8$	0.3854	$8 5/8$	0.7188
$11/16$	0.0573	$4 11/16$	0.3906	$8 11/16$	0.7240
$3/4$	0.0625	$4 3/4$	0.3958	$8 3/4$	0.7292
$13/16$	0.0677	$4 13/16$	0.4010	$8 13/16$	0.7344
$7/8$	0.0729	$4 7/8$	0.4063	$8 7/8$	0.7396
$15/16$	0.0781	$4 15/16$	0.4115	$8 15/16$	0.7448
1	0.0833	5	0.4167	9	0.7500
$1 1/16$	0.0885	$5 1/16$	0.4219	$9 1/16$	0.7552
$1 1/8$	0.0938	$5 1/8$	0.4271	$9 1/8$	0.7604
$1 3/16$	0.0990	$5 3/16$	0.4323	$9 3/16$	0.7656
$1 1/4$	0.1042	$5 1/4$	0.4375	$9 1/4$	0.7708
$1 5/16$	0.1094	$5 5/16$	0.4427	$9 5/16$	0.7760
$1 3/8$	0.1146	$5 3/8$	0.4479	$9 3/8$	0.7813
$1 7/16$	0.1198	$5 7/16$	0.4531	$9 7/16$	0.7865
$1 1/2$	0.1250	$5 1/2$	0.4583	$9 1/2$	0.7917
$1 9/16$	0.1302	$5 9/16$	0.4635	$9 9/16$	0.7969
$1 5/8$	0.1354	$5 5/8$	0.4688	$9 5/8$	0.8021
$1 11/16$	0.1406	$5 11/16$	0.4740	$9 11/16$	0.8073
$1 3/4$	0.1458	$5 3/4$	0.4792	$9 3/4$	0.8125
$1 13/16$	0.1510	$5 13/16$	0.4844	$9 13/16$	0.8177
$1 7/8$	0.1563	$5 7/8$	0.4896	$9 7/8$	0.8229
$1 15/16$	0.1615	$5 15/16$	0.4948	$9 15/16$	0.8281
2	0.1667	6	0.5000	10	0.8333
$2 1/16$	0.1719	$6 1/16$	0.5052	$10 1/16$	0.8385
$2 1/8$	0.1771	$6 1/8$	0.5104	$10 1/8$	0.8438
$2 3/16$	0.1823	$6 3/16$	0.5156	$10 3/16$	0.8490
$2 1/4$	0.1875	$6 1/4$	0.5208	$10 1/4$	0.8542
$2 5/16$	0.1927	$6 5/16$	0.5260	$10 5/16$	0.8594
$2 3/8$	0.1979	$6 3/8$	0.5313	$10 3/8$	0.8646
$2 7/16$	0.2031	$6 7/16$	0.5365	$10 7/16$	0.8698
$2 1/2$	0.2083	$6 1/2$	0.5417	$10 1/2$	0.8750
$2 9/16$	0.2135	$6 9/16$	0.5469	$10 9/16$	0.8802
$2 5/8$	0.2188	$6 5/8$	0.5521	$10 5/8$	0.8854
$2 11/16$	0.2240	$6 11/16$	0.5573	$10 11/16$	0.8906
$2 3/4$	0.2292	$6 3/4$	0.5625	$10 3/4$	0.8958
$2 13/16$	0.2344	$6 13/16$	0.5677	$10 13/16$	0.9010
$2 7/8$	0.2396	$6 7/8$	0.5729	$10 7/8$	0.9063
$2 15/16$	0.2448	$6 15/16$	0.5781	$10 15/16$	0.9115
3	0.2500	7	0.5833	11	0.9167
$3 1/16$	0.2552	$7 1/16$	0.5885	$11 1/16$	0.9219
$3 1/8$	0.2604	$7 1/8$	0.5938	$11 1/8$	0.9271
$3 3/16$	0.2656	$7 3/16$	0.5990	$11 3/16$	0.9323
$3 1/4$	0.2708	$7 1/4$	0.6042	$11 1/4$	0.9375
$3 5/16$	0.2760	$7 5/16$	0.6094	$11 5/16$	0.9427
$3 3/8$	0.2813	$7 3/8$	0.6146	$11 3/8$	0.9479
$3 7/16$	0.2865	$7 7/16$	0.6198	$11 7/16$	0.9531
$3 1/2$	0.2917	$7 1/2$	0.6250	$11 1/2$	0.9583
$3 9/16$	0.2969	$7 9/16$	0.6302	$11 9/16$	0.9635
$3 5/8$	0.3021	$7 5/8$	0.6354	$11 5/8$	0.9688
$3 11/16$	0.3073	$7 11/16$	0.6406	$11 11/16$	0.9740
$3 3/4$	0.3125	$7 3/4$	0.6458	$11 3/4$	0.9792
$3 13/16$	0.3177	$7 13/16$	0.6510	$11 13/16$	0.9844
$3 7/8$	0.3229	$7 7/8$	0.6563	$11 7/8$	0.9896
$3 15/16$	0.3281	$7 15/16$	0.6615	$11 15/16$	0.9948
4	0.3333	8	0.6667	12	1.0000

POWER AND ROOT TABLE

Number	Powers		Roots		Number	Powers		Roots	
	Square	Cube	Square	Cube		Square	Cube	Square	Cube
1	1	1	1.000	1.000	51	2601	132,651	7.141	3.708
2	4	8	1.414	1.260	52	2704	140,608	7.211	3.733
3	9	27	1.732	1.442	53	2809	148,877	7.280	3.756
4	16	64	2.000	1.587	54	2916	157,464	7.348	3.780
5	25	125	2.236	1.710	55	3025	166,375	7.416	3.803
6	36	216	2.449	1.817	56	3136	175,616	7.483	3.826
7	49	343	2.646	1.913	57	3249	185,193	7.550	3.849
8	64	512	2.828	2.000	58	3364	195,112	7.616	3.871
9	81	729	3.000	2.080	59	3481	205,379	7.681	3.893
10	100	1000	3.162	2.154	60	3600	216,000	7.746	3.915
11	121	1331	3.317	2.224	61	3721	226,981	7.810	3.936
12	144	1728	3.464	2.289	62	3844	238,328	7.874	3.958
13	169	2197	3.606	2.351	63	3969	250,047	7.937	3.979
14	196	2744	3.742	2.410	64	4096	262,144	8.000	4.000
15	225	3375	3.873	2.466	65	4225	274,625	8.062	4.021
16	256	4096	4.000	2.520	66	4356	287,496	8.124	4.041
17	289	4913	4.123	2.571	67	4489	300,763	8.185	4.062
18	324	5832	4.243	2.621	68	4624	314,432	8.246	4.082
19	361	6859	4.359	2.668	69	4761	328,509	8.307	4.102
20	400	8000	4.472	2.714	70	4900	343,000	8.367	4.121
21	441	9261	4.583	2.759	71	5041	357,911	8.426	4.141
22	484	10,648	4.690	2.802	72	5184	373,248	8.485	4.160
23	529	12,167	4.796	2.844	73	5329	389,017	8.544	4.179
24	576	13,824	4.899	2.884	74	5476	405,224	8.602	4.198
25	625	15,625	5.000	2.924	75	5625	421,875	8.660	4.217
26	676	17,576	5.099	2.962	76	5776	438,976	8.718	4.236
27	729	19,683	5.196	3.000	77	5929	456,533	8.775	4.254
28	784	21,952	5.292	3.037	78	6084	474,552	8.832	4.273
29	841	24,389	5.385	3.072	79	6241	493,039	8.888	4.291
30	900	27,000	5.477	3.107	80	6400	512,000	8.944	4.309
31	961	29,791	5.568	3.141	81	6561	531,441	9.000	4.327
32	1024	32,798	5.657	3.175	82	6724	551,368	9.055	4.344
33	1089	35,937	5.745	3.208	83	6889	571,787	9.110	4.362
34	1156	39,304	5.831	3.240	84	7056	592,704	9.165	4.380
35	1225	42,875	5.916	3.271	85	7225	614,125	9.220	4.397
36	1296	46,656	6.000	3.302	86	7396	636,056	9.274	4.414
37	1369	50,653	6.083	3.332	87	7569	658,503	9.327	4.481
38	1444	54,872	6.164	3.362	88	7744	681,472	9.381	4.448
39	1521	59,319	6.245	3.391	89	7921	704,969	9.434	4.465
40	1600	64,000	6.325	3.420	90	8100	729,000	9.487	4.481
41	1681	68,921	6.403	3.448	91	8281	753,571	9.539	4.498
42	1764	74,088	6.481	3.476	92	8464	778,688	9.592	4.514
43	1849	79,507	6.557	3.503	93	8649	804,357	9.644	4.531
44	1936	85,184	6.633	3.530	94	8836	830,584	9.695	4.547
45	2025	91,125	6.708	3.557	95	9025	857,375	9.747	4.563
46	2116	97,336	6.782	3.583	96	9216	884,736	9.798	4.579
47	2209	103,823	6.856	3.609	97	9409	912,673	9.849	4.595
48	2304	110,592	6.928	3.634	98	9604	941,192	9.900	4.610
49	2401	117,649	7.000	3.659	99	9801	970,299	9.950	4.626
50	2500	125,000	7.071	3.684	100	10,000	1,000,000	10.000	4.642

LINEAR MEASURE
12 inches = 1 foot
3 feet = 1 yard
36 inches = 1 yard
5.5 yards = 1 rod
16.5 feet = 1 rod
40 rods = 1 furlong
660 feet = 1 furlong
8 furlongs = 1 mile
320 rods = 1 mile
1760 yards = 1 mile
5290 feet = 1 mile

ANGULAR MEASURE
60 seconds = 1 minute
60 minutes = 1 degree
57.3 degrees = 1 radian
180 degrees = π radians
360 degrees = 2π radians

TIME MEASURE
60 seconds = 1 minute
60 minutes = 1 hour
24 hours = 1 day
7 days = 1 week
52 weeks = 1 year
365.26 days = 1 year

AVOIRDUPOIS WEIGHT
437.5 grains = 1 ounce
16 ounces = 1 pound
100 pounds = 1 hundredweight
1000 pounds = 1 kip
2 kips = 1 ton
2000 pounds = 1 ton
2240 pounds = 1 long ton

DRY MEASURE
2 pints = 1 quart
4 quarts = 1 gallon
2 gallons = 1 peck
8 quarts = 1 peck
4 pecks = 1 bushel

LIQUID MEASURE
4 gills = 1 pint
2 pints = 1 quart
57.75 cubic inches = 1 quart
4 quarts = 1 gallon
231 cubic inches = 1 gallon
31.5 gallons = 1 barrel

SQUARE MEASURE
144 square inches = 1 square foot
9 square feet = 1 square yard
1296 square inches = 1 square yard
30.25 square yards = 1 square rod
160 square rods = 1 acre
4840 square yards = 1 acre
43,560 square feet = 1 acre
640 acres = 1 square mile

CUBIC MEASURE
7.48 gallons = 1 cubic foot
1728 cubic inches = 1 cubic foot
27 cubic feet = 1 cubic yard
202 gallons = 1 cubic yard
128 cubic feet = 1 cord

SURVEYOR'S LINEAR MEASURE
7.92 inches = 1 link
16.5 feet = 1 rod
25 links = 1 rod
4 rods = 1 chain
66 feet = 1 chain
100 links = 1 chain
80 chains = 1 mile

SURVEYOR'S SQUARE MEASURE
625 square links = 1 square rod
16 square rods = 1 square chain
10 square chains = 1 acre
640 acres = 1 square mile
1 square mile = 1 section
36 square miles = 1 township
36 sections = 1 township

TEMPERATURE CONVERSIONS
°F = (9/5 × °C) + 32
°C = 5/9 (°F − 32)

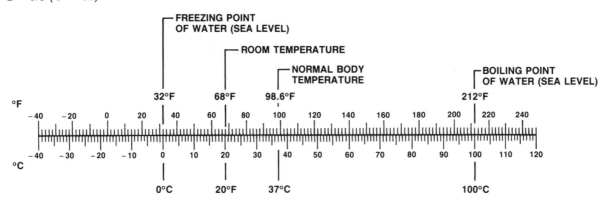

APPENDIX H—CSI FORMAT

DIVISION 0—BIDDING AND CONTRACT REQUIREMENTS

Reference
Number
00010 Pre-bid Information
00100 Instructions to Bidders
00200 Information Available to Bidders
00300 Bid/Tender Forms
00400 Supplements to Bid/Tender Forms
00500 Agreement Forms
00600 Bonds and Certificates
00700 General Conditions of the Contract
00800 Supplementary Conditions
00900 Addenda and Modifications
00950 Drawings Index

DIVISION 1—GENERAL REQUIREMENTS
01010 General Contractors
01020 Allowances
01030 Special Project Procedures
01040 Coordination
01050 Field Engineering
01060 Regulatory Requirements
01070 Abbreviations and Symbols
01080 Identification Systems
01100 Alternates/Alternatives
01150 Measurement and Payment
01200 Project Meetings
01300 Submittals
01400 Quality Control
01500 Construction Facilities and Temporary Controls
01600 Material and Equipment
01650 Starting of Systems
01660 Testing, Adjusting, and Balancing of Systems
01700 Contract Closeout
01800 Maintenance Materials

DIVISION 2—SITEWORK
02010 Subsurface Investigation
02050 Demolition
02100 Site Preparation
02150 Underpinning
02200 Earthwork
02300 Tunneling
02350 Piles, Caissons, and Cofferdams
02400 Drainage
02440 Site Improvements
02480 Landscaping
02500 Paving and Surfacing
02580 Bridges
02590 Ponds and Reservoirs
02700 Piped Utilities
02800 Power and Commmunication Utilities
02850 Railroad Work
02880 Marine Work

DIVISION 3—CONCRETE
03010 Concrete Materials
03050 Concreting Procedures
03100 Concrete Formwork
03150 Forms

03180 Form Ties and Accessories
03200 Concrete Reinforcement
03250 Concrete Accessories
03300 Cast-in-place Concrete
03350 Special Concrete Finishes
03360 Specially Placed Concrete
03370 Concrete Curing
03400 Precast Concrete
03500 Cementitious Decks
03700 Concrete Restoration and Cleaning

DIVISION 4—MASONRY
04050 Masonry Procedures
04100 Mortar
04150 Masonry Accessories
04200 Unit Masonry
04400 Stone
04500 Masonry Restoration and Cleaning
04550 Refractories
04600 Corrosion-resistant Masonry

DIVISION 5—METALS
05010 Metal Materials and Methods
05050 Metal Fastening
05100 Structural Metal Framing
05200 Metal Joists
05300 Metal Decking
05400 Cold-formed Metal Framing
05500 Metal Fabrications
05700 Ornamental Metal
05800 Expansion Control
05900 Metal Finishes

DIVISION 6—WOOD AND PLASTICS
06050 Fasteners and Supports
06100 Rough Carpentry
06130 Heavy Timber Construction
06150 Wood-Metal Systems
06170 Prefabricated Structural Wood
06200 Finish Carpentry
06300 Wood Treatment
06400 Architectural Woodwork
06500 Prefabricated Structural Plastics
06600 Plastic Fabrications

DIVISION 7—THERMAL AND MOISTURE PROTECTION
07100 Waterproofing
07150 Dampproofing
07200 Insulation
07250 Fireproofing
07300 Shingles and Roofing Tiles
07400 Preformed Roofing and Siding
07500 Membrane Roofing
07570 Traffic Topping
07600 Flashing and Sheet Metal
07800 Roof Accessories
07900 Sealants

DIVISION 8—DOORS AND WINDOWS

08100 Metal Doors and Frames
08200 Wood and Plastic Doors
08250 Door Opening Assemblies
08300 Special Doors
08400 Entrances and Storefronts
08500 Metal Windows
08600 Wood and Plastic Windows
08650 Window Covering
08700 Hardware
08800 Glazing
08900 Glazed Curtain Walls

DIVISION 9—FINISHES

09100 Metal Support Systems
09200 Lath and Plaster
09230 Aggregate Coatings
09300 Tile
09400 Terrazzo
09500 Acoustical Treatment
09550 Wood Flooring
09600 Stone and Brick Flooring
09650 Resilient Flooring
09680 Carpeting
09700 Special Flooring
09760 Floor Treatment
09800 Special Coatings
09900 Painting
09950 Wall Covering

DIVISION 10—SPECIALITIES

10100 Chalkboards and Tackboards
10150 Compartments and Cubicles
10200 Louvers and Vents
10240 Grills and Screens
10230 Service Wall Systems
10260 Wall and Corner Guards
10270 Access Flooring
10280 Speciality Modules
10290 Pest Control
10300 Fireplaces and Stoves
10340 Prefabricated Steeples, Spires, and Cupolas
10350 Flagpoles
10400 Identifying Devices
10450 Pedestrian Control Devices
10500 Lockers
10520 Fire Extinguishers, Cabinets, and Accessories
10530 Protective Covers
10550 Postal Specialities
10600 Partitions
10650 Scales
10670 Storage Shelving
10700 Exterior Sun Control Devices
10750 Telephone Enclosures
10800 Toilet and Bath Accessories
10900 Wardrobe Specialities

DIVISION 11—EQUIPMENT

11010 Maintenance Equipment
11020 Security and Vault Equipment
11030 Checkroom Equipment
11040 Ecclesiastical Equipment
11050 Library Equipment
11060 Theater and Stage Equipment
11070 Musical Equipment
11080 Registration Equipment
11100 Mercantile Equipment
11110 Commercial Laundry and Dry Cleaning Equipment
11120 Vending Equipment
11130 Audio-Visual Equipment
11140 Service Station Equipment
11150 Parking Equipment
11160 Loading Dock Equipment
11170 Waste Handling Equipment
11190 Detention Equipment
11200 Water Supply and Treatment Equipment
11300 Fluid Waste Disposal and Treatment Equipment
11400 Food Service Equipment
11450 Residential Equipment
11460 Unit Kitchens
11470 Darkroom Equipment
11480 Athletic, Recreational, and Therapeutic Equipment
11500 Industrial and Process Equipment
11600 Laboratory Equipment
11650 Planetarium and Observatory Equipment
11700 Medical Equipment
11780 Mortuary Equipment
11800 Telecommunication Equipment
11850 Navigation Equipment

DIVISION 12—FURNISHINGS

12100 Artwork
12300 Manufactured Cabinets and Casework
12500 Window Treatment
12550 Fabrics
12600 Furniture and Accessories
12670 Rugs and Mats
12700 Multiple Seating
12800 Interior Plants and Plantings

DIVISION 13—SPECIAL CONSTRUCTION

13010 Air-supported Structures
13020 Integrated Assemblies
13030 Audiometric Rooms
13040 Clean Rooms
13050 Hyperbaric Rooms
13060 Insulated Rooms
13070 Integrated Ceilings
13080 Sound, Vibration, and Seismic Control
13090 Radiation Protection
13100 Nuclear Reactors
13110 Observatories
13120 Pre-engineered Structures
13130 Special-purpose Rooms and Buildings
13140 Vaults
13150 Pools
13160 Ice Rinks
13170 Kennels and Animal Shelters
13200 Seismographic Instrumentation
13210 Stress Recording Instrumentation
13220 Solar and Wind Instrumentation
13410 Liquid and Gas Storage Tanks
13510 Restoration of Underground Pipelines
13520 Filter Underdrains and Media
13530 Digestion Tank Covers and Appurtenances

13540 Oxygenation Systems
13550 Thermal Sludge Conditioning Systems
13560 Site Constructed Incinerators
13600 Utility Control Systems
13700 Industrial and Process Control Systems
13800 Oil and Gas Refining Installations and Control Systems
13900 Transportation Instrumentation
13940 Building Automation Systems
13970 Fire Suppression and Supervisory Systems
13980 Solar Energy Systems
13990 Wind Energy Systems

DIVISION 14—CONVEYING SYSTEMS
14100 Dumbwaiters
14200 Elevators
14300 Hoists and Cranes
14400 Lifts
14500 Material Handling Systems
14600 Turntables
14700 Moving Stairs and Walks
14800 Powered Scaffolding
14900 Transportation Systems

DIVISION 15—MECHANICAL
15050 HVAC and Piping Contractors
15200 Noise, Vibration, and Seismic Control
15250 Insulation
15300 Special Piping Systems
15400 Plumbing Systems
15450 Plumbing Fixtures and Trim
15500 Fire Protection
15600 Power or Heat Generation
15650 Refrigeration
15700 Liquid Heat Transfer
15800 Air Distribution
15900 Controls and Instrumentation

DIVISION 16—ELECTRICAL
16050 Electrical Contractors
16200 Power Generation
16300 Power Transmission
16400 Service and Distribution
16500 Lighting
16600 Special Systems
16700 Communications
16850 Heating and Cooling
16900 Controls and Instrumentation

APPENDIX I—TRADE ORGANIZATIONS

ASA **Acoustical Society of America**
500 Sunnyside Blvd.
Woodberry, NY 11797

ASC **Adhesive and Sealant Council, Inc.**
1500 Wilson Blvd., Suite 515
Arlington, VA 22209-2495

ARI **Air-Conditioning and Refrigeration Institute**
1501 Wilson Blvd.
Arlington, VA 22209

ACCA **Air Conditioning Contractors of America**
1513 16th St.
Washington, DC 20036

ACEA **Allied Construction Employers Association**
180 N. Executive Drive
Brookfield, WI 53008

AA **Aluminum Association**
900 19th St., NW
Washington, DC 20006

AAA **American Arbitration Association**
140 W. 51st St.
New York, NY 10020

AASHTO **American Association of State Highway and Transportation Officials**
444 N. Capitol St., NW, Suite 225
Washington, DC 20001

ABCA **American Building Contractors Association**
11100 Valley Blvd., Suite 120
El Monte, CA 91731

ACI **American Concrete Institute**
22400 W. Seven Mile Rd.
Detroit, MI 48219

ACPA **American Concrete Pavement Association**
3800 N. Wilke Rd., Suite 490
Arlington Heights, IL 60004

ACPA **American Concrete Pipe Association**
8320 Old Courthouse Rd.
Vienna, VA 22180

ACPA **American Concrete Pumping Association**
P.O. Box 4307
1034 Tennessee St.
Vallejo, CA 94590

ACEC **American Consulting Engineers Council**
1015 15th St., NW, Suite 802
Washington, DC 20005

AGA **American Gas Association, Inc.**
1515 Wilson Blvd.
Arlington, VA 22209

AHA **American Hardboard Association**
520 N. Hicks Rd.
Palatine, IL 60067

AHMA **American Hardware Manufacturers Association**
931 N. Plum Grove Rd.
Schaumburg, IL 60173

AHLI **American Home Lighting Institute**
435 N. Michigan Ave., Suite 1717
Chicago, IL 60611

AIA **American Institute of Architects**
1735 New York Ave., NW
Washington, DC 20006

AIC **American Institute of Constructors**
20 S. Front St.
Columbus, OH 43215

ASID **American Society of Interior Designers**
200 Lexington Ave.
New York, NY 10016

AISC **American Institute of Steel Construction, Inc.**
400 N. Michigan Ave.
Chicago, IL 60611

AITC **American Institute of Timber Construction**
11818 S.E. Mill Plain Blvd.
Vancouver, WA 98684

AISI **American Iron and Steel Institute**
1133 15th St., NW, Suite 300
Washington, DC 20005

ALSC **American Lumber Standards Committee**
P.O. Box 210
Germantown, MD 20874

ANSI **American National Standards Institute**
1430 Broadway
New York, NY 10018

APFA **American Pipe Fitting Association**
8136 Old Keene Mill Rd. #B-311
Springfield, VA 22152

APA **American Plywood Association**
P.O. Box 11700
Tacoma, WA 98411

ARTBA **American Road and Transportation Builders Association**
525 School St., SW
Washington, DC 20024

ASTM **American Society for Testing and Materials**
1916 Race St.
Philadelphia, PA 19103

ASCE **American Society of Civil Engineers**
345 E. 47th St.
New York, NY 10017

ASCC **American Society of Concrete Construction**
426 S. Westgate
Addison, IL 60101

ASHRAE **American Society of Heating, Refrigerating, and Air-Conditioning Engineers Inc.**
1791 Tullie Circle, NE
Atlanta, GA 30329

ASID **American Society of Interior Designers**
1430 Broadway
New York, NY 10018

ASME	**American Society of Mechanical Engineers** United Engineering Center 345 E. 47th St. New York, NY 10017	**BHMA**	**Builder's Hardware Manufacturers Association, Inc.** 60 E. 42nd St., Rm. 511 New York, NY 10165
ASPE	**American Society of Professional Estimators** 3617 Thousand Oaks Blvd., Suite 210 Westlake, CA 91362	**BOCA**	**Building Officials and Code Administrators International** 4051 W. Flossmoor Rd. Country Club Hills, IL 60477
ASSE	**American Society of Sanitary Engineers** P.O. Box 40362 Bay Village, OH 44140		**Building Materials Research Institute, Inc.** 501 5th Ave., #1402 New York, NY 10017
ASA	**American Subcontractors Association** 1004 Duke St. Alexandria, VA 22314	**BRB**	**Building Research Board** 2101 Constitution Ave., NW Washington, DC 20418
AWS	**American Welding Society, Inc.** 550 N.W. LeJeune Rd. Miami, FL 33126	**BSI**	**Building Stone Institute** 420 Lexington Ave., Suite 2800 New York, NY 10170
AAMA	**Architectural Aluminum Manufacturers Association** 2700 River Rd., Suite 118 Des Plaines, IL 60018	**BSC**	**Building Systems Council** 15th and M St., NW Washington, DC 20005
APA	**Architectural Precast Association** 825 E. 64th St. Indianapolis, IN 46220	**CRA**	**California Redwood Association** 405 Enfrente Dr., Suite 200 Nevato, CA 94949
AWI	**Architectural Woodwork Institute** 2310 S. Walter Reed Dr. Arlington, VA 22206	**CRI**	**Carpet and Rug Institute** P.O. Box 2048 Dalton, GA 30722-2048
AIA/NA	**Asbestos Information Association/North America** 1745 Jefferson Davis Hwy., Suite 509 Arlington, VA 22202	**CISPI**	**Cast Iron Soil Pipe Institute** 1499 Chain Bridge Rd., Suite 203 McLean, VA 22101
AI	**Asphalt Institute** Asphalt Institute Building College Park, MD 20740	**CISCA**	**Ceilings and Interior Systems Construction Association** 104 Wilmot, Suite 201 Deerfield, IL 60015
ABC	**Associated Builders and Contractors, Inc.** 729 15th St., NW Washington, DC 20005	**CTI**	**Ceramic Tile Institute** 700 N. Virgil Ave. Los Angeles, CA 90029
AGC	**Associated General Contractors of America** 1957 East St., NW Washington, DC 20006	**CLFMI**	**Chain Link Fence Manufacturers Institute** 1776 Massachusetts Ave., NW, Suite 500 Washington, DC 20036
SMACNA	**Associated Sheet Metal Contractors, Inc.** 3121 W. Hallandale Beach Blvd., Suite 114 Hallandale, FL 33009	**CRSI**	**Concrete Reinforcing Steel Institute** 933 N. Plum Grove Rd. Schaumburg, IL 60195
ASC	**Associated Speciality Contractors** 7315 Wisconsin Ave. Bethesda, MD 20814	**CIEA**	**Construction Industry Employers Association** 625 Ensminger Rd. Tonawanda, NY 14150
ABC	**Association of Bituminous Contractors** 2020 K St., NW, Suite 800 Washington, DC 20006	**CIMA**	**Construction Industry Manufacturers Association** 111 E. Wisconsin Ave., Suite 940 Milwaukee, WI 53202-4879
AWCI	**Association of the Wall and Ceiling Industries International** 25 K St., NE, Suite 300 Washington, DC 20002	**CPMA**	**Construction Products Manufacturing Council** P.O. Box 21008 Washington, DC 20009-0508
BIA	**Brick Institute of America** 11490 Commerce Park Dr., Suite 300 Reston, VA 22091	**CSI**	**Construction Specifications Institute** 601 Madison St. Alexandria, VA 22314

CDA	**Copper Development Association, Inc.** Greenwich Office Park 2 51 Weaver St. Grant, CT 06836	IHEA	**Industrial Heating Equipment Association** 1901 N. Moore St. Arlington, VA 22209
	Corps of Engineers/U.S. Department of the Army 20 Massachusetts Ave., NW Washington, DC 20314	IEEE	**Institute of Electrical and Electronics Engineers** 345 E. 47th St. New York, NY 10017
CABO	**Council of American Building Officials** 5203 Leesburg Pike, Suite 708 Falls Church, VA 22041	ICAA	**Insulation Contractors Association of America** 15819 Crabbs Branch Way Rockville, MD 20855
DFI	**Deep Foundations Institute** P.O. Box 359 Springfield, NJ 07081		**International Association of Bridge, Structural and Ornamental Iron Workers** 1750 New York Avenue, NW, Suite 400 Washington, DC 20006
DHI	**Door and Hardware Institute** 7711 Old Springhouse Rd. McLean, VA 22102-3474	IALD	**International Association of Lighting Designers** 18 E. 16th St., Suite 208 New York, NY 10003
DIPRA	**Ductile Iron Pipe Research Association** 245 Riverchase Parkway E., Suite 0 Birmingham, AL 35244	IAPMO	**International Association of Plumbing and Mechanical Officials** 20001 Walnut Dr., S Walnut, CA 91789
EPA	**Environmental Protection Agency** 401 M St., SW Washington, DC 20460	IBB	**International Brotherhood of Boilermakers, Iron Ship Builders, Blacksmiths, Forgers and Helpers** 753 State Avenue, Suite 565 Kansas City, KS 66101
ESCSI	**Expanded Shale, Clay and Slate Institute** 6218 Montrose Rd. Rockville, MD 20852	IBW	**International Brotherhood of Electrical Workers** 1125 15th Street, NW Washington, DC 20005
FTI	**Facing Tile Institute** P.O. Box 8880 Canton, OH 44711	IBPAT	**International Brotherhood of Painters and Allied Trades** 1750 New York Avenue, NW Washington, DC 20006
FHA	**Federal Housing Administration** 451 7th St., SW, Rm. 3158 Washington, DC 20410	ICBO	**International Council of Building Officials** 5360 S. Workman Mill Rd. Whittier, CA 90601
FPRS	**Forest Products Research Society** 2801 Marshall Ct. Madison, WI 53705	IILP	**International Institute for Lath and Plaster** 795 Raymond Ave. St. Paul, MN 55114
GBCA	**General Building Contractors Association** 36 S. 18th St. P.O. Box 15959 Philadelphia, PA 19103	IMI	**International Masonry Institute** 823 15th St., NW, Suite 1001 Washington, DC 20005
GA	**Gypsum Association** 1603 Orrington Ave., Suite 1210 Evanston, IL 60201	IRF	**International Road Federation** 525 School St., SW Washington, DC 20024
HPMA	**Hardwood Plywood Manufacturers Association** P.O. Box 2789 Reston, VA 22090	IUBAC	**International Union of Bricklayers and Allied Craftsmen** Bowen Building 815 15th St., NW Washington, DC 20005
IESNA	**Illuminating Engineering Society of North America** 345 E. 47th St. New York, NY 10017	IUEC	**International Union of Elevator Constructors** Clark Building 5565 Sterrett Place, Suite 530 Columbia, Maryland 21044
ILIA	**Indiana Limestone Institute of America** Stone City Bank Building, Suite 400 Bedford, IN 47421	IUOE	**International Union of Operating Engineers** 1125 17th Street, NW Washington, DC 20036
IFI	**Industrial Fasteners Institute** 1505 E. Ohio Building Cleveland, OH 44114		

LIUNA	**Laborers' International Union of North America** 905 16th Street, NW Washington, DC 20006-1765	NARSC	**National Association of Reinforcing Steel Contractors** 10382 Main St. P.O. Box 225 Fairfax, VA 22030

	Manufacturers Standardization Society of the Valve and Fittings Industry 127 Park St., NE Vienna, VA 22180
MFMA	**Maple Flooring Manufacturers Association** 60 Revere Dr., Suite 500 Northbrook, IL 60062
MIA	**Marble Institute of America** 33505 State St. Farmington, MI 48024
MCAA	**Mason Contractors Association of America** 17W 601 14th St. Oakbrook Terrace, IL 60181
MCAA	**Mechanical Contractors Association of America** 5410 Grosvenor, Suite 120 Bethesda, MD 20814
MBMA	**Metal Building Manufacturers Association** 1230 Keith Building Cleveland, OH 44115
MLSFA	**Metal Lath/Steel Framing Association** 600 S. Federal, Suite 400 Chicago, IL 60605
NAPA	**National Asphalt Pavement Association** 6811 Kenilworth Ave., Suite 620 P.O. Box 517 Riverdale, MD 20737
NADC	**National Association of Demolition Contractors** 4415 W. Harrison St. Hillside, IL 60162
NADC	**National Association of Dredging Contractors** 1625 I St., NW, Suite 321 Washington, DC 20006
NAEC	**National Association of Elevator Contractors** 4053 LaVista Rd., Suite 120 Tucker, GA 30084
NAFCD	**National Association of Floor Covering Distributors** 13-126 Merchandise Mart Chicago, IL 60654
NAHB	**National Association of Home Builders** 15th and M St., NW Washington, DC 20005
NAHRO	**National Association of Housing Redevelopment Officials** 1320 18th St., NW Washington, DC 20036
NAPHCC	**National Association of Plumbing, Heating, and Cooling Contractors** P.O. Box 6808 Falls Church, VA 22046

NAWIC	**National Association of Women in Construction** 327 S. Adams St. Fort Worth, TX 76104
NBMA	**National Building Manufacturers Association** 142 Lexington Ave. New York, NY 10016
	National Building Material Distributors Association 1701 Lake Ave., Suite 170 Glenview, IL 60025
NCMA	**National Concrete Masonry Association** P.O. Box 781 Herndon, VA 22070
NCSBCS	**National Conference of States on Building Codes and Standards** 481 Carlisle Dr. Herndon, VA 22070
NCA	**National Constructors Association** 1101 15th St., NW, Suite 1000 Washington, DC 20005
NCRP	**National Council on Radiation Protection and Measurement** 7910 Woodmont Ave., Suite 800 Bethesda, MD 20814
NECA	**National Electrical Contractors Association** 7315 Wisconsin Ave. 13th Floor, West Building Bethesda, MD 20814
NEMA	**National Electrical Manufacturers Association** 2101 L St., NW, Suite 300 Washington, DC 20037
NFPA	**National Fire Protection Association** Batterymarch Park Quincy, MA 02269
NFPA	**National Forest Products Association** 1250 Connecticut Ave., NW, Suite 200 Washington, DC 20036
NGA	**National Glass Association** 8200 Greensboro Dr., Suite 302 McLean, VA 22102
	National Housing Rehabilitation Association 1726 18th St., NW Washington, DC 20009
NKCA	**National Kitchen Cabinet Association** P.O. Box 6830 Falls Church, VA 22046
NLA	**National Lime Association** 3601 N. Fairfax Dr. Arlington, VA 22201

NLBMDA	**National Lumber and Building Material Dealers Association** 40 Ivy St., SE Washington, DC 20003	PDI	**Plumbing and Drainage Institute** 1106 W. 77th St., S. Dr. Indianapolis, IN 46260
NOFMA	**National Oak Flooring Manufacturers Association** P.O. Box 3009 Memphis, TN 38173-0009	PHCIB	**Plumbing-Heating-Cooling Information Bureau** 303 E. Wacker Dr., Suite 711 Chicago, IL 60601
NPCA	**National Paint and Coatings Association** 1500 Rhode Island Ave., NW Washington, DC 20005	PMI	**Plumbing Manufacturers Institute** 800 Roosevelt Rd., Building C, Suite 20 Glen Ellyn, IL 60137
	National Particleboard Association 2306 Perkins Pl. Silver Spring, MD 20910	PTI	**Post-Tensioning Institute** 1717 W. Northern Ave., Suite 218 Pheonix, AZ 85021
NPCA	**National Precast Concrete Association** 825 E. 64th St. Indianapolis, IN 46220	PCA	**Portland Cement Association** 5420 Old Orchard Rd. Skokie, IL 60077
NRMCA	**National Ready Mixed Concrete Association** 900 Spring St. Silver Spring, MD 20910	PCI	**Prestressed Concrete Institute** 175 W. Jackson Blvd., Suite 1859 Chicago, IL 60604
NRCA	**National Roofing Contractors Association** 1 O'Hare Center 6250 River Rd. Rosemont, IL 60018	RCSHSB	**Red Cedar Shingle and Handsplit Shake Bureau** 515 116th Ave., NE, Suite 275 Bellevue, WA 98004
NSPE	**National Society of Professional Engineers** 1420 King St. Alexandria, VA 22314	RCRC	**Reinforced Concrete Research Council** 5420 Old Orchard Rd. Skokie, IL 60077
NSA	**National Stone Association** 1415 Elliot Pl., NW Washington, DC 20007	RFCI	**Resilient Floor Covering Institute** 966 Hungerford Dr., Suite 12B Rockville, MD 20850
NTMA	**National Terrazzo and Mosiac Association** 3166 Des Plaines Ave., Suite 132 Des Plaines, IL 60018	RFCA	**Resilient Flooring and Carpet Association, Inc.** 14570 E. 14th St., Suite 511 San Landro, CA 94578
	National Wood Window and Door Association 205 Touhy Ave. Park Ridge, IL 60068	SSFI	**Scaffolding, Shoring, and Forming Institute, Inc.** 1230 Keith Building Cleveland, OH 44115
NWMA	**National Woodwork Manufacturers' Association** 400 W. Madison St. Chicago, IL 60606	SMA	**Screen Manufacturers Association** 655 Irving Park, Suite 201 Chicago, IL 60613-3198
OPCMIA	**Operative Plasterers' and Cement Masons' International Association of the United States and Canada** 1125 17th St., NW, 6th Floor Washington, DC 20036	SWI	**Sealant and Waterproofers Institute** 3101 Broadway, Suite 300 Kansas City, MO 64111
PDCA	**Painting and Decorating Contractors of America** 7223 Lee Hwy. Falls Church, VA 22046	SIGMA	**Sealed Insulating Glass Manufacturers Association** 111 E. Wacker Dr., Suite 600 Chicago, IL 60601
PSIC	**Passive Solar Industries Council** 2836 Duke St. Alexandria, VA 22314	SMACNA	**Sheet Metal and Air Conditioning Contractors National Association, Inc.** 8224 Old Courthouse Rd. Vienna, VA 22180
	Pipe Line Contractors Association 4100 First City Center 1700 Pacific Ave. Dallas, TX 75201	SMWIA	**Sheet Metal Workers International Association** 1750 New York Ave., NW Washington, DC 20006
PPI	**Plastics Pipe Institute** 355 Lexington Ave. New York, NY 10017		

SBCCI	**Southern Building Code Congress International, Inc.** 900 Montclair Rd. Birmingham, AL 35213
SDI	**Steel Deck Institute** P.O. Box 9506 Canton, OH 44711
SDI	**Steel Door Institute** 712 Lakewood Center N 14600 Detroit Ave. Cleveland, OH 44107
SJI	**Steel Joist Institute** 1205 48th Ave., N, Suite A Myrtle Beach, SC 29577
SSPC	**Steel Structures Painting Council** 4400 5th Ave. Pittsburgh, PA 15213
SWI	**Steel Window Institute** 1230 Keith Building Cleveland, OH 44115
SMA	**Stucco Manufacturers Association** 14006 Ventura Blvd. Sherman Oaks, CA 91423
SBA	**Systems Builders Association** P.O. Box 117 West Milton, OH 45383
TIMA	**Thermal Insulation Manufacturers Association** 29 Bank St. Stanford, CT 06901
TCAA	**Tile Contractors Association of America, Inc.** 112 N. Alfred St. Alexandria, VA 22314
TCA	**Tile Council of America** P.O. Box 2222 Princeton, NJ 08542
TCA	**Tilt-up Concrete Association** 5420 Old Orchard Rd. Skokie, IL 60077

UL	**Underwriters' Laboratories, Inc.** 333 Pfingsten Rd. Northbrook, IL 60062
	United Association of Journeymen and Apprentices of the Plumbing and Pipe Fitting Industry of the United States and Canada 901 Massachusetts Ave., NW Washington, DC 20001
UBC	**United Brotherhood of Carpenters and Joiners of America** 101 Constitution Ave., NW Washington, DC 20001
	U.S. Department of Labor/Occupational Safety and Health Administration 200 Constitution Ave., NW Washington, DC 20210
	U.S. Forest Products Laboratory One Gifford Pinchot Dr. Madison, WI 53705-2398
UURWAW	**United Union of Roofers, Waterproofers and Allied Workers** 1125 17th Street, NW, 5th Floor Washington, DC 20036
VMA	**Valve Manufacturers Association of America** 1050 17th St., NW, Suite 701 Washington, DC 20036
WMA	**Wallcovering Manufacturers Association** 355 Lexington Ave. New York, NY 10017
WWPA	**Western Wood Products Association** Yeon Building 522 S.W. 5th Ave. Portland, OR 97204
WRI	**Wire Reinforcement Institute** 8361-A Greensboro Dr. McLean, VA 22102

APPENDIX J—PRINT COMPONENTS

ABBREVIATIONS OF TERMS

Punctutation is not utilized except for the slant (/) and hyphen (-). A period (.) is added to an abbreviation if it does not clearly represent an abbreviation in context.

Term	Drawing Abbreviation	Term	Drawing Abbreviation
A		alignment	ALIGN
abbreviate	ABBR	alkaline	ALK
above	ABV	allowance	ALLOW
above water	AW	alloy	ALY
abrasive	ABRSV	alloy steel	ALY STL
abrasive-resistant	ABRSV RES	alteration	ALTRN
absolute	ABS	alternate	ALTN
absorption	ABSORB	alternating current	AC
abstract	ABSTR	aluminum	AL
accelerate	ACCEL	ambient	AMB
acceptable quality level	AQL	American National Standard	AMER NATL STD
access	ACS	American Steel Wire Gauge	ASWG
access opening	AO	American Wire Gauge	AWG
accessory	ACCESS	ammonia	AMNA
access panel	AP	amount	AMT
account	ACCT	ampere	AMP
accumulate	ACCUM	amplitude	AMPTD
accumulator	ACC	anchor	AHR
acetate	ACTT	anchor bolt	AB
acetylene	ACET	anchored filament	ANCFIL
acidproof	AP	angle stop valve	ASV
acoustic	ACST	angular	ANLR
acoustical plaster ceiling	APC	annular	ANLR
acoustical tile	AT or ACT	annunciator	ANN
across	ACR	anodize	ANDZ
activate	ACTVT	antenna	ANT
activator	ACTVTR	apartment	APT
activity	ACT	aperture	APERT
actual	ACTL	appearance	APP
actuate	ACTE	appendix	APPX
actuating	ACTG	application	APPL
actuator	ACTR	approach	APRCH
adapter	ADPTR	approval	APPVL
addition	ADD	approve	APPV
additive	ADDT	approved	APVD
adhesive	ADH	approximate	APPROX
adjacent	ADJ	architecture	ARCH
adjustable	ADJT or A	arc weld	ARCW
adsorbent	ADSORB	area	A
advance	ADV	area drain	AD
advance material request	AMR	armature	ARM
agent	AGT	armored	ARMD
aileron	AIL	armored cable	ARM CA
air blast circuit breaker	ABCB	arrester	ARSR
air-break switch	AB SW	artificial	ARTF
air circulating	ACIRC	asbestos	ASB
air cleaner	AIRCLNR	as drawn	AD
air-condition	AIR COND	asphalt	ASPH
air-conditioner	AIR COND	asphalt roof shingles	ASPHRS
air-cooled	ACLD	asphalt tile	AT
air induction	AINDTN	asphalt-tile base	ATB
air lock	AL	asphalt-tile floor	ATF
air passage	AP	aspirator	ASPRTR
airtight	AT	as required	AR
alcohol	ALC	astragal	A

Term	Drawing Abbreviation	Term	Drawing Abbreviation
attachment	ATCH	bituminous	BITUM
attenuation	ATTEN	black iron	BI
attenuator	ATTEN	blackout door	BOD
automatic	AUTO	blade	BL
automatic overload	AUTO OVLD	blank	BLK
automatic sprinkler	AS	blanket	BLKT
automatic starter	AUTOSTRT	bleeder	BLDR
automatic stop and check valve	AUTO S&CV	block	BLK
auxiliary	AUX	blocking	BLKG
auxiliary power unit	APU	blower	BLO
auxiliary register	AUXR	blowout	BLWT
auxiliary switch	ASW	board	BD
auxiliary switch (breaker) normally open	ASO	board foot	BF
avenue	AVE	boiler	BLR
average diameter	AVG DIA	bolster	BOLS
avoirdupois	AVDP	bolt	BLT
awning	AWN	bolt circle	BC
axial flow	AX FL	bolted plate	BP
axial pitch	AXP	bonded	BND
axial pressure angle	APA	bonding	BNDG
azimuth	AZ	bonnet	BNT
		bookcase	BC
B		book shelves	BK SH
back end	BE	borrowed light	BLT
backface	BF	both faces	BF
background	BKGD	both sides	BS
back pressure	BP	bottom	BOT
back-to-back	B TO B	bottom chord	BC
ball bearing	BBRG	bottom face	BF
ball stop	BSP	bottom layer	BL
barrel tile roof	BTR	boulevard	BLVD
barrier	BARR	boundary	BDY
barrier, moisture vapor-proof	BMVP	bow door	BDO
barrier, waterproof	BWP	bow light	BW LT
base	B	bowstring	BWSTRN
base line	BL	bracket	BRKT
basement	BSMT	braided	BRD
base plate	BP	branch	BR
basic	BSC	break	BRK
basic contour line	BCL	breaker	BRKR
batch mixer	BMXR	brick	BRK
bathroom	B	bridge	BRDG
bathtub	BT	bright	BRT
batten	BATT	brilliance	BRIL
batter	BAT	Brinell hardness	BH
beam	BM	Brinell hardness number	BHN
bearing	BRG	British thermal unit	Btu
bearing plate	BPL or BRG PL	broom closet	BCL
bedroom	BR	brush	BR
bell and spigot	B&S	brush holder	BRH
below	BLW	building	BLDG
bench	BNCH	building line	BL
bench mark	BM	built-in	BLTIN
bend line	BL	built-up roofing	BUR
bend radius	BR	bulkhead	BHD
between	BETW	bulldozer	BDZR
between centers	BC	bullnose	BN
between perpendiculars	BP	burner	BNR
bevel	BEV	burnish	BNSH
beveled wood siding	BWS	bushel	BU
billet	BL	bushing	BSHG
billet steel	BL STL	butterfly	BTFL
bit	B	button head	BTNHD

Term	Drawing Abbreviation	Term	Drawing Abbreviation
butt weld	BTWLD	chain	CH
bypass	BYP	chamfer	CHAM
		change order	CO
C		channel	CHAN
cabinet	CAB	charge	CHG
cable	CA	check bit	CHKB
cable duct	CD	check valve	CV
caliber	CAL	chimney	CHM
calibrate	CAL	chopper	CHP
caliper	CLPR	chord	CHD
calorie	CAL	circle	CIR
camber	CAM	circuit	CKT
canopy	CAN	circuit breaker	CB
cantilever	CANTIL	circuit interrupter	CI
canvas	CANV	circular	CIRC
capacitor	CAP	circulate	CRCLT
capacity	CAP	circulating water pump	CWP
capital	CAP	circumference	CRCMF
capped	CPPD	clamp	CLP
cap screw	CAP SCR	clamp screw	CLP SCR
carbide	CBD	clay pipe	CP
carbon steel	CS	cleanout	CO
carpenter	CPNTR	clear	CLR
carpet	CARP	clearance	CL
carriage	CRG	clear glass	CL GL
case	CS	clear glazed structural facing units	CGSFU
cased opening	CO	clear glazed structural unit base	CGSUB
casement	CSMT	cleat	CLT
case harden	CH	clinometer	CLN
casing	CSG	close	CL
castellate	CSTL	closed end	CLE
cast iron	CI	closet	CLO
cast-iron pipe	CIP	closure	CLOS
castellated nut	CAS NUT	coaming	COAM
cast steel	CS	coarse	CRS
cast stone	CS	coat and hat hooks	C&H
catalyst	CTLST	coated	CTD
catch basin	CB	coated metal	CMET
catwalk	CTWALK	coating	CTG
caulked joint	CLKJ	coaxial	COAX
cavity	CAV	coefficient	COEF
ceiling	CLG	cofferdam	COFF
cellar	CEL	cold air	CA
Celsius	°C	cold-drawn	CD
cement	CEM	cold-drawn steel	CDS
cement base	CB	cold-rolled	CR
cement floor	CF	cold-rolled steel	CRS
cement mortar	CEM MORT	cold water	CW
cement plaster	CPL	collar	CLR
cement plaster ceiling	CPC	collector	COLL
center	CTR	color code	CC
centering	CTRG	column	COL
center line	CL	combination	COMB
center punch	CP	combustible	COMBL
center-to-center	C TO C	combustion	COMB
central	CTL	combustor	CMBSTR
ceramic	CER	commercial	COML
ceramic glazed structural facing units	CGSFU	common	COM
ceramic glazed structural unit base	CGSUB	compacted	COMP
ceramic tile	CT	compass	CMPS
ceramic-tile base	CTB	complete	COMPL
ceramic-tile floor	CTF	composite	CMPST
ceramic-to-metal (seal)	CERMET	composition	CMPSN

Term	Drawing Abbreviation	Term	Drawing Abbreviation
composition floor	COMPF	cross section	XSECT
composition roof	COMPR	crowfoot	CRFT
compressed air	COMPA	crown	CRN
compression	CPRSN	crushed stone	CRSHD STN
compressor	CPRSR	cubic	CU
concave	CNCV	cubic foot per minute	CFM
concealed	CNCL	cubic foot per second	CFS
concrete	CONC	cubic inch	CU IN
concrete block	CCB	cubic yard	CU YD
concrete floor	CCF	culvert	CULV
concrete pipe	CP	current	CUR
concrete splash block	CSB	curve	CRV
condensate	CNDS	cutoff	CO
condenser	COND	cutoff valve	COV
conductor	CNDCT	cutout	CO
conduit	CND	cutout valve	COV
construction	CONSTR	cut stone	CUTS
construction joint	CJ	cycle	C
continuous	CONT	cylinder lock	CYLL
continuous window	CONTW	cylindrical	CYL
contour	CTR		
contract	CONTR	**D**	
contractor	CONTR	damper	DMPR
control joint	CLJ	dampproofing	DP
control switch	CS	datum	DAT
conventional	CVNTL	daylight opening	DLO
convex	CVX	dead load	DL
conveyor	CNVR	decibel	DB
coolant	COOL	decimal	DEC
cooler	CLR	deck	DK
cooling fan	CF	decrease	DECR
cord	CD	deep	D
corner	COR	degree	DEG
corner bead	COR BD	dehumidify	DHMY
cornice	COR	delay	DLY
corrosion	CRSN	demagnetize	DMGZ
corrosion-resistant	CRE	demolition	DML
corrosion-resistant steel	CRES	denatured	DNTRD
corrosive	CRSV	density	DENS
corrugate	CORR	depression	DEPR
corrugated fiberboard	CORR FBD	depth	DP
corrugated metal pipe	CMP	derrick	DRK
corrugated wire glass	CWG	design	DSGN
cotter	COT	design specification	DSPEC
cotter pin	COT	detached	DTCH
counter	CNTR	detail	DET
counterbalance	CBAL	develop	DVL
counterbore	CBORE	developed length	DEV LG
counterdrill	CDRILL	development	DEV
counterflashing	CFLG	dew point	DP
countershaft	CTSHFT	diagonal	DIAG
countersink	CSK	diallyl phthalate	DIAL PHTH
countersink other side	CSKO	differential	DIFF
counterweight	CTWT	diffusing	DIFFUS
county	CO	dilute	DIL
coupling	CPLG	dimension	DIM
courses	C	dimmer	DMR
cover	COV	dining	DNG
crane	CRN	dining room	DNG R or DR
crane load	CL	direct	DIR
crank	CRK	direct current	DC
critical path method	CPM	direct drive	DDR
critical path technique	CPT	direction	DIR

Term	Drawing Abbreviation	Term	Drawing Abbreviation
discharge	DISCH	**E**	
disconnect	DISC	each	EA
disconnecting device	DD	each face	EF
disconnect switch	DS	east	E
dishwater	DW	eccentric	ECC
distance	DIST	edge thickness	ET
distribution box	DB	edgewise	EDGW
distribution panel	DPNL	effective	EFF
divide	DIV	efficiency	EFF
division	DIV	ejection	EJN
door	DR	ejector	EJCTR
door closer	DCL	elastic	ELAS
door stop	DST	elasticity	ELAS
door switch	DSW	elbow	ELB
dormer	DRM	electric	ELEC
double	DBL	electrical metallic tubing	EMT
double-acting	DBL ACT	electric contact	ELCTC
double-acting door	DAD	electric panel	EP
double end	DE	electric power distribution	EPD
double extra strong	XX STR	electric water cooler	EWC
double face	DBLF	electrode	ELCTD
double-hung windows	DHW	electroluminescent	EL
double-pole double-throw	DPDT	electrolytic	ELCTLT
double-pole double-throw switch	DPDT SW	electromagnetic	EM
double-pole single-throw	DPST	electromechanical	ELMCH
double-pole single-throw switch	DPST SW	electromotive force	EMF
double-pole switch	DP SW	electron (n)-type semiconductor material	N
double strength glass	DSG	electronic	ELEK
double-wall	DBLW	electronic checkout	ECO
double weight	DW	electronics	ELEX
douglas fir	DF	electropneumatic	ELPNEU
dovetail	DVTL	electrostatic	ES
dowel	DWL	elevate	ELEV
down	DN	elevation	EL
downdraft	DNDFT	elevator	ELEV
downspout	DS	elliptical	ELP
dozen	DOZ	emboss	EMB
draft	DFT	emergency	EMER
draft stop	DS	emergency power	EPWR
drain	DR	emulsion	EMUL
drain tile	DT	enamel	ENAM
drawer	DWR	encased	ENCSD
drawing	DWG	enclose	ENCL
dressed (lumber)	DRS	enclosure	ENCL
dressed and matched	D&M	end-to-end	E TO E
drift	DFT	energize	ENRGZ
drill	DR	enlarge	ENLG
drinking fountain	DF	enlarged	ENLGD
dripproof	DP	entrance	ENTR
drop forge	DF	environment	ENVIR
dry bulb	DB	equal	EQL
dryer	D	equally spaced	EQL SP
dry pipe valve	DPV	equal section	ES
drywall	DW	equipment	EQPT
dual-purpose	DP	equivalent	EQUIV
dumbwaiter	DW	erecting	ERCG
duplex	DX	erection	ERECT
duplicate	DUP	erector	ERCR
dust-tight	DT	escutcheon	ESC
dutch door	DD	estimate	EST
duty cycle	DTY CY	estimated completion date	ECD
dwelling	DWEL	evaluation	EVLTN
dynamite	DYNMT		

Term	Drawing Abbreviation	Term	Drawing Abbreviation
evaporator	EVAP	fire alarm bell	FABL
example	EX	fire alarm box	FABX
except	EXC	firebrick	FBCK
exchange	EXCH	fireclay	FC
exhaust	EXH	fired	FIR
exhaust vent	EXHV	fire door	FDR
existing	EXST	fire extinguisher	FEXT
expand	EXP	fire hose	FH
expanded metal	EM	fire hose cabinet	FHC
expansion	EXP	fire hose rack	FHR
expansion joint	EXP JT	fire hydrant	FHY
explode	EXPLD	fireplace	FP
explosion-proof	EP	fire plug	FPL
exposed	EXP	fireproof	FPRF
exposure	EXPSR	fire-resistant	FRES
extension	EXT	fire wall	FW
exterior	EXT	fitting	FTG
exterior grade	EXT GR	fixed	FXD
external pipe thread	EPT	fixed transom	FTR
extinguish	EXT	fixed window	FX WDW
extinguisher	EXT	fixture	FXTR
extra	EX	flagstone	FLGSTN
extractor	EXTR	flame	FLM
extra fine (threads)	EF	flameproof	FLMPRF
extra heavy	XHVY	flammable	FLMB
extra strong	XSTR	flange	FLG
extrude	EXTD	flared	FLRD
eyelet	EYLT	flaring	FLRG
eyepiece	EYPC	flashing	FL
		flat	FL
F		flat bar	FB
fabricate	FAB	flat fillister head	FFILH
face brick	FB	flat head	FLH
faceplate	FP	flat point	FP
face or field of drawing	F/D	flat tile roof	FTR
face-to-face	F TO F	flat washer	FLW
facing	FCG	floating	FLTG
Fahrenheit	°F	floodlight	FLDT
fastener	FSTNR	floor	FL
feeder	FDR	floor drain	FD
female	FEM	flooring	FLG
female flared	FFL	floor line	FL
female thread	FTHRD	flow	FL
fiber	FBR	flow rate	FLRT
fiberboard	FBRBD	flow switch	FLSW
fiberboard, corrugated	FBDC	fluid	FL
fiberboard, double wall	FDWL	fluid flow	FDFL
fiberboard, solid	FBDS	fluid pressure line	FDPL
fibrous	FBRS	fluorescent	FLUOR
field	FLD	flush	FL
figure	FIG	flush mount	FLMT
filament	FIL	folding	FLDG
filler	FLR	foot	FT
fillet	FIL	footcandle	FC
fillister	FIL	footing	FTG
fillister head	FILH	foot per minute	FPM
filter	FLTR	foot per second	FPS
finish	FNSH	foot pound force	FT LB
finish all over	FAO	forced-draft	FD
finished floor	FNSH FL	forced-draft blower	FDB
finish grade	FG	foreman	FMAN
finish one side	F1S	forged	FGD
finish two sides	F2S	forged steel	FST

Term	Drawing Abbreviation	Term	Drawing Abbreviation
foundation	FDN	grade line	GL
four-pole	4P	gradient	GRAD
four-pole double-throw switch	4PDT SW	graduation	GRDTN
four-pole single-throw switch	4PST SW	grain	GR
four-pole switch	4PSW	granite	GRAN
four-way	4WAY	granulated	GNLTD
four-wire	4W	grating	GRTG
fragmentation	FRAG	gravel	GVL
framework	FRWK	grease trap	GT
freezer	FRZR	grill	GRL
freezing point	FP	grommet	GROM
fresh water	FW	groove	GRV
fresh water pump	FWP	grooved	GRVD
from	FR	groover	GRVR
front	FR	grooving	GRVG
front view	FV	gross	GR
fuel line	FLN	gross vehicle weight	GVW
fuel tank	FTK	gross weight	GRWT
full scale	FSC	ground	GND
full voltage	FV	grounded (outlet)	G
funnel	FUNL	ground glass	GGL
furnish	FURN	guard	GD
furred ceiling	FC	guard rail	GDR
furring	FUR	gutter	GUT
fuse	FU	gymnasium	GYM
fuse block	FB	gypsum	GYP
fuse box	FUBX	gypsum-plaster ceiling	GPC
fuse holder	FUHLR	gypsum-plaster wall	GPW
fusible	FSBL	gypsum sheathing board	GSB
future	FUT	gypsum wallboard	GWB
G		**H**	
gauge	GA	half-round	½RD
gauge board	GABD	hand (comb form)	HD
gallery	GALL	hand rail	HNDRL
gallon	GAL	handwheel	HNDWL
gallon per hour	GPH	hanging	HNG
gallon per minute	GPM	hard	HD
galvanize	GALV	hardboard	HBD
galvanized iron	GALVI	hard-drawn	HD DRN
galvanized steel	GALVS	hardware	HDW
garage	GAR	hardwood	HDWD
garbage	GBG	hazardous	HAZ
gas	G	head	HD
gasket	GSKT	header	HDR
gate valve	GTV	headless	HDLS
gear	GR	heat	HT
gear rack	GRK	heated	HTD
general	GENL	heater	HTR
general contractor	GEN CONT	heat exchange	HE
general-purpose	GP	heating	HTG
generator	GEN	heating, ventilating and	
girder	G	air conditioning	HVAC
glass	GL	heat-resisting	HT RES
glass block	GLB	heat shield	HT SHLD
glaze	GLZ	heat-treat	HT TR
glazed facing unit	GFU	heavy	HVY
glazed structural facing unit	GSFU	heavy-duty	HD
glazed wall tile	GWT	height	HGT
globe stop valve	GSV	helical	HLCL
globe valve	GLV	herringbone	HGBN
grab rod	GR	hertz	Hz
grade	GR	hexagon	HEX

Term	Drawing Abbreviation	Term	Drawing Abbreviation
hexagonal head	HEX HD	inlet	INL
high	H	inlet and outlet	I&O
high carbon	HC	insecticide	ICTCD
high carbon steel	HCS	insert	INSR
high carbon steel, heat-treated	HCSHT	inside	INS
high humidity	HI HUM	inside diameter	ID
high impact	HIMP	inside radius	IR
high intensity	HINT	install	INSTL
high point	HPT	installation	INSTL
high pressure	HP	installation and maintenance	I&M
high-pressure steam	HPS	insulate	INSUL
high tensile	HTNSL	insulation	INSUL
high tensile strength	HTS	integral	INT
high tension	HT	integrated circuit	IC
high voltage	HV	interior	INTR
high-voltage regulator	HVR	intermediate	INTMD
high-water line	HWL	intermittent	INTMT
highway	HWY	internal	INTL
hinge	HNG	internal diameter	ID
hoist	HST	internal pipe thread	IPT
hollow	HOL	interrupt	INTRPT
hollow core	HC	iron	I
hollow metal	HM	iron pipe	IP
hollow tile	HT	iron pipe size	IPS
honeycomb	HNYCMB	iron pipe thread	IPT
hook rail	HR	isometric	ISO
horizontal	HOR		
horizontal center line	HCL	**J**	
horsepower	HP	jack	JK
hose bibb	HB	jackscrew	JKSCR
hose clamp	HC	jamb	JB or JMB
hose rack	HR	janitor's closet	JC
hose thread	HSTH	japanned	JAP
hot-galvanize	H GALV	joiner	J
hot water	HW	joint	JT
housing	HSG	joist	J
hot air	HA	joists and planks	J&P
hovering	HVRNG	junction	JCT
humidity	HMD	junction field-effect transistor	JFET
hydrated	HYDTD		
hydraulic	HYDR	**K**	
hydroelectric	HYDRELC	Kalamein	KAL
		Kalamein door	Kald
I		keel	K
idler	IDL	keene's cement plaster	KCP
illuminate	ILLUM	keyway	KWY
impedance	IMP	kick plate	KPL
incandescent	INCAND	kiln-dried	KD
inch	IN	kilovolt-ampere hour	KVAH
inch per second	IPS	kilovolt-ampere hour meter	KVAHM
inch-pound	IN LB	kilovolt-ampere meter	KVAM
include	INCL	kinescope	KINE
inclusive	INCL	kitchen	K or KIT
incomplete	INCOMP	knee brace	KB
increase	INCR	knife switch	KN SW
indirect waste	IW	knock down	KD
induced draft	ID	knockout	KO
induction	IND	knot	KN
inductor	IDCTR		
industry	IND	**L**	
inert	INRT	label	LBL
inert gas	INRTG	lacquer	LAQ
infrared	IR	ladder	LAD

Term	Drawing Abbreviation	Term	Drawing Abbreviation
ladder rung	LR	low tension	LT
lagging	LAG	low torque	LTQ
laminate	LAM	low voltage	LV
landing	LDG	lumber	LBR
lanyard	LNYD	lumen per watt	LPW
lapping	LPG	lunar excursion module	LEM
large	LGE		
latch	LCH	**M**	
lateral	LATL	machine	MACH
lath	LTH	machined surface	MASU
lattice	LTC	machinery	MCHRY
laundry	LAU	machine screw	MSCR
laundry chute	LC	machine steel	MST
laundry tray	LT	magnesium	MAG
lavatory	LAV	magnetic	MAG
layer	LYR	magnetic armature	MG
leader	LDR	mahogany	MAH
left	L	mailbox	MB
left hand	LH	main	MN
left side	LS	makeup	MKUP
length	LG	male pipe thread	MPT
level	LVL	male threaded	MTHRD
library	LBRY	malleable	MAL
lift-up door	LUD	malleable iron	MI
light	LT	mandrel	MDRL
lighting	LTG	manhole	MH
lightning arrester	LA	manhole cover	MC
lightproof louver	LPL	manifold	MANF
lightproof vent	LPV	manually operated	MNL OPR
light switch	LTSW	manual overload	MAN OVLD
lightweight concrete	LWC	manufacture	MFR
lightweight insulating concrete	LWIC	manufactured	MFD
limestone	LS	marble	MR
limit switch	LIM SW	marble floor	MRF
linear; lineal	LIN	marble threshold	MRT
linen closet	LCL or LCLO	marker	MKR
line-of-sight	LOS	masonry	MSNRY
lining	LNG	masonry opening	MO
link	LK	master	MA
linoleum	LINOL	master switch	MSW
linoleum floor	LF	masthead	MHD
lintel	LNTL	mastic	MSTC
list of material	LM	mastic floor	MF
live load	LL	mastic joint	MJ
living room	LR	matched	MTCHD
local	LCL	material	MATL
locking	LKG	material list	ML
locknut	LKNT	matrix	MAT
locknut pipe thread	NPSL	matt-finish structural facing units	MFSFU
lock washer	LK WASH	mattress	MTRS
long	L	maximum	MAX
lookout	LKT	maximum design meter	MDM
louver	LVR	maximum permissible exposure	MPE
louvered door	LVD	maximum torque	MT
louver opening	LVO	maximum working pressure	MWP
low	L	maximum working voltage	MWV
low-alloy steel	LAS	mean sea level	MSL
low carbon	LC	measure	MEAS
lower	LWR	median	MDN
lowest	LWST	medical	MED
lowest usable frequency	LUF	medicine cabinet	MC
low pressure	LP	medium	MDM
low temperature	LTEMP	melamine	MEL

Term	Drawing Abbreviation	Term	Drawing Abbreviation
melting point	MP	National Electrical Safety Code	NESC
member	MBR	National extra fine (thread)	NEF
membrane	MEMB	National fine (thread)	NF
membrane waterproofing	MWP	National pipe thread	NP
meridian	MER	National special (thread)	NS
metal	MET	National taper pipe (thread)	NPT
metal anchor	MA	natural	NAT
metal anchor slots	MAS	negative	(–) or NEG
metal awning-type window	MATW	neoprene	NPRN
metal base	METB	net weight	NTWT
metal-covered door	MCD	neutral	NEUT
metal door	METD	nipple	NIP
metal flashing	METF	nominal	NOM
metal grill	METG	noncombustible	NCOMBL
metal jalousie	METJ	noncorrosive metal	NCM
metal lath and plaster	MLP	nonflammable	NONFLMB
metallic	MTLC	nonmagnetic	NMAG
metal partition	METP	nonmetallic	NM
metal rolling door	MRD	nonreinforced-concrete pipe	NRCP
metal roof	METR	nonslip tread	NST
metal strip	METS	normal	NORM
metal threshold	MT	north	N
metal through-wall flashing	MTWF	nosing	NOS
mezzanine	MEZZ	not applicable	NA
middle	MDL	notched	NCH
mildew-resistant thread	MRT	not in contract	NIC
mile	MI	not to scale	NTS
mile per gallon	MPG	nozzle	NOZ
mile per hour	MPH	number	NO
mineral	MNRL	nylon	NYL
mineral-surface roof	MSR		
minimum	MIN	**O**	
minor	MIN		
mirror	MIR	objective	OBJV
miscellaneous	MISC	obscure	OB
miter	MIT	obscure glass	OGL
mixture	MXT	observation window	OBW
mobile	MBL	obsolete	OBS
model block	MB	occupy	OCC
modification	MOD	octagon	OCT
modification work order	MWO	office	OFCE
modify	MOD	ohmmeter	OHM
mogul	MGL	oil circuit breaker	OCB
molded	MLD	oil-cooled	OCLD
molding	MLDG	oil-insulated	OI
monolithic	ML	on center	OC
monument	MON	one-pole	SP
mortar	MOR	one-stage	1STG
mosaic	MOS	one-way	1/W
motor direct-connected	MDC	opaque	OPA
motor-driven	MTRDN	open end	OE
motorized	MTZ	opening	OPNG
motor-operated	MO	open web joist	OJ or OW J
mount	MT	open-window unit	OWU
movable	MVBL	operate	OPR
multiple	MULT	operating steam pressure	OSP
		operation	OPN
N		operational	OPNL
		operator	OP
nameplate	NPL	operator table	OPRT
narrow	NAR	opposite	OPP
National Bureau of Standards	NBS	optimum	OPT
National coarse (thread)	NC	ordinance	ORD
National Electrical Code	NEC	origin	ORIG

Term	Drawing Abbreviation	Term	Drawing Abbreviation
original	ORIG	piece	PC
ounce	OZ	pigment	PGMT
outlet	OUT	pilaster	P
output	OUT	piling	PLG
outside	OUT	pillar	PLR
outside circumference	OC	pillow block	PLBLK
outside diameter	OD	pilot	PLT
outside face	OF	pinion	PIN
outside radius	OR	pint	PT
out-to-out	O TO O	pipeline	PPLN
oval head	OVH	pipe rail	PR
over	OV	pipe sleeve	PSL
overall	OA	pipe tap	PT
overcurrent	OC	piping	PP
overcurrent relay	OCR	pitch	P
overflow	OVFL	pitch diameter	PD
overhanging	OVHG	pivot	PVT
overhaul	OVHL	pivoted door	PD
overhead	OVHD	place	PL
overload	OVLD	plain	PL
overload relay	ORLY	plain washer	PW
override	OVRD	plank	PLK
oxygen	OXY	plan view	PV
		plaster	PLAS
P		plaster mockup	PMU
packing	PKG	plastic	PLSTC
padlock	PL	plate	PL
page	P	plate (electron tube)	P
paint	PNT	plated	PLD
painted	PTD	plate glass	PLGL
painted base	PB	platform	PLATF
paints and oils	P&O	plating	PLTG
pair	PR	pliers	PLR
pan head	PNH	plug	PL
panel	PNL	plumbing	PLMB
panel point	PP	plunger	PLGR
panic bolt	PANB	plywood	PLYWD
pantry	PAN	pneumatic	PNEU
paperboard	PBD	pocket	PKT
parallel	PRL	point	PT
parkway	PKWY	point of beginning	POB
partial	PART	point of compound curve	PCC
partition	PTN	point of intersection	PI
passage	PASS	polarity	PLRT
passing window	PW	pole	P
passive	PSIV	polyester	POLYEST
peak	PK	polystyrene	PS
peak-to-peak	P-P	polyvinyl chloride	PVC
pedestal	PED	popping	POP
penetration	PEN	porcelain	PORC
penny (nails, etc)	d	porch	P
pentagon	PNTGN	portable	PORT
perfect	PERF	position switch	POSN SW
perforate	PERF	posts and timbers	P&T
perimeter	PERIM	potable water	POTW
permanent	PERM	potential	POT
permeability	PERMB	potential difference	PD
perpendicular	PERP	potential switch	PSW
personnel carrier	PC	pound	LB
phase	PH	pound-foot	LB FT
phase modulation	PM	power	PWR
phenolic	PHEN	power supply	PWR SPLY
phillips head	PHH	power supply unit	PSU

Term	Drawing Abbreviation	Term	Drawing Abbreviation
precast	PRCST	rattail	RTTL
prefabricated	PREFAB	reactive	REAC
preferred	PFD	reactive volt-ampere meter	RVA
prefinished	PFN	reactive voltmeter	RVM
preheater	PHR	reamer	RMR
preservative	PSVTV	rear view	RV
pressboard	PBD	receiving	RCVG
pressed metal	PRSD MET	receptacle	RCPT
pressure	PRESS	reception	RCPTN
pressure gauge	PG	recess	REC
pressure-reducing valve	PRV	recharger	RECHRG
primary	PRI	recirculate	RECIRC
primer	PRMR	rectangle	RECT
priming	PRM	rectifier	RECT
projected window	PW	redrawn	REDWN
propelling	PROP	reduce	RDC
property	PROP	reducer	RDCR
property line	PL	reduction	RDCN
protective	PROT	redwood	RWD
publication	PUBN	reel	RE
pull box	PB	reference	REF
pull button	PLB	reference line	REFL
pull-button switch	PB SW	reflected	REFLD
pulley	PUL	reflective insulation	RI
pull switch	PS	refractory	RFRC
pulse-width modulation	PWM	refrigerant	RFGT
pulverized	PLVRZD	refrigerator	REFR
pump	PMP	register	RGTR
punch	PCH	regulator	RGLTR
push button	PB	reinforce	REINF
push-pull	PP	reinforced concrete	RC
		reinforced-concrete culvert pipe	RCP
Q		reinforced-concrete pipe	RCP
quadrant	QDRNT	reinforced tile lintel	RTL
quality control	QC	reinforcing steel	RST
quarry	QRY	relief	RLF
quarry tile	QT	relief valve	RV
quarry-tile base	QTB	Remote Control System	RCS
quarry-tile floor	QTF	removable cover	REM COV
quarry-tile roof	QTR	renewable	RNWBL
quart	QT	replace	REPL
quarter	QTR	request for price quotation	RPQ
quarter-phase	¼PH	required	REQD
quarter-round	¼RD	requirement	REQT
quick-acting	QA	resilient	RESIL
quotation	QUOT	resistor	RES
		retaining	RTNG
R		retard	RTD
rabbet	RAB	return	RTN
radial	RDL	reverse	RVS
radiator	RDTR	reverse-acting	RACT
radioactive	RAACT	reverse-current device	RCD
adius	RAD	reversible	RVSBL
rail	R	revolution	REV
railing	RLG	revolution per minute	RPM
raintight	RT	revolution per second	RPS
rainwater conductor	RWC	revolve	RVLV
raised	RSD	revolving	RVLG
raised face	RF	rheostat	RHEO
random	RNDM	ribbed	RIB
range	RNG	ribbon	RBN
ratchet	RCHT	ridge	RDG
rate of change	RC	rigging	RGNG

Term	Drawing Abbreviation	Term	Drawing Abbreviation
right	R	separate	SEP
right angle	RTANG	service	SVCE
right hand	RH	service sink	SS
right-of-way	R/W	set screw	SSCR
rigid	RGD	sewage	SEW
riser	R	sewer	SEW
rivet	RVT	shackle	SH
road	RD	shaft	SFT
Rockwell hardness	RH	shake	SHK
rolled	RLD	shank	SHK
roller	RLR	shear plate	SP
roll roofing	RR	sheathing	SHTHG
roof	RF	sheave	SHV
roof drain	RD	sheet	SH
roofing	RFG	sheeting	SH
roof leader	RL	sheet metal	SM
room	RM	shelf and rod	SH&RD
rotor	RTR	shell	SHL
rough	RGH	shellac	SHL
rough opening	RO	shelving	SHELV
rough sawn	RS	shielding	SHLD
round	RND	shingle	SHGL
round head	RDH	short circuit	SHORT
rubber	RBR	shoulder	SHLDR
rubber base	RB	shower	SH
rubber insulation	RINSUL	shower and toilet	SH & T
rubber-tile floor	RTF	shower drain	SD
rubble stone	RS	shroud	SHRD
rust preventative	RPVNTV	shutoff valve	SOV
rustproof	RSTPF	shutter	SHTR
		side light	SI LT
S		siding	SDG
saddle	SDL	signal	SIG
safety	SAF	sill cock	SC
safety valve	SV	silver solder	SILS
safe working pressure	SWP	similar	SIM
salt-glazed structural facing units	SGSFU	single	SGL
sanitary	SAN	single conductor	1/C
saturate	SAT	single-conductor	
scale	SC	cable	SCC
schedule	SCHED	single-end	SE
scraper	SRPR	single-face	SIF
screen	SCRN	single-phase	1PH
screen door	SCD	single-pole	SP
screen gate	SCR GT	single-pole double-throw	SPDT
screw	SCR	single-pole double-throw switch	SPDT SW
screwdriver	SCDR	single-pole single-throw	SPST
scupper	SCUP	single-pole single-throw switch	SPST SW
scuttle	S	single-pole switch	SP SW
sealed	SLD	single strength glass	SSG
seamless	SMLS	single-throw	ST
seamless steel tubing	SSTU	sink	SK
second	SEC	siphon	SPHN
secondary	SEC	skirt	SKT
section	SECT	skylight	SLT
security guard window	SGW	slate	SLT or S
security window screen and guard	SWSG	sliding	SL
select	SEL	sliding door	SLD
self-cleaning	SLFCLN	sliding expansion joint	SEJ
self-closing	SELF CL	slip joint	SJ
self-locking	SLFLKG	slip ring	SR
self-sealing	SLFSE	slope	SLP
self-tapping	SLFTPG	slop sink	SS

Term	Drawing Abbreviation	Term	Drawing Abbreviation
smooth contour	SC	stringer	STGR
smooth-surface built-up roof	SSBR	strip	STP
socket	SKT	strongback	STRBK
socket head	SCH	structural	STRL
soffit	SF	structural carbon steel, hard	SCSH
soil pipe	SP	structural carbon steel, medium	SCSM
soil stack	SSK	structural carbon steel, soft	SCSS
solder	SLDR	structural clay tile	SCT
soldering	SLDR	structural glass	SG
solenoid	SOL	structure	STRUCT
solid core	SC	submergence	SUBMG
solid fiberboard	SFB	subsoil drain	SSD
solvent	SLVT	substitute	SUBST
soundproof	SNDPRF	substructure	SUBSTR
south	S	superintendent	SUPT
space heater	SPH	superstructure	SUPERSTR
spacer	SPCR	supply	SPLY
special	SPCL	surface	SURF
special equipment	SE	surfaced or dress four sides	S4S
special-purpose	SP	surfaced or dressed one side	S1S
specific	SP	surfaced or dressed one side and	
specification	SPEC	one edge	S1S1E
spike	SPK	surfaced or dressed two sides	S2S
spiral	SPL	survey	SURV
splash block	SB	suspend	SUSP
splashproof	SP	suspended acoutical-plaster ceiling	SAPC
splice	SPLC	suspended acoustical-tile ceiling	SATC
spline	SPLN	supsended plaster ceiling	SPC
split ring	SR	suspended sprayed acoustical ceiling	SSAC
spotlight	SLT	suspension	SPNSN
spot-weld	SW	swage	SWG
sprayed acoustical ceiling	SAC	sweat	SWT
spreader	SPRDR	swinging door	SWGD
sprinkler	SPR	switch	SW
square	SQ		
square foot	SQ FT	**T**	
square head	SQH	tackle	TKL
square inch	SQ IN	tailpiece	TLPC
square yard	SQ YD	tandem	TDM
stack	STK	taper	TPR
stained	STN	taper pipe thread	NPT
stainless	STNLS	tapping	TPG
stainless steel	SST	tarpaulin	TARP
stairs	ST	T-bar (structural shape)	T
stairway	STWY	telephone	TEL
standard	STD	telephone booth	TELB
standoff	STDF	television	TV
standpipe	SP	temper	TEM
starter	START	temperature	TEMP
static pressure	ST PR	tempered	TMPD
steel	STL	template	TEMPL
steel basement window	SBW	temporary	TEMP
stiffener	STIF	temporary construction hole	TCH
stirrup	STIR	tensile	TNSL
stone	STN	tensile strength	TS
storage	STOR	tension	TNSN
storm water	STW	terra cotta	TC
stove bolt	SB	terazzo	TER
straight	STR	terazzo base	TERB
straight thread (pipe couplings)	NPSC	thermal	THRM
strainer	STR	thermocouple	TC
street	ST	thermostat	THERMO
strength	STR	thick	THK

Term	Drawing Abbreviation	Term	Drawing Abbreviation
thickness	THKNS	tubing	TBG
thread	THD	tumbler	TBLR
thread both ends	TBE	tunnel	TNL
thread cutting	TC	turnbuckle	TRNBKL
three-conductor	3/C	turned	TRND
three-phase	3PH	twisted	TW
three-pole	3P	two-conductor	2/C
three-pole double-throw	3PDT	two-phase	2PH
three-pole single-throw	3PST	two-pole	DP
three-way	3WAY	two-pole double-throw	DPDT
three-wire	3W	two-pole single-throw	DPST
threshold	TH	two-stage	2STG
throat	THRT	two-way	2WAY
tight joints	NPTF	typical	TYP
tile base	TB		
tile drain	TD	**U**	
tile floor	TF	ultimate	ULT
tile-shingle roof	TSR	ultraviolet	UV
tile threshold	TT	under	UND
tile wainscot	TW	underground	UGND
timber	TMBR	undersize	US
toggle	TGL	underwater	UWTR
toilet	T	unexcavated	UNEXC
toilet-paper holder	TH	unfinished	UNFIN
tongue	TNG	unglazed ceramic mosaic tile	UCMT
tongue and groove	T&G	unglazed structural facing units	USFU
tooth	T	unglazed structural unit base	USUB
top	T	Unified coarse thread	UNC
top and bottom	T&B	Unified extra fine thread	UNEF
top and bottom bolt	T&BB	Unified fine thread	UNF
top chord	TC	Unified special thread	UNS
top of frame	T/FR	union	UN
topping	TOPG	unit heater	UH
torque	TRQ	universal	UNIV
torsion	TRSN	unless otherwise specified	UOS
total	TOT	untreated	UTRTD
total load	TLLD	untreated hard-pressed fiberboard	UHPFB
towel rack or rod	TR	updraft	UPDFT
track	TRK	upper	UPR
trailer	TRLR	upper and lower	U&L
trailing edge	TE	urinal	UR
transducer	XDCR	urinal water closet	URWC
transformer	XFMR	utility	UTIL
transistor	XSTR	utility room	UR
transom	TR		
transparent	TRANS	**V**	
transverse	TRANSV	vacuum	VAC
traversing	TRAV	vacuum tube	VT
treated	TRTD	valley	VAL
triangle	TRNGL	valve	V
trimmer	TRMR	valve stem	VSTM
triple	TPL	vaporproof	VAP PRF
triple-pole	3P	vaportight	VT
triple-pole double-throw	3PDT	variance	VAR
triple-pole double-throw switch	3PDT SW	variation	VAR
triple-pole single-throw	3PST	varnish	VARN
triple-pole single-throw switch	3PST SW	vent	V
triple-pole switch	3P SW	vent hole	VH
triple-throw	3T	vent pipe	VP
triple wall	TPLW	vent stack	VS
triplex	TRX	ventilate	VENT
truss	TR	ventilating equipment	VE
truss head	TRH	ventilator	VENT

Term	Drawing Abbreviation	Term	Drawing Abbreviation
vernier	VERN	weep hole	WH
vertical	VERT	welded	WLD
vestibule	VEST	welded wire fabric	WWF
vibration	VSBL	welder	WLDR
vinyl tile	VT or VTILE	weldless	WLDS
visible	VSBL	west	W
vitreous	VIT	white pine	WP
vitrified clay	VC	wide	W
vitrified clay tile	VCT	wide flange	WF
void	VD	width	WD
volt	V	winch	WN
voltage	V	winder	WNDR
voltage drop	VD	wind load	WL
voltage regulator	VR	window	WDO
voltage relay	VRLY	window guard	WG
voltmeter	VM	window unit	WU
voltmeter switch	VS	wire (comb form)	W
volt per meter	V/M	wire gauge	WG
volume	VOL	wire glass	WGL
volute	VLT	wire mesh	WM
vulcanize	VULC	wire rope	WR
		wireway	WW
W		wiring	WRG
wainscot	WA	with (comb form)	W/
walk in closet	WIC	without	W/O
wall	W	wood	WD
wallboard	WLB	wood awning-type window	WATW
wall receptacle	WR	wood base	WB
wall vent	WV	wood black floor	WBF
warehouse	WHSE	wood blocking	WBL
warm air	WA	wood casement window	WCW
washer	WSHR	wood door	WD
washing machine	WM	wood door and frame	WDF
washroom	WR	wood furring strips	WFS
waste	W	wood jalousie	WJ
waste pipe	WP	wood panel	WDP
waste stack	WS	wood-shingle roof	WSR
water	WTR	wood threshold	WT
water closet	WC	working pressure	WPR
water heater	WH	working steam pressure	WSP
water meter	WM	working voltage	WV
waterproof	WTRPRF	warm gear	WMGR
waterproof fan-cooled	WPFC	wrench	WR
waterproofing	WPG	wrought	WRT
water resistant	WR	wrought iron	WI
watertight	WTRTT		
watt	W	**Y**	
watthour	WH	yard	YD
weatherproof	WTHPRF	yellow pine	YP
weatherproof (insul)	WP		
weather-resistant	WR	**Z**	
weather seal	WSL	zone	Z
weather stripping	WS		

ALPHABET OF LINES

LINE	REPRESENTATION		USE
Object line		THICK	Define shape.
Hidden line	→‖←─1/8″ →‖←1/32″	THIN	Show hidden features, future or existing construction to be removed.
Center line	←──3/4″−11/2″──→ ‖←1/8″ →‖←1/16″	THIN	Locate center points of arcs and circles, exterior elevation lines, and projections.
Dimension line	4′-0″ 4′-0″ 4′-0″	THIN	Show size or location.
Extension line		THIN	Define size or location.
Leader		THIN	Indicate specific features.
Long−break line	←─3/4″−11/2″─→	THIN	Indicate long breaks.
Short−break line		THIN	Indicate short breaks.
Cutting plane	←3/4″−11/2″→ 1/16″→‖← ←1/8″→‖←	THICK	Show internal features.
Section line	1/16″	THICK	Indicate internal features.
Phantom line	←3/4″−11/2″→ 1/8″→‖← →‖←1/16″	THICK	Indicate movement, property and boundary lines.

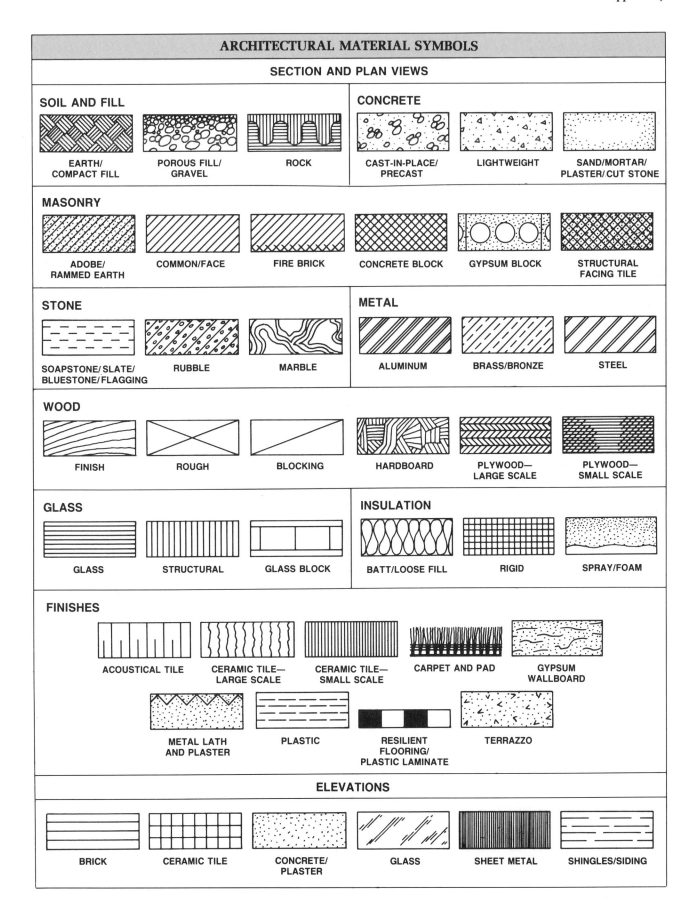

ARCHITECTURAL MATERIAL SYMBOLS

SECTION AND PLAN VIEWS

SOIL AND FILL
- EARTH/COMPACT FILL
- POROUS FILL/GRAVEL
- ROCK

CONCRETE
- CAST-IN-PLACE/PRECAST
- LIGHTWEIGHT
- SAND/MORTAR/PLASTER/CUT STONE

MASONRY
- ADOBE/RAMMED EARTH
- COMMON/FACE
- FIRE BRICK
- CONCRETE BLOCK
- GYPSUM BLOCK
- STRUCTURAL FACING TILE

STONE
- SOAPSTONE/SLATE/BLUESTONE/FLAGGING
- RUBBLE
- MARBLE

METAL
- ALUMINUM
- BRASS/BRONZE
- STEEL

WOOD
- FINISH
- ROUGH
- BLOCKING
- HARDBOARD
- PLYWOOD—LARGE SCALE
- PLYWOOD—SMALL SCALE

GLASS
- GLASS
- STRUCTURAL
- GLASS BLOCK

INSULATION
- BATT/LOOSE FILL
- RIGID
- SPRAY/FOAM

FINISHES
- ACOUSTICAL TILE
- CERAMIC TILE—LARGE SCALE
- CERAMIC TILE—SMALL SCALE
- CARPET AND PAD
- GYPSUM WALLBOARD
- METAL LATH AND PLASTER
- PLASTIC
- RESILIENT FLOORING/PLASTIC LAMINATE
- TERRAZZO

ELEVATIONS
- BRICK
- CERAMIC TILE
- CONCRETE/PLASTER
- GLASS
- SHEET METAL
- SHINGLES/SIDING

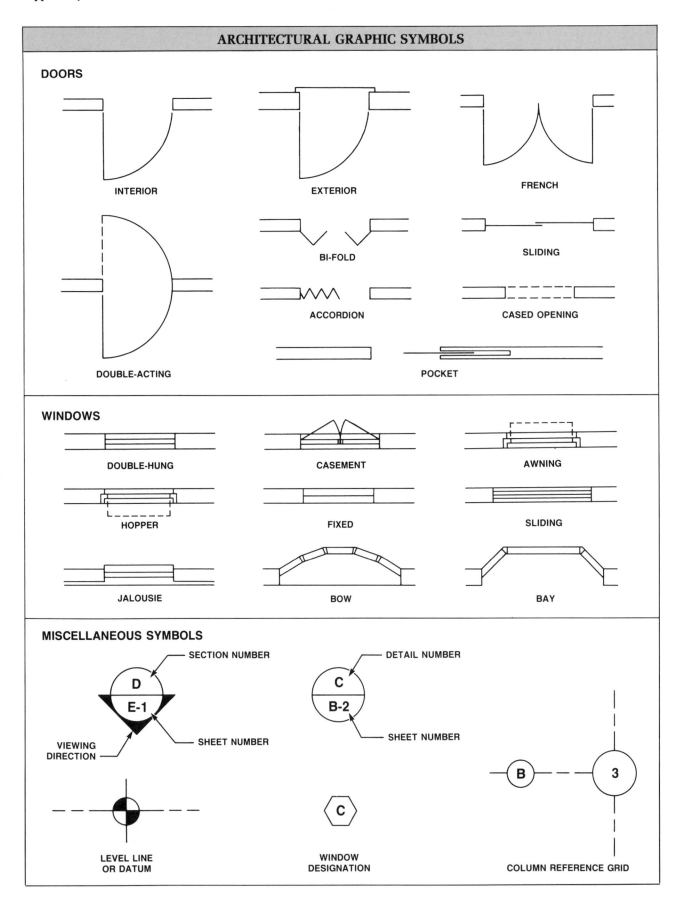

ARCHITECTURAL GRAPHIC SYMBOLS

DOORS

INTERIOR

EXTERIOR

FRENCH

DOUBLE-ACTING

BI-FOLD

SLIDING

ACCORDION

CASED OPENING

POCKET

WINDOWS

DOUBLE-HUNG

CASEMENT

AWNING

HOPPER

FIXED

SLIDING

JALOUSIE

BOW

BAY

MISCELLANEOUS SYMBOLS

SECTION NUMBER

D

E-1

VIEWING
DIRECTION

SHEET NUMBER

DETAIL NUMBER

C

B-2

SHEET NUMBER

B — — — 3

LEVEL LINE
OR DATUM

WINDOW
DESIGNATION

COLUMN REFERENCE GRID

ELECTRICAL SYMBOLS

LIGHTING OUTLETS

CEILING, WALL

OUTLET BOX AND INCANDESCENT LIGHTING FIXTURE.

INCANDESCENT TRACK LIGHTING

BLANKED OUTLET

DROP CORD

EXIT LIGHT AND OUTLET BOX. SHADED AREAS DENOTE FACES.

OUTDOOR POLE-MOUNTED FIXTURES

JUNCTION BOX

LAMPHOLDER WITH PULL SWITCH

MULTIPLE FLOODLIGHT ASSEMBLY

EMERGENCY BATTERY PACK WITH CHARGER

INDIVIDUAL FLUORESCENT FIXTURE

OUTLET BOX AND FLUORESCENT LIGHTING TRACK FIXTURE

CONTINUOUS FLUORESCENT FIXTURE

SURFACE-MOUNTED FLUORESCENT FIXTURE

PANELBOARDS

FLUSH-MOUNTED PANELBOARD AND CABINET

SURFACE-MOUNTED PANELBOARD AND CABINET

CONVENIENCE OUTLETS

SINGLE RECEPTACLE OUTLET

DUPLEX RECEPTACLE OUTLET

TRIPLEX RECEPTACLE OUTLET

SPLIT-WIRED DUPLEX RECEPTACLE OUTLET

SPLIT-WIRED TRIPLEX RECEPTACLE OUTLET

SINGLE SPECIAL-PURPOSE RECEPTACLE OUTLET

DUPLEX SPECIAL-PURPOSE RECEPTACLE OUTLET

RANGE OUTLET — R

SPECIAL-PURPOSE CONNECTION — DW

CLOSED-CIRCUIT TELEVISION CAMERA

CLOCK HANGER RECEPTACLE — C

FAN HANGER RECEPTACLE — F

FLOOR SINGLE RECEPTACLE OUTLET

FLOOR DUPLEX RECEPTACLE OUTLET

FLOOR SPECIAL-PURPOSE OUTLET

UNDERFLOOR DUCT AND JUNCTION BOX FOR TRIPLE, DOUBLE, OR SINGLE DUCT SYSTEM AS INDICATED BY NUMBER OF PARALLEL LINES

BUSDUCTS AND WIREWAYS

SERVICE, FEEDER, OR PLUG-IN BUSWAY — B B B

CABLE THROUGH LADDER OR CHANNEL — C C C

WIREWAY — W W W

SWITCH OUTLETS

SINGLE-POLE SWITCH — S

DOUBLE-POLE SWITCH — S_2

THREE-WAY SWITCH — S_3

FOUR-WAY SWITCH — S_4

AUTOMATIC DOOR SWITCH — S_D

KEY-OPERATED SWITCH — S_K

CIRCUIT BREAKER — S_{CB}

WEATHERPROOF CIRCUIT BREAKER — S_{WCB}

DIMMER — S_{DM}

REMOTE CONTROL SWITCH — S_{RC}

WEATHERPROOF SWITCH — S_{WP}

FUSED SWITCH — S_F

WEATHERPROOF FUSED SWITCH — S_{WF}

TIME SWITCH — S_T

CEILING PULL SWITCH

SWITCH AND SINGLE RECEPTACLE — $\ominus_S$

SWITCH AND DOUBLE RECEPTACLE — $\ominus_S$

ANY STANDARD SYMBOL WITH THE ADDITION OF A LOWERCASE SUBSCRIPT LETTER MAY BE USED TO DESIGNATE A VARIATION IN STANDARD EQUIPMENT.

$\bigcirc_{a,b}$

$\ominus_{a,b}$

$S_{a,b}$

ELECTRICAL SYMBOLS (continued)

COMMERCIAL AND INDUSTRIAL SYSTEMS

PAGING SYSTEM DEVICE

FIRE ALARM SYSTEM DEVICE

COMPUTER DATA SYSTEM DEVICE

PRIVATE TELEPHONE SYSTEM DEVICE

SOUND SYSTEM

FIRE ALARM CONTROL PANEL — FACP

SIGNALING SYSTEM OUTLETS FOR RESIDENTIAL SYSTEMS

PUSH BUTTON

BUZZER

BELL

BELL AND BUZZER COMBINATION

COMPUTER DATA OUTLET

BELL RINGING TRANSFORMER — BT

ELECTRIC DOOR OPENER — D

CHIME — CH

TELEVISION OUTLET — TV

THERMOSTAT — T

UNDERGROUND ELECTRICAL DISTRIBUTION OR ELECTRICAL LIGHTING SYSTEMS

MANHOLE — M

HANDHOLE — H

TRANSFORMER-MANHOLE OR VAULT — TM

TRANSFORMER PAD — TP

UNDERGROUND DIRECT BURIAL CABLE

UNDERGROUND DUCT LINE

STREET LIGHT STANDARD FED FROM UNDERGROUND CIRCUIT

ABOVE-GROUND ELECTRICAL DISTRIBUTION OR LIGHTING SYSTEMS

POLE

STREET LIGHT AND BRACKET

PRIMARY CIRCUIT

SECONDARY CIRCUIT

DOWN GUY

HEAD GUY

SIDEWALK GUY

SERVICE WEATHERHEAD

PANEL CIRCUITS AND MISCELLANEOUS

LIGHTING PANEL

POWER PANEL

WIRING—CONCEALED IN CEILING OR WALL

WIRING—CONCEALED IN FLOOR

WIRING EXPOSED

HOME RUN TO PANEL BOARD.
Indicate number of circuits by number of arrows. Any circuit without such designation indicates a two-wire circuit. For a greater number of wires indicate as follows:
—/// (3 wires) —//// (4 wires), etc.

FEEDERS
Use heavy lines and designate by number corresponding to listing in feeder schedule.

WIRING TURNED UP

WIRING TURNED DOWN

GENERATOR — G

MOTOR — M

INSTRUMENT (SPECIFY) — I

TRANSFORMER — T

CONTROLLER

EXTERNALLY-OPERATED DISCONNECT SWITCH

PULL BOX

PIPE FITTING AND VALVE SYMBOLS

	FLANGED	SCREWED	BELL & SPIGOT		FLANGED	SCREWED	BELL & SPIGOT		FLANGED	SCREWED	BELL & SPIGOT
BUSHING				REDUCING FLANGE				AUTOMATIC BY-PASS VALVE			
CAP				BULL PLUG							
REDUCING CROSS				PIPE PLUG				AUTOMATIC REDUCING VALVE			
STRAIGHT-SIZE CROSS				CONCENTRIC REDUCER				STRAIGHT CHECK VALVE			
CROSSOVER				ECCENTRIC REDUCER				COCK			
45° ELBOW				SLEEVE				DIAPHRAGM VALVE			
90° ELBOW				STRAIGHT-SIZE TEE				FLOAT VALVE			
ELBOW— TURNED DOWN				TEE—OUTLET UP				GATE VALVE			
ELBOW— TURNED UP				TEE—OUTLET DOWN				MOTOR-OPERATED GATE VALVE			
BASE ELBOW				DOUBLE-SWEEP TEE				GLOBE VALVE			
DOUBLE-BRANCH ELBOW				REDUCING TEE				MOTOR-OPERATED GLOBE VALVE			
LONG-RADIUS ELBOW				SINGLE-SWEEP TEE				ANGLE HOSE VALVE			
REDUCING ELBOW				SIDE OUTLET TEE— OUTLET DOWN				GATE VALVE			
SIDE OUTLET ELBOW— OUTLET DOWN				SIDE OUTLET TEE— OUTLET UP				GLOBE VALVE			
SIDE OUTLET ELBOW— OUTLET UP				UNION				LOCKSHIELD VALVE			
STREET ELBOW				ANGLE CHECK VALVE				QUICK-OPENING VALVE			
CONNECTING PIPE JOINT				ANGLE GATE VALVE— ELEVATION				SAFETY VALVE			
EXPANSION JOINT				ANGLE GATE VALVE—PLAN							
LATERAL				ANGLE GLOBE VALVE— ELEVATION				GOVERNOR- OPERATED AUTOMATIC VALVE			
ORIFICE FLANGE				ANGLE GLOBE VALVE—PLAN							

The American Society of Mechanical Engineers

PLUMBING SYMBOLS

FIXTURES

STANDARD BATHTUB	
OVAL BATHTUB	
WHIRLPOOL BATH	
SHOWER STALL	
SHOWER HEAD	
TANK-TYPE WATER CLOSET	
WALL-MOUNTED WATER CLOSET	
FLOOR-MOUNTED WATER CLOSET	
LOW-PROFILE WATER CLOSET	
BIDET	
WALL-MOUNTED URINAL	
FLOOR-MOUNTED URINAL	
TROUGH-TYPE URINAL	
WALL-MOUNTED LAVATORY	
PEDESTAL LAVATORY	
BUILT-IN LAVATORY	
WHEELCHAIR LAVATORY	
CORNER LAVATORY	
FLOOR DRAIN	
FLOOR SINK	

FIXTURES (continued)

LAUNDRY TRAY	
BUILT-IN SINK	
DOUBLE OR TRIPLE BUILT-IN SINK	
COMMERCIAL KITCHEN SINK	
SERVICE SINK	SS
CLINIC SERVICE SINK	
FLOOR-MOUNTED SERVICE SINK	
DRINKING FOUNTAIN	DF
WATER COOLER	
HOT WATER TANK	HWT
WATER HEATER	WH
METER	M
HOSE BIBB	HB
GAS OUTLET	G
GREASE SEPARATOR	G
GARAGE DRAIN	
FLOOR DRAIN WITH BACKWATER VALVE	

PIPING

SOIL, WASTE, OR LEADER— ABOVE GRADE	———————
SOIL, WASTE, OR LEADER— BELOW GRADE	— — — —
VENT	– – – – – –
COMBINATION WASTE AND VENT	—— SV ——
STORM DRAIN	—— S ——
COLD WATER	—– · —– · —– ·

PIPING (continued)

CHILLED DRINKING WATER SUPPLY	—— DWS ——
CHILLED DRINKING WATER RETURN	—— DWR ——
HOT WATER	— ·· — ·· —
HOT WATER RETURN	— – – · — – – · —
SANITIZING HOT WATER SUPPLY (180°F)	⫫ ·· ⫫ ·· ⫫
SANITIZING HOT WATER RETURN (180°F)	⫫ ··· — ⫫ ··
DRY STANDPIPE	—— DSP ——
COMBINATION STANDPIPE	—— CSP ——
MAIN SUPPLIES SPRINKLER	—— S ——
BRANCH AND HEAD SPRINKLER	—o—————o—
GAS—LOW PRESSURE	— G — G —
GAS—MEDIUM PRESSURE	—— MG ——
GAS—HIGH PRESSURE	—— HG ——
COMPRESSED AIR	—— A ——
OXYGEN	—— O ——
NITROGEN	—— N ——
HYDROGEN	—— H ——
HELIUM	—— HE ——
ARGON	—— AR ——
LIQUID PETROLEUM GAS	—— LPG ——
INDUSTRIAL WASTE	—— INW ——
CAST IRON	—— CI ——
CULVERT PIPE	—— CP ——
CLAY TILE	—— CT ——
DUCTILE IRON	—— DI ——
REINFORCED CONCRETE	—— RCP ——
DRAIN—OPEN TILE OR AGRICULTURAL TILE	== == ==

WELDING SYMBOLS

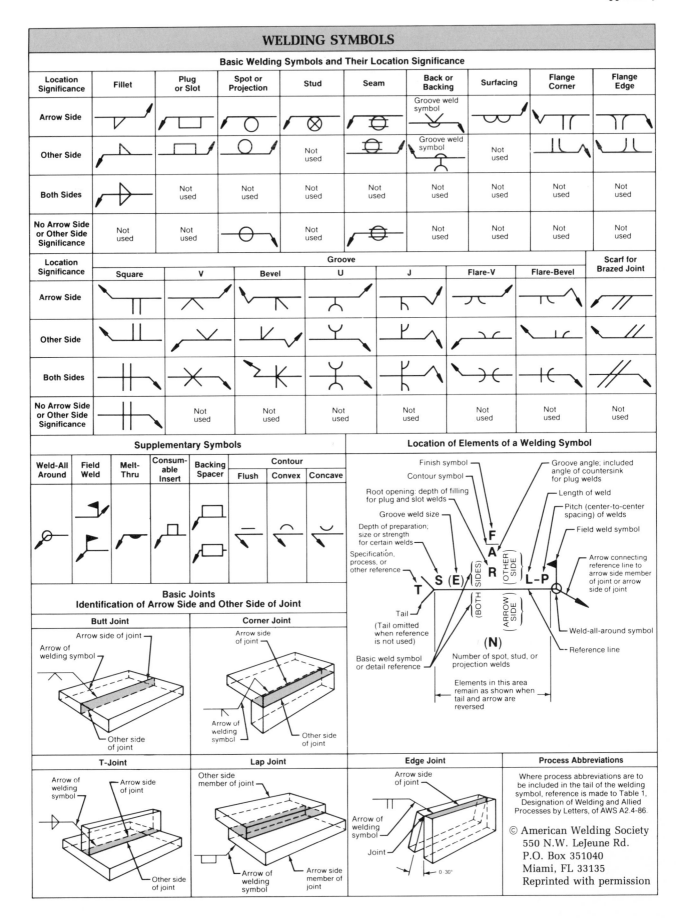

WELDING SYMBOLS (continued)

Typical Welding Symbols

Double-Fillet Welding Symbol	Chain Intermittant Fillet Welding Symbol	Staggered Intermittant Fillet Welding Symbol
Weld size — Length — 1/4 6 — 1/16 4 — Omission indicates that weld extends between abrupt changes in direction or as dimensioned	Pitch (distance between centers) of increments — 5/16 2-5 — 7/16 2-6 — Size (length of leg) — Length of increments	Pitch (distance between centers) of increments — 1/2 3-5 — 1/2 3-5 — Size (length of leg) — Length of increments

Plug Welding Symbol	Back Welding Symbol	Backing Welding Symbol
Include angle of countersink — Pitch (distance between centers) of welds — Size (diameter of hole at root) — 30° 1 3/4 4 — Depth of filling in inches (omission indicates filling is complete)	Back weld — or — 2nd operation — 1st operation	Backing weld — or — 1st operation — 2nd operation

Spot Welding Symbol	Stud Welding Symbol	Seam Welding Symbol
Size or strength — Number of welds — Pitch — 0.25 (5) 4 — RSW — Process	1/2 ⊗ 6 — (7) — Pitch — Size — Number of studs	Size or strength — Increment length — Pitch — 0.30 3-9 — RSEW — Process

Square-Groove Welding Symbol	Single-V Groove Welding Symbol	Double-Bevel-Groove Welding Symbol
(3/16) 1 4 — Weld size — Root opening	Depth of preparation — 1/2 (1/2) 2 60° — Root opening — Weld size — Groove angle	Weld size — (1) (1-1/4) — Weld size — Arrow points toward member to be prepared

Symbol with Backgouging	Flare-V Groove Welding Symbol	Flare-Bevel-Groove Welding Symbol
Depth of preparation — 3/8 — Back gouge	(1/4) — Weld size	Weld size — (3/4)

Multiple Reference Lines	Complete Penetration	Edge Flange Welding Symbol
1st operation on line nearest arrow — 2nd operation — 3rd operation	Indicates complete penetration regardless of type of weld or joint preparation — CJP	Radius — 3/64 · 1/16 — 1/16 — Weld size — Height above point of tangency

Flash or Upset Welding Symbol	Melt-Thru Symbol	Joint with Backing
Process reference — FW	1/32	R — 'R' indicates backing removed after welding

Joint with Spacer	Flush Contour Symbol	Convex Contour Symbol
With modified groove weld symbol — Double bevel groove		G

HVAC SYMBOLS

EQUIPMENT SYMBOLS	DUCTWORK	HEATING PIPING

EQUIPMENT SYMBOLS

EXPOSED RADIATOR

RECESSED RADIATOR

FLUSH ENCLOSED RADIATOR

PROJECTING ENCLOSED RADIATOR

UNIT HEATER (PROPELLER)—PLAN

UNIT HEATER (CENTRIFUGAL)—PLAN

UNIT VENTILATOR— PLAN

STEAM

DUPLEX STRAINER

PRESSURE REDUCING VALVE

AIR LINE VALVE

STRAINER

THERMOMETER

PRESSURE GAUGE AND COCK

RELIEF VALVE

AUTOMATIC 3-WAY VALVE

AUTOMATIC 2-WAY VALVE

SOLENOID VALVE

DUCTWORK

DUCT (1ST FIGURE, WIDTH; 2ND FIGURE, DEPTH) 12 × 20

DIRECTION OF FLOW

FLEXIBLE CONNECTION

DUCTWORK WITH ACOUSTICAL LINING

FIRE DAMPER WITH ACCESS DOOR FD AD

MANUAL VOLUME DAMPER VD

AUTOMATIC VOLUME DAMPER

EXHAUST, RETURN OR OUTSIDE AIR DUCT— SECTION 20 × 12

SUPPLY DUCT—SECTION 20 × 12

CEILING DIFFUSER SUPPLY OUTLET 20″ DIA. CD 1000 CFM

CEILING DIFFUSER SUPPLY OUTLET 20 × 12 CD 700 CFM

LINEAR DIFFUSER 96 × 6-LD 400 CFM

FLOOR REGISTER 20 × 12 FR 700 CFM

TURNING VANES

FAN AND MOTOR WITH BELT GUARD

LOUVER OPENING 20 × 12-L 700 CFM

HEATING PIPING

HIGH PRESSURE STEAM ——HPS——

MEDIUM PRESSURE STEAM ——MPS——

LOW PRESSURE STEAM ——LPS——

HIGH PRESSURE RETURN ——HPR——

MEDIUM PRESSURE RETURN ——MPR——

LOW PRESSURE RETURN ——LPR——

BOILER BLOW OFF ——BD——

CONDENSATE OR VACUUM PUMP DISCHARGE ——VPD——

FEEDWATER PUMP DISCHARGE ——PPD——

MAKE UP WATER ——MU——

AIR RELIEF LINE ——V——

FUEL OIL SUCTION ——FOS——

FUEL OIL RETURN ——FOR——

FUEL OIL VENT ——FOV——

COMPRESSED AIR ——A——

HOT WATER HEATING SUPPLY ——HW——

HOT WATER HEATING RETURN ——HWR——

AIR CONDITIONING PIPING

REFRIGERANT LIQUID ——RL——

REFRIGERANT DISCHARGE ——RD——

REFRIGERANT SUCTION ——RS——

CONDENSER WATER SUPPLY ——CWS——

CONDENSER WATER RETURN ——CWR——

CHILLED WATER SUPPLY ——CHWS——

CHILLED WATER RETURN ——CHWR——

MAKE UP WATER ——MU——

HUMIDIFICATION LINE ——H——

DRAIN ——D——

REFRIGERATION SYMBOLS

THERMOSTAT, SELF-CONTAINED	COMBINATION STRAINER AND DRYER	EVAPORATOR, FORCED CONVECTION
PRESSURE SWITCH	SIGHT GLASS	IMMERSION COOLING UNIT
EXPANSION VALVE, HAND	FLOAT VALVE, HIGH SIDE	EVAPORATIVE CONDENSOR
EXPANSION VALVE, AUTOMATIC	FLOAT VALVE, LOW SIDE	HEAT EXCHANGER
EXPANSION VALVE, THERMOSTATIC	GAUGE	AIR-COOLED CONDENSING UNIT
CONSTANT PRESSURE VALVE, SUCTION	COOLING TOWER	WATER-COOLED CONDENSING UNIT
THERMAL BULB		
SCALE TRAP		
DRYER	EVAPORATOR, FINNED TYPE, NATURAL CONVECTION	
FILTER AND STRAINER		